Studies in Systems, Decision and Control

Volume 190

Series editor

Janusz Kacprzyk, Polish Academy of Sciences, Warsaw, Poland
e-mail: kacprzyk@ibspan.waw.pl

The series "Studies in Systems, Decision and Control" (SSDC) covers both new developments and advances, as well as the state of the art, in the various areas of broadly perceived systems, decision making and control–quickly, up to date and with a high quality. The intent is to cover the theory, applications, and perspectives on the state of the art and future developments relevant to systems, decision making, control, complex processes and related areas, as embedded in the fields of engineering, computer science, physics, economics, social and life sciences, as well as the paradigms and methodologies behind them. The series contains monographs, textbooks, lecture notes and edited volumes in systems, decision making and control spanning the areas of Cyber-Physical Systems, Autonomous Systems, Sensor Networks, Control Systems, Energy Systems, Automotive Systems, Biological Systems, Vehicular Networking and Connected Vehicles, Aerospace Systems, Automation, Manufacturing, Smart Grids, Nonlinear Systems, Power Systems, Robotics, Social Systems, Economic Systems and other. Of particular value to both the contributors and the readership are the short publication timeframe and the world-wide distribution and exposure which enable both a wide and rapid dissemination of research output.

More information about this series at http://www.springer.com/series/13304

George A. Anastassiou

Ordinary and Fractional Approximation by Non-additive Integrals: Choquet, Shilkret and Sugeno Integral Approximators

Springer

George A. Anastassiou
Department of Mathematical Sciences
University of Memphis
Memphis, TN, USA

ISSN 2198-4182 ISSN 2198-4190 (electronic)
Studies in Systems, Decision and Control
ISBN 978-3-030-04286-8 ISBN 978-3-030-04287-5 (eBook)
https://doi.org/10.1007/978-3-030-04287-5

Library of Congress Control Number: 2018961216

This Springer imprint is published by the registered company Springer Nature Switzerland AG
The registered company address is: Gewerbestrasse 11, 6330 Cham, Switzerland

Friendship is a function and operation of similarity, and confidence of a person is an expression of his/her internal cumulative strengths

George A. Anastassiou

Dedicated to My Family

Preface

Nonlinear mathematics extend the linear mathematics to meet various needs and demands of pure and applied mathematics, among others to cover a great variety of applications in the real world. Ordinary and fractional approximations by non-additive integrals, especially by integral approximators of Choquet, Silkret and Sugeno types are a new trend in approximation theory. These integrals are only subadditive and only the first two are positive linear, and they produce very fast and flexible approximations based on limited data. We present both the univariate and multivariate cases. The Choquet integral is based on Choquet capacities, the Shilkret integral is based on maxitive measures and the Sugeno fuzzy integral is based on fuzzy measures. All these set functions are much weaker forms of the Lebesgue measure and they were conceived to fulfil needs of economic theories and other applied sciences. This monograph is the natural evolution of recent author's research work put in a book form for the first time. The presented approaches are original, and chapters are self-contained and can be read independently. This monograph is suitable to be used in related graduate classes and research projects. We exhibit to the maximum our approximation methods to all possible directions. For the history, G. Choquet (1953) introduced the capacities and his integral. Initially these were applied to statistical mechanics and potential theory, and they gave rise to the study of non-additive measure theory. Slowly but steadily these ideas of Choquet started to attract economists especially after the very important work of Shapley (1953) in the study of cooperative games. Capacities and Choquet integrals became main stream to Decision theorists since 1989 when D. Schmeidler was the first to use them in a model of choice with non-additive beliefs. The expected utility results are strengthened by the use of Choquet capacities instead of probability measures. Nowadays, Choquet integral has wide applications, among others, to decision making under risk and uncertainty, in finance, in economics, in portfolio problems and in insurance. The motivation to write this monograph came also by the following: various issues related to the modeling and analysis of ordinary and fractional order systems have gained an increased popularity, as witnessed by many books and volumes in Springer's program: http://www.springer.com/gp/search?query=fractional&submit=Prze%C5%9Blij and the purpose of our book is to provide a deeper formal analysis on some issues that are relevant to many areas for instance: decision making, complex processes, systems modeling and control and related areas. The above are deeply embedded in the fields of mathematics, engineering, computer science, physics, economics, social and life sciences.

The list of presented topics are as follows:

- Quantitative Approximation by Kantorovich-Choquet quasi-interpolation neural network operators
- Quantitative Approximation by Perturbed Kantorovich-Choquet Neural Network Operators
- Approximation by Shift Invariant Univariate Sublinear-Choquet Operators
- Quantitative Approximation by Shift Invariant Multivariate Sublinear-Choquet Operators
- Hardy type inequalities for Choquet integrals
- Approximation by Choquet integrals
- Conformable Fractional Approximation by Choquet integrals
- Multivariate and Convex Approximation by Choquet integrals
- Caputo and Canavati fractional Approximation by Choquet integrals
- Mixed Conformable and Iterated fractional Approximation by Choquet integrals
- Quantitative Approximation by a Kantorovich-Shilkret quasi-interpolation neural network operator
- Multivariate Approximation with Rates by Perturbed Kantorovich-Shilkret Neural Network Operators
- Approximation by Shift Invariant Univariate Sublinear-Shilkret Operators
- Approximation by Shift Invariant Multivariate Sublinear-Shilkret Operators
- Ostrowski-Sugeno Fuzzy inequalities
- Fractional Ostrowski-Sugeno Fuzzy univariate Inequalities
- Multivariate Ostrowski-Sugeno Fuzzy Inequalities

An extensive list of references is given per chapter.

The book's results are expected to find applications in many areas of pure and applied mathematics, especially in approximation theory, numerical analysis and mathematical economics in both ordinary and fractional sense. As such this monograph is suitable for researchers, graduate students and seminars of the above disciplines, also to be in all science and engineering libraries.

The preparation of the book took place during 2017–2018 at the University of Memphis.

The author likes to thank Prof. Alina Lupas of University of Oradea, Romania, for checking and reading the manuscript.

Memphis, TN, USA
September 2018

George A. Anastassiou

Contents

Chapter 1
Approximation with Rates by Kantorovich–Choquet Quasi-interpolation Neural Network Operators

In this chapter we present univariate and multivariate basic approximation by Kantorovich–Choquet type quasi-interpolation neural network operators with respect to supremum norm. This is done with rates using the first univariate and multivariate moduli of continuity. We approximate continuous and bounded functions on $\mathbb{R}^N$, $N \in \mathbb{N}$. When they are also uniformly continuous we have pointwise and uniform convergences. It follows [11].

1.1 Introduction

The author in [2, 3], see Chapters 2–5, was the first to establish neural network approximations to continuous functions with rates by very specifically defined neural network operators of Cardaliagnet–Euvrard and "Squashing" types, by employing the modulus of continuity of the engaged function or its high order derivative, and producing very tight Jackson type inequalities. He treats there both the univariate and multivariate cases. The defining these operators "bell-shaped" and "squashing" functions are assumed to be compact support. Also in [3] he gives the Nth order asymptotic expansion for the error of weak approximation of these two operators to a special natural class of smooth functions, see Chapters 4–5 there.

The author inspired by [13], continued his studies on neural networks approximation by introducing and using the proper quasi-interpolation operators of sigmoidal and hyperbolic tangent type which resulted into [4–8], by treating both the univariate and multivariate cases. He did also the corresponding fractional case [9].

The author here performs univariate and multivariate error function based neural network approximations to continuous functions over the whole $\mathbb{R}^N$, $N \in \mathbb{N}$, then he extends his results to complex valued functions. He also finds similar results when the activation function is induced by the sigmoidal and hyperbolic tangent function.

G. A. Anastassiou, *Ordinary and Fractional Approximation by Non-additive Integrals: Choquet, Shilkret and Sugeno Integral Approximators*, Studies in Systems, Decision and Control 190, https://doi.org/10.1007/978-3-030-04287-5_1

All convergences here are with rates expressed via the modulus of continuity of the involved function and given by very tight Jackson type inequalities.

The author comes up with the "right" precisely defined flexible quasi-interpolation, Baskakov–Choquet type integral coefficient neural networks operators associated with: the error function, sigmoidal and hyperbolic tangent functions. In preparation to prove our results we establish important properties of the basic density functions defining our operators.

Feed-forward neural networks (FNNs) with one hidden layer, the only type of networks we deal with in this chapter, are mathematically expressed as

$$N_n(x) = \sum_{j=0}^{n} c_j \sigma\left(\left\langle a_j \cdot x\right\rangle + b_j\right), \; x \in \mathbb{R}^s, s \in \mathbb{N},$$

where for $0 \leq j \leq n$, $b_j \in \mathbb{R}$ are the thresholds, $a_j \in \mathbb{R}^s$ are the connection weights, $c_j \in \mathbb{R}$ are the coefficients, $\left\langle a_j \cdot x\right\rangle$ is the inner product of a_j and x, and σ is the activation function of the network. In many fundamental neural network models, the activation functions are the error, sigmoidal and hyperbolic tangent functions. About neural networks in general read [17–19]. We have been greatly inspired by [16].

1.2 Background

Next we present briefly about the Choquet integral.

We make

Definition 1.1 Consider $\Omega \neq \varnothing$ and let $\mathcal{C}$ be a σ-algebra of subsets in Ω.

(i) (see, e.g., [20], p. 63) The set function $\mu : \mathcal{C} \to [0, +\infty]$ is called a monotone set function (or capacity) if $\mu(\varnothing) = 0$ and $\mu(A) \leq \mu(B)$ for all $A, B \in \mathcal{C}$, with $A \subset B$. Also, μ is called submodular if

$$\mu(A \cup B) + \mu(A \cap B) \leq \mu(A) + \mu(B), \quad \text{for all } A, B \in \mathcal{C}.$$

μ is called bounded if $\mu(\Omega) < +\infty$ and normalized if $\mu(\Omega) = 1$.

(ii) (see, e.g., [20], p. 233, or [14]) If μ is a monotone set function on $\mathcal{C}$ and if $f : \Omega \to \mathbb{R}$ is $\mathcal{C}$-measurable (that is, for any Borel subset $B \subset \mathbb{R}$ it follows $f^{-1}(B) \in \mathcal{C}$), then for any $A \in \mathcal{C}$, the Choquet integral is defined by

$$(C)\int_A f d\mu = \int_0^{+\infty} \mu\left(F_\beta(f) \cap A\right) d\beta + \int_{-\infty}^0 \left[\mu\left(F_\beta(f) \cap A\right) - \mu(A)\right] d\beta,$$

where we used the notation $F_\beta(f) = \{\omega \in \Omega : f(\omega) \geq \beta\}$. Notice that if $f \geq 0$ on A, then in the above formula we get $\int_{-\infty}^0 = 0$.

The integrals on the right-hand side are the usual Riemann integral.

The function f will be called Choquet integrable on A if $(C)\int_A f d\mu \in \mathbb{R}$.

Next we list some well known properties of the Choquet integral.

Remark 1.2 If $\mu : \mathcal{C} \to [0, +\infty]$ is a monotone set function, then the following properties hold:

(i) For all $a \geq 0$ we have $(C)\int_A af d\mu = a \cdot (C)\int_A f d\mu$ (if $f \geq 0$ then see, e.g., [20], Theorem 11.2, (5), p. 228 and if f is arbitrary sign, then see, e.g., [15], p. 64, Proposition 5.1, (ii)).

(ii) For all $c \in \mathbb{R}$ and f of arbitrary sign, we have (see, e.g., [20], pp. 232–233, or [15], p. 65) $(C)\int_A (f + c)\, d\mu = (C)\int_A f d\mu + c \cdot \mu(A)$.

If μ is submodular too, then for all f, g of arbitrary sign and lower bounded, we have (see, e.g., [15], p. 75, Theorem 6.3)

$$(C)\int_A (f + g)\, d\mu \leq (C)\int_A f d\mu + (C)\int_A g d\mu.$$

(iii) If $f \leq g$ on A then $(C)\int_A f d\mu \leq (C)\int_A g d\mu$ (see, e.g., [20], p. 228, Theorem 11.2, (3) if $f, g \geq 0$ and p. 232 if f, g are of arbitrary sign).

(iv) Let $f \geq 0$. If $A \subset B$ then $(C)\int_A f d\mu \leq (C)\int_B f d\mu$. In addition, if μ is finitely subadditive, then

$$(C)\int_{A\cup B} f d\mu \leq (C)\int_A f d\mu + (C)\int_B f d\mu.$$

(v) It is immediate that $(C)\int_A 1 \cdot d\mu(t) = \mu(A)$.

(vi) The formula $\mu(A) = \gamma(M(A))$, where $\gamma : [0, 1] \to [0, 1]$ is an increasing and concave function, with $\gamma(0) = 0$, $\gamma(1) = 1$ and M is a probability measure (or only finitely additive) on a σ-algebra on Ω (that is, $M(\emptyset) = 0$, $M(\Omega) = 1$ and M is countably additive), gives simple examples of normalized, monotone and submodular set functions (see, e.g., [15], pp. 16–17, Example 2.1). Such of set functions μ are also called distorsions of countably normalized, additive measures (or distorted measures). For a simple example, we can take $\gamma(t) = \frac{2t}{1+t}$, $\gamma(t) = \sqrt{t}$.

If the above γ function is increasing, concave and satisfies only $\gamma(0) = 0$, then for any bounded Borel measure m, $\mu(A) = \gamma(m(A))$ gives a simple example of bounded, monotone and submodular set function.

(vii) If μ is a countably additive bounded measure, then the Choquet integral $(C)\int_A f d\mu$ reduces to the usual Lebesgue type integral (see, e.g., [15], p. 62, or [20], p. 226).

(viii) If $f \geq 0$, then $(C)\int_A f d\mu \geq 0$.

(ix) Let $\mu = \sqrt{M}$, where M is the Lebesgue measure on $[0, +\infty)$, then μ is a monotone and submodular set function, furthermore μ is strictly positive, see [16].

(x) If $\Omega = \mathbb{R}^N$, $N \in \mathbb{N}$, we call μ strictly positive if $\mu(A) > 0$, for any open subset $A \subseteq \mathbb{R}^N$.

1.3 Results - I

We consider here the (Gauss) error special function [1, 12]

$$erf\,(x) = \frac{2}{\sqrt{\pi}} \int_0^x e^{-t^2} dt,\ x \in \mathbb{R}, \tag{1.1}$$

which is a sigmoidal type function and a strictly increasing function.

It has the basic properties

$$erf\,(0) = 0,\ erf\,(-x) = -erf\,(x)\,,\ erf\,(+\infty) = 1,\ erf\,(-\infty) = -1, \tag{1.2}$$

and

$$(erf\,(x))' = \frac{2}{\sqrt{\pi}} e^{-x^2},\ \ x \in \mathbb{R}, \tag{1.3}$$

$$\int erf\,(x)\,dx = x\,erf\,(x) + \frac{e^{-x^2}}{\sqrt{\pi}} + C, \tag{1.4}$$

where C is a constant.

The error function is related to the cumulative probability distribution function of the standard normal distribution

$$\Phi\,(x) = \frac{1}{2} + \frac{1}{2} erf\left(\frac{x}{\sqrt{2}}\right).$$

We consider the activation function

$$\chi\,(x) = \frac{1}{4}\,(erf\,(x+1) - erf\,(x-1))\,,\ x \in \mathbb{R}, \tag{1.5}$$

and we notice that

$$\chi\,(-x) = \frac{1}{4}\,(erf\,(-x+1) - erf\,(-x-1)) =$$

$$\frac{1}{4}\,(erf\,(-(x-1)) - erf\,(-(x+1))) = \frac{1}{4}\,(-erf\,(x-1) + erf\,(x+1)) = \chi\,(x)\,, \tag{1.6}$$

thus χ is an even function.

Since $x+1 > x-1$, then $erf\,(x+1) > erf\,(x-1)$, and $\chi\,(x) > 0$, all $x \in \mathbb{R}$.

We see that

$$\chi\,(0) = \frac{erf\,(1)}{2} \simeq \frac{0.843}{2} = 0.4215. \tag{1.7}$$

Let $x > 0$, we have

$$\chi'(x) = \frac{1}{4}\left(\frac{2}{\sqrt{\pi}}e^{-(x+1)^2} - \frac{2}{\sqrt{\pi}}e^{-(x-1)^2}\right) =$$

$$\frac{1}{2\sqrt{\pi}}\left(\frac{1}{e^{(x+1)^2}} - \frac{1}{e^{(x-1)^2}}\right) = \frac{1}{2\sqrt{\pi}}\left(\frac{e^{(x-1)^2} - e^{(x+1)^2}}{e^{(x+1)^2}e^{(x-1)^2}}\right) < 0, \tag{1.8}$$

proving $\chi'(x) < 0$, for $x > 0$.

That is χ is strictly decreasing on $[0, \infty)$ and is strictly increasing on $(-\infty, 0]$, and $\chi'(0) = 0$.

Clearly the x-axis is the horizontal asymptote on χ.

Conclusion, χ is a bell symmetric function with maximum $\chi(0) \simeq 0.4215$.

We further need

Theorem 1.3 ([10]) *We have that*

$$\sum_{i=-\infty}^{\infty} \chi(x-i) = 1, \quad \text{all} \quad x \in \mathbb{R}. \tag{1.9}$$

Thus

$$\sum_{i=-\infty}^{\infty} \chi(nx-i) = 1, \ \forall\, n \in \mathbb{N}, \forall\, x \in \mathbb{R}. \tag{1.10}$$

Furthermore we get:

Since χ is even it holds $\sum_{i=-\infty}^{\infty} \chi(i-x) = 1$, for any $x \in \mathbb{R}$.

Hence $\sum_{i=-\infty}^{\infty} \chi(i+x) = 1, \forall\, x \in \mathbb{R}$, and $\sum_{i=-\infty}^{\infty} \chi(x+i) = 1, \forall\, x \in \mathbb{R}$.

Theorem 1.4 ([10]) *It holds*

$$\int_{-\infty}^{\infty} \chi(x)\, dx = 1. \tag{1.11}$$

So $\chi(x)$ is a density function on $\mathbb{R}$.

Theorem 1.5 ([10]) *Let* $0 < \alpha < 1$, *and* $n \in \mathbb{N}$ *with* $n^{1-\alpha} \geq 3$. *It holds*

$$\sum_{\substack{k=-\infty \\ : |nx-k| \geq n^{1-\alpha}}}^{\infty} \chi(nx-k) < \frac{1}{2\sqrt{\pi}\left(n^{1-\alpha}-2\right)e^{\left(n^{1-\alpha}-2\right)^2}}. \tag{1.12}$$

We give

Definition 1.6 Let $\mathcal{L}$ be the Lebesgue σ-algebra on $\mathbb{R}$, and the set function $\mu : \mathcal{L} \to [0, +\infty)$, which is assumed to be monotone, submodular and strictly positive. Let $f \in C_B^+(\mathbb{R})$ (the set of continuous and bounded functions from $\mathbb{R} \to \mathbb{R}_+$). We define the univariate Kantorovich–Choquet type neural network operator

$$C_n^\mu(f, x) := \sum_{k=-\infty}^{\infty} \left(\frac{(C)\int_0^{\frac{1}{n}} f\left(t + \frac{k}{n}\right) d\mu(t)}{\mu\left(\left[0, \frac{1}{n}\right]\right)} \right) \chi(nx - k), \quad \forall\, x \in \mathbb{R}, n \in \mathbb{N}. \tag{1.13}$$

Clearly here $\mu\left(\left[0, \frac{1}{n}\right]\right) > 0, \forall\, n \in \mathbb{N}$.

Above we notice that

$$\left\| C_n^\mu(f) \right\|_\infty \leq \| f \|_\infty, \tag{1.14}$$

so that $C_n^\mu(f, x)$ is well-defined.

We make

Remark 1.7 Let $t \in \left[0, \frac{1}{n}\right]$ and $x \in \mathbb{R}$, then

$$f\left(t + \frac{k}{n}\right) = f\left(t + \frac{k}{n}\right) - f(x) + f(x) \leq \left| f\left(t + \frac{k}{n}\right) - f(x) \right| + f(x),$$

hence

$$(C)\int_0^{\frac{1}{n}} f\left(t + \frac{k}{n}\right) d\mu(t) \leq$$

$$(C)\int_0^{\frac{1}{n}} \left| f\left(t + \frac{k}{n}\right) - f(x) \right| d\mu(t) + (C)\int_0^{\frac{1}{n}} f(x)\, d\mu(t) =$$

$$(C)\int_0^{\frac{1}{n}} \left| f\left(t + \frac{k}{n}\right) - f(x) \right| d\mu(t) + f(x)\, \mu\left(\left[0, \frac{1}{n}\right]\right).$$

That is

$$(C)\int_0^{\frac{1}{n}} f\left(t + \frac{k}{n}\right) d\mu(t) - f(x)\, \mu\left(\left[0, \frac{1}{n}\right]\right) \leq (C)\int_0^{\frac{1}{n}} \left| f\left(t + \frac{k}{n}\right) - f(x) \right| d\mu(t). \tag{1.15}$$

Similarly we have

$$f(x) = f(x) - f\left(t + \frac{k}{n}\right) + f\left(t + \frac{k}{n}\right) \leq \left| f\left(t + \frac{k}{n}\right) - f(x) \right| + f\left(t + \frac{k}{n}\right),$$

hence

$$(C)\int_0^{\frac{1}{n}} f(x)\, \mu(dt) \leq$$

$$(C)\int_0^{\frac{1}{n}}\left|f\left(t+\frac{k}{n}\right)-f(x)\right|d\mu(t)+(C)\int_0^{\frac{1}{n}}f\left(t+\frac{k}{n}\right)\mu(dt).$$

That is

$$f(x)\mu\left(\left[0,\frac{1}{n}\right]\right)\leq$$

$$(C)\int_0^{\frac{1}{n}}\left|f\left(t+\frac{k}{n}\right)-f(x)\right|d\mu(t)+(C)\int_0^{\frac{1}{n}}f\left(t+\frac{k}{n}\right)\mu(dt),$$

and

$$f(x)\mu\left(\left[0,\frac{1}{n}\right]\right)-(C)\int_0^{\frac{1}{n}}f\left(t+\frac{k}{n}\right)\mu(dt)\leq(C)\int_0^{\frac{1}{n}}\left|f\left(t+\frac{k}{n}\right)-f(x)\right|d\mu(t).\tag{1.16}$$

By (1.15) and (1.16) we derive that

$$\left|(C)\int_0^{\frac{1}{n}}f\left(t+\frac{k}{n}\right)\mu(dt)-f(x)\mu\left(\left[0,\frac{1}{n}\right]\right)\right|\leq$$

$$(C)\int_0^{\frac{1}{n}}\left|f\left(t+\frac{k}{n}\right)-f(x)\right|d\mu(t).\tag{1.17}$$

In particular, it holds

$$\left|\left(\frac{(C)\int_0^{\frac{1}{n}}f\left(t+\frac{k}{n}\right)\mu(dt)}{\mu\left(\left[0,\frac{1}{n}\right]\right)}\right)-f(x)\right|\leq\frac{(C)\int_0^{\frac{1}{n}}\left|f\left(t+\frac{k}{n}\right)-f(x)\right|d\mu(t)}{\mu\left(\left[0,\frac{1}{n}\right]\right)}.\tag{1.18}$$

We define

$$\omega_1(f,h):=\sup_{\substack{x,y\in\mathbb{R}\\|x-y|\leq h}}|f(x)-f(y)|,\quad h>0,$$

the first modulus of continuity of $f\in C_B^+(\mathbb{R})$.

We present

Theorem 1.8 *Let $f\in C_B^+(\mathbb{R})$, $0<\alpha<1$, $x\in\mathbb{R}$, $n\in\mathbb{N}$ with $n^{1-\alpha}\geq 3$. Then*
(i)

$$\sup_\mu\left|C_n^\mu(f,x)-f(x)\right|\leq\omega_1\left(f,\frac{1}{n}+\frac{1}{n^\alpha}\right)+\frac{\|f\|_\infty}{\sqrt{\pi}\left(n^{1-\alpha}-2\right)e^{\left(n^{1-\alpha}-2\right)^2}}=:\rho_{1n},\tag{1.19}$$

(ii)

$$\sup_{\mu} \left\| C_n^{\mu}(f) - f \right\|_{\infty} \leq \rho_{1n}. \tag{1.20}$$

For $f \in \left(C_B^+(\mathbb{R}) \cap C_u^+(\mathbb{R})\right)$ *we get* $\lim_{n\to\infty} C_n^{\mu}(f) = f$*, pointwise and uniformly.*

Here $C_u^+(\mathbb{R})$ *is the set of functions that are uniformly continuous from* $\mathbb{R}$ *into* $\mathbb{R}_+$*.*

Proof We observe that

$$\left|C_n^{\mu}(f,x) - f(x)\right| = \left|\sum_{k=-\infty}^{\infty}\left(\left(\frac{(C)\int_0^{\frac{1}{n}} f\left(t+\frac{k}{n}\right)d\mu(t)}{\mu\left(\left[0,\frac{1}{n}\right]\right)}\right) - f(x)\right)\chi(nx-k)\right| \leq$$

$$\sum_{k=-\infty}^{\infty}\left|\left(\frac{(C)\int_0^{\frac{1}{n}} f\left(t+\frac{k}{n}\right)d\mu(t)}{\mu\left(\left[0,\frac{1}{n}\right]\right)}\right) - f(x)\right|\chi(nx-k) \overset{(1.18)}{\leq}$$

$$\sum_{k=-\infty}^{\infty}\left(\frac{(C)\int_0^{\frac{1}{n}} \left|f\left(t+\frac{k}{n}\right) - f(x)\right| d\mu(t)}{\mu\left(\left[0,\frac{1}{n}\right]\right)}\right)\chi(nx-k) = \tag{1.21}$$

$$\sum_{\substack{k=-\infty \\ \left|x-\frac{k}{n}\right| \leq \frac{1}{n^{\alpha}}}}^{\infty}\left(\frac{(C)\int_0^{\frac{1}{n}} \left|f\left(t+\frac{k}{n}\right) - f(x)\right| d\mu(t)}{\mu\left(\left[0,\frac{1}{n}\right]\right)}\right)\chi(nx-k) +$$

$$\sum_{\substack{k=-\infty \\ \left|x-\frac{k}{n}\right| \geq \frac{1}{n^{\alpha}}}}^{\infty}\left(\frac{(C)\int_0^{\frac{1}{n}} \left|f\left(t+\frac{k}{n}\right) - f(x)\right| d\mu(t)}{\mu\left(\left[0,\frac{1}{n}\right]\right)}\right)\chi(nx-k) \leq \tag{1.22}$$

$$\sum_{\substack{k=-\infty \\ \left|x-\frac{k}{n}\right| \leq \frac{1}{n^{\alpha}}}}^{\infty}\left(\frac{(C)\int_0^{\frac{1}{n}} \omega_1\left(f,\left|t+\frac{k}{n}-x\right|\right) d\mu(t)}{\mu\left(\left[0,\frac{1}{n}\right]\right)}\right)\chi(nx-k) +$$

$$2\left\|f\right\|_{\infty}\left(\sum_{\substack{k=-\infty \\ |nx-k| \geq n^{1-\alpha}}}^{\infty} \chi(|nx-k|)\right) \overset{(1.12)}{\leq}$$

$$\sum_{\substack{k=-\infty \\ |nx-k| \le n^{1-\alpha}}}^{\infty} \left(\frac{(C) \int_0^{\frac{1}{n}} \omega_1 \left(f, |t| + \frac{1}{n^{\alpha}}\right) \mu(dt)}{\mu\left(\left[0, \frac{1}{n}\right]\right)} \right) \chi(nx-k) +$$

$$\frac{\|f\|_{\infty}}{\sqrt{\pi}\left(n^{1-\alpha}-2\right) e^{\left(n^{1-\alpha}-2\right)^2}} \le \tag{1.23}$$

$$\omega_1\left(f, \frac{1}{n} + \frac{1}{n^{\alpha}}\right) \left(\sum_{\substack{k=-\infty \\ |nx-k| \le n^{1-\alpha}}}^{\infty} \chi(nx-k) \right) +$$

$$\frac{\|f\|_{\infty}}{\sqrt{\pi}\left(n^{1-\alpha}-2\right) e^{\left(n^{1-\alpha}-2\right)^2}} \overset{(1.10)}{\le}$$

$$\omega_1\left(f, \frac{1}{n} + \frac{1}{n^{\alpha}}\right) + \frac{\|f\|_{\infty}}{\sqrt{\pi}\left(n^{1-\alpha}-2\right) e^{\left(n^{1-\alpha}-2\right)^2}},$$

proving the claim. ■

Additionally we give

Definition 1.9 Here $i = \sqrt{-1}$. Denote $C_B^{+}(\mathbb{R}, \mathbb{C}) = \{f : \mathbb{R} \to \mathbb{C} | f = f_1 + i f_2$, where $f_1, f_2 \in C_B^{+}(\mathbb{R})\}$. We set for $f \in C_B^{+}(\mathbb{R}, \mathbb{C})$ that

$$C_n^{\mu}(f, x) := C_n^{\mu}(f_1, x) + i C_n^{\mu}(f_2, x), \quad \forall\, n \in \mathbb{N}, x \in \mathbb{R}. \tag{1.24}$$

Theorem 1.10 *Let* $f \in C_B^{+}(\mathbb{R}, \mathbb{C})$, $f = f_1 + i f_2$, $0 < \alpha < 1$, $n \in \mathbb{N}$, $n^{1-\alpha} \ge 3$, $x \in \mathbb{R}$. *Then*

(i)

$$\sup_{\mu} \left| C_n^{\mu}(f, x) - f(x) \right| \le \left(\omega_1\left(f_1, \frac{1}{n} + \frac{1}{n^{\alpha}}\right) + \omega_1\left(f_2, \frac{1}{n} + \frac{1}{n^{\alpha}}\right) \right) \tag{1.25}$$

$$+ \frac{\left(\|f_1\|_{\infty} + \|f_2\|_{\infty}\right)}{\sqrt{\pi}\left(n^{1-\alpha}-2\right) e^{\left(n^{1-\alpha}-2\right)^2}} =: \rho_{2n},$$

and

(ii)

$$\sup_{\mu} \left\| C_n^{\mu}(f) - f \right\|_{\infty} \le \rho_{2n}. \tag{1.26}$$

Proof We have that

$$\left|C_n^{\mu}(f,x)-f(x)\right|=\left|C_n^{\mu}(f_1,x)+iC_n^{\mu}(f_2,x)-f_1(x)-if_2(x)\right|=$$

$$\left|\left(C_n^{\mu}(f_1,x)-f_1(x)\right)+i\left(C_n^{\mu}(f_2,x)-f_2(x)\right)\right|\leq$$

$$\left|C_n^{\mu}(f_1,x)-f_1(x)\right|+\left|C_n^{\mu}(f_2,x)-f_2(x)\right|\overset{(1.19)}{\leq}$$

$$\left(\omega_1\left(f_1,\frac{1}{n}+\frac{1}{n^{\alpha}}\right)+\frac{\|f_1\|_{\infty}}{\sqrt{\pi}\left(n^{1-\alpha}-2\right)e^{\left(n^{1-\alpha}-2\right)^2}}\right)+ \tag{1.27}$$

$$\left(\omega_1\left(f_2,\frac{1}{n}+\frac{1}{n^{\alpha}}\right)+\frac{\|f_2\|_{\infty}}{\sqrt{\pi}\left(n^{1-\alpha}-2\right)e^{\left(n^{1-\alpha}-2\right)^2}}\right),$$

proving the claim. ■

We make

Remark 1.11 We introduce

$$Z(x_1,...,x_N):=Z(x):=\prod_{i=1}^{N}\chi(x_i),\ x=(x_1,...,x_N)\in\mathbb{R}^N,\ N\in\mathbb{N}. \tag{1.28}$$

It has the properties:

(i)

$$Z(x)>0,\ \ \forall\, x\in\mathbb{R}^N, \tag{1.29}$$

(ii)

$$\sum_{k=-\infty}^{\infty}Z(x-k):=\sum_{k_1=-\infty}^{\infty}\sum_{k_2=-\infty}^{\infty}\cdots\sum_{k_N=-\infty}^{\infty}Z(x_1-k_1,...,x_N-k_N)=1, \tag{1.30}$$

where $k:=(k_1,...,k_n)\in\mathbb{Z}^N$, $\forall\, x\in\mathbb{R}^N$,

hence

(iii)

$$\sum_{k=-\infty}^{\infty}Z(nx-k):=\sum_{k_1=-\infty}^{\infty}\sum_{k_2=-\infty}^{\infty}\cdots\sum_{k_N=-\infty}^{\infty}Z(nx_1-k_1,...,nx_N-k_N)=1, \tag{1.31}$$

$\forall\, x\in\mathbb{R}^N;\ n\in\mathbb{N}$,

and

(iv)

$$\int_{\mathbb{R}^N} Z(x)\,dx = 1, \tag{1.32}$$

that is Z is a multivariate density function.

Here $\|x\|_\infty := \max\{|x_1|, ..., |x_N|\}$, $x \in \mathbb{R}^N$, also set $\infty := (\infty, ..., \infty)$, $-\infty = (-\infty, ..., -\infty)$ upon the multivariate context.

It is also clear that (see (1.12))

(v)

$$\sum_{\substack{k=-\infty \\ \left\| \frac{k}{n} - x \right\|_\infty > \frac{1}{n^\beta}}}^{\infty} Z(nx-k) \le \frac{1}{2\sqrt{\pi}\left(n^{1-\beta}-2\right) e^{\left(n^{1-\beta}-2\right)^2}}, \tag{1.33}$$

$0 < \beta < 1$, $n \in \mathbb{N} : n^{1-\beta} \ge 3$, $x \in \mathbb{R}^N$.

For $f \in C_B^+\left(\mathbb{R}^N\right)$ (continuous and bounded functions from $\mathbb{R}^N$ into $\mathbb{R}_+$), we define the first modulus of continuity

$$\omega_1(f,h) := \sup_{\substack{x,y \in \mathbb{R}^N \\ \|x-y\|_\infty \le h}} |f(x) - f(y)|, \quad h > 0. \tag{1.34}$$

Given that $f \in C_U^+\left(\mathbb{R}^N\right)$ (uniformly continuous from $\mathbb{R}^N$ into $\mathbb{R}_+$), we have that

$$\lim_{h \to 0} \omega_1(f,h) = 0. \tag{1.35}$$

We make

Definition 1.12 Let $\mathcal{L}^*$ be the Lebesgue σ-algebra on $\mathbb{R}^N$, $N \in \mathbb{N}$, and the set function $\mu^* : \mathcal{L}^* \to [0, +\infty)$, which is assumed to be monotone, submodular and strictly positive. For $f \in C_B^+\left(\mathbb{R}^N\right)$, we define the multivariate Kantorovich–Choquet type neural network operator for any $x \in \mathbb{R}^N$:

$$\overline{C}_n^{\mu^*}(f,x) = \overline{C}_n^{\mu^*}(f, x_1, ..., x_N) := \tag{1.36}$$

$$\sum_{k=-\infty}^{\infty} \left(\frac{(C)\int_{\left[0,\frac{1}{n}\right]^N} f\left(t + \frac{k}{n}\right) d\mu^*(t)}{\mu^*\left(\left[0,\frac{1}{n}\right]^N\right)} \right) Z(nx-k) =$$

$$\sum_{k_1=-\infty}^{\infty} \sum_{k_2=-\infty}^{\infty} \cdots \sum_{k_N=-\infty}^{\infty} \left(\frac{(C)\int_0^{\frac{1}{n}} \cdots \int_0^{\frac{1}{n}} f\left(t_1 + \frac{k_1}{n}, t_2 + \frac{k_2}{n}, ..., t_N + \frac{k_N}{n}\right) d\mu^*(t_1, ..., t_N)}{\mu^*\left(\left[0,\frac{1}{n}\right]^N\right)} \right)$$

$$\cdot\left(\prod_{i=1}^{N} Z\left(nx_i - k_i\right)\right),$$

where $x = (x_1, ..., x_N) \in \mathbb{R}^N$, $k = (k_1, ..., k_N)$, $t = (t_1, ..., t_N)$, $n \in \mathbb{N}$.

Clearly here $\mu^*\left(\left[0, \frac{1}{n}\right]^N\right) > 0$, $\forall\, n \in \mathbb{N}$.

Above we notice that

$$\left\|\overline{C}_n^{\mu^*}(f)\right\|_\infty \leq \|f\|_\infty, \tag{1.37}$$

so that $\overline{C}_n^{\mu^*}(f, x)$ is well-defined.

Remark 1.13 Acting as in the proof of (1.18) we derive

$$\left|\left(\frac{(C)\int_{\left[0,\frac{1}{n}\right]^N} f\left(t + \frac{k}{n}\right) d\mu^*(t)}{\mu^*\left(\left[0, \frac{1}{n}\right]^N\right)}\right) - f(x)\right| \leq \frac{(C)\int_{\left[0,\frac{1}{n}\right]^N} \left|f\left(t + \frac{k}{n}\right) - f(x)\right| d\mu^*(t)}{\mu^*\left(\left[0, \frac{1}{n}\right]^N\right)}. \tag{1.38}$$

We present

Theorem 1.14 *Let $f \in C_B^+\left(\mathbb{R}^N\right)$, $0 < \beta < 1$, $x \in \mathbb{R}^N$; $N, n \in \mathbb{N}$ with $n^{1-\beta} \geq 3$. Then*

(i)

$$\sup_{\mu^*}\left|\overline{C}_n^{\mu^*}(f, x) - f(x)\right| \leq \omega_1\left(f, \frac{1}{n} + \frac{1}{n^\beta}\right) + \frac{\|f\|_\infty}{\sqrt{\pi}\left(n^{1-\beta} - 2\right) e^{\left(n^{1-\beta}-2\right)^2}} =: \rho_{3n}, \tag{1.39}$$

(ii)

$$\sup_{\mu^*}\left\|\overline{C}_n^{\mu^*}(f) - f\right\|_\infty \leq \rho_{3n}. \tag{1.40}$$

Given that $f \in \left(C_U^+\left(\mathbb{R}^N\right) \cap C_B^+\left(\mathbb{R}^N\right)\right)$, we obtain $\lim_{n\to\infty} \overline{C}_n^{\mu^*}(f) = f$, uniformly.

Proof We observe that

$$\left|\overline{C}_n^{\mu^*}(f, x) - f(x)\right| =$$

$$\left|\sum_{k=-\infty}^{\infty}\left(\frac{(C)\int_{\left[0,\frac{1}{n}\right]^N} f\left(t + \frac{k}{n}\right) d\mu^*(t)}{\mu^*\left(\left[0, \frac{1}{n}\right]^N\right)}\right) Z(nx - k) - \sum_{k=-\infty}^{\infty} f(x)\, Z(nx - k)\right|$$

$$= \left|\sum_{k=-\infty}^{\infty}\left(\left(\frac{(C)\int_{\left[0,\frac{1}{n}\right]^N} f\left(t + \frac{k}{n}\right) d\mu^*(t)}{\mu^*\left(\left[0, \frac{1}{n}\right]^N\right)}\right) - f(x)\right) Z(nx - k)\right| \leq \tag{1.41}$$

$$\sum_{k=-\infty}^{\infty}\left|\left(\frac{(C)\int_{[0,\frac{1}{n}]^N} f\left(t+\frac{k}{n}\right)d\mu^*(t)}{\mu^*\left(\left[0,\frac{1}{n}\right]^N\right)}\right)-f(x)\right| Z(nx-k) \overset{(1.38)}{\leq}$$

$$\sum_{k=-\infty}^{\infty}\left(\frac{(C)\int_{[0,\frac{1}{n}]^N}\left|f\left(t+\frac{k}{n}\right)-f(x)\right|d\mu^*(t)}{\mu^*\left(\left[0,\frac{1}{n}\right]^N\right)}\right) Z(nx-k) =$$

$$\sum_{\begin{cases} k=-\infty \\ \left\|\frac{k}{n}-x\right\|_\infty \leq \frac{1}{n^\beta}\end{cases}}^{\infty}\left(\frac{(C)\int_{[0,\frac{1}{n}]^N}\left|f\left(t+\frac{k}{n}\right)-f(x)\right|d\mu^*(t)}{\mu^*\left(\left[0,\frac{1}{n}\right]^N\right)}\right) Z(nx-k)+$$

$$\sum_{\begin{cases} k=-\infty \\ \left\|\frac{k}{n}-x\right\|_\infty > \frac{1}{n^\beta}\end{cases}}^{\infty}\left(\frac{(C)\int_{[0,\frac{1}{n}]^N}\left|f\left(t+\frac{k}{n}\right)-f(x)\right|d\mu^*(t)}{\mu^*\left(\left[0,\frac{1}{n}\right]^N\right)}\right) Z(nx-k) \leq \tag{1.42}$$

$$\sum_{\begin{cases} k=-\infty \\ \left\|\frac{k}{n}-x\right\|_\infty \leq \frac{1}{n^\beta}\end{cases}}^{\infty}\left(\frac{(C)\int_{[0,\frac{1}{n}]^N}\omega_1\left(f,\|t\|_\infty+\left\|\frac{k}{n}-x\right\|_\infty\right)d\mu^*(t)}{\mu^*\left(\left[0,\frac{1}{n}\right]^N\right)}\right) Z(nx-k)+$$

$$2\|f\|_\infty\left(\sum_{\begin{cases} k=-\infty \\ \left\|\frac{k}{n}-x\right\|_\infty > \frac{1}{n^\beta}\end{cases}}^{\infty} Z(|nx-k|)\right) \overset{(1.33)}{\leq}$$

$$\omega_1\left(f,\frac{1}{n}+\frac{1}{n^\beta}\right)+\frac{\|f\|_\infty}{\sqrt{\pi}\left(n^{1-\beta}-2\right)e^{\left(n^{1-\beta}-2\right)^2}}, \tag{1.43}$$

proving the claim. ■

Additionally we give

Definition 1.15 Denote by $C_B^+\left(\mathbb{R}^N,\mathbb{C}\right)=\{f:\mathbb{R}^N\to\mathbb{C}\,|\,f=f_1+if_2$, where f_1, $f_2\in C_B^+\left(\mathbb{R}^N\right)$, $N\in\mathbb{N}\}$. We set for $f\in C_B^+\left(\mathbb{R}^N,\mathbb{C}\right)$ that

$$\overline{C}_n^{\mu^*}(f,x):=\overline{C}_n^{\mu^*}(f_1,x)+i\overline{C}_n^{\mu^*}(f_2,x),\quad\forall\, n\in\mathbb{N},\ x\in\mathbb{R}^N. \tag{1.44}$$

Theorem 1.16 *Let* $f \in C_B^+(\mathbb{R}^N, \mathbb{C})$, $f = f_1 + i f_2$, $N \in \mathbb{N}$, $0 < \beta < 1$, $x \in \mathbb{R}^N$; $n \in \mathbb{N}$ *with* $n^{1-\beta} \geq 3$. *Then*

(i)

$$\sup_{\mu^*} \left| \overline{C}_n^{\mu^*}(f, x) - f(x) \right| \leq \left(\omega_1 \left(f_1, \frac{1}{n} + \frac{1}{n^\beta} \right) + \omega_1 \left(f_2, \frac{1}{n} + \frac{1}{n^\beta} \right) \right) + \tag{1.45}$$

$$\frac{\left(\|f_1\|_\infty + \|f_2\|_\infty \right)}{\sqrt{\pi} \left(n^{1-\beta} - 2 \right) e^{\left(n^{1-\beta} - 2 \right)^2}} =: \rho_{4n},$$

and

(ii)

$$\sup_{\mu^*} \left\| \overline{C}_n^{\mu^*}(f) - f \right\|_\infty \leq \rho_{4n}. \tag{1.46}$$

Proof Similar to Theorem 1.10 by applying twice (1.39). ■

1.4 Results - II

We make

Remark 1.17 We consider here the sigmoidal function of logarithmic type

$$s_i(x_i) = \frac{1}{1 + e^{-x_i}}, \; x_i \in \mathbb{R}, i = 1, ..., N; \; x := (x_1, ..., x_N) \in \mathbb{R}^N,$$

each has the properties $\lim_{x_i \to +\infty} s_i(x_i) = 1$ and $\lim_{x_i \to -\infty} s_i(x_i) = 0$, $i = 1, ..., N$.

These functions play the role of activation functions in the hidden layer of neural networks, also have applications in biology, demography, etc.

As in [13], we consider

$$\Phi_i(x_i) := \frac{1}{2} \left(s_i(x_i + 1) - s_i(x_i - 1) \right), \; x_i \in \mathbb{R}, i = 1, ..., N. \tag{1.47}$$

We notice the following properties:

(i) $\Phi_i(x_i) > 0$, $\forall \, x_i \in \mathbb{R}$,

(ii) $\sum_{k_i = -\infty}^{\infty} \Phi_i(x_i - k_i) = 1$, $\forall \, x_i \in \mathbb{R}$,

(iii) $\sum_{k_i = -\infty}^{\infty} \Phi_i(n x_i - k_i) = 1$, $\forall \, x_i \in \mathbb{R}$; $n \in \mathbb{N}$,

(iv) $\int_{-\infty}^{\infty} \Phi_i(x_i) \, dx_i = 1$,

(v) Φ_i is a density function,

(vi) Φ_i is even: $\Phi_i(-x_i) = \Phi_i(x_i)$, $x_i \geq 0$, for $i = 1, ..., N$.

We see that [8]

$$\Phi_i(x_i) = \left(\frac{e^2-1}{2e^2}\right)\frac{1}{\left(1+e^{x_i-1}\right)\left(1+e^{-x_i-1}\right)}, \quad i = 1, ..., N.$$

(vii) Φ_i is decreasing on $\mathbb{R}_+$, and increasing on $\mathbb{R}_-$, $i = 1, ..., N$.
Let $0 < \beta < 1$, $n \in \mathbb{N}$. Then as in [8] we get
(viii)

$$\sum_{\begin{cases} k_i = -\infty \\ : |nx_i - k_i| > n^{1-\alpha} \end{cases}}^{\infty} \Phi_i(nx_i - k_i) = \sum_{\begin{cases} k_i = -\infty \\ : |nx_i - k_i| > n^{1-\alpha} \end{cases}}^{\infty} \Phi_i(|nx_i - k_i|)$$

$$\leq 3.1992 e^{-n^{(1-\beta)}}, \; i = 1, ..., N.$$

We use here the complete multivariate activation function [7]

$$\Phi(x_1, ..., x_N) := \Phi(x) := \prod_{i=1}^{N} \Phi_i(x_i), \; x \in \mathbb{R}^N. \tag{1.48}$$

It has the properties [7]:
(i)' $\Phi(x) > 0, \forall\, x \in \mathbb{R}^N$.
We see that

$$\sum_{k_1=-\infty}^{\infty} \sum_{k_2=-\infty}^{\infty} ... \sum_{k_N=-\infty}^{\infty} \Phi(x_1 - k_1, x_2 - k_2, ..., x_N - k_N) =$$

$$\sum_{k_1=-\infty}^{\infty} \sum_{k_2=-\infty}^{\infty} ... \sum_{k_N=-\infty}^{\infty} \prod_{i=1}^{N} \Phi_i(x_i - k_i) = \prod_{i=1}^{N} \left(\sum_{k_i=-\infty}^{\infty} \Phi_i(x_i - k_i) \right) = 1. \tag{1.49}$$

That is
(ii)'

$$\sum_{k=-\infty}^{\infty} \Phi(x-k) := \sum_{k_1=-\infty}^{\infty} \sum_{k_2=-\infty}^{\infty} ... \sum_{k_N=-\infty}^{\infty} \Phi(x_1 - k_1, x_2 - k_2, ..., x_N - k_N) = 1, \tag{1.50}$$

$k := (k_1, ..., k_N)$, $\forall\, x \in \mathbb{R}^N$.
(iii)'

$$\sum_{k=-\infty}^{\infty} \Phi(nx - k) :=$$

$$\sum_{k_1=-\infty}^{\infty}\sum_{k_2=-\infty}^{\infty}\dots\sum_{k_N=-\infty}^{\infty}\Phi\left(nx_1-k_1,nx_2-k_2,...,nx_N-k_N\right)=1, \tag{1.51}$$

$\forall\, x\in\mathbb{R}^N;\ n\in\mathbb{N}$.

(iv)'

$$\int_{\mathbb{R}^N}\Phi\left(x\right)dx=1, \tag{1.52}$$

that is Φ is a multivariate density function.

Here $\|x\|_\infty:=\max\{|x_1|,...,|x_n|\}$, $x\in\mathbb{R}^N$, also set $\infty:=(\infty,...,\infty)$, $-\infty:=(-\infty,...,-\infty)$ upon the multivariate context.

For $0<\beta<1$ and $n\in\mathbb{N}$, fixed $x\in\mathbb{R}^N$, we have proved [7]

(v)'

$$\sum_{\begin{cases} k=-\infty \\ \left\|\frac{k}{n}-x\right\|_\infty>\frac{1}{n^\beta}\end{cases}}^{\infty}\Phi\left(nx-k\right)\leq 3.1992e^{-n^{(1-\beta)}}. \tag{1.53}$$

We give

Definition 1.18 Let $\mathcal{L}^*$ be the Lebesgue σ-algebra on $\mathbb{R}^N$, $N\in\mathbb{N}$, and the set function $\mu^*:\mathcal{L}^*\to[0,+\infty)$, which is assumed to be monotone, submodular and strictly positive. For $f\in C_B^+\left(\mathbb{R}^N\right)$, we define the multivariate Kantorovich–Choquet type neural network operator for any $x\in\mathbb{R}^N$:

$$K_n^{\mu^*}\left(f,x\right)=K_n^{\mu^*}\left(f,x_1,...,x_N\right):=$$

$$\sum_{k=-\infty}^{\infty}\left(\frac{(C)\int_{\left[0,\frac{1}{n}\right]^N}f\left(t+\frac{k}{n}\right)d\mu^*\left(t\right)}{\mu^*\left(\left[0,\frac{1}{n}\right]^N\right)}\right)\Phi\left(nx-k\right)= \tag{1.54}$$

$$\sum_{k_1=-\infty}^{\infty}\sum_{k_2=-\infty}^{\infty}\dots\sum_{k_N=-\infty}^{\infty}\left(\frac{(C)\int_0^{\frac{1}{n}}\dots\int_0^{\frac{1}{n}}f\left(t_1+\frac{k_1}{n},t_2+\frac{k_2}{n},...,t_N+\frac{k_N}{n}\right)d\mu^*(t_1,...,t_N)}{\mu^*\left(\left[0,\frac{1}{n}\right]^N\right)}\right)$$

$$\cdot\left(\prod_{i=1}^N\Phi_i\left(nx_i-k_i\right)\right),$$

where $x=(x_1,...,x_N)\in\mathbb{R}^N$, $k=(k_1,...,k_N)$, $t=(t_1,...,t_N)$, $n\in\mathbb{N}$.

Clearly here $\mu^*\left(\left[0,\frac{1}{n}\right]^N\right)>0$, $\forall\, n\in\mathbb{N}$.

Above we notice that

$$\left\|K_n^{\mu^*}\left(f\right)\right\|_\infty\leq\|f\|_\infty, \tag{1.55}$$

so that $K_n^{\mu^*}\left(f,x\right)$ is well-defined.

We make

Remark 1.19 We consider also here the hyperbolic tangent function $\tanh x$, $x \in \mathbb{R}$ (see also [5])

$$\tanh x := \frac{e^x - e^{-x}}{e^x + e^{-x}}. \tag{1.56}$$

It has the properties $\tanh 0 = 0$, $-1 < \tanh x < 1$, $\forall \ x \in \mathbb{R}$, and $\tanh (-x) = -\tanh x$. Furthermore $\tanh x \to 1$ as $x \to \infty$, and $\tanh x \to -1$, as $x \to -\infty$, and it is strictly increasing on $\mathbb{R}$.

This function plays the role of an activation function in the hidden layer of neural networks.

We further consider [5]

$$\Psi (x) := \frac{1}{4} (\tanh (x+1) - \tanh (x-1)) > 0, \ \forall x \in \mathbb{R}. \tag{1.57}$$

We easily see that $\Psi (-x) = \Psi (x)$, that is Ψ is even on $\mathbb{R}$. Obviously Ψ is differentiable, thus continuous.

It follows

Proposition 1.20 ([5]) *$\Psi (x)$ for $x \geq 0$ is strictly decreasing.*

Obviously $\Psi (x)$ is strictly increasing for $x \leq 0$. Also it holds $\lim_{x \to -\infty} \Psi (x) = 0 = \lim_{x \to \infty} \Psi (x)$.

Infact Ψ has the bell shape with horizontal asymptote the x-axis. So the maximum of Ψ is zero, $\Psi (0) = 0.3809297$.

Theorem 1.21 ([5]) *We have that $\sum_{i=-\infty}^{\infty} \Psi (x-i) = 1, \ \forall x \in \mathbb{R}$.*

Thus

$$\sum_{i=-\infty}^{\infty} \Psi (nx - i) = 1, \quad \forall n \in \mathbb{N}, \ \forall x \in \mathbb{R}.$$

Also it holds

$$\sum_{i=-\infty}^{\infty} \Psi (x + i) = 1, \ \forall x \in \mathbb{R}.$$

Theorem 1.22 ([5]) *It holds $\int_{-\infty}^{\infty} \Psi (x) \, dx = 1$.*

So $\Psi (x)$ is a density function on $\mathbb{R}$.

Theorem 1.23 ([5]) *Let $0 < \alpha < 1$ and $n \in \mathbb{N}$. It holds*

$$\sum_{\substack{k=-\infty \\ : |nx-k| \geq n^{1-\alpha}}}^{\infty} \Psi (nx - k) \leq e^4 \cdot e^{-2n^{(1-\alpha)}}.$$

Remark 1.24 In this chapter we also use the complete multivariate activation function

$$\Theta(x_1, ..., x_N) := \Theta(x) := \prod_{i=1}^{N} \Psi(x_i), \; x = (x_1, ..., x_N) \in \mathbb{R}^N, \; N \in \mathbb{N}. \quad (1.58)$$

It has the properties (see [6]):

(i) $\Theta(x) > 0$, $\forall \, x \in \mathbb{R}^N$,

(ii)

$$\sum_{k=-\infty}^{\infty} \Theta(x-k) := \sum_{k_1=-\infty}^{\infty} \sum_{k_2=-\infty}^{\infty} \cdots \sum_{k_N=-\infty}^{\infty} \Theta(x_1 - k_1, x_2 - k_2, ..., x_N - k_N) = 1, \quad (1.59)$$

where $k := (k_1, ..., k_N)$, $\forall \, x \in \mathbb{R}^N$.

(iii)

$$\sum_{k=-\infty}^{\infty} \Theta(nx - k) :=$$

$$\sum_{k_1=-\infty}^{\infty} \sum_{k_2=-\infty}^{\infty} \cdots \sum_{k_N=-\infty}^{\infty} \Theta(nx_1 - k_1, nx_2 - k_2, ..., nx_N - k_N) = 1, \quad (1.60)$$

$\forall \, x \in \mathbb{R}^N$; $n \in \mathbb{N}$.

(iv)

$$\int_{\mathbb{R}^N} \Theta(x) \, dx = 1, \quad (1.61)$$

that is Θ is a multivariate density function.

By [6] we get

(v)

$$\sum_{\substack{k=-\infty \\ \left\| \frac{k}{n} - x \right\|_{\infty} > \frac{1}{n^{\beta}}}}^{\infty} \Theta(nx - k) \leq e^4 \cdot e^{-2n^{(1-\beta)}}, \quad (1.62)$$

$0 < \beta < 1$; $n \in \mathbb{N}$, $x \in \mathbb{R}^N$.

We make

Definition 1.25 Let $\mathcal{L}^*$ be the Lebesgue σ-algebra on $\mathbb{R}^N$, $N \in \mathbb{N}$, and the set function $\mu^* : \mathcal{L}^* \to [0, +\infty)$, which is assumed to be monotone, submodular and strictly positive. For $f \in C_B^+(\mathbb{R}^N)$, we define the multivariate Kantorovich–Choquet type neural network operator for any $x \in \mathbb{R}^N$:

$$L_n^{\mu^*}(f, x) = L_n^{\mu^*}(f, x_1, ..., x_N) :=$$

$$\sum_{k=-\infty}^{\infty}\left(\frac{(C)\int_{[0,\frac{1}{n}]^N} f\left(t+\frac{k}{n}\right)d\mu^*(t)}{\mu^*\left(\left[0,\frac{1}{n}\right]^N\right)}\right)\Theta(nx-k)= \tag{1.63}$$

$$\sum_{k_1=-\infty}^{\infty}\sum_{k_2=-\infty}^{\infty}\cdots\sum_{k_N=-\infty}^{\infty}\left(\frac{(C)\int_0^{\frac{1}{n}}\dots\int_0^{\frac{1}{n}} f\left(t_1+\frac{k_1}{n},t_2+\frac{k_2}{n},\dots,t_N+\frac{k_N}{n}\right)d\mu^*(t_1,\dots,t_N)}{\mu^*\left(\left[0,\frac{1}{n}\right]^N\right)}\right)$$

$$\cdot\left(\prod_{i=1}^{N}\Psi(nx_i-k_i)\right),$$

where $x=(x_1,\dots,x_N)\in\mathbb{R}^N$, $k=(k_1,\dots,k_N)$, $t=(t_1,\dots,t_N)$, $n\in\mathbb{N}$.

Clearly here $\mu^*\left(\left[0,\frac{1}{n}\right]^N\right)>0$, $\forall\, n\in\mathbb{N}$.

Above we notice that

$$\left\|L_n^{\mu^*}(f)\right\|_\infty\le\|f\|_\infty, \tag{1.64}$$

so that $L_n^{\mu^*}(f,x)$ is well-defined.

We give

Theorem 1.26 *Let $f\in C_B^+(\mathbb{R}^N)$, $0<\beta<1$, $x\in\mathbb{R}^N$; $n,N\in\mathbb{N}$. Then*

(i)

$$\sup_{\mu^*}\left|K_n^{\mu^*}(f,x)-f(x)\right|\le\omega_1\left(f,\frac{1}{n}+\frac{1}{n^\beta}\right)+(6.3984)\|f\|_\infty e^{-n^{(1-\beta)}}=:\lambda_{1n}, \tag{1.65}$$

(ii)

$$\sup_{\mu^*}\left\|K_n^{\mu^*}(f)-f\right\|_\infty\le\lambda_{1n}. \tag{1.66}$$

Given that $f\in\left(C_U^+(\mathbb{R}^N)\cap C_B^+(\mathbb{R}^N)\right)$, we obtain $\lim_{n\to\infty}K_n^{\mu^*}(f)=f$, uniformly.

Proof As similar to Theorem 1.14 is omitted. ∎

We also present

Theorem 1.27 *Let $f\in C_B^+(\mathbb{R}^N)$, $0<\beta<1$, $x\in\mathbb{R}^N$; $n,N\in\mathbb{N}$. Then*

(i)

$$\sup_{\mu^*}\left|L_n^{\mu^*}(f,x)-f(x)\right|\le\omega_1\left(f,\frac{1}{n}+\frac{1}{n^\beta}\right)+\|f\|_\infty 2e^4e^{-2n^{(1-\beta)}}=:\lambda_{2n}, \tag{1.67}$$

(ii)

$$\sup_{\mu^*}\left\|L_n^{\mu^*}(f)-f\right\|_\infty\le\lambda_{2n}. \tag{1.68}$$

Given that $f \in \left(C_U^+\left(\mathbb{R}^N\right) \cap C_B^+\left(\mathbb{R}^N\right)\right)$, *we obtain* $\lim_{n\to\infty} L_n^{\mu^*}(f) = f$, *uniformly.*

Proof As similar to Theorem 1.14 is omitted. ■

We need

Definition 1.28 We set for $f \in C_B^+\left(\mathbb{R}^N, \mathbb{C}\right)$ that

$$K_n^{\mu^*}(f, x) := K_n^{\mu^*}(f_1, x) + i K_n^{\mu^*}(f_2, x), \tag{1.69}$$

$$L_n^{\mu^*}(f, x) := L_n^{\mu^*}(f_1, x) + i L_n^{\mu^*}(f_2, x),$$

$\forall\, n \in \mathbb{N}, x \in \mathbb{R}^N$.

We finish with

Theorem 1.29 *Let* $f \in C_B^+\left(\mathbb{R}^N, \mathbb{C}\right)$, $f = f_1 + i f_2$, $0 < \beta < 1$, $x \in \mathbb{R}^N$, $n, N \in \mathbb{N}$. *Then*

(i)

$$\sup_{\mu^*}\left|K_n^{\mu^*}(f, x) - f(x)\right| \leq \left(\omega_1\left(f_1, \frac{1}{n} + \frac{1}{n^\beta}\right) + \omega_1\left(f_2, \frac{1}{n} + \frac{1}{n^\beta}\right)\right) + \tag{1.70}$$

$$\left(\|f_1\|_\infty + \|f_2\|_\infty\right)(6.3984)\, e^{-n^{(1-\beta)}} =: \lambda_{3n},$$

(ii)

$$\sup_{\mu^*}\left\|K_n^{\mu^*}(f) - f\right\|_\infty \leq \lambda_{3n}, \tag{1.71}$$

(iii)

$$\sup_{\mu^*}\left|L_n^{\mu^*}(f, x) - f(x)\right| \leq \left(\omega_1\left(f_1, \frac{1}{n} + \frac{1}{n^\beta}\right) + \omega_1\left(f_2, \frac{1}{n} + \frac{1}{n^\beta}\right)\right) + \tag{1.72}$$

$$\left(\|f_1\|_\infty + \|f_2\|_\infty\right) 2e^4 e^{-2n^{(1-\beta)}} =: \lambda_{4n},$$

(iv)

$$\sup_{\mu^*}\left\|L_n^{\mu^*}(f) - f\right\|_\infty \leq \lambda_{4n}. \tag{1.73}$$

Proof Similar to Theorem 1.10. ■

References

1. M. Abramowitz, I.A. Stegun (eds.), *Handbook of Mathematical Functions with Formulas, Graphs, and Mathematical Tables* (Dover Publications, New York, 1972)

2. G.A. Anastassiou, Rate of convergence of some neural network operators to the unit-univariate case. J. Math. Anal. Appl. **212**, 237–262 (1997)
3. G.A. Anastassiou, *Quantitative Approximations* (Chapman & Hall/CRC, New York, 2001)
4. G.A. Anastassiou, *Intelligent Systems: Approximation by Artificial Neural Networks, Intelligent Systems Reference Library*, vol. 19 (Springer, Heidelberg, 2011)
5. G.A. Anastassiou, Univariate hyperbolic tangent neural network approximation. Math. Comput. Model. **53**, 1111–1132 (2011)
6. G.A. Anastassiou, Multivariate hyperbolic tangent neural network approximation. Comput. Math. **61**, 809–821 (2011)
7. G.A. Anastassiou, Multivariate sigmoidal neural network approximation. Neural Netw. **24**, 378–386 (2011)
8. G.A. Anastassiou, Univariate sigmoidal neural network approximation. J. Comput. Anal. Appl. **14**(4), 659–690 (2012)
9. G.A. Anastassiou, Fractional neural network approximation. Comput. Math. Appl. **64**, 1655–1676 (2012)
10. G.A. Anastassiou, Univariate error function based neural network approximation. Indian J. Math. **57**(2), 243–291 (2015)
11. G.A. Anastassiou, Quantitative approximation by Kantorovich–Choquet quasi-interpolation neural network operators. Acta Mathematica Universitatis Comenianae (2018). Accepted for publication
12. L.C. Andrews, *Special Functions of Mathematics for Engineers*, 2nd edn. (Mc Graw-Hill, New York, 1992)
13. Z. Chen, F. Cao, The approximation operators with sigmoidal functions. Comput. Math. Appl. **58**, 758–765 (2009)
14. G. Choquet, Theory of capacities. Ann. Inst. Fourier (Grenoble) **5**, 131–295 (1954)
15. D. Denneberg, *Non-additive Measure and Integral* (Kluwer, Dordrecht, 1994)
16. S. Gal, Uniform and pointwise quantitative approximation by Kantorovich–Choquet type integral operators with respect to monotone and submodular set functions. Mediterr. J. Math. **14**(5), 12 pp. (2017). Art. 205
17. I.S. Haykin, *Neural Networks: A Comprehensive Foundation*, 2nd edn. (Prentice Hall, New York, 1998)
18. W. McCulloch, W. Pitts, A logical calculus of the ideas immanent in nervous activity. Bull. Math. Biophys. **7**, 115–133 (1943)
19. T.M. Mitchell, *Machine Learning* (WCB-McGraw-Hill, New York, 1997)
20. Z. Wang, G.J. Klir, *Generalized Measure Theory* (Springer, New York, 2009)

Chapter 2
Approximation with Rates by Perturbed Kantorovich–Choquet Neural Network Operators

This chapter deals with the determination of the rate of convergence to the unit of Perturbed Kantorovich–Choquet univariate and multivariate normalized neural network operators of one hidden layer. These are given through the univariate and multivariate moduli of continuity of the involved univariate or multivariate function or its high order derivatives and that appears in the right-hand side of the associated univariate and multivariate Jackson type inequalities. The activation function is very general, especially it can derive from any univariate or multivariate sigmoid or bell-shaped function. The right hand sides of our convergence inequalities do not depend on the activation function. The sample functionals are of Kantorovich–Choquet type. We give applications for the first derivatives of the involved function. It follows [1].

2.1 Introduction

Feed-forward neural networks (FNNs) with one hidden layer, the only type of networks we deal with in this chapter are mathematically expressed as

$$N_n(x) = \sum_{j=0}^{n} c_j \sigma\left(\langle a_j \cdot x\rangle + b_j\right), \; x \in \mathbb{R}^s, \; s \in \mathbb{N},$$

where for $0 \leq j \leq n$, $b_j \in \mathbb{R}$ are the thresholds, $a_j \in \mathbb{R}^s$ are the connection weights, $c_j \in \mathbb{R}$ are the coefficients, $\langle a_j \cdot x\rangle$ is the inner product of a_j and x, and σ is the activation function of the network. In many fundamental network models, the activation functions is the sigmoid function of logistic type or other sigmoid function or bell-shaped function.

It is well known that FNN's are universal approximators. Theoretically, any continuous function defined on a compact set can be approximated to any desired degree of accuracy by increasing the number of hidden neurons. It was proved by Cybenko

G. A. Anastassiou, *Ordinary and Fractional Approximation by Non-additive Integrals: Choquet, Shilkret and Sugeno Integral Approximators*, Studies in Systems, Decision and Control 190, https://doi.org/10.1007/978-3-030-04287-5_2

[16] and Funahashi [19], that any continuous function can be approximated on a compact set with uniform topology by a network of the form $N_n(x)$, using any continuous, sigmoid activation function. Hornik et al. in [23], have shown that any measurable function can be approached with such a network. Furthermore, these authors proved in [24], that any function of the Sobolev spaces can be approached with all derivatives. A variety of density results on FNN approximations to multivariate functions were later established by many authors using different methods, for more or less general situations: [25] by Leshno et al. [29] by Mhaskar and Micchelli, [13] by Chui and Li, [11] by Chen and Chen, [21] by Hahm and Hong, etc.

Usually these results only give theorems about the existence of an approximation. A related and important problem is that of complexity: determining the number of neurons required to guarantee that all functions belonging to a space can be approximated to the prescribed degree of accuracy ϵ.

Barron [7] shows that if the function is supposed to satisfy certain conditions expressed in terms of its Fourier transform, and if each of the neurons evaluates a sigmoid activation function, then at most $O\left(\epsilon^{-2}\right)$ neurons are needed to achieve the order of approximation ϵ. Some other authors have published similar results on the complexity of FNN approximations: Mhaskar and Micchelli [30], Suzuki [32], Maiorov and Meir [26], Makovoz [27], Ferrari and Stengel [18], Xu and Cao [34], Cao et al. [8], etc.

P. Cardaliaguet and G. Euvrard were the first, see [9], to describe precisely and study neural network approximation operators to the unit operator. Namely they showed that: be given $f : \mathbb{R} \to \mathbb{R}$ a continuous bounded function and b a centered bell-shaped function, then the functions

$$F_n(x) = \sum_{k=-n^2}^{n^2} \frac{f\left(\frac{k}{n}\right)}{In^{\alpha}} b\left(n^{1-\alpha}\left(x - \frac{k}{n}\right)\right),$$

where $I := \int_{-\infty}^{\infty} b(t)\,dt$, $0 < \alpha < 1$, converge uniformly on compacta to f.

We see above that the weights $\frac{f\left(\frac{k}{n}\right)}{In^{\alpha}}$ are explicitly known, for the first time shown in [9].

Furthermore the authors in [9] proved that: let $f : \mathbb{R}^p \to \mathbb{R}$, $p \in \mathbb{N}$, be a continuous bounded function and b a p-dimensional bell-shaped function. Then the functions

$$G_n(x) =$$

$$\sum_{k_1=-n^2}^{n^2} \cdots \sum_{k_p=-n^2}^{n^2} \frac{f\left(\frac{k_1}{n}, \ldots, \frac{k_p}{n}\right)}{In^{\alpha}} b\left(n^{1-\alpha}\left(x_1 - \frac{k_1}{n}\right), \ldots, n^{1-\alpha}\left(x_p - \frac{k_p}{n}\right)\right),$$

where I is the interval of b on $\mathbb{R}^p$ and $0 < \alpha < 1$, converge uniformly on compacta to f.

Still the work [9] is qualitative and non quantitative.

The author in [2–4], see Chapters 2–5, was the first to establish neural network approximations to continuous functions with rates, that is quantitative works, by very specifically defined neural network operators of Cardaliaguet-Euvrard and "Squashing" types, by employing the modulus of continuity of the engaged function or its high order derivative or partial derivatives, and producing very tight Jackson type inequalities. He treats there both the univariate and multivariate cases. The defining these operators "bell-shaped" and "squashing" functions are assumed to be of compact support. Also in [4] gave the Nth order asymptotic expansion for the error of weak approximation of these two operators to a special natural class of smooth functions, see Chaps. 4 and 5 there.

Though the work in [2–4], was quantitative, the rate of convergence is not calculated.

Finally the author in [5, 6], by normalizing his operators he achieved to determine the exact rates of convergence.

In this chapter the author continues and extends his related work, by introducing the Baskakov-Choquet univariate and multivariate perturbed neural network operators of Cardaliaguet-Euvrard type.

The same coefficient $f\left(\frac{k}{n}\right)$ is replaced by as suitable natural perturbations via a Choquet integral, what is actually happens in reality of neural network operations.

The calculation of $f\left(\frac{k}{n}\right)$ at the neurons many times are not calculated as such, but rather in a distored way.

Next we justify why we take here the univariate and multivariate activation functions to be of compact support, of course it helps us to conduct our study.

The activation function, same as transfer function or learning rule, is connected and associated to firing of neurons. Firing, which sends electric pulses or an output signal to other neurons, occurs when the activation level is above the threshold level set by the learning rule.

Each Neural Network firing is essentially of finite time duration. Essentially the firing in time decays, but in practice sends positive energy over a finite time interval.

Thus by using an activation function of compact support, in practice we do not alter much of the good results of our approximation.

To be more precise, in the multivariate case, we may take the compact support to be a large symmetric to the origin box. This is what happens in real time with the firing of neurons.

For more about neural network in general we refer to [10, 14, 15, 22, 28, 31].

2.2 Background

Next we present briefly about the Choquet integral.

We define the following

Definition 2.1 Consider $\Omega \neq \varnothing$ and let $\mathcal{C}$ be a σ-algebra of subsets in Ω.

(i) (see, e.g., [33], p. 63) The set function $\mu : \mathcal{C} \to [0, +\infty]$ is called a monotone set function (or capacity) if $\mu(\emptyset) = 0$ and $\mu(A) \leq \mu(B)$ for all $A, B \in \mathcal{C}$, with $A \subset B$. Also, μ is called submodular if

$$\mu(A \cup B) + \mu(A \cap B) \leq \mu(A) + \mu(B), \text{ for all } A, B \in \mathcal{C}.$$

μ is called bounded if $\mu(\Omega) < +\infty$ and normalized if $\mu(\Omega) = 1$.

(ii) (see, e.g., [33], p. 233, or [12]) If μ is a monotone set function on $\mathcal{C}$ and if $f : \Omega \to \mathbb{R}$ is $\mathcal{C}$-measurable (that is, for any Borel subset $B \subset \mathbb{R}$ it follows $f^{-1}(B) \in \mathcal{C}$), the for any $A \in \mathcal{C}$, the Choquet integral is defined by

$$(C)\int_A f d\mu = \int_0^{+\infty} \mu\left(F_\beta(f) \cap A\right) d\beta + \int_{-\infty}^0 \left[\mu\left(F_\beta(f) \cap A\right) - \mu(A)\right] d\beta,$$

where we used the notation $F_\beta(f) = \{\omega \in \Omega : f(\omega) \geq \beta\}$. Notice that if $f \geq 0$ on A, then in the above formula we get $\int_{-\infty}^0 = 0$.

The integrals on the right-hand side are the usual Riemann integral.

The function f will be called Choquet integrable on A if $(C)\int_A f d\mu \in \mathbb{R}$.

Next we list some well known properties of the Choquet integral.

Remark 2.2 If $\mu : \mathcal{C} \to [0, +\infty]$ is a monotone set function, then the following properties hold:

(i) For all $a \geq 0$ we have $(C)\int_A af d\mu = a \cdot (C)\int_A f d\mu$ (if $f \geq 0$ then see, e.g., [33], Theorem 11.2, (5), p. 228 and if f is arbitrary sign, then see, e.g., [17], p. 64, Proposition 5.1, (ii)).

(ii) For all $c \in \mathbb{R}$ and f of arbitrary sign, we have (see, e.g., [33], pp. 232–233, or [17], p. 65) $(C)\int_A (f + c) d\mu = (C)\int_A f d\mu + c \cdot \mu(A)$.

If μ is submodular too, then for all f, g of arbitrary sign and lower bounded, we have (see, e.g., [17], p. 75, Theorem 6.3)

$$(C)\int_A (f + g) d\mu \leq (C)\int_A f d\mu + (C)\int_A g d\mu.$$

(iii) If $f \leq g$ on A then $(C)\int_A f d\mu \leq (C)\int_A g d\mu$ (see, e.g., [33], p. 228, Theorem 11.2, (3) if $f, g \geq 0$ and p. 232 if f, g are of arbitrary sign).

(iv) Let $f \geq 0$. If $A \subset B$ then $(C)\int_A f d\mu \leq (C)\int_B f d\mu$. In addition, if μ is finitely subadditive, then

$$(C)\int_{A \cup B} f d\mu \leq (C)\int_A f d\mu + (C)\int_B f d\mu.$$

(v) It is immediate that $(C)\int_A 1 \cdot d\mu(t) = \mu(A)$.

(vi) The formula $\mu(A) = \gamma(M(A))$, where $\gamma : [0, 1] \to [0, 1]$ is an increasing and concave function, with $\gamma(0) = 0$, $\gamma(1) = 1$ and M is a probability measure (or only finitely additive) on a σ-algebra on Ω (that is, $M(\emptyset) = 0$, $M(\Omega) = 1$ and M is countably additive), gives simple examples of normalized, monotone and submodular set functions (see, e.g., [17], pp. 16-17, Example 2.1). Such of set functions μ are also called distorsions of countably normalized, additive measures (or distorted measures). For a simple example, we can take $\gamma(t) = \frac{2t}{1+t}$, $\gamma(t) = \sqrt{t}$.

If the above γ function is increasing, concave and satisfies only $\gamma(0) = 0$, then for any bounded Borel measure m, $\mu(A) = \gamma(m(A))$ gives a simple example of bounded, monotone and submodular set function.

(vii) If μ is a countably additive bounded measure, then the Choquet integral $(C)\int_A f d\mu$ reduces to the usual Lebesgue type integral (see, e.g., [17], p. 62, or [33], p. 226).

(viii) If $f \geq 0$, then $(C)\int_A f d\mu \geq 0$.

(ix) Let $\mu = \sqrt{M}$, where M is the Lebesgue measure on $[0, +\infty)$, then μ is a monotone and submodular set function, furthermore μ is strictly positive, see [20].

(x) If $\Omega = \mathbb{R}^N$, $N \in \mathbb{N}$, we call μ strictly positive if $\mu(A) > 0$, for any open subset $A \subseteq \mathbb{R}^N$.

2.3 Univariate Results

Here the activation function $b : \mathbb{R} \to \mathbb{R}_+$ is of compact support $[-T, T]$, $T > 0$. That is, $b(x) > 0$ for any $x \in [-T, T]$, and clearly b may have jump discontinuities. Also the shape of the graph of b could be anything. Typically in neural networks approximation we take b as a sigmoidal function or bell-shaped function, of course here of compact support $[-T, T]$, $T > 0$.

Example 2.3 (i) b can be the characteristic function on $[-1, 1]$,

(ii) b can be that hat function over $[-1+\varepsilon, 1-\varepsilon]$, where $\varepsilon > 0$ is small i.e.,

$$b(x) = \begin{cases} 1+x, & -1 \leq x \leq 0, \\ 1-x, & 0 < x \leq 1, \\ 0, & \text{elsewhere,} \end{cases}$$

(iii) the truncated sigmoidals

$$b(x) = \begin{cases} \frac{1}{1+e^{-x}} \text{ or } \tanh x \text{ or } erf(x), & \text{for } x \in [-T, T], \text{ with large } T > 0, \\ 0, \ x \in \mathbb{R} - [-T, T], \end{cases}$$

(iv) the truncated Gompertz function

$$b(x) = \begin{cases} e^{-\alpha e^{-\beta x}}, \ x \in [-T, T]; \ \alpha, \beta > 0; \ \text{large } T > 0, \\ 0, \ x \in \mathbb{R} - [-T, T], \end{cases}$$

The Gompertz functions are also sigmoidal functions, with wide applications to many applied fields, e.g. demography and tumor growth modeling, etc.

So the general function b we will be using here covers all kinds of activation functions in neural network approximations.

Here we consider functions $f : \mathbb{R} \to \mathbb{R}_+$ that are either continuous and bounded, or uniformly continuous.

Let here the parameters $\overline{\mu}, \overline{\nu} \geq 0$; $\mu_i, \nu_i \geq 0$, $i = 1, ..., r \in \mathbb{N}$; $w_i \geq 0 : \sum_{i=1}^{r} w_i = 1$; $0 < \alpha < 1$, $x \in \mathbb{R}$, $n \in \mathbb{N}$.

We use here the first modulus of continuity

$$\omega_1(f, \delta) := \sup_{\substack{x, y \in \mathbb{R} \\ |x-y| \leq \delta}} |f(x) - f(y)|,$$

and given that f is uniformly continuous we get $\lim_{\delta \to 0} \omega_1(f, \delta) = 0$.

In this chapter mainly we study the pointwise convergence with rates over $\mathbb{R}$, to the unit operator, of the following defined one hidden layer normalized neural network perturbed operators:

Definition 2.4 Let $\mathcal{L}$ be the Lebesgue σ-algebra on $\mathbb{R}$, and the set function $\mu : \mathcal{L} \to [0, +\infty)$, which is assumed to be monotone, submodular and strictly positive. We define the univariate Kantorovich–Choquet type neural network operators ($n \in \mathbb{N}$, $x \in \mathbb{R}$):

$$\left(K_n^{\mu}(f)\right)(x) =$$

$$\frac{\left[\sum_{k=-n^2}^{n^2} \left(\sum_{i=1}^{r} w_i \frac{1}{\mu\left(\left[0, \frac{1}{n+\nu_i}\right]\right)} (C) \int_0^{\frac{1}{n+\nu_i}} f\left(t + \frac{k+\mu_i}{n+\nu_i}\right) d\mu(t) \right) b\left(n^{1-\alpha}\left(x - \frac{k}{n}\right)\right)\right]}{\sum_{k=-n^2}^{n^2} b\left(n^{1-\alpha}\left(x - \frac{k}{n}\right)\right)}. \tag{2.1}$$

For f continuous and bounded we get

$$\left\| K_n^{\mu}(f) \right\|_{\infty} \leq \|f\|_{\infty}, \text{ plus } K_n^{\mu}(1) = 1.$$

The flexible operator K_n^{μ} in the corresponding signal processing context, represents the natural called "time-jitter" error, where the sample information is calculated in a perturbed neighborhood of $\frac{k+\overline{\mu}}{n+\overline{\nu}}$ rather than exactly at the node $\frac{k}{n}$.

The terms in the ratio of sums (2.1) are nonzero, if and only if

$$\left| n^{1-\alpha}\left(x - \frac{k}{n}\right) \right| \leq T, \text{ i.e. } \left| x - \frac{k}{n} \right| \leq \frac{T}{n^{1-\alpha}} \tag{2.2}$$

iff

$$nx - Tn^{\alpha} \leq k \leq nx + Tn^{\alpha}. \tag{2.3}$$

In order to have the desired order of the numbers

$$-n^2 \leq nx - Tn^{\alpha} \leq nx + Tn^{\alpha} \leq n^2, \tag{2.4}$$

it is sufficiently enough to assume that

$$n \geq T + |x|. \tag{2.5}$$

When $x \in [-T, T]$ it is enough to assume $n \geq 2T$, which implies (2.4).

Proposition 2.5 ([2]) *Let $a \leq b$, $a, b \in \mathbb{R}$. Let $card(k)$ (≥ 0) be the maximum number of integers contained in $[a, b]$. Then*

$$\max(0, (b-a)-1) \leq card(k) \leq (b-a)+1. \tag{2.6}$$

Note 2.6 We would like to establish a lower bound on $card(k)$ over the interval $[nx - Tn^{\alpha}, nx + Tn^{\alpha}]$. From Proposition 2.5 we get that

$$card(k) \geq \max(2Tn^{\alpha} - 1, 0). \tag{2.7}$$

We obtain card $(k) \geq 1$, if

$$2Tn^{\alpha} - 1 \geq 1 \text{ iff } n \geq T^{-\frac{1}{\alpha}}. \tag{2.8}$$

So to have the desired order (2.4) and $card(k) \geq 1$ over $[nx - Tn^{\alpha}, nx + Tn^{\alpha}]$, we need to consider

$$n \geq \max\left(T + |x|, T^{-\frac{1}{\alpha}}\right). \tag{2.9}$$

Also notice that $card(k) \to +\infty$, as $n \to +\infty$.

Denote by $[\cdot]$ the integral part of a number and by $\lceil \cdot \rceil$ its ceiling.

So under assumption (2.9), $K_n^{\mu}(f)$ collapses to:

$$\left(K_n^{\mu}(f)\right)(x) =$$

$$\frac{\left[\sum_{k=\lceil nx-Tn^{\alpha}\rceil}^{[nx+Tn^{\alpha}]} \left(\sum_{i=1}^{r} w_i \frac{1}{\mu\left(\left[0, \frac{1}{n+\nu_i}\right]\right)} (C) \int_0^{\frac{1}{n+\nu_i}} f\left(t + \frac{k+\mu_i}{n+\nu_i}\right) d\mu(t)\right) b\left(n^{1-\alpha}\left(x - \frac{k}{n}\right)\right)\right]}{\sum_{k=\lceil nx-Tn^{\alpha}\rceil}^{[nx+Tn^{\alpha}]} b\left(n^{1-\alpha}\left(x - \frac{k}{n}\right)\right)}. \tag{2.10}$$

We make

Remark 2.7 Let k as in (2.3). We observe that

$$\left|\frac{k+\overline{\mu}}{n+\overline{\nu}}-x\right| \leq \left|\frac{k}{n+\overline{\nu}}-x\right| + \frac{\overline{\mu}}{n+\overline{\nu}}. \tag{2.11}$$

Next we see

$$\left|\frac{k}{n+\overline{\nu}}-x\right| \leq \left|\frac{k}{n+\overline{\nu}}-\frac{k}{n}\right| + \left|\frac{k}{n}-x\right| \overset{(2.2)}{\leq} \frac{\overline{\nu}\,|k|}{n\,(n+\overline{\nu})} + \frac{T}{n^{1-\alpha}} \tag{2.12}$$

(by $|k| \leq \max\left(|nx - Tn^{\alpha}|, |nx + Tn^{\alpha}|\right) \leq n\,|x| + Tn^{\alpha}$)

$$\leq \left(\frac{\overline{\nu}}{n+\overline{\nu}}\right)\left(|x| + \frac{T}{n^{1-\alpha}}\right) + \frac{T}{n^{1-\alpha}}. \tag{2.13}$$

Consequently it holds

$$\left|\frac{k+\overline{\mu}}{n+\overline{\nu}}-x\right| \leq \left(\frac{\overline{\nu}}{n+\overline{\nu}}\right)\left(|x| + \frac{T}{n^{1-\alpha}}\right) + \frac{T}{n^{1-\alpha}} + \frac{\overline{\mu}}{n+\overline{\nu}} = \tag{2.14}$$

$$\left(\frac{\overline{\nu}\,|x| + \overline{\mu}}{n+\overline{\nu}}\right) + \left(1 + \frac{\overline{\nu}}{n+\overline{\nu}}\right)\frac{T}{n^{1-\alpha}}.$$

Hence we obtain

$$\omega_1\left(f, \left|\frac{k+\overline{\mu}}{n+\overline{\nu}}-x\right|\right) \overset{(2.14)}{\leq} \omega_1\left(f, \left(\frac{\overline{\nu}\,|x| + \overline{\mu}}{n+\overline{\nu}}\right) + \left(1 + \frac{\overline{\nu}}{n+\overline{\nu}}\right)\frac{T}{n^{1-\alpha}}\right), \tag{2.15}$$

where $\overline{\mu}, \overline{\nu} \geq 0$, $0 < \alpha < 1$, so that the dominant speed above is $\frac{1}{n^{1-\alpha}}$.

We present the following

Theorem 2.8 *Let $f : \mathbb{R} \to \mathbb{R}_+$ continuous and bounded or uniformly continuous. Let $x \in \mathbb{R}$, $T > 0$ and $n \in \mathbb{N}$ such that $n \geq \max\left(T + |x|, T^{-\frac{1}{\alpha}}\right)$. Then*

$$\sup_{\mu}\left|\left(K_n^{\mu}(f)\right)(x) - f(x)\right| \leq \tag{2.16}$$

$$\max_{i\in\{1,\ldots,r\}}\left\{\omega_1\left(f, \left(\frac{\nu_i\,|x| + \mu_i + 1}{n+\nu_i}\right) + \left(1 + \frac{\nu_i}{n+\nu_i}\right)\frac{T}{n^{1-\alpha}}\right)\right\}.$$

If f is uniformly continuous, then $\lim_{n\to+\infty} K_n^{\mu}(f)(x) = f(x)$.

Proof We notice the following:

$$f\left(t + \frac{k+\mu_i}{n+\nu_i}\right) = f\left(t + \frac{k+\mu_i}{n+\nu_i}\right) - f(x) + f(x) \leq$$

$$\left|f\left(t+\frac{k+\mu_i}{n+\nu_i}\right)-f(x)\right|+f(x). \tag{2.17}$$

Hence

$$\frac{1}{\mu\left(\left[0,\frac{1}{n+\nu_i}\right]\right)}(C)\int_0^{\frac{1}{n+\nu_i}} f\left(t+\frac{k+\mu_i}{n+\nu_i}\right)d\mu(t) \le$$

$$\frac{1}{\mu\left(\left[0,\frac{1}{n+\nu_i}\right]\right)}(C)\int_0^{\frac{1}{n+\nu_i}} \left|f\left(t+\frac{k+\mu_i}{n+\nu_i}\right)-f(x)\right|d\mu(t)+f(x). \tag{2.18}$$

That is

$$\frac{1}{\mu\left(\left[0,\frac{1}{n+\nu_i}\right]\right)}(C)\int_0^{\frac{1}{n+\nu_i}} f\left(t+\frac{k+\mu_i}{n+\nu_i}\right)d\mu(t)-f(x) \le$$

$$\frac{1}{\mu\left(\left[0,\frac{1}{n+\nu_i}\right]\right)}(C)\int_0^{\frac{1}{n+\nu_i}} \left|f\left(t+\frac{k+\mu_i}{n+\nu_i}\right)-f(x)\right|d\mu(t). \tag{2.19}$$

Similarly, we have

$$f(x)=f(x)-f\left(t+\frac{k+\mu_i}{n+\nu_i}\right)+f\left(t+\frac{k+\mu_i}{n+\nu_i}\right)\le$$

$$\left|f\left(t+\frac{k+\mu_i}{n+\nu_i}\right)-f(x)\right|+f\left(t+\frac{k+\mu_i}{n+\nu_i}\right). \tag{2.20}$$

Hence

$$f(x)\le\frac{1}{\mu\left(\left[0,\frac{1}{n+\nu_i}\right]\right)}(C)\int_0^{\frac{1}{n+\nu_i}} \left|f\left(t+\frac{k+\mu_i}{n+\nu_i}\right)-f(x)\right|d\mu(t)+$$

$$\frac{1}{\mu\left(\left[0,\frac{1}{n+\nu_i}\right]\right)}(C)\int_0^{\frac{1}{n+\nu_i}} f\left(t+\frac{k+\mu_i}{n+\nu_i}\right)d\mu(t). \tag{2.21}$$

That is,

$$f(x)-\frac{1}{\mu\left(\left[0,\frac{1}{n+\nu_i}\right]\right)}(C)\int_0^{\frac{1}{n+\nu_i}} f\left(t+\frac{k+\mu_i}{n+\nu_i}\right)d\mu(t)\le \tag{2.22}$$

$$\frac{1}{\mu\left(\left[0, \frac{1}{n+\nu_i}\right]\right)} (C) \int_0^{\frac{1}{n+\nu_i}} \left| f\left(t + \frac{k+\mu_i}{n+\nu_i}\right) - f(x) \right| d\mu(t).$$

We have proved that

$$\left| \frac{1}{\mu\left(\left[0, \frac{1}{n+\nu_i}\right]\right)} (C) \int_0^{\frac{1}{n+\nu_i}} f\left(t + \frac{k+\mu_i}{n+\nu_i}\right) d\mu(t) - f(x) \right| \leq \tag{2.23}$$

$$\frac{1}{\mu\left(\left[0, \frac{1}{n+\nu_i}\right]\right)} (C) \int_0^{\frac{1}{n+\nu_i}} \left| f\left(t + \frac{k+\mu_i}{n+\nu_i}\right) - f(x) \right| d\mu(t).$$

We call

$$\delta_{n,k}(f) := \sum_{i=1}^{r} w_i \frac{1}{\mu\left(\left[0, \frac{1}{n+\nu_i}\right]\right)} (C) \int_0^{\frac{1}{n+\nu_i}} f\left(t + \frac{k+\mu_i}{n+\nu_i}\right) d\mu(t). \tag{2.24}$$

We observe that

$$\delta_{n,k}(f) - f(x) = \sum_{i=1}^{r} w_i \frac{1}{\mu\left(\left[0, \frac{1}{n+\nu_i}\right]\right)} (C) \int_0^{\frac{1}{n+\nu_i}} f\left(t + \frac{k+\mu_i}{n+\nu_i}\right) d\mu(t) - f(x) \tag{2.25}$$

$$= \sum_{i=1}^{r} w_i \left[\frac{1}{\mu\left(\left[0, \frac{1}{n+\nu_i}\right]\right)} (C) \int_0^{\frac{1}{n+\nu_i}} f\left(t + \frac{k+\mu_i}{n+\nu_i}\right) d\mu(t) - f(x) \right].$$

Hence

$$\left| \delta_{n,k}(f) - f(x) \right| \leq$$

$$\sum_{i=1}^{r} w_i \left| \frac{1}{\mu\left(\left[0, \frac{1}{n+\nu_i}\right]\right)} (C) \int_0^{\frac{1}{n+\nu_i}} f\left(t + \frac{k+\mu_i}{n+\nu_i}\right) d\mu(t) - f(x) \right| \overset{(2.23)}{\leq}$$

$$\sum_{i=1}^{r} w_i \frac{1}{\mu\left(\left[0, \frac{1}{n+\nu_i}\right]\right)} (C) \int_0^{\frac{1}{n+\nu_i}} \left| f\left(t + \frac{k+\mu_i}{n+\nu_i}\right) - f(x) \right| d\mu(t) \leq \tag{2.26}$$

$$\sum_{i=1}^{r} w_i \frac{1}{\mu\left(\left[0, \frac{1}{n+\nu_i}\right]\right)} (C) \int_0^{\frac{1}{n+\nu_i}} \omega_1\left(f, |t| + \left| \frac{k+\mu_i}{n+\nu_i} - x \right| \right) d\mu(t) \leq$$

$$\sum_{i=1}^{r} w_i \omega_1 \left(f, \frac{1}{n+\mu_i} + \left| \frac{k+\mu_i}{n+\nu_i} - x \right| \right) \overset{(2.15)}{\leq}$$

$$\sum_{i=1}^{r} w_i \omega_1 \left(f, \left(\frac{\nu_i |x| + \mu_i + 1}{n+\nu_i} \right) + \left(1 + \frac{\nu_i}{n+\nu_i} \right) \frac{T}{n^{1-\alpha}} \right) \leq \tag{2.27}$$

$$\max_{i \in \{1,\ldots,r\}} \left\{ \omega_1 \left(f, \left(\frac{\nu_i |x| + \mu_i + 1}{n+\nu_i} \right) + \left(1 + \frac{\nu_i}{n+\nu_i} \right) \frac{T}{n^{1-\alpha}} \right) \right\}.$$

We have proved that

$$\left| \delta_{n,k}(f) - f(x) \right| \leq$$

$$\max_{i \in \{1,\ldots,r\}} \left\{ \omega_1 \left(f, \left(\frac{\nu_i |x| + \mu_i + 1}{n+\nu_i} \right) + \left(1 + \frac{\nu_i}{n+\nu_i} \right) \frac{T}{n^{1-\alpha}} \right) \right\}. \tag{2.28}$$

Therefore by (2.10) and (2.24) we get that

$$\left(K_n^{\mu}(f) \right)(x) - f(x) = \frac{\sum_{k=\lceil nx - Tn^{\alpha} \rceil}^{[nx+Tn^{\alpha}]} \delta_{n,k}(f)\, b\left(n^{1-\alpha}\left(x - \frac{k}{n} \right) \right)}{\sum_{k=\lceil nx - Tn^{\alpha} \rceil}^{[nx+Tn^{\alpha}]} b\left(n^{1-\alpha}\left(x - \frac{k}{n} \right) \right)} - f(x) =$$

$$\frac{\sum_{k=\lceil nx - Tn^{\alpha} \rceil}^{[nx+Tn^{\alpha}]} \delta_{n,k}(f)\, b\left(n^{1-\alpha}\left(x - \frac{k}{n} \right) \right) - f(x) \sum_{k=\lceil nx - Tn^{\alpha} \rceil}^{[nx+Tn^{\alpha}]} b\left(n^{1-\alpha}\left(x - \frac{k}{n} \right) \right)}{\sum_{k=\lceil nx - Tn^{\alpha} \rceil}^{[nx+Tn^{\alpha}]} b\left(n^{1-\alpha}\left(x - \frac{k}{n} \right) \right)} =$$

$$\frac{\sum_{k=\lceil nx - Tn^{\alpha} \rceil}^{[nx+Tn^{\alpha}]} \left(\delta_{n,k}(f) - f(x) \right) b\left(n^{1-\alpha}\left(x - \frac{k}{n} \right) \right)}{\sum_{k=\lceil nx - Tn^{\alpha} \rceil}^{[nx+Tn^{\alpha}]} b\left(n^{1-\alpha}\left(x - \frac{k}{n} \right) \right)}. \tag{2.29}$$

Hence it holds

$$\left| \left(K_n^{\mu}(f) \right)(x) - f(x) \right| \leq \frac{\sum_{k=\lceil nx - Tn^{\alpha} \rceil}^{[nx+Tn^{\alpha}]} \left| \delta_{n,k}(f) - f(x) \right| b\left(n^{1-\alpha}\left(x - \frac{k}{n} \right) \right)}{\sum_{k=\lceil nx - Tn^{\alpha} \rceil}^{[nx+Tn^{\alpha}]} b\left(n^{1-\alpha}\left(x - \frac{k}{n} \right) \right)} \overset{(2.28)}{\leq}$$

$$\frac{\left(\sum_{k=\lceil nx-Tn^{\alpha}\rceil}^{[nx+Tn^{\alpha}]} b\left(n^{1-\alpha}\left(x-\frac{k}{n}\right)\right)\right)\max_{i\in\{1,\dots,r\}}\left\{\omega_1\left(f,\left(\frac{\nu_i|x|+\mu_i+1}{n+\nu_i}\right)+\left(1+\frac{\nu_i}{n+\nu_i}\right)\frac{T}{n^{1-\alpha}}\right)\right\}}{\sum_{k=\lceil nx-Tn^{\alpha}\rceil}^{[nx+Tn^{\alpha}]} b\left(n^{1-\alpha}\left(x-\frac{k}{n}\right)\right)}$$

$$=\max_{i\in\{1,\dots,r\}}\left\{\omega_1\left(f,\left(\frac{\nu_i|x|+\mu_i+1}{n+\nu_i}\right)+\left(1+\frac{\nu_i}{n+\nu_i}\right)\frac{T}{n^{1-\alpha}}\right)\right\}, \tag{2.30}$$

proving the claim. ■

Corollary 2.9 *In the assumptions of Theorem 2.8, let $x\in[-T^*,T^*]$, $T^*>0$. Then*

$$\sup_{\mu}\left\|K_n^{\mu}(f)-f\right\|_{\infty,[-T^*,T^*]}\leq \tag{2.31}$$

$$\max_{i\in\{1,\dots,r\}}\left\{\omega_1\left(f,\left(\frac{\nu_iT^*+\mu_i+1}{n+\nu_i}\right)+\left(1+\frac{\nu_i}{n+\nu_i}\right)\frac{T}{n^{1-\alpha}}\right)\right\}.$$

Proof By (2.16) we get the desired result. ■

Notice the right hand sides of (2.16) and (2.31) do not depend on b or μ.

We continue with

Theorem 2.10 *Let $x\in\mathbb{R}$, $T>0$, $0<\alpha<1$, $n\in\mathbb{N}:n\geq\max\left(T+|x|,T^{-\frac{1}{\alpha}}\right)$. Let $f\in C^N(\mathbb{R},\mathbb{R}_+)$, $N\in\mathbb{N}$, such that $f^{(N)}$ is uniformly continuous or continuous and bounded. Then*

$$\sup_{\mu}\left|\left(K_n^{\mu}(f)\right)(x)-f(x)\right|\leq\sum_{j=1}^{N}\frac{\left|f^{(j)}(x)\right|}{(j)!}\cdot$$

$$\left(\sum_{i=1}^{r}w_i\left(\left(\frac{\nu_i|x|+\mu_i+1}{n+\nu_i}\right)+\left(1+\frac{\nu_i}{n+\nu_i}\right)\frac{T}{n^{1-\alpha}}\right)^j\right)+ \tag{2.32}$$

$$\sum_{i=1}^{r}w_i\left\{\omega_1\left(f^{(N)},\left(\frac{\nu_i|x|+\mu_i+1}{n+\nu_i}\right)+\left(1+\frac{\nu_i}{n+\nu_i}\right)\frac{T}{n^{1-\alpha}}\right)\cdot\right.$$

$$\left.\frac{\left(\left(\frac{\nu_i|x|+\mu_i+1}{n+\nu_i}\right)+\left(1+\frac{\nu_i}{n+\nu_i}\right)\frac{T}{n^{1-\alpha}}\right)^N}{N!}\right\}.$$

Inequality (2.32) implies the pointwise convergence with rates of $\left(K_n^{\mu}(f)\right)(x)$ to $f(x)$, as $n\to\infty$, at the speed $\frac{1}{n^{1-\alpha}}$.

Proof Let k as in (2.3). We observe that

$$f\left(t+\frac{k+\mu_i}{n+\nu_i}\right)-f(x)=\sum_{j=1}^{N}\frac{f^{(j)}(x)}{j!}\left(t+\frac{k+\mu_i}{n+\nu_i}-x\right)^j+ \tag{2.33}$$

$$\int_x^{t+\frac{k+\mu_i}{n+\nu_i}}\left(f^{(N)}(z)-f^{(N)}(x)\right)\frac{\left(t+\frac{k+\mu_i}{n+\nu_i}-z\right)^{N-1}}{(N-1)!}dz,$$

$i=1,\ldots,r$.

Above we take $0\leq t\leq\frac{1}{n+\nu_i}$.

Next we estimate the remainder of (2.33).

We distinguish the cases:

(i) if $t+\frac{k+\mu_i}{n+\nu_i}\geq x$, then

$$\theta_i:=\left|\int_x^{t+\frac{k+\mu_i}{n+\nu_i}}\left|f^{(N)}(z)-f^{(N)}(x)\right|\frac{\left|t+\frac{k+\mu_i}{n+\nu_i}-z\right|^{N-1}}{(N-1)!}dz\right|=$$

$$\int_x^{t+\frac{k+\mu_i}{n+\nu_i}}\left|f^{(N)}(z)-f^{(N)}(x)\right|\frac{\left(t+\frac{k+\mu_i}{n+\nu_i}-z\right)^{N-1}}{(N-1)!}dz\leq$$

$$\omega_1\left(f^{(N)},|t|+\left|\frac{k+\mu_i}{n+\nu_i}-x\right|\right)\frac{\left(|t|+\left|\frac{k+\mu_i}{n+\nu_i}-x\right|\right)^N}{N!}\overset{(2.14)}{\leq} \tag{2.34}$$

$$\omega_1\left(f^{(N)},\left(\frac{\nu_i|x|+\mu_i+1}{n+\nu_i}\right)+\left(1+\frac{\nu_i}{n+\nu_i}\right)\frac{T}{n^{1-\alpha}}\right)\cdot$$

$$\frac{\left(\left(\frac{\nu_i|x|+\mu_i+1}{n+\nu_i}\right)+\left(1+\frac{\nu_i}{n+\nu_i}\right)\frac{T}{n^{1-\alpha}}\right)^N}{N!}.$$

That is, if $t+\frac{k+\mu_i}{n+\nu_i}\geq x$, we proved that

$$\theta_i\leq\omega_1\left(f^{(N)},\left(\frac{\nu_i|x|+\mu_i+1}{n+\nu_i}\right)+\left(1+\frac{\nu_i}{n+\nu_i}\right)\frac{T}{n^{1-\alpha}}\right)\cdot \tag{2.35}$$

$$\frac{\left(\left(\frac{\nu_i|x|+\mu_i+1}{n+\nu_i}\right)+\left(1+\frac{\nu_i}{n+\nu_i}\right)\frac{T}{n^{1-\alpha}}\right)^N}{N!}.$$

(ii) if $t+\frac{k+\mu_i}{n+\nu_i}<x$, then

$$\theta_i := \int_{t+\frac{k+\mu_i}{n+\nu_i}}^{x} \left| f^{(N)}(z) - f^{(N)}(x) \right| \frac{\left(z - \left(t + \frac{k+\mu_i}{n+\nu_i} \right) \right)^{N-1}}{(N-1)!} dz \leq$$

$$\omega_1 \left(f^{(N)}, |t| + \left| \frac{k+\mu_i}{n+\nu_i} - x \right| \right) \frac{\left(|t| + \left| \frac{k+\mu_i}{n+\nu_i} - x \right| \right)^N}{N!} \overset{(2.14)}{\leq} \tag{2.36}$$

$$\omega_1 \left(f^{(N)}, \left(\frac{\nu_i |x| + \mu_i + 1}{n+\nu_i} \right) + \left(1 + \frac{\nu_i}{n+\nu_i} \right) \frac{T}{n^{1-\alpha}} \right) \cdot$$

$$\frac{\left(\left(\frac{\nu_i |x| + \mu_i + 1}{n+\nu_i} \right) + \left(1 + \frac{\nu_i}{n+\nu_i} \right) \frac{T}{n^{1-\alpha}} \right)^N}{N!},$$

the same estimate as in (2.35).

Hence it holds ($i = 1, ..., r$)

$$\left| f \left(t + \frac{k+\mu_i}{n+\nu_i} \right) - f(x) \right| \overset{(2.33)}{\leq} \sum_{j=1}^{N} \frac{\left| f^{(j)}(x) \right|}{j!} \left| t + \frac{k+\mu_i}{n+\nu_i} - x \right|^j +$$

$$\left| \int_x^{t+\frac{k+\mu_i}{n+\nu_i}} \left(f^{(N)}(z) - f^{(N)}(x) \right) \frac{\left(t + \frac{k+\mu_i}{n+\nu_i} - z \right)^{N-1}}{(N-1)!} dz \right| \leq \tag{2.37}$$

$$\sum_{j=1}^{N} \frac{\left| f^{(j)}(x) \right|}{j!} \left(|t| + \left| \frac{k+\mu_i}{n+\nu_i} - x \right| \right)^j +$$

$$\left| \int_x^{t+\frac{k+\mu_i}{n+\nu_i}} \left| f^{(N)}(z) - f^{(N)}(x) \right| \frac{\left| t + \frac{k+\mu_i}{n+\nu_i} - z \right|^{N-1}}{(N-1)!} dz \right| \overset{\text{(by (2.14) and (2.35), (2.36))}}{\leq}$$

$$\sum_{j=1}^{N} \frac{\left| f^{(j)}(x) \right|}{j!} \left(\left(\frac{\nu_i |x| + \mu_i + 1}{n+\nu_i} \right) + \left(1 + \frac{\nu_i}{n+\nu_i} \right) \frac{T}{n^{1-\alpha}} \right)^j +$$

$$\omega_1 \left(f^{(N)}, \left(\frac{\nu_i |x| + \mu_i + 1}{n+\nu_i} \right) + \left(1 + \frac{\nu_i}{n+\nu_i} \right) \frac{T}{n^{1-\alpha}} \right) \cdot$$

$$\frac{\left(\left(\frac{\nu_i |x| + \mu_i + 1}{n+\nu_i} \right) + \left(1 + \frac{\nu_i}{n+\nu_i} \right) \frac{T}{n^{1-\alpha}} \right)^N}{N!}. \tag{2.38}$$

We have proved so far that ($i = 1, ..., r$)

$$\left| f\left(t + \frac{k+\mu_i}{n+\nu_i} \right) - f(x) \right| \leq$$

$$\sum_{j=1}^{N} \frac{\left| f^{(j)}(x) \right|}{j!} \left(\left(\frac{\nu_i |x| + \mu_i + 1}{n+\nu_i} \right) + \left(1 + \frac{\nu_i}{n+\nu_i} \right) \frac{T}{n^{1-\alpha}} \right)^j +$$

$$\omega_1 \left(f^{(N)}, \left(\frac{\nu_i |x| + \mu_i + 1}{n+\nu_i} \right) + \left(1 + \frac{\nu_i}{n+\nu_i} \right) \frac{T}{n^{1-\alpha}} \right) \cdot$$

$$\frac{\left(\left(\frac{\nu_i |x| + \mu_i + 1}{n+\nu_i} \right) + \left(1 + \frac{\nu_i}{n+\nu_i} \right) \frac{T}{n^{1-\alpha}} \right)^N}{N!}. \tag{2.39}$$

Let $\delta_{n,k}(f)$ be as in (2.24).

As in the proof of Theorem 2.8 we have

$$\left| \delta_{n,k}(f) - f(x) \right| \leq$$

$$\sum_{i=1}^{r} w_i \frac{1}{\mu\left(\left[0, \frac{1}{n+\nu_i} \right] \right)} (C) \int_0^{\frac{1}{n+\nu_i}} \left| f\left(t + \frac{k+\mu_i}{n+\nu_i} \right) - f(x) \right| d\mu(t) \overset{(2.39)}{\leq}$$

$$\sum_{i=1}^{r} w_i \left\{ \sum_{j=1}^{N} \frac{\left| f^{(j)}(x) \right|}{j!} \left(\left(\frac{\nu_i |x| + \mu_i + 1}{n+\nu_i} \right) + \left(1 + \frac{\nu_i}{n+\nu_i} \right) \frac{T}{n^{1-\alpha}} \right)^j + \right.$$

$$\omega_1 \left(f^{(N)}, \left(\frac{\nu_i |x| + \mu_i + 1}{n+\nu_i} \right) + \left(1 + \frac{\nu_i}{n+\nu_i} \right) \frac{T}{n^{1-\alpha}} \right) \cdot$$

$$\left. \frac{\left(\left(\frac{\nu_i |x| + \mu_i + 1}{n+\nu_i} \right) + \left(1 + \frac{\nu_i}{n+\nu_i} \right) \frac{T}{n^{1-\alpha}} \right)^N}{N!} \right\}. \tag{2.40}$$

As in the proof of Theorem 2.8 we have

$$\left| \left(K_n^{\mu}(f) \right)(x) - f(x) \right| \leq \frac{\sum_{k=\lceil nx - Tn^{\alpha} \rceil}^{[nx+Tn^{\alpha}]} \left| \delta_{n,k}(f) - f(x) \right| b\left(n^{1-\alpha} \left(x - \frac{k}{n} \right) \right)}{\sum_{k=\lceil nx - Tn^{\alpha} \rceil}^{[nx+Tn^{\alpha}]} b\left(n^{1-\alpha} \left(x - \frac{k}{n} \right) \right)} \leq$$

$$\sum_{i=1}^{r} w_i \left\{ \sum_{j=1}^{N} \frac{\left| f^{(j)}(x) \right|}{j!} \left(\left(\frac{\nu_i |x| + \mu_i + 1}{n + \nu_i} \right) + \left(1 + \frac{\nu_i}{n + \nu_i} \right) \frac{T}{n^{1-\alpha}} \right)^j + \right. \tag{2.41}$$

$$\omega_1 \left(f^{(N)}, \left(\frac{\nu_i |x| + \mu_i + 1}{n + \nu_i} \right) + \left(1 + \frac{\nu_i}{n + \nu_i} \right) \frac{T}{n^{1-\alpha}} \right) \cdot$$

$$\left. \frac{\left(\left(\frac{\nu_i |x| + \mu_i + 1}{n + \nu_i} \right) + \left(1 + \frac{\nu_i}{n + \nu_i} \right) \frac{T}{n^{1-\alpha}} \right)^N}{N!} \right\},$$

proving the claim. ■

Corollary 2.11 *All as in Theorem 2.10 hold, plus* $f^{(j)}(x) = 0$, $j = 1, ..., N$. *Then*

$$\sup_{\mu} \left| \left(K_n^{\mu}(f) \right)(x) - f(x) \right| \leq$$

$$\sum_{i=1}^{r} w_i \left\{ \omega_1 \left(f^{(N)}, \left(\frac{\nu_i |x| + \mu_i + 1}{n + \nu_i} \right) + \left(1 + \frac{\nu_i}{n + \nu_i} \right) \frac{T}{n^{1-\alpha}} \right) \cdot \right.$$

$$\left. \frac{\left(\left(\frac{\nu_i |x| + \mu_i + 1}{n + \nu_i} \right) + \left(1 + \frac{\nu_i}{n + \nu_i} \right) \frac{T}{n^{1-\alpha}} \right)^N}{N!} \right\}. \tag{2.42}$$

Proof By (2.32) we get the desired result. ■

In (2.42) notice the extremely high speed of convergence $\frac{1}{n^{(1-\alpha)(N+1)}}$.
The uniform convergence with rates follows from the following

Corollary 2.12 *Let* $x \in [-T^*, T^*]$, $T^* > 0$; $T > 0$ *and* $n \in \mathbb{N}$ *such that* $n \geq \max\left(T + T^*, T^{-\frac{1}{\alpha}} \right)$, $0 < \alpha < 1$. *Let* $f \in C^N(\mathbb{R}, \mathbb{R}_+)$, $N \in \mathbb{N}$, *such that* $f^{(N)}$ *is uniformly continuous or is continuous and bounded. Then*

$$\sup_{\mu} \left\| K_n^{\mu}(f) - f \right\|_{\infty,[-T^*,T^*]} \leq \sum_{j=1}^{N} \frac{\left\| f^{(j)} \right\|_{\infty,[-T^*,T^*]}}{j!} \cdot$$

$$\left(\sum_{i=1}^{r} w_i \left[\left(\frac{\nu_i T^* + \mu_i + 1}{n + \nu_i} \right) + \left(1 + \frac{\nu_i}{n + \nu_i} \right) \frac{T}{n^{1-\alpha}} \right]^j \right) + \tag{2.43}$$

$$\sum_{i=1}^{r} w_i \left\{ \omega_1 \left(f^{(N)}, \left(\frac{\nu_i T^* + \mu_i + 1}{n + \nu_i} \right) + \left(1 + \frac{\nu_i}{n + \nu_i} \right) \frac{T}{n^{1-\alpha}} \right) \cdot \right.$$

$$\left.\frac{\left(\left(\frac{\nu_i T^*+\mu_i+1}{n+\nu_i}\right)+\left(1+\frac{\nu_i}{n+\nu_i}\right)\frac{T}{n^{1-\alpha}}\right)^N}{N!}\right\}.$$

Proof By (2.32) we get the desired result. ■

Corollary 2.13 *All as in Theorem 2.10, case of $N=1$ hold. It holds*

$$\sup_{\mu}\left|\left(K_n^{\mu}(f)\right)(x)-f(x)\right|\le$$

$$\left|f'(x)\right|\left(\sum_{i=1}^r w_i\left(\left(\frac{\nu_i|x|+\mu_i+1}{n+\nu_i}\right)+\left(1+\frac{\nu_i}{n+\nu_i}\right)\frac{T}{n^{1-\alpha}}\right)\right)+ \tag{2.44}$$

$$\sum_{i=1}^r w_i\left\{\omega_1\left(f',\left(\frac{\nu_i|x|+\mu_i+1}{n+\nu_i}\right)+\left(1+\frac{\nu_i}{n+\nu_i}\right)\frac{T}{n^{1-\alpha}}\right)\cdot\right.$$

$$\left.\left(\left(\frac{\nu_i|x|+\mu_i+1}{n+\nu_i}\right)+\left(1+\frac{\nu_i}{n+\nu_i}\right)\frac{T}{n^{1-\alpha}}\right)\right\}.$$

Proof By (2.32) we get the desired result. ■

2.4 Multivariate Results

Here the activation function $b:\mathbb{R}^d\to\mathbb{R}_+$, $d\in\mathbb{N}$, is of compact support $B:=\prod_{i=1}^d[-T_i,T_i]$, $T_i>0$, $i=1,...,d$. That is $b(x)>0$ for any $x\in B$, and clearly b may have jump discontinuities. Also the shape of the graph of b is immaterial.

Typically in neural networks approximation we take b to be a d -dimensional bell-shaped function (i.e. per coordinate is a centered bell-shaped function), or a product of univariate centered bell-shaped functions, or a product of sigmoid functions, in our case all of them are of compact support B.

Example 2.14 Take $b(x)=\beta(x_1)\beta(x_2)...\beta(x_d)$, where β is any of the following functions, $j=1,...,d$:

(i) $\beta(x_j)$ is the characteristic function on $[-1,1]$,

(ii) $\beta(x_j)$ is the hat function over $[-1+\varepsilon,1-\varepsilon]$, where $\varepsilon>0$ *is small*, that is,

$$\beta(xj)=\begin{cases}1+x_j, & -1\le x_j\le 0,\\ 1-x_j, & 0<x_j\le 1,\\ 0, & \text{elsewhere},\end{cases}$$

(iii) the truncated sigmoids

$$\beta(x_j) = \begin{cases} \frac{1}{1+e^{-x_j}} \text{ or } \tanh x_j \text{ or } erf(x_j), & \text{for } x_j \in [-T_j, T_j], \text{ with large } T_j > 0, \\ 0, \ x_j \in \mathbb{R} - [-T_j, T_j], \end{cases}$$

(iv) the truncated Gompertz function

$$\beta(x_j) = \begin{cases} e^{-\alpha e^{-\beta x_j}}, \ x_j \in [-T_j, T_j]; \ \alpha, \beta > 0; \text{ large } T_j > 0, \\ 0, \ x_j \in \mathbb{R} - [-T_j, T_j], \end{cases}$$

Thus the general activation function b we will be using here includes all kinds of activation functions in neural network approximations.

Here we consider functions $f : \mathbb{R}^d \to \mathbb{R}_+$ that either continuous and bounded, or uniformly continuous.

Let here the parameters: $0 < \alpha < 1$, $x = (x_1, ..., x_d) \in \mathbb{R}^d$, $n \in \mathbb{N}$; $r = (r_1, ..., r_d) \in \mathbb{N}^d$, $i = (i_1, ..., i_d) \in \mathbb{N}^d$, with $i_j = 1, 2, ..., r_j$, $j = 1, ..., d$; also let $w_i = w_{i_1,...,i_d} \geq 0$, such that $\sum_{i_1=1}^{r_1} \sum_{i_2=1}^{r_2} ... \sum_{i_d=1}^{r_d} w_{i_1,...,i_d} = 1$, in brief written as $\sum_{i=1}^{r} w_i = 1$. We further consider the parameters $k = (k_1, ..., k_d) \in \mathbb{Z}^d$; $\mu_i = (\mu_{i_1}, ..., \mu_{i_d}) \in \mathbb{R}^d_+$, $\nu_i = (\nu_{i_1}, ..., \nu_{i_d}) \in \mathbb{R}^d_+$; and $\lambda_i = \lambda_{i_1,...,i_d}$, $\rho_i = \rho_{i_1,...,i_d} \geq 0$.

We use here the first modulus of continuity, with $\delta > 0$,

$$\omega_1(f, \delta) := \sup_{\substack{x, y \in \mathbb{R}^d \\ \|x - y\|_\infty \leq \delta}} |f(x) - f(y)|,$$

where $\|x\|_\infty = \max(|x_1|, ..., |x_d|)$.

Given that f is uniformly continuous we get $\lim_{\delta \to 0} \omega_1(f, \delta) = 0$.

So in this section mainly we study the pointwise convergence with rates over $\mathbb{R}^d$, to the unit operator, of the following one hidden layer multivariate normalized neural network perturbed operator:

Definition 2.15 Let $\mathcal{L}^*$ be the Lebesgue σ-algebra on $\mathbb{R}^N$, $N \in \mathbb{N}$, and the set function $\mu^* : \mathcal{L}^* \to [0, +\infty)$, which is assumed to be monotone, submodular and strictly positive. We define the multivariate Kantorovich–Choquet type neural network operators for any $x \in \mathbb{R}^N$, $n \in \mathbb{N}$:

$$\left(K_n^{\mu^*}(f)\right)(x) =$$

$$\frac{\left[\sum_{k=-n^2}^{n^2} \left(\sum_{i=1}^{r} w_i \frac{1}{\mu^*\left(\left[0, \frac{1}{n+\rho_i}\right]^d\right)} (C) \int_0^{\frac{1}{n+\rho_i}} f\left(t + \frac{k+\lambda_i}{n+\rho_i}\right) d\mu^*(t) \right) b\left(n^{1-\alpha}\left(x - \frac{k}{n}\right)\right)\right]}{\sum_{k=-n^2}^{n^2} b\left(n^{1-\alpha}\left(x - \frac{k}{n}\right)\right)} \tag{2.45}$$

$$= \sum_{k_1=-n^2}^{n^2} \cdots \sum_{k_d=-n^2}^{n^2} \left(\sum_{i_1=1}^{r_1} \cdots \sum_{i_d=1}^{r_d} w_{i_1,\ldots,i_d} \frac{1}{\mu^* \left(\left[0, \frac{1}{n+\rho_{i_1,\ldots,i_d}} \right]^d \right)} \right.$$

$$\left. (C) \int \cdots \int_{\left[0, \frac{1}{n+\rho_{i_1,\ldots,i_d}}\right]^d} \cdots \int f\left(t_1 + \frac{k_1 + \lambda_{i_1,\ldots,i_d}}{n + \rho_{i_1,\ldots,i_d}}, \ldots, t_d + \frac{k_d + \lambda_{i_1,\ldots,i_d}}{n + \rho_{i_1,\ldots,i_d}} \right) d\mu^* (t_1, \ldots, t_d) \right)$$

$$\frac{b\left(n^{1-\alpha}\left(x_1 - \frac{k_1}{n}\right), \ldots, n^{1-\alpha}\left(x_d - \frac{k_d}{n}\right)\right)}{\sum_{k_1=-n^2}^{n^2} \cdots \sum_{k_d=-n^2}^{n^2} b\left(n^{1-\alpha}\left(x_1 - \frac{k_1}{n}\right), \ldots, n^{1-\alpha}\left(x_d - \frac{k_d}{n}\right)\right)}.$$

For f continuous and bounded we get:

$$\left\| K_n^{\mu^*}(f) \right\|_\infty \leq \|f\|_\infty, \quad \text{plus} \quad K_n^{\mu^*}(1) = 1.$$

Remark 2.16 The terms in the ratio of sums in (2.45) can be nonzero, if and only if simultaneously

$$\left| n^{1-\alpha} \left(x_j - \frac{k_j}{n} \right) \right| \leq T_j, \quad \text{all } j = 1, \ldots, d, \tag{2.46}$$

i.e., $\left| x_j - \frac{k_j}{n} \right| \leq \frac{T_j}{n^{1-\alpha}}$, all $j = 1, \ldots, d$, iff

$$n x_j - T_j n^\alpha \leq k_j \leq n x_j + T_j n^\alpha, \quad \text{all } j = 1, \ldots, d. \tag{2.47}$$

To have the order

$$-n^2 \leq n x_j - T_j n^\alpha \leq k_j \leq n x_j + T_j n^\alpha \leq n^2, \tag{2.48}$$

we need $n \geq T_j + |x_j|$, all $j = 1, \ldots, d$. So (2.48) is true when we take

$$n \geq \max_{j \in \{1,\ldots,d\}} \left(T_j + |x_j| \right). \tag{2.49}$$

When $x \in B$ in order to have (2.48) it is enough to assume that $n \geq 2T^*$, where $T^* := \max\{T_1, \ldots, T_d\} > 0$. Consider

$$\widetilde{I}_j := \left[n x_j - T_j n^\alpha, n x_j + T_j n^\alpha \right], \; j = 1, \ldots, d, \; n \in \mathbb{N}.$$

The length of $\widetilde{I}_j$ is $2T_jn^{\alpha}$. By Proposition 2.5, we get that the cardinality of $k_j \in \mathbb{Z}$ that belong to $\widetilde{I}_j := card\left(k_j\right) \geq \max\left(2T_jn^{\alpha} - 1, 0\right)$, any $j \in \{1, ..., d\}$. In order to have $card\left(k_j\right) \geq 1$, we need $2T_jn^{\alpha} - 1 \geq 1$ if and only if $n \geq T_j^{-\frac{1}{\alpha}}$, any $j \in \{1, ..., d\}$.

Therefore, a sufficient condition in order to obtain the order (2.48) along with the interval $\widetilde{I}_j$ to contain at least one integer for all $j = 1, ..., d$ is that

$$n \geq \max_{j\in\{1,...,d\}} \left\{T_j + \left|x_j\right|, T_j^{-\frac{1}{\alpha}}\right\}. \tag{2.50}$$

Clearly as $n \to +\infty$ we get that $card\left(k_j\right) \to +\infty$, all $j = 1, ..., d$. Also notice that $card\left(k_j\right)$ equals to the cardinality of integers in $\left[\left\lceil nx_j - T_jn^{\alpha}\right\rceil, \left[nx_j + T_jn^{\alpha}\right]\right]$ for all $j = 1, ..., d$. Here $[\cdot]$ denotes the integral part of the number while $\lceil\cdot\rceil$ denotes its ceiling.

From now on, in this section we assume (2.50).

We shall denote by $T = (T_1, ..., T_d)$, $[nx + Tn^{\alpha}] = ([nx_1 + T_1n^{\alpha}], ..., [nx_d + T_dn^{\alpha}])$, and $\lceil nx - Tn^{\alpha}\rceil = (\lceil nx_1 - T_1n^{\alpha}\rceil, ..., \lceil nx_d - T_dn^{\alpha}\rceil)$. Furthermore it holds

$$\left(K_n^{\mu^*}(f)\right)(x) =$$

$$\frac{\sum_{k=\lceil nx-Tn^{\alpha}\rceil}^{[nx+Tn^{\alpha}]}\left(\sum_{i=1}^{r} w_i \frac{1}{\mu^*\left(\left[0,\frac{1}{n+\rho_i}\right]^d\right)} (C)\int_0^{\frac{1}{n+\rho_i}} f\left(t+\frac{k+\lambda_i}{n+\rho_i}\right)d\mu^*(t)\right) b\left(n^{1-\alpha}\left(x-\frac{k}{n}\right)\right)}{\sum_{k=\lceil nx-Tn^{\alpha}\rceil}^{[nx+Tn^{\alpha}]} b\left(n^{1-\alpha}\left(x-\frac{k}{n}\right)\right)} \tag{2.51}$$

$$= \sum_{k_1=\lceil nx_1-T_1n^{\alpha}\rceil}^{[nx_1+T_1n^{\alpha}]} \cdots \sum_{k_d=\lceil nx_d-T_dn^{\alpha}\rceil}^{[nx_d+T_dn^{\alpha}]} \left(\sum_{i_1=1}^{r_1} \cdots \sum_{i_d=1}^{r_d} w_{i_1,...,i_d} \frac{1}{\mu^*\left(\left[0,\frac{1}{n+\rho_{i_1,...,i_d}}\right]^d\right)}\right.$$

$$\left.(C)\int\ldots\int_{\left[0,\frac{1}{n+\rho_{i_1,...,i_d}}\right]^d}\ldots\int f\left(t_1+\frac{k_1+\lambda_{i_1,...,i_d}}{n+\rho_{i_1,...,i_d}}, ..., t_d+\frac{k_d+\lambda_{i_1,...,i_d}}{n+\rho_{i_1,...,i_d}}\right)d\mu^*(t_1, ..., t_d)\right)$$

$$\frac{b\left(n^{1-\alpha}\left(x_1-\frac{k_1}{n}\right), ..., n^{1-\alpha}\left(x_d-\frac{k_d}{n}\right)\right)}{\sum_{k_1=\lceil nx_1-T_1n^{\alpha}\rceil}^{[nx_1+T_1n^{\alpha}]} \cdots \sum_{k_d=\lceil nx_d-T_dn^{\alpha}\rceil}^{[nx_d+T_dn^{\alpha}]} b\left(n^{1-\alpha}\left(x_1-\frac{k_1}{n}\right), ..., n^{1-\alpha}\left(x_d-\frac{k_d}{n}\right)\right)}.$$

So if $\left|n^{1-\alpha}\left(x_j - \frac{k_j}{n}\right)\right| \leq T_j$, all $j = 1, ..., d$, we get that

$$\left\| x - \frac{k}{n} \right\|_{\infty} \leq \frac{T^*}{n^{1-\alpha}}. \tag{2.52}$$

For convinience we call

$$V(x) := \sum_{k=\lceil nx - Tn^{\alpha} \rceil}^{[nx+Tn^{\alpha}]} b\left(n^{1-\alpha}\left(x - \frac{k}{n}\right)\right) =$$

$$\sum_{k_1=\lceil nx_1 - T_1 n^{\alpha} \rceil}^{[nx_1+T_1 n^{\alpha}]} \cdots \sum_{k_d=\lceil nx_d - T_d n^{\alpha} \rceil}^{[nx_d+T_d n^{\alpha}]} b\left(n^{1-\alpha}\left(x_1 - \frac{k_1}{n}\right), \ldots, n^{1-\alpha}\left(x_d - \frac{k_d}{n}\right)\right). \tag{2.53}$$

We make

Remark 2.17 Here always k is as in (2.48).

We have for

$$0 \leq t_j \leq \frac{1}{n + \rho_{i_1,\ldots,i_d}}, \quad j = 1, \ldots, d, \tag{2.54}$$

that

$$\left\| t + \frac{k + \lambda_{i_1,\ldots,i_d}}{n + \rho_{i_1,\ldots,i_d}} - x \right\|_{\infty} \leq \frac{1}{n + \rho_{i_1,\ldots,i_d}} + \left\| \frac{k + \lambda_{i_1,\ldots,i_d}}{n + \rho_{i_1,\ldots,i_d}} - x \right\|_{\infty} \leq \tag{2.55}$$

$$\frac{1 + \lambda_{i_1,\ldots,i_d}}{n + \rho_{i_1,\ldots,i_d}} + \left\| \frac{k}{n + \rho_{i_1,\ldots,i_d}} - x \right\|_{\infty} \leq$$

$$\frac{1 + \lambda_{i_1,\ldots,i_d}}{n + \rho_{i_1,\ldots,i_d}} + \left\| \frac{k}{n + \rho_{i_1,\ldots,i_d}} - \frac{k}{n} \right\|_{\infty} + \left\| \frac{k}{n} - x \right\|_{\infty} \leq$$

$$\frac{1 + \lambda_{i_1,\ldots,i_d}}{n + \rho_{i_1,\ldots,i_d}} + \frac{T^*}{n^{1-\alpha}} + \frac{\rho_{i_1,\ldots,i_d} \|k\|_{\infty}}{(n + \rho_{i_1,\ldots,i_d})\, n} \leq$$

$$\frac{1 + \lambda_{i_1,\ldots,i_d}}{n + \rho_{i_1,\ldots,i_d}} + \frac{T^*}{n^{1-\alpha}} + \frac{\rho_{i_1,\ldots,i_d}}{n\left(n + \rho_{i_1,\ldots,i_d}\right)} \left(n \|x\|_{\infty} + T^* n^{\alpha}\right) = \tag{2.56}$$

$$\frac{1 + \lambda_{i_1,\ldots,i_d}}{n + \rho_{i_1,\ldots,i_d}} + \frac{T^*}{n^{1-\alpha}} + \left(\frac{\rho_{i_1,\ldots,i_d}}{n + \rho_{i_1,\ldots,i_d}}\right)\left(\|x\|_{\infty} + \frac{T^*}{n^{1-\alpha}}\right).$$

We have found that

$$\left\| t + \frac{k + \lambda_{i_1,\ldots,i_d}}{n + \rho_{i_1,\ldots,i_d}} - x \right\|_{\infty} \leq$$

$$\left(\frac{\rho_{i_1,\ldots,i_d} \|x\|_{\infty} + \lambda_{i_1,\ldots,i_d} + 1}{n + \rho_{i_1,\ldots,i_d}}\right) + \left(1 + \frac{\rho_{i_1,\ldots,i_d}}{n + \rho_{i_1,\ldots,i_d}}\right) \frac{T^*}{n^{1-\alpha}}. \tag{2.57}$$

So when $0 \le t_j \le \frac{1}{n+\rho_{i_1,\dots,i_d}}$, $j = 1, \dots, d$, we get that

$$\omega_1\left(f, \left\| t + \frac{k+\lambda_{i_1,\dots,i_d}}{n+\rho_{i_1,\dots,i_d}} - x \right\|_\infty\right) \le$$

$$\omega_1\left(f, \left(\frac{\rho_{i_1,\dots,i_d} \|x\|_\infty + \lambda_{i_1,\dots,i_d} + 1}{n+\rho_{i_1,\dots,i_d}}\right) + \left(1 + \frac{\rho_{i_1,\dots,i_d}}{n+\rho_{i_1,\dots,i_d}}\right)\frac{T^*}{n^{1-\alpha}}\right), \tag{2.58}$$

with dominant speed $\frac{1}{n^{1-\alpha}}$.

We present the following

Theorem 2.18 *Let $f : \mathbb{R}^d \to \mathbb{R}_+$ be continuous and bounded or uniformly continuous. Let $x \in \mathbb{R}^d$ and $n \in \mathbb{N}$ such that $n \ge \max_{j\in\{1,\dots,d\}} \left(T_j + |x_j|, T_j^{-\frac{1}{\alpha}}\right)$, $T_j > 0$, $0 < \alpha < 1$. Then*

$$\sup_{\mu^*} \left|\left(K_n^{\mu^*}(f)\right)(x) - f(x)\right| \le$$

$$\sum_{i=1}^{r} w_i \omega_1\left(f, \left(\frac{\rho_i \|x\|_\infty + \lambda_i + 1}{n+\rho_i}\right) + \left(1 + \frac{\rho_i}{n+\rho_i}\right)\frac{T^*}{n^{1-\alpha}}\right) = \tag{2.59}$$

$$\sum_{i_1=1}^{r_1} \dots \sum_{i_d=1}^{r_d} w_{i_1,\dots,i_d} \omega_1\left(f, \left(\frac{\rho_{i_1,\dots,i_d} \|x\|_\infty + \lambda_{i_1,\dots,i_d} + 1}{n+\rho_{i_1,\dots,i_d}}\right) + \right.$$

$$\left.\left(1 + \frac{\rho_{i_1,\dots,i_d}}{n+\rho_{i_1,\dots,i_d}}\right)\frac{T^*}{n^{1-\alpha}}\right).$$

Proof Here holds the analogous inequality to (2.23) in multidimensions, same proof. We observe the following

$$\left|\left(K_n^{\mu^*}(f)\right)(x) - f(x)\right| \overset{(2.51)}{=}$$

$$\left| \sum_{k=\lceil nx - Tn^\alpha \rceil}^{[nx+Tn^\alpha]} \left(\sum_{i=1}^{r} w_i \frac{1}{\mu^*\left(\left[0, \frac{1}{n+\rho_i}\right]^d\right)} (C) \int_0^{\frac{1}{n+\rho_i}} f\left(t + \frac{k+\lambda_i}{n+\rho_i}\right) d\mu^*(t) \right) \cdot \right. \tag{2.60}$$

$$\left. \frac{b\left(n^{1-\alpha}\left(x - \frac{k}{n}\right)\right)}{V(x)} - f(x) \right| =$$

$$\left| \sum_{k=\lceil nx-Tn^{\alpha}\rceil}^{[nx+Tn^{\alpha}]} \left(\sum_{i=1}^{r} w_i \frac{1}{\mu^*\left(\left[0,\frac{1}{n+\rho_i}\right]^d\right)} (C) \int_0^{\frac{1}{n+\rho_i}} f\left(t+\frac{k+\lambda_i}{n+\rho_i}\right) d\mu^*(t) \right) \cdot \right.$$

$$\left. \frac{b\left(n^{1-\alpha}\left(x-\frac{k}{n}\right)\right)}{V(x)} - \frac{f(x)\,V(x)}{V(x)} \right| =$$

$$\left| \frac{\sum_{k=\lceil nx-Tn^{\alpha}\rceil}^{[nx+Tn^{\alpha}]} \left[\left(\sum_{i=1}^{r} w_i \frac{1}{\mu^*\left(\left[0,\frac{1}{n+\rho_i}\right]^d\right)} (C) \int_0^{\frac{1}{n+\rho_i}} f\left(t+\frac{k+\lambda_i}{n+\rho_i}\right) d\mu^*(t) \right) - f(x) \right]}{V(x)} \right.$$

$$\left. \cdot b\left(n^{1-\alpha}\left(x-\frac{k}{n}\right)\right) \right| \leq$$

$$\frac{\sum_{k=\lceil nx-Tn^{\alpha}\rceil}^{[nx+Tn^{\alpha}]} \left[\sum_{i=1}^{r} w_i \left| \frac{1}{\mu^*\left(\left[0,\frac{1}{n+\rho_i}\right]^d\right)} (C) \int_0^{\frac{1}{n+\rho_i}} f\left(t+\frac{k+\lambda_i}{n+\rho_i}\right) d\mu^*(t) - f(x) \right| \right]}{V(x)}$$

$$\cdot b\left(n^{1-\alpha}\left(x-\frac{k}{n}\right)\right) \leq \tag{2.61}$$

$$\frac{\sum_{k=\lceil nx-Tn^{\alpha}\rceil}^{[nx+Tn^{\alpha}]} \left[\sum_{i=1}^{r} w_i \left[\frac{1}{\mu^*\left(\left[0,\frac{1}{n+\rho_i}\right]^d\right)} (C) \int_0^{\frac{1}{n+\rho_i}} \left| f\left(t+\frac{k+\lambda_i}{n+\rho_i}\right) - f(x) \right| d\mu^*(t) \right] \right]}{V(x)}$$

$$\cdot b\left(n^{1-\alpha}\left(x-\frac{k}{n}\right)\right) \leq$$

$$\frac{\sum_{k=\lceil nx-Tn^{\alpha}\rceil}^{[nx+Tn^{\alpha}]} \left[\sum_{i=1}^{r} w_i \left[\frac{1}{\mu^*\left(\left[0,\frac{1}{n+\rho_i}\right]^d\right)} (C) \int_0^{\frac{1}{n+\rho_i}} \omega_1\left(f, \left\| t+\frac{k+\lambda_i}{n+\rho_i} - x \right\|_{\infty}\right) d\mu^*(t) \right] \right]}{V(x)}$$

$$\cdot b\left(n^{1-\alpha}\left(x-\frac{k}{n}\right)\right) \overset{\text{(by (2.58))}}{\leq}$$

$$\frac{\sum_{k=\lceil nx-Tn^{\alpha}\rceil}^{[nx+Tn^{\alpha}]}\left[\sum_{i=1}^{r} w_i\left[\frac{1}{\mu^*\left(\left[0,\frac{1}{n+\rho_i}\right]^d\right)}(C)\int_0^{\frac{1}{n+\rho_i}}\omega_1\left(f,\left(\frac{\rho_i\|x\|_\infty+\lambda_i+1}{n+\rho_i}\right)+\right.\right.\right.}{V(x)}$$

$$\left.\left.\left.\left(1+\frac{\rho_i}{n+\rho_i}\right)\frac{T^*}{n^{1-\alpha}}\right)d\mu^*(t)\right]\right]\cdot b\left(n^{1-\alpha}\left(x-\frac{k}{n}\right)\right)=$$

$$\frac{\sum_{k=\lceil nx-Tn^{\alpha}\rceil}^{[nx+Tn^{\alpha}]}\left[\sum_{i=1}^{r} w_i\omega_1\left(f,\left(\frac{\rho_i\|x\|_\infty+\lambda_i+1}{n+\rho_i}\right)+\left(1+\frac{\rho_i}{n+\rho_i}\right)\frac{T^*}{n^{1-\alpha}}\right)\right]}{V(x)}$$

$$\cdot b\left(n^{1-\alpha}\left(x-\frac{k}{n}\right)\right)=$$

$$\sum_{i=1}^{r} w_i\omega_1\left(f,\left(\frac{\rho_i\|x\|_\infty+\lambda_i+1}{n+\rho_i}\right)+\left(1+\frac{\rho_i}{n+\rho_i}\right)\frac{T^*}{n^{1-\alpha}}\right), \tag{2.62}$$

proving the claim. ■

Corollary 2.19 *Let $f:\mathbb{R}^d\to\mathbb{R}_+$ continuous and bounded or uniformly continuous function. Let $x\in\prod_{j=1}^{d}[-\gamma_j,\gamma_j]\subset\mathbb{R}^d$, $\gamma_j>0$, $\gamma^*=\max\{\gamma_1,\ldots,\gamma_d\}$ and $n\in\mathbb{N}$: $n\geq\max_{j\in\{1,\ldots,d\}}\left\{T_j+\gamma_j,T_j^{-\frac{1}{\alpha}}\right\}$. Then*

$$\sup_{\mu^*}\left\|K_n^{\mu^*}(f)-f\right\|_{\infty,\prod_{j=1}^{d}[-\gamma_j,\gamma_j]}\leq$$

$$\sum_{i=1}^{r} w_i\omega_1\left(f,\left(\frac{\rho_i\gamma^*+\lambda_i+1}{n+\rho_i}\right)+\left(1+\frac{\rho_i}{n+\rho_i}\right)\frac{T^*}{n^{1-\alpha}}\right)= \tag{2.63}$$

$$\sum_{i_1=1}^{r_1}\cdots\sum_{i_d=1}^{r_d} w_{i_1,\ldots,i_d}\omega_1\left(f,\left(\frac{\rho_{i_1,\ldots,i_d}\gamma^*+\lambda_{i_1,\ldots,i_d}+1}{n+\rho_{i_1,\ldots,i_d}}\right)+\left(1+\frac{\rho_{i_1,\ldots,i_d}}{n+\rho_{i_1,\ldots,i_d}}\right)\frac{T^*}{n^{1-\alpha}}\right).$$

Proof By (2.59) we get the desired result. ■

Next here that the right hand sides of (2.59) and (2.63) do not depend on μ^* and b.

We continue with the higher order of approximation results based on the high order differentiability of the approximated function.

Theorem 2.20 *Let* $x \in \mathbb{R}^d$ *and* $n \in \mathbb{N}$ *such that* $n \geq \max_{j \in \{1,...,d\}} \left(T_j + |x_j|, T_j^{-\frac{1}{\alpha}} \right)$, $T_j > 0$, $0 < \alpha < 1$. *Let also* $f \in C^N(\mathbb{R}^d, \mathbb{R}_+)$, $N \in \mathbb{N}$, *such that all of its partial derivatives* $f_{\widetilde{\alpha}}$ *of order* N, $\widetilde{\alpha} : |\widetilde{\alpha}| = \sum_{j=1}^{d} \alpha_j = N$, *are uniformly continuous or continuous and bounded. Then*

$$\sup_{\mu^*} \left| \left(K_n^{\mu^*}(f) \right)(x) - f(x) \right| \leq$$

$$\sum_{l=1}^{N} \frac{1}{l!} \left(\sum_{i=1}^{r} w_i \left[\left(\frac{\rho_i \|x\|_\infty + \lambda_i + 1}{n + \rho_i} \right) + \left(1 + \frac{\rho_i}{n + \rho_i} \right) \frac{T^*}{n^{1-\alpha}} \right]^l \right) \cdot$$

$$\left(\left(\sum_{j=1}^{d} \left| \frac{\partial}{\partial x_j} \right| \right)^l f(x) \right) + \tag{2.64}$$

$$\frac{d^N}{N!} \sum_{i=1}^{r} w_i \left[\left(\frac{\rho_i \|x\|_\infty + \lambda_i + 1}{n + \rho_i} \right) + \left(1 + \frac{\rho_i}{n + \rho_i} \right) \frac{T^*}{n^{1-\alpha}} \right]^N \cdot$$

$$\max_{\widetilde{\alpha}:|\widetilde{\alpha}|=N} \omega_1 \left(f_{\widetilde{\alpha}}, \left[\left(\frac{\rho_i \|x\|_\infty + \lambda_i + 1}{n + \rho_i} \right) + \left(1 + \frac{\rho_i}{n + \rho_i} \right) \frac{T^*}{n^{1-\alpha}} \right] \right).$$

Inequality (2.64) implies the pointwise convergence with rates of $\left(K_n^{\mu^*}(f) \right)(x)$ *to* $f(x)$, *as* $n \to \infty$, *at speed* $\frac{1}{n^{1-\alpha}}$.

Proof As in the proof of Theorem 2.18 we get:

$$\left| \left(K_n^{\mu^*}(f) \right)(x) - f(x) \right| \leq$$

$$\frac{\sum_{k=\lceil nx - Tn^\alpha \rceil}^{[nx+Tn^\alpha]} \left[\sum_{i=1}^{r} w_i \left[\frac{1}{\mu^*\left(\left[0, \frac{1}{n+\rho_i} \right]^d \right)} (C) \int_0^{\frac{1}{n+\rho_i}} \left| f\left(t + \frac{k+\lambda_i}{n+\rho_i} \right) - f(x) \right| d\mu^*(t) \right] \right]}{V(x)} \tag{2.65}$$

$$\cdot b\left(n^{1-\alpha} \left(x - \frac{k}{n} \right) \right).$$

Here we consider $0 \leq t_j \leq \frac{1}{n+\rho_{i_1,...,i_d}}$, $j = 1, ..., d$.

Set

$$g_{t+\frac{k+\lambda_i}{n+\rho_i}}(\lambda^*) = f\left(x + \lambda^* \left(\frac{k+\lambda_i}{n+\rho_i} - x \right) \right), \quad 0 \leq \lambda^* \leq 1. \tag{2.66}$$

Then we have

$$g^{(l)}_{t+\frac{k+\lambda_i}{n+\rho_i}}(\lambda^*) =$$

$$\left[\left(\sum_{j=1}^{d}\left(t_j + \frac{k_j+\lambda_i}{n+\rho_i} - x_j\right)\frac{\partial}{\partial x_j}\right)^l f\right]\left(x + \lambda^*\left(t + \frac{k+\lambda_i}{n+\rho_i} - x\right)\right), \quad (2.67)$$

and

$$g_{t+\frac{k+\lambda_i}{n+\rho_i}}(0) = f(x). \quad (2.68)$$

By multivariate Taylor's formula, we get

$$f\left(t + \frac{k+\lambda_i}{n+\rho_i}\right) - f(x) = g_{t+\frac{k+\lambda_i}{n+\rho_i}}(1) - f(x) =$$

$$\sum_{l=1}^{N}\frac{g^{(l)}_{t+\frac{k+\lambda_i}{n+\rho_i}}(0)}{l!} + R_N\left(t + \frac{k+\lambda_i}{n+\rho_i}, 0\right), \quad (2.69)$$

where

$$R_N\left(t + \frac{k+\lambda_i}{n+\rho_i}, 0\right) =$$

$$\int_0^1\left(\int_0^{\lambda_1^*}\cdots\left(\int_0^{\lambda_{N-1}^*}\left(g^{(N)}_{t+\frac{k+\lambda_i}{n+\rho_i}}(\lambda_N^*) - g^{(N)}_{t+\frac{k+\lambda_i}{n+\rho_i}}(0)\right)d\lambda_N^*\right)\cdots\right)d\lambda_1^*. \quad (2.70)$$

Here we denote by

$$f_{\widetilde{\alpha}} := \frac{\partial^{\widetilde{\alpha}} f}{\partial x^{\widetilde{\alpha}}},\ \widetilde{\alpha} := (\alpha_1, ..., \alpha_d), \alpha_j \in \mathbb{Z}^+,\ j = 1, ..., d, \quad (2.71)$$

such that $|\widetilde{\alpha}| = \sum_{j=1}^{d}\alpha_j = N$.

Notice that

$$\left|g^{(l)}_{t+\frac{k+\lambda_i}{n+\rho_i}}(0)\right| \overset{(2.57)}{\leq} \left(\left(\sum_{j=1}^{d}\left|\frac{\partial}{\partial x_j}\right|\right)^l f(x)\right) \cdot$$

$$\left[\left(\frac{\rho_i\|x\|_\infty + \lambda_i + 1}{n+\rho_i}\right) + \left(1 + \frac{\rho_i}{n+\rho_i}\right)\frac{T^*}{n^{1-\alpha}}\right]^l, \quad (2.72)$$

We observe that $(0 \leq \lambda_N^* \leq 1)$

$$\left| g^{(N)}_{t+\frac{k+\lambda_i}{n+\rho_i}}\left(\lambda_N^*\right) - g^{(N)}_{t+\frac{k+\lambda_i}{n+\rho_i}}(0) \right| =$$

$$| \left[\left(\sum_{j=1}^{d} \left(t_j + \frac{k_j+\lambda_i}{n+\rho_i} - x_j \right) \frac{\partial}{\partial x_j} \right)^N f \right] \left(x + \lambda_N^* \left(t + \frac{k+\lambda_i}{n+\rho_i} - x \right) \right) -$$

$$\left[\left(\sum_{j=1}^{d} \left(t_j + \frac{k_j+\lambda_i}{n+\rho_i} - x_j \right) \frac{\partial}{\partial x_j} \right)^N f \right] (x) | \overset{\text{(by (2.57), (2.58))}}{\leq} \tag{2.73}$$

$$d^N \left[\left(\frac{\rho_i \|x\|_\infty + \lambda_i + 1}{n+\rho_i} \right) + \left(1 + \frac{\rho_i}{n+\rho_i} \right) \frac{T^*}{n^{1-\alpha}} \right]^N \cdot$$

$$\max_{\widetilde{\alpha}:|\widetilde{\alpha}|=N} \omega_1 \left(f_{\widetilde{\alpha}}, \left[\left(\frac{\rho_i \|x\|_\infty + \lambda_i + 1}{n+\rho_i} \right) + \left(1 + \frac{\rho_i}{n+\rho_i} \right) \frac{T^*}{n^{1-\alpha}} \right] \right).$$

Thus we find

$$\left| R_N \left(t + \frac{k+\lambda_i}{n+\rho_i}, 0 \right) \right| \leq$$

$$\int_0^1 \left(\int_0^{\lambda_1^*} \ldots \left(\int_0^{\lambda_{N-1}^*} \left| g^{(N)}_{t+\frac{k+\lambda_i}{n+\rho_i}}\left(\lambda_N^*\right) - g^{(N)}_{t+\frac{k+\lambda_i}{n+\rho_i}}(0) \right| d\lambda_N^* \right) \ldots \right) d\lambda_1^* \leq \tag{2.74}$$

$$\frac{d^N}{N!} \left[\left(\frac{\rho_i \|x\|_\infty + \lambda_i + 1}{n+\rho_i} \right) + \left(1 + \frac{\rho_i}{n+\rho_i} \right) \frac{T^*}{n^{1-\alpha}} \right]^N \cdot$$

$$\max_{\widetilde{\alpha}:|\widetilde{\alpha}|=N} \omega_1 \left(f_{\widetilde{\alpha}}, \left[\left(\frac{\rho_i \|x\|_\infty + \lambda_i + 1}{n+\rho_i} \right) + \left(1 + \frac{\rho_i}{n+\rho_i} \right) \frac{T^*}{n^{1-\alpha}} \right] \right).$$

Therefore it holds

$$\left| f\left(t + \frac{k+\lambda_i}{n+\rho_i} \right) - f(x) \right| \overset{(2.69)}{\leq}$$

$$\sum_{l=1}^{N} \frac{\left| g^{(l)}_{t+\frac{k+\lambda_i}{n+\rho_i}}(0) \right|}{l!} + \left| R_N \left(t + \frac{k+\lambda_i}{n+\rho_i}, 0 \right) \right| \overset{\text{(by (2.72), (2.74))}}{\leq}$$

$$\sum_{l=1}^{N} \frac{1}{l!} \left(\left(\sum_{j=1}^{d} \left| \frac{\partial}{\partial x_j} \right| \right)^l f(x) \right) \left[\left(\frac{\rho_i \|x\|_\infty + \lambda_i + 1}{n+\rho_i} \right) + \left(1 + \frac{\rho_i}{n+\rho_i} \right) \frac{T^*}{n^{1-\alpha}} \right]^l \tag{2.75}$$

$$+\frac{d^N}{N!}\left[\left(\frac{\rho_i\left\|x\right\|_\infty+\lambda_i+1}{n+\rho_i}\right)+\left(1+\frac{\rho_i}{n+\rho_i}\right)\frac{T^*}{n^{1-\alpha}}\right]^N \cdot$$

$$\max_{\widetilde{\alpha}:|\widetilde{\alpha}|=N}\omega_1\left(f_{\widetilde{\alpha}},\left[\left(\frac{\rho_i\left\|x\right\|_\infty+\lambda_i+1}{n+\rho_i}\right)+\left(1+\frac{\rho_i}{n+\rho_i}\right)\frac{T^*}{n^{1-\alpha}}\right]\right).$$

Hence we have that

$$\sum_{i=1}^r w_i\left[\frac{1}{\mu^*\left(\left[0,\frac{1}{n+\rho_i}\right]^d\right)}(C)\int_0^{\frac{1}{n+\rho_i}}\left|f\left(t+\frac{k+\lambda_i}{n+\rho_i}\right)-f(x)\right|d\mu^*(t)\right]\overset{(2.75)}{\leq}$$

$$\sum_{i=1}^r w_i\left\{\sum_{l=1}^N\frac{1}{l!}\left(\left(\sum_{j=1}^d\left|\frac{\partial}{\partial x_j}\right|\right)^l f(x)\right)\cdot\right. \tag{2.76}$$

$$\left[\left(\frac{\rho_i\left\|x\right\|_\infty+\lambda_i+1}{n+\rho_i}\right)+\left(1+\frac{\rho_i}{n+\rho_i}\right)\frac{T^*}{n^{1-\alpha}}\right]^l$$

$$+\frac{d^N}{N!}\left[\left(\frac{\rho_i\left\|x\right\|_\infty+\lambda_i+1}{n+\rho_i}\right)+\left(1+\frac{\rho_i}{n+\rho_i}\right)\frac{T^*}{n^{1-\alpha}}\right]^N \cdot$$

$$\left.\max_{\widetilde{\alpha}:|\widetilde{\alpha}|=N}\omega_1\left(f_{\widetilde{\alpha}},\left[\left(\frac{\rho_i\left\|x\right\|_\infty+\lambda_i+1}{n+\rho_i}\right)+\left(1+\frac{\rho_i}{n+\rho_i}\right)\frac{T^*}{n^{1-\alpha}}\right]\right)\right\}.$$

Consequently we derive

$$\left|\left(K_n^{\mu^*}(f)\right)(x)-f(x)\right|\overset{((2.65),(2.76))}{\leq}$$

$$\sum_{k=\lceil nx-Tn^\alpha\rceil}^{[nx+Tn^\alpha]}\left[\sum_{l=1}^N\frac{1}{l!}\sum_{i=1}^r w_i\left[\left(\frac{\rho_i\left\|x\right\|_\infty+\lambda_i+1}{n+\rho_i}\right)+\left(1+\frac{\rho_i}{n+\rho_i}\right)\frac{T^*}{n^{1-\alpha}}\right]^l\cdot\right.$$

$$\left(\left(\sum_{j=1}^d\left|\frac{\partial}{\partial x_j}\right|\right)^l f(x)\right)+\frac{d^N}{N!}\sum_{i=1}^r w_i\left[\left(\frac{\rho_i\left\|x\right\|_\infty+\lambda_i+1}{n+\rho_i}\right)+\left(1+\frac{\rho_i}{n+\rho_i}\right)\frac{T^*}{n^{1-\alpha}}\right]^N\cdot \tag{2.77}$$

$$\frac{\left.\max_{\widetilde{\alpha}:|\widetilde{\alpha}|=N}\omega_1\left(f_{\widetilde{\alpha}},\left[\left(\frac{\rho_i\|x\|_\infty+\lambda_i+1}{n+\rho_i}\right)+\left(1+\frac{\rho_i}{n+\rho_i}\right)\frac{T^*}{n^{1-\alpha}}\right]\right)\right]b\left(n^{1-\alpha}\left(x-\frac{k}{n}\right)\right)}{V(x)}=$$

$$\sum_{l=1}^{N}\frac{1}{l!}\sum_{i=1}^{r}w_i\left[\left(\frac{\rho_i\|x\|_\infty+\lambda_i+1}{n+\rho_i}\right)+\left(1+\frac{\rho_i}{n+\rho_i}\right)\frac{T^*}{n^{1-\alpha}}\right]^l\cdot$$

$$\left(\left(\sum_{j=1}^{d}\left|\frac{\partial}{\partial x_j}\right|\right)^l f(x)\right)+\frac{d^N}{N!}\sum_{i=1}^{r}w_i\left[\left(\frac{\rho_i\|x\|_\infty+\lambda_i+1}{n+\rho_i}\right)+\left(1+\frac{\rho_i}{n+\rho_i}\right)\frac{T^*}{n^{1-\alpha}}\right]^N\cdot \tag{2.78}$$

$$\max_{\widetilde{\alpha}:|\widetilde{\alpha}|=N}\omega_1\left(f_{\widetilde{\alpha}},\left[\left(\frac{\rho_i\|x\|_\infty+\lambda_i+1}{n+\rho_i}\right)+\left(1+\frac{\rho_i}{n+\rho_i}\right)\frac{T^*}{n^{1-\alpha}}\right]\right),$$

proving the claim. ■

We have

Corollary 2.21 *Let all as in Theorem 2.20 hold. Additionally, assume that* $f_{\widetilde{\alpha}}(x)=0$, $\widetilde{\alpha}:|\widetilde{\alpha}|=\rho$, $1\le\rho\le N$. *Then*

$$\sup_{\mu^*}\left|\left(K_n^{\mu^*}(f)\right)(x)-f(x)\right|\le$$

$$\frac{d^N}{N!}\sum_{i=1}^{r}w_i\left[\left(\frac{\rho_i\|x\|_\infty+\lambda_i+1}{n+\rho_i}\right)+\left(1+\frac{\rho_i}{n+\rho_i}\right)\frac{T^*}{n^{1-\alpha}}\right]^N\cdot \tag{2.79}$$

$$\max_{\widetilde{\alpha}:|\widetilde{\alpha}|=N}\omega_1\left(f_{\widetilde{\alpha}},\left[\left(\frac{\rho_i\|x\|_\infty+\lambda_i+1}{n+\rho_i}\right)+\left(1+\frac{\rho_i}{n+\rho_i}\right)\frac{T^*}{n^{1-\alpha}}\right]\right).$$

Inequality (2.79) implies the pointwise convergence with rates of $\left(K_n^{\mu^*}(f)\right)(x)$ *to* $f(x)$, *as* $n\to\infty$, *at speed* $\frac{1}{n^{(1-\alpha)(N+1)}}$.

Proof By (2.64) we get the desired result. ■

The uniform convergence with rates follows from the following

Corollary 2.22 *All as in Theorem 2.20. Let* $x\in G:=\prod_{j=1}^{d}[-\gamma_j,\gamma_j]\subset\mathbb{R}^d$, $\gamma_j>0$, $\gamma^*:=\max\{\gamma_1,\ldots,\gamma_d\}$ *and* $n\in\mathbb{N}:n\ge\max_{j\in\{1,\ldots,d\}}\left(T_j+\gamma_j,T_j^{-\frac{1}{\alpha}}\right)$. *Then*

$$\sup_{\mu^*}\left\|K_n^{\mu^*}(f)-f\right\|_{\infty,\prod_{j=1}^{d}[-\gamma_j,\gamma_j]}\le\sum_{l=1}^{N}\frac{1}{l!}\left(\left(\sum_{j=1}^{d}\left|\frac{\partial}{\partial x_j}\right|\right)^l\|f\|_{\infty,\prod_{j=1}^{d}[-\gamma_j,\gamma_j]}\right)$$

$$\left[\sum_{i=1}^{r}w_i\left[\left(\frac{\rho_i\gamma^*+\lambda_i+1}{n+\rho_i}\right)+\left(1+\frac{\rho_i}{n+\rho_i}\right)\frac{T^*}{n^{1-\alpha}}\right]^l\right]+ \tag{2.80}$$

$$\frac{d^N}{N!}\sum_{i=1}^{r} w_i \left[\left(\frac{\rho_i \gamma^* + \lambda_i + 1}{n+\rho_i}\right) + \left(1 + \frac{\rho_i}{n+\rho_i}\right)\frac{T^*}{n^{1-\alpha}}\right]^N \cdot$$

$$\max_{\widetilde{\alpha}:|\widetilde{\alpha}|=N} \omega_1 \left(f_{\widetilde{\alpha}}, \left[\left(\frac{\rho_i \gamma^* + \lambda_i + 1}{n+\rho_i}\right) + \left(1 + \frac{\rho_i}{n+\rho_i}\right)\frac{T^*}{n^{1-\alpha}}\right]\right).$$

Inequality (2.80) implies the pointwise convergence with rates of $K_n^{\mu^*}(f)$ *to* f *on* G*, as* $n \to \infty$*, at speed* $\frac{1}{n^{1-\alpha}}$.

Proof By (2.64) we get the desired result. ■

Corollary 2.23 *All as in Theorem 2.20 with* $N = 1$ *hold. Then*

$$\sup_{\mu^*}\left|\left(K_n^{\mu^*}(f)\right)(x) - f(x)\right| \leq \left(\sum_{j=1}^{d}\left|\frac{\partial f(x)}{\partial x_j}\right|\right) \cdot$$

$$\left[\sum_{i=1}^{r} w_i \left[\left(\frac{\rho_i \|x\|_\infty + \lambda_i + 1}{n+\rho_i}\right) + \left(1 + \frac{\rho_i}{n+\rho_i}\right)\frac{T^*}{n^{1-\alpha}}\right]\right] + \tag{2.81}$$

$$d\sum_{i=1}^{r} w_i \left[\left(\frac{\rho_i \|x\|_\infty + \lambda_i + 1}{n+\rho_i}\right) + \left(1 + \frac{\rho_i}{n+\rho_i}\right)\frac{T^*}{n^{1-\alpha}}\right] \cdot$$

$$\max_{\widetilde{\alpha}:|\widetilde{\alpha}|=N} \omega_1 \left(f_{\widetilde{\alpha}}, \left[\left(\frac{\rho_i \|x\|_\infty + \lambda_i + 1}{n+\rho_i}\right) + \left(1 + \frac{\rho_i}{n+\rho_i}\right)\frac{T^*}{n^{1-\alpha}}\right]\right).$$

Inequality (2.81) implies the pointwise convergence with rates of $\left(K_n^{\mu^*}(f)\right)(x)$ *to* $f(x)$*, as* $n \to \infty$*, at speed* $\frac{1}{n^{1-\alpha}}$.

Proof By (2.64) we get the desired result. ■

We finish with the following

Note 2.24 Notice here that the right hand sides of (2.64), (2.79), (2.80) and (2.81), do not depend on μ^* and b.

References

1. G.A. Anastassiou, *Quantitative Approximation by Perturbed Kantorovich-Choquet Neural Network Operators* (Fisicasy Naturales. Serie A. Mathematicas, accepted for publication, Revista de la Real Academia de Ciencias Exactas, 2018)
2. G.A. Anastassiou, Rate of convergence of some neural network operators to the unit-univariate case. J. Math. Anal. Appl. **212**, 237–262 (1997)

3. G.A. Anastassiou, Rate of convergence of some multivariate neural network operators to the unit. J. Comput. Math. Appl. **40**, 1–19 (2000)
4. G.A. Anastassiou, *Quantitative Approximations* (Chapman and Hall/CRC, New York, 2001)
5. G.A. Anastassiou, Rate of convergence of some neural network operators to the unit-univariate case, revisited. Vesnik **65**(4), 511–518 (2013)
6. G.A. Anastassiou, Rate of convergence of some multivariate neural network operators to the unit, revisited. J. Comput. Anal. Appl. **15**(7), 1300–1309 (2013)
7. A.R. Barron, Universal approximation bounds for superpositions of a sigmoidal function. IEEE Trans. Inform. Theory **39**, 930–945 (1993)
8. F.L. Cao, T.F. Xie, Z.B. Xu, The estimate for approximation error of neural networks: a constructive approach. Neurocomputing **71**, 626–630 (2008)
9. P. Cardaliaguet, G. Euvrard, Approximation of a function and its derivative with a neural network. Neural Netw. **5**, 207–220 (1992)
10. Z. Chen, F. Cao, The approximation operators with sigmoidal functions. Comput. Math. Appl. **58**, 758–765 (2009)
11. T.P. Chen, H. Chen, Universal approximation to nonlinear operators by neural networks with arbitrary activation functions and its applications to a dynamic system. IEEE Trans. Neural Netw. **6**, 911–917 (1995)
12. G. Choquet, Theory of capacities. Ann. Inst. Fourier (Grenoble) **5**, 131–295 (1954)
13. C.K. Chui, X. Li, Approximation by ridge functions and neural networks with one hidden layer. J. Approx. Theory **70**, 131–141 (1992)
14. D. Costarelli, R. Spigler, Approximation results for neural network operators activated by sigmoidal functions. Neural Netw. **44**, 101–106 (2013)
15. D. Costarelli, R. Spigler, Multivariate neural network operators with sigmoidal activation functions. Neural Netw. **48**, 72–77 (2013)
16. G. Cybenko, Approximation by superpositions of sigmoidal function. Math. Control Signals Syst. **2**, 303–314 (1989)
17. D. Denneberg, *Non-additive Meas. Integral* (Kluwer, Dordrecht, 1994)
18. S. Ferrari, R.F. Stengel, Smooth function approximation using neural networks. IEEE Trans. Neural Netw. **16**, 24–38 (2005)
19. K.I. Funahashi, On the approximate realization of continuous mappings by neural networks. Neural Netw. **2**, 183–192 (1989)
20. S. Gal, Uniform and Pointwise Quantitative Approximation by Kantorovich-Choquet type Integral Operators with respect to monotone and submodular set functions. Mediter. J. Math. **14**(5), 12 (2017). Art. 205
21. N. Hahm, B.I. Hong, An approximation by neural networks with a fixed weight. Comput. Math. Appl. **47**, 1897–1903 (2004)
22. S. Haykin, *Neural Netw.: A Compr. Found.*, 2nd edn. (Prentice Hall, New York, 1998)
23. K. Hornik, M. Stinchombe, H. White, Multilayer feedforward networks are universal approximations. Neural Netw. **2**, 359–366 (1989)
24. K. Hornik, M. Stinchombe, H. White, Universal approximation of an unknown mapping and its derivatives using multilayer feedforward networks. Neural Netw. **3**, 551–560 (1990)
25. M. Leshno, V.Y. Lin, A. Pinks, S. Schocken, Multilayer feedforward networks with a nonpolynomial activation function can approximate any function. Neural Netw. **6**, 861–867 (1993)
26. V. Maiorov, R.S. Meir, Approximation bounds for smooth functions in $C\left(R^d\right)$ by neural and mixture networks. IEEE Trans. Neural Netw. **9**, 969–978 (1998)
27. Y. Makovoz, Uniform approximation by neural networks. J. Approx. Theory **95**, 215–228 (1998)
28. W. McCulloch, W. Pitts, A logical calculus of the ideas immanent in nervous activity. Bull. Math. Biophys. **7**, 115–133 (1943)
29. H.N. Mhaskar, C.A. Micchelli, Approximation by superposition of a sigmoidal function. Adv. Appl. Math. **13**, 350–373 (1992)
30. H.N. Mhaskar, C.A. Micchelli, Degree of approximation by neural networks with a single hidden layer. Adv. Appl. Math. **16**, 151–183 (1995)

31. T.M. Mitchell, *Machine Learning* (WCB-McGraw-Hill, New York, 1997)
32. S. Suzuki, Constructive function approximation by three-layer artificial neural networks. Neural Netw. **11**, 1049–1058 (1998)
33. Z. Wang, G.J. Klir, *Generalized Measure Theory* (Springer, New York, 2009)
34. Z.B. Xu, F.L. Cao, The essential order of approximation for neural networks. Sci. China (Ser. F) **47**, 97–112 (2004)

Chapter 3
Approximation with Rates by Shift Invariant Univariate Sublinear-Choquet Operators

A very general positive sublinear Choquet integral type operator is given through a convolution-like iteration of another general positive sublinear operator with a scaling type function. For it sufficient conditions are given for shift invariance, preservation of global smoothness, convergence to the unit with rates. Furthermore, two examples of very general specialized operators are presented fulfilling all the above properties, the higher order of approximation of these operators is also studied. It follows [3].

3.1 Introduction

Let X, Y be function spaces of functions from $\mathbb{R}$ into $\mathbb{R}_+$. Let $L_N : X \to Y$, $N \in \mathbb{N}$, be a sequence of operators with the following properties:

(i) (positive homogeneous)

$$L_N(\alpha f) = \alpha L_N(f), \forall \alpha \geq 0, \forall f \in X.$$

(ii) (Monotonicity) if $f, g \in X$ satisfy $f \leq g$, then $L_N(f) \leq L_N(g)$, $\forall N \in \mathbb{N}$, and

(iii) (Subadditivity)

$$L_N(f+g) \leq L_N(f) + L_N(g), \forall f, g \in X.$$

We call L_N positive sublinear operators.

In this chapter we deal with sequences of Choquet positive sublinear operators that are constructed, with the help of Choquet integral. Our functions spaces are continuous functions from $\mathbb{R}$ into $\mathbb{R}_+$. The sequence of operators is generated by a basic operator via dilated translations of convolution type using the Choquet integral. We prove that our operators possess the following properties: of shift invariance, of

G. A. Anastassiou, *Ordinary and Fractional Approximation by Non-additive Integrals: Choquet, Shilkret and Sugeno Integral Approximators*, Studies in Systems, Decision and Control 190, https://doi.org/10.1007/978-3-030-04287-5_3

global smoothness preservation, of convergence to the unit operator with rates. Then we apply our results to two specific families of such Choquet type operators.

We continue with the higher order of approximation study of these specific operators, and all results are quantitative.

Earlier similar studies have been done by the author, see [5], Chapters. 10–17, and [2], Chapters. 16, 17. These serve as motivation and inspiration to our current work.

3.2 Background

Next we present briefly about the Choquet integral.

We make

Definition 3.1 Consider $\Omega \neq \emptyset$ and let $\mathcal{C}$ be a σ-algebra of subsets in Ω.

(i) (see, e.g., [9], p. 63) The set function $\mu : \mathcal{C} \to [0, +\infty]$ is called a monotone set function (or capacity) if $\mu(\emptyset) = 0$ and $\mu(A) \leq \mu(B)$ for all $A, B \in \mathcal{C}$, with $A \subset B$. Also, μ is called submodular if

$$\mu(A \cup B) + \mu(A \cap B) \leq \mu(A) + \mu(B), \text{ for all } A, B \in \mathcal{C}.$$

μ is called bounded if $\mu(\Omega) < +\infty$ and normalized if $\mu(\Omega) = 1$.

(ii) (see, e.g., [9], p. 233, or [6]) If μ is a monotone set function on $\mathcal{C}$ and if $f : \Omega \to \mathbb{R}$ is $\mathcal{C}$-measurable (that is, for any Borel subset $B \subset \mathbb{R}$ it follows $f^{-1}(B) \in \mathcal{C}$), then for any $A \in \mathcal{C}$, the Choquet integral is defined by

$$(C)\int_A f d\mu = \int_0^{+\infty} \mu\left(F_\beta(f) \cap A\right) d\beta + \int_{-\infty}^0 \left[\mu\left(F_\beta(f) \cap A\right) - \mu(A)\right] d\beta,$$

where we used the notation $F_\beta(f) = \{\omega \in \Omega : f(\omega) \geq \beta\}$. Notice that if $f \geq 0$ on A, then in the above formula we get $\int_{-\infty}^0 = 0$.

The integrals on the right-hand side are the usual Riemann integral.

The function f will be called Choquet integrable on A if $(C)\int_A f d\mu \in \mathbb{R}$.

Next we list some well known properties of the Choquet integral.

Remark 3.2 If $\mu : \mathcal{C} \to [0, +\infty]$ is a monotone set function, then the following properties hold:

(i) For all $a \geq 0$ we have $(C)\int_A af d\mu = a \cdot (C)\int_A f d\mu$ (if $f \geq 0$ then see, e.g., [9], Theorem 11.2, (5), p. 228 and if f is arbitrary sign, then see, e.g., [7], p. 64, Proposition 5.1, (ii)).

(ii) For all $c \in \mathbb{R}$ and f of arbitrary sign, we have (see, e.g., [9], pp. 232–233, or [7], p. 65) $(C)\int_A (f + c) d\mu = (C)\int_A f d\mu + c \cdot \mu(A)$.

If μ is submodular too, then for all f, g of arbitrary sign and lower bounded, we have (see, e.g., [7], p. 75, Theorem 6.3)

$$(C)\int_A (f+g)\,d\mu \leq (C)\int_A f\,d\mu + (C)\int_A g\,d\mu.$$

(iii) If $f \leq g$ on A then $(C)\int_A f d\mu \leq (C)\int_A g d\mu$ (see, e.g., [9], p. 228, Theorem 11.2, (3) if $f, g \geq 0$ and p. 232 if f, g are of arbitrary sign).

(iv) Let $f \geq 0$. If $A \subset B$ then $(C)\int_A f d\mu \leq (C)\int_B f d\mu$. In addition, if μ is finitely subadditive, then

$$(C)\int_{A\cup B} f\,d\mu \leq (C)\int_A f\,d\mu + (C)\int_B f\,d\mu.$$

(v) It is immediate that $(C)\int_A 1 \cdot d\mu(t) = \mu(A)$.

(vi) The formula $\mu(A) = \gamma(M(A))$, where $\gamma : [0, 1] \to [0, 1]$ is an increasing and concave function, with $\gamma(0) = 0$, $\gamma(1) = 1$ and M is a probability measure (or only finitely additive) on a σ-algebra on Ω (that is, $M(\emptyset) = 0$, $M(\Omega) = 1$ and M is countably additive), gives simple examples of normalized, monotone and submodular set functions (see, e.g., [7], pp. 16–17, Example 2.1). Such of set functions μ are also called distorsions of countably normalized, additive measures (or distorted measures). For a simple example, we can take $\gamma(t) = \frac{2t}{1+t}$, $\gamma(t) = \sqrt{t}$.

If the above γ function is increasing, concave and satisfies only $\gamma(0) = 0$, then for any bounded Borel measure m, $\mu(A) = \gamma(m(A))$ gives a simple example of bounded, monotone and submodular set function.

(vii) If μ is a countably additive bounded measure, then the Choquet integral $(C)\int_A f d\mu$ reduces to the usual Lebesgue type integral (see, e.g., [7], p. 62, or [9], p. 226).

(viii) If $f \geq 0$, then $(C)\int_A f d\mu \geq 0$.

(ix) Let $\mu = \sqrt{M}$, where M is the Lebesgue measure, then μ is a monotone and submodular set function, furthermore μ is strictly positive, see [8]

(x) If $\Omega = \mathbb{R}^N$, $N \in \mathbb{N}$, we call μ strictly positive if $\mu(A) > 0$, for any open subset $A \subseteq \mathbb{R}^N$.

3.3 Univariate Theory

This section is motivated and inspired by [5], and Chapter 10, [4]. In a way it generalizes that material.

Let $\mathcal{L}$ be the Lebesgue $\sigma-$ algebra on $\mathbb{R}$, and the set function $\mu : \mathcal{L} \to [0, +\infty]$, which is assumed to be monotone and submodular. Let $C_U(\mathbb{R}, \mathbb{R}_+)$ be the space of uniformly continuous functions from $\mathbb{R}$ into $\mathbb{R}_+$, and $C(\mathbb{R}, \mathbb{R}_+)$ the space of continuous functions from $\mathbb{R}$ into $\mathbb{R}_+$. For any $f \in C_U(\mathbb{R}, \mathbb{R}_+)$ we have $\omega_1(f, \delta) < +\infty$, $\delta > 0$, where

$$\omega_1(f, \delta) := \sup_{\substack{x, y \in \mathbb{R}: \\ |x-y| \leq \delta}} |f(x) - f(y)|, \ \delta > 0,$$

is the first modulus of continuity.

Let $\{l_k\}_{k \in \mathbb{Z}}$ be a sequence of positive sublinear operators that map $C_U(\mathbb{R}, \mathbb{R}_+)$ into $C(\mathbb{R}, \mathbb{R}_+)$ with the property

$$(l_k(f))(x) := l_0\left(f\left(2^{-k}\cdot\right)\right)(x), \forall x \in \mathbb{R}, \forall f \in C_U(\mathbb{R}, \mathbb{R}_+). \tag{3.1}$$

For a fixed $a > 0$ we assume that

$$\sup_{\substack{u,y \in \mathbb{R}: \\ |u-y| \leq a}} |l_0(f, u) - f(y)| \leq \omega_1\left(f, \frac{ma+n}{2^r}\right), \forall f \in C_U(\mathbb{R}, \mathbb{R}_+), \tag{3.2}$$

where $m \in \mathbb{N}$, $n \in \mathbb{Z}_+$, $r \in \mathbb{Z}$.

Let $\varphi : \mathbb{R} \to \mathbb{R}_+$ which is Lebesgue measurable, such that

$$(C)\int_{-a}^{a} \varphi(u)\, d\mu(u) = 1. \tag{3.3}$$

Example 3.3 Let $[c, d] \subset \mathbb{R}$ such that $\left[-\frac{1}{2}, \frac{1}{2}\right] \subset [c, d]$. Let the characteristic function

$$\varphi(x) := \chi_{[c,d]}(x) = \begin{cases} 1, & x \in [c, d] \\ 0, & \text{elsewhere}, \end{cases} \tag{3.4}$$

and λ is the Lebesgue measurable on $\mathbb{R}$, where $a = \frac{1}{2}$. Here we take $\mu := \sqrt{\lambda}$, which is a monotone and submodular set function (see Remark 3.2 (ix)).

We notice that

$$(C)\int_{-\frac{1}{2}}^{\frac{1}{2}} \chi_{[c,d]}(u)\, d\sqrt{\lambda(u)} = (C)\int_{-\frac{1}{2}}^{\frac{1}{2}} 1 d\sqrt{\lambda(u)} = \sqrt{\lambda\left(\left[-\frac{1}{2}, \frac{1}{2}\right]\right)} = 1, \tag{3.5}$$

fulfilling (3.3).

We define the positive sublinear-Choquet operators

$$(\mathcal{L}_0(f))(x) := (C)\int_{-a}^{a} (l_0 f)(x-u)\, \varphi(u)\, d\mu(u), \tag{3.6}$$

and

$$(\mathcal{L}_k(f))(x) := \left(\mathcal{L}_0\left(f\left(2^{-k}\cdot\right)\right)\right)\left(2^k x\right), \forall k \in \mathbb{Z}, \forall x \in \mathbb{R}. \tag{3.7}$$

Therefore it holds

$$(\mathcal{L}_k(f))(x) = (C)\int_{-a}^{a} \left(l_0\left(f\left(2^{-k}\cdot\right)\right)\right)\left(2^k x - u\right) \varphi(u)\, d\mu(u) = \tag{3.8}$$

$$(C)\int_{-a}^{a} (l_k(f))\left(2^k x - u\right) \varphi(u)\, d\mu(u),$$

$\forall\, x \in \mathbb{R}, \forall\, k \in \mathbb{Z}$.

Indeed here we have

$$(\mathcal{L}_k(f))(x) \leq (C)\int_{-a}^{a} \left\| l_k(f)\left(2^k x - \cdot\right)\right\|_{\infty,[-a,a]} \varphi(u)\, d\mu(u) =$$

$$\left\| l_k(f)\left(2^k x - \cdot\right)\right\|_{\infty,[-a,a]} \left((C)\int_{-a}^{a} \varphi(u)\, d\mu(u)\right) =$$

$$\left\| l_k(f)\left(2^k x - \cdot\right)\right\|_{\infty,[-a,a]} < +\infty.$$

Hence $(\mathcal{L}_k(f))(x) \in \mathbb{R}_+$ is well-defined.

Let $f, g \in \mathcal{M}(\mathbb{R}, \mathbb{R}_+)$ (Lebesgue measurable functions) where $x \in A$, $A \subset \mathbb{R}$ is a Lebesgue measurable set.

We observe that

$$f(x) = f(x) - g(x) + g(x) \leq |f(x) - g(x)| + g(x),$$

hence

$$(C)\int_A f(x)\, d\mu(x) \leq (C)\int_A |f(x) - g(x)|\, d\mu(x) + (C)\int_A g(x)\, d\mu(x),$$

and

$$(C)\int_A f(x)\, d\mu(x) - (C)\int_A g(x)\, d\mu(x) \leq (C)\int_A |f(x) - g(x)|\, d\mu(x).$$

Similarly, it holds

$$(C)\int_A g(x)\, d\mu(x) - (C)\int_A f(x)\, d\mu(x) \leq (C)\int_A |f(x) - g(x)|\, d\mu(x).$$

Therefore we derive

$$\left|(C)\int_A f(x)\, d\mu(x) - (C)\int_A g(x)\, d\mu(x)\right| \leq (C)\int_A |f(x) - g(x)|\, d\mu(x). \tag{3.9}$$

We need

Definition 3.4 Let $f_\alpha(\cdot) := f(\cdot + \alpha)$, $\alpha \in \mathbb{R}$, and Φ be an operator. If $\Phi(f_\alpha) = (\Phi f)_\alpha$, then Φ is called a shift invariant operator.

We give

Theorem 3.5 *Assume that*

$$\left(l_0\left(f\left(2^{-k} \cdot + \alpha\right)\right)\right)\left(2^k u\right) = \left(l_0\left(f\left(2^{-k} \cdot\right)\right)\right)\left(2^k(u + \alpha)\right), \tag{3.10}$$

for all $k \in \mathbb{Z}$, $\alpha \in \mathbb{R}$ fixed, all $u \in \mathbb{R}$ and any $f \in C_U(\mathbb{R}, \mathbb{R}_+)$. Then $\mathcal{L}_k$ is a shift invariant operator for all $k \in \mathbb{Z}$.

Proof We have that

$$\left(\mathcal{L}_k\left(f\left(\cdot+\alpha\right)\right)\right)(x) = \left(\mathcal{L}_k\left(f_\alpha\right)\right)(x) \overset{(3.8)}{=}$$

$$(C)\int_{-a}^{a}\left(l_0\left(f_\alpha\left(2^{-k}\cdot\right)\right)\right)\left(2^k x - u\right)\varphi(u)\,d\mu(u) =$$

$$(C)\int_{-a}^{a}\left(l_0\left(f\left(2^{-k}\cdot+\alpha\right)\right)\right)\left(2^k x - u\right)\varphi(u)\,d\mu(u) =$$

$$(C)\int_{-a}^{a}\left(l_0\left(f\left(2^{-k}\cdot+\alpha\right)\right)\right)\left(2^k\left(x - 2^{-k}u\right)\right)\varphi(u)\,d\mu(u) \overset{(3.10)}{=} \tag{3.11}$$

$$(C)\int_{-a}^{a}\left(l_0\left(f\left(2^{-k}\cdot\right)\right)\right)\left(2^k\left(x - 2^{-k}u+\alpha\right)\right)\varphi(u)\,d\mu(u) =$$

$$(C)\int_{-a}^{a}\left(l_0\left(f\left(2^{-k}\cdot\right)\right)\right)\left(2^k(x+\alpha) - u\right)\varphi(u)\,d\mu(u) \overset{(3.8)}{=} \left(\mathcal{L}_k(f)\right)(x+\alpha),$$

that is

$$\mathcal{L}_k\left(f_\alpha\right) = \left(\mathcal{L}_k(f)\right)_\alpha, \tag{3.12}$$

proving the claim. ■

It follows the global smoothness of the operators $\mathcal{L}_k$.

Theorem 3.6 *For any $f \in C_U(\mathbb{R}, \mathbb{R}_+)$ assume that, for all $u \in \mathbb{R}$,*

$$\left|\left(l_0(f)\right)(x-u) - \left(l_0(f)\right)(y-u)\right| \leq \omega_1\left(f, |x-y|\right), \tag{3.13}$$

for any $x, y \in \mathbb{R}$. Then

$$\omega_1\left(\mathcal{L}_k f, \delta\right) \leq \omega_1(f, \delta), \ \forall\, \delta > 0. \tag{3.14}$$

Proof We observe that

$$\left|\left(\mathcal{L}_0(f)\right)(x) - \left(\mathcal{L}_0(f)\right)(y)\right| =$$

$$\left|(C)\int_{-a}^{a}\left(l_0 f\right)(x-u)\varphi(u)\,d\mu(u) - (C)\int_{-a}^{a}\left(l_0 f\right)(y-u)\varphi(u)\,d\mu(u)\right| \overset{(3.9)}{\leq} \tag{3.15}$$

$$(C)\int_{-a}^{a}\left|\left(l_0 f\right)(x-u) - \left(l_0 f\right)(y-u)\right|\varphi(u)\,d\mu(u) \overset{(3.13)}{\leq}$$

$$\omega_1(f, |x-y|)\left((C)\int_{-a}^{a}\varphi(u)\,d\mu(u)\right) \overset{(3.3)}{=} \omega_1(f, |x-y|).$$

So that

$$|(\mathcal{L}_0(f))(x) - (\mathcal{L}_0(f))(y)| \leq \omega_1(f, |x-y|). \tag{3.16}$$

From (3.7), (3.16) we get

$$|(\mathcal{L}_k(f))(x) - (\mathcal{L}_k(f))(y)| \overset{(3.7)}{=}$$

$$\left|\left(\mathcal{L}_0\left(f\left(2^{-k}\cdot\right)\right)\right)\left(2^k x\right) - \left(\mathcal{L}_0\left(f\left(2^{-k}\cdot\right)\right)\right)\left(2^k y\right)\right| \leq \tag{3.17}$$

$$\omega_1\left(f\left(2^{-k}\cdot\right), 2^k|x-y|\right) = \omega_1(f, |x-y|),$$

i.e. global smoothness for $\mathcal{L}_k$ has been proved. ■

The convergence of $\mathcal{L}_k$ to the unit operator, as $k \to +\infty$, k with rates follows:

Theorem 3.7 *For $f \in C_U(\mathbb{R}, \mathbb{R}_+)$, under the assumption (3.2), we have*

$$|(\mathcal{L}_k(f))(x) - f(x)| \leq \omega_1\left(f, \frac{ma+n}{2^{k+r}}\right), \tag{3.18}$$

where $m \in \mathbb{N}$, $n \in \mathbb{Z}_+$, $k, r \in \mathbb{Z}$.

Proof We notice that

$$|(\mathcal{L}_k(f))(x) - f(x)| \overset{(3.7)}{=} \left|\left(\mathcal{L}_0\left(f\left(2^{-k}\cdot\right)\right)\right)\left(2^k x\right) - f(x)\right| \overset{(3.8)}{=}$$

$$\left|(C)\int_{-a}^{a}\left(l_0\left(f\left(2^{-k}\cdot\right)\right)\right)\left(2^k x - u\right)\varphi(u)\,d\mu(u) - f(x)\right| \overset{(3.3)}{=}$$

$$\left|(C)\int_{-a}^{a}\left(l_0\left(f\left(2^{-k}\cdot\right)\right)\right)\left(2^k x - u\right)\varphi(u)\,d\mu(u) - (C)\int_{-a}^{a} f(x)\varphi(u)\,d\mu(u)\right| \overset{(3.9)}{\leq}$$

$$(C)\int_{-a}^{a}\left|\left(l_0\left(f\left(2^{-k}\cdot\right)\right)\right)\left(2^k x - u\right) - f(x)\right|\varphi(u)\,d\mu(u) = \tag{3.19}$$

$$(C)\int_{-a}^{a}\left|\left(l_0\left(f\left(2^{-k}\cdot\right)\right)\right)\left(2^k x - u\right) - f\left(2^{-k}\cdot\right)\left(2^k x\right)\right|\varphi(u)\,d\mu(u) \overset{(3.2)}{\leq}$$

(here $\left|\left(2^k x - u\right) - 2^k x\right| = |u| \leq a$)

$$\omega_1\left(f\left(2^{-k}\cdot\right), \frac{ma+n}{2^r}\right)\left((C)\int_{-a}^{a}\varphi(u)\,d\mu(u)\right) =$$

$$\omega_1\left(f\left(2^{-k}\cdot\right),\frac{ma+n}{2^r}\right)\cdot 1=\omega_1\left(f,\frac{ma+n}{2^{k+r}}\right),\tag{3.20}$$

proving the claim. ■

We give some applications.

For each $k\in\mathbb{Z}$, we define

(i)

$$\left(B_k f\right)(x):=(C)\int_{-a}^{a} f\left(x-\frac{u}{2^k}\right)\varphi(u)\,d\mu(u),\tag{3.21}$$

i.e., here

$$\begin{gathered}\left(l_k(f)\right)(u)=f\left(\tfrac{u}{2^k}\right),\\ \text{and}\\ \left(l_0(f)\right)(u)=f(u),\end{gathered}\tag{3.22}$$

are continuous in $u\in\mathbb{R}$.

Also for $k\in\mathbb{Z}$, we define

(ii)

$$\left(\Gamma_k(f)\right)(x):=(C)\int_{-a}^{a}\gamma_k^f\left(2^k x-u\right)\varphi(u)\,d\mu(u),\tag{3.23}$$

where

$$\left(l_k(f)\right)(u)=\gamma_k^f(u):=\sum_{j=0}^{n} w_j f\left(\frac{u}{2^k}+\frac{j}{2^k n}\right),\tag{3.24}$$

$n\in\mathbb{N}$, $w_j\geq 0$, $\sum_{j=0}^{n} w_j=1$,

is continuous in $u\in\mathbb{R}$.

Notice here that

$$\left(l_0(f)\right)(u)=\gamma_0^f(u)=\sum_{j=0}^{n} w_j f\left(u+\frac{j}{n}\right)\tag{3.25}$$

is also continuous in $u\in\mathbb{R}$.

Indeed we have

$$\left(\Gamma_k(f)\right)(x)=(C)\int_{-a}^{a}\left[\sum_{j=0}^{n} w_j f\left(\left(x-\frac{u}{2^k}\right)+\frac{j}{2^k n}\right)\right]\varphi(u)\,d\mu(u).\tag{3.26}$$

Clearly here we have

$$\begin{aligned}(B_k(f))(x) &= \left(B_0\left(f\left(2^{-k}\cdot\right)\right)\right)\left(2^k x\right),\\ &\text{and}\\ (\Gamma_k(f))(x) &= \left(\Gamma_0\left(f\left(2^{-k}\cdot\right)\right)\right)\left(2^k x\right),\end{aligned} \tag{3.27}$$

$\forall\, k \in \mathbb{Z}, \forall\, x \in \mathbb{R}$.

We give

Proposition 3.8 *B_k, Γ_k are shift invariant operators.*

Proof (i) For B_k operators: Here $l_0 f = f$. Hence

$$\left(l_0\left(f\left(2^{-k}\cdot+\alpha\right)\right)\right)\left(2^k u\right) = f\left(2^{-k}2^k u+\alpha\right) = f(u+\alpha) = \tag{3.28}$$

$$\left(l_0\left(f\left(2^{-k}\cdot\right)\right)\right)\left(2^k(u+\alpha)\right).$$

(ii) Foe Γ_k operators:

$$(l_0(f))(u) = \sum_{j=0}^{n} w_j f\left(u+\frac{j}{n}\right).$$

Hence

$$\left(l_0\left(f\left(2^{-k}\cdot+\alpha\right)\right)\right)\left(2^k u\right) = \sum_{j=0}^{n} w_j f\left(2^{-k}\left(2^k u+\frac{j}{n}\right)+\alpha\right) =$$

$$\sum_{j=0}^{n} w_j f\left(2^{-k}\left(2^k(u+\alpha)+\frac{j}{n}\right)\right) = \left(l_0\left(f\left(2^{-k}\cdot\right)\right)\right)\left(2^k(u+\alpha)\right), \tag{3.29}$$

proving the claim. ∎

Next we show that the operators B_k, Γ_k possess the property of global smoothness preservation.

Theorem 3.9 *For all $f \in C_U(\mathbb{R}, \mathbb{R}_+)$ and all $\delta > 0$ we have*

$$\begin{aligned}\omega_1(B_k f, \delta) &\le \omega_1(f, \delta),\\ &\textit{and}\\ \omega_1(\Gamma_k f, \delta) &\le \omega_1(f, \delta).\end{aligned} \tag{3.30}$$

Proof (i) For B_k operators: Here $l_0 f = f$, therefore

$$|(l_0(f))(x-u) - (l_0(f))(y-u)| = |f(x-u) - f(y-u)| \le \omega_1(f, |x-y|). \tag{3.31}$$

(ii) For Γ_k operators: We observe that

$$|(l_0(f))(x-u)-(l_0(f))(y-u)| = \left|\gamma_0^f(x-u)-\gamma_0^f(y-u)\right| =$$

$$\left|\sum_{j=0}^{n} w_j\left(f\left(x-u+\frac{j}{n}\right)-f\left(y-u+\frac{j}{n}\right)\right)\right| \leq$$

$$\sum_{j=0}^{n} w_j\left|f\left(x-u+\frac{j}{n}\right)-f\left(y-u+\frac{j}{n}\right)\right| \leq$$

$$\omega_1(f,|x-y|)\left(\sum_{j=0}^{n} w_j\right) = \omega_1(f,|x-y|), \tag{3.32}$$

proving the claim. ∎

The operators B_k, Γ_k, $k \in \mathbb{Z}$, converge to the unit operator with rates presented next.

Theorem 3.10 *For $k \in \mathbb{Z}$,*

$$\begin{gathered}|(B_k(f))(x)-f(x)| \leq \omega_1\left(f,\tfrac{a}{2^k}\right), \\ \text{and} \\ |(\Gamma_k(f))(x)-f(x)| \leq \omega_1\left(f,\tfrac{a+1}{2^k}\right).\end{gathered} \tag{3.33}$$

Proof (i) For B_k operators: Here $(l_0(f))(u) = f(u)$ and

$$\sup_{\substack{u,y\in\mathbb{R}\\|u-y|\leq a}}|(l_0(f))(u)-f(y)| = \sup_{\substack{u,y\in\mathbb{R}\\|u-y|\leq a}}|f(u)-f(y)| = \omega_1(f,a), \tag{3.34}$$

and we use Theorem 3.7.

(ii) For Γ_k operators: Here we see that

$$\sup_{\substack{u,y\in\mathbb{R}\\|u-y|\leq a}}|(l_0(f))(u)-f(y)| = \sup_{\substack{u,y\in\mathbb{R}\\|u-y|\leq a}}\left|\sum_{j=0}^{n} w_j f\left(u+\frac{j}{n}\right)-f(y)\right| \leq$$

$$\sup_{\substack{u,y\in\mathbb{R}\\|u-y|\leq a}}\sum_{j=0}^{n} w_j\left|f\left(u+\frac{j}{n}\right)-f(y)\right| \leq \sup_{\substack{u,y\in\mathbb{R}\\|u-y|\leq a}}\sum_{j=0}^{n} w_j\omega_1\left(f,\left|u+\frac{j}{n}-y\right|\right) \leq \tag{3.35}$$

$$\sup_{\substack{u,y\in\mathbb{R}\\|u-y|\leq a}}\sum_{j=0}^{n}w_j\omega_1\left(f,\frac{j}{n}+|u-y|\right)\leq\left(\sum_{j=0}^{n}w_j\right)\omega_1\left(f,1+\alpha\right)=\omega_1\left(f,\alpha+1\right).$$

By (3.18) we are done. ■

3.4 Higher Order of Approximation

Here all are as in Sect. 3.3. See also earlier our work [1], and [2], Chapter 16.

We give

Theorem 3.11 *Let $f\in C^N(\mathbb{R},\mathbb{R}_+)$, $N\geq 1$. Consider the Choquet-sublinear operators*

$$(B_kf)(x)=(C)\int_{-a}^{a}f\left(x-\frac{u}{2^k}\right)\varphi(u)\,d\mu(u),$$

$\forall\, k\in\mathbb{Z}$, $\forall\, x\in\mathbb{R}$. Then

$$|(B_kf)(x)-f(x)|\leq\sum_{i=1}^{N}\frac{\left|f^{(i)}(x)\right|}{i!}\frac{a^i}{2^{ki}}+\frac{a^N}{2^{kN}N!}\omega_1\left(f^{(N)},\frac{a}{2^k}\right).\tag{3.36}$$

If $f^{(N)}$ is uniformly continuous or bounded and continuous, then as $k\to+\infty$ we obtain that $(B_kf)(x)\to f(x)$ pointwise with rates.

Proof Since $f\in C^N(\mathbb{R},\mathbb{R}_+)$, $N\geq 1$, by Taylor's formula we have

$$f\left(x-\frac{u}{2^k}\right)-f(x)=\sum_{i=1}^{N}\frac{f^{(i)}(x)}{i!}\left(-\frac{u}{2^k}\right)^i+\tag{3.37}$$

$$\int_x^{x-\frac{u}{2^k}}\left(f^{(N)}(t)-f^{(N)}(x)\right)\frac{\left(x-\frac{u}{2^k}-t\right)^{N-1}}{(N-1)!}dt.$$

Call

$$\Gamma_u(x):=\left|\int_x^{x-\frac{u}{2^k}}\left(f^{(N)}(t)-f^{(N)}(x)\right)\frac{\left(x-\frac{u}{2^k}-t\right)^{N-1}}{(N-1)!}dt\right|.\tag{3.38}$$

Next we estimate $\Gamma_u(x)$, where $u\in[-a,a]$.

(i) Case of $-a\leq u\leq 0$, then $x\leq x-\frac{u}{2^k}$. Then

$$\Gamma_u(x)\leq\int_x^{x-\frac{u}{2^k}}\left|f^{(N)}(t)-f^{(N)}(x)\right|\frac{\left(x-\frac{u}{2^k}-t\right)^{N-1}}{(N-1)!}dt\leq$$

$$\int_{x}^{x-\frac{u}{2^k}} \omega_1\left(f^{(N)}, |t-x|\right) \frac{\left(x-\frac{u}{2^k}-t\right)^{N-1}}{(N-1)!} dt \leq$$

$$\omega_1\left(f^{(N)}, \frac{|u|}{2^k}\right) \int_{x}^{x-\frac{u}{2^k}} \frac{\left(x-\frac{u}{2^k}-t\right)^{N-1}}{(N-1)!} dt \leq \tag{3.39}$$

$$\omega_1\left(f^{(N)}, \frac{a}{2^k}\right) \frac{\left(-\frac{u}{2^k}\right)^N}{N!} \leq \omega_1\left(f^{(N)}, \frac{a}{2^k}\right) \frac{a^N}{2^{kN} N!}.$$

That is, when $-a \leq u \leq 0$, then

$$\Gamma_u(x) \leq \omega_1\left(f^{(N)}, \frac{a}{2^k}\right) \frac{a^N}{2^{kN} N!}. \tag{3.40}$$

(ii) Case of $0 \leq u \leq a$, then $x \geq x - \frac{u}{2^k}$. Then

$$\Gamma_u(x) = \left| \int_{x-\frac{u}{2^k}}^{x} \left(f^{(N)}(t) - f^{(N)}(x)\right) \frac{\left(t-x+\frac{u}{2^k}\right)^{N-1}}{(N-1)!} dt \right| \leq$$

$$\int_{x-\frac{u}{2^k}}^{x} \left|f^{(N)}(t) - f^{(N)}(x)\right| \frac{\left(t-x+\frac{u}{2^k}\right)^{N-1}}{(N-1)!} dt \leq \tag{3.41}$$

$$\int_{x-\frac{u}{2^k}}^{x} \omega_1\left(f^{(N)}, |t-x|\right) \frac{\left(t-x+\frac{u}{2^k}\right)^{N-1}}{(N-1)!} dt \leq$$

$$\omega_1\left(f^{(N)}, \frac{|u|}{2^k}\right) \int_{x-\frac{u}{2^k}}^{x} \frac{\left(t-x+\frac{u}{2^k}\right)^{N-1}}{(N-1)!} dt \leq$$

$$\omega_1\left(f^{(N)}, \frac{a}{2^k}\right) \frac{\left(\frac{u}{2^k}\right)^N}{N!} \leq \omega_1\left(f^{(N)}, \frac{a}{2^k}\right) \frac{a^N}{2^{kN} N!}. \tag{3.42}$$

That is, when $0 \leq u \leq a$, then

$$\Gamma_u(x) \leq \omega_1\left(f^{(N)}, \frac{a}{2^k}\right) \frac{a^N}{2^{kN} N!}. \tag{3.43}$$

We proved that

$$\Gamma_u(x) \leq \omega_1\left(f^{(N)}, \frac{a}{2^k}\right) \frac{a^N}{2^{kN} N!} := \rho \geq 0, \tag{3.44}$$

$\forall\, k \in \mathbb{Z}, \forall\, x \in \mathbb{R}, |u| \leq a.$

By (3.37) we get that ($|u| \leq a$)

$$\left|f\left(x-\frac{u}{2^k}\right)-f(x)\right| \leq \sum_{i=1}^{N} \frac{\left|f^{(i)}(x)\right|}{i!} \frac{a^i}{2^{ki}}+\rho. \tag{3.45}$$

We observe that

$$|(B_k f)(x)-f(x)| =$$

$$\left|(C)\int_{-a}^{a} f\left(x-\frac{u}{2^k}\right)\varphi(u)\,d\mu(u)-(C)\int_{-a}^{a} f(x)\,\varphi(u)\,d\mu(u)\right| \overset{(3.9)}{\leq} \tag{3.46}$$

$$(C)\int_{-a}^{a}\left|f\left(x-\frac{u}{2^k}\right)-f(x)\right|\varphi(u)\,d\mu(u) \leq$$

$$\left(\sum_{i=1}^{N} \frac{\left|f^{(i)}(x)\right|}{i!} \frac{a^i}{2^{ki}}+\rho\right)\left((C)\int_{-a}^{a}\varphi(u)\,d\mu(u)\right) =$$

$$\left(\sum_{i=1}^{N} \frac{\left|f^{(i)}(x)\right|}{i!} \frac{a^i}{2^{ki}}+\rho\right)\cdot 1 = \tag{3.47}$$

$$\sum_{i=1}^{N} \frac{\left|f^{(i)}(x)\right|}{i!} \frac{a^i}{2^{ki}}+\frac{a^N}{2^{kN}N!}\omega_1\left(f^{(N)},\frac{a}{2^k}\right),$$

proving the claim. ∎

Corollary 3.12 *Let $f \in C^1(\mathbb{R},\mathbb{R}_+)$. Then*

$$|(B_k f)(x)-f(x)| \leq \frac{a}{2^k}\left(\left|f'(x)\right|+\omega_1\left(f',\frac{a}{2^k}\right)\right), \tag{3.48}$$

$\forall\, k \in \mathbb{Z}$, $\forall\, x \in \mathbb{R}$.

Proof By (3.36) for $N = 1$. ∎

We also present

Theorem 3.13 *Let $f \in C^N(\mathbb{R},\mathbb{R}_+)$, $N \geq 1$. Consider the Choquet-sublinear operators*

$$(\Gamma_k(f))(x) = (C)\int_{-a}^{a}\left[\sum_{j=0}^{n} w_j f\left(\left(x-\frac{u}{2^k}\right)+\frac{j}{2^k n}\right)\right]\varphi(u)\,d\mu(u), \tag{3.49}$$

$\forall\, k \in \mathbb{Z}$, $\forall\, x \in \mathbb{R}$. *Then*

$$\left|\left(\Gamma_k f\right)(x) - f(x)\right| \leq \sum_{i=1}^{N} \frac{\left|f^{(i)}(x)\right|}{i!} \frac{(a+1)^i}{2^{ki}} + \frac{(a+1)^N}{N! 2^{kN}} \omega_1\left(f^{(N)}, \frac{a+1}{2^k}\right). \tag{3.50}$$

If $f^{(N)}$ is uniformly continuous or bounded and continuous, then as $k \to +\infty$ we obtain that $\left(\Gamma_k f\right)(x) \to f(x)$, pointwise with rates.

Corollary 3.14 *Let $f \in C^1(\mathbb{R}, \mathbb{R}_+)$. Then*

$$\left|\left(\Gamma_k f\right)(x) - f(x)\right| \leq \frac{(a+1)}{2^k}\left[\left|f'(x)\right| + \omega_1\left(f', \frac{a+1}{2^k}\right)\right], \tag{3.51}$$

$\forall\, k \in \mathbb{Z}, \forall\, x \in \mathbb{R}$.

Proof By (3.50) for $N = 1$. ■

Proof (*of Theorem* 3.13)
Since $f \in C^N(\mathbb{R})$, $N \geq 1$, by Taylor's formula we get

$$\sum_{j=0}^{n} w_j f\left(\left(x - \frac{u}{2^k}\right) + \frac{j}{2^k n}\right) - f(x) =$$

$$\sum_{i=1}^{N} \frac{f^{(i)}(x)}{i!} \sum_{j=0}^{n} w_j \left(-\frac{u}{2^k} + \frac{j}{2^k n}\right)^i + \tag{3.52}$$

$$\sum_{j=0}^{n} w_j \int_x^{\left(x - \frac{u}{2^k}\right) + \frac{j}{2^k n}} \left(f^{(N)}(t) - f^{(N)}(x)\right) \frac{\left(\left(x - \frac{u}{2^k}\right) + \frac{j}{2^k n} - t\right)^{N-1}}{(N-1)!} dt.$$

Call

$$\varepsilon(x, u, j) := \int_x^{\left(x - \frac{u}{2^k}\right) + \frac{j}{2^k n}} \left(f^{(N)}(t) - f^{(N)}(x)\right) \frac{\left(\left(x - \frac{u}{2^k}\right) + \frac{j}{2^k n} - t\right)^{N-1}}{(N-1)!} dt. \tag{3.53}$$

We estimate $\varepsilon(x, u, j)$. Here $|u| \leq a$.
(i) case of $u \leq \frac{j}{n}$, iff $\frac{u}{2^k} \leq \frac{j}{2^k n}$, iff $x \leq x - \frac{u}{2^k} + \frac{j}{2^k n}$.
Hence

$$\left|\varepsilon(x, u, j)\right| \leq \int_x^{\left(x - \frac{u}{2^k}\right) + \frac{j}{2^k n}} \left|f^{(N)}(t) - f^{(N)}(x)\right| \frac{\left(\left(x - \frac{u}{2^k}\right) + \frac{j}{2^k n} - t\right)^{N-1}}{(N-1)!} dt \leq \tag{3.54}$$

$$\int_x^{\left(x - \frac{u}{2^k}\right) + \frac{j}{2^k n}} \omega_1\left(f^{(N)}, |t - x|\right) \frac{\left(\left(x - \frac{u}{2^k}\right) + \frac{j}{2^k n} - t\right)^{N-1}}{(N-1)!} dt \leq$$

$$\omega_1\left(f^{(N)},\left[\frac{j}{2^k n}-\frac{u}{2^k}\right]\right)\int_x^{\left(x-\frac{u}{2^k}\right)+\frac{j}{2^k n}}\frac{\left(\left(x-\frac{u}{2^k}\right)+\frac{j}{2^k n}-t\right)^{N-1}}{(N-1)!}dt\leq$$

$$\omega_1\left(f^{(N)},\frac{a+1}{2^k}\right)\frac{\left(\frac{j}{2^k n}-\frac{u}{2^k}\right)^N}{N!}\leq\omega_1\left(f^{(N)},\frac{a+1}{2^k}\right)\frac{(a+1)^N}{2^{kN}N!}. \tag{3.55}$$

For $u\leq\frac{j}{n}$, we have proved that

$$|\varepsilon(x,u,j)|\leq\omega_1\left(f^{(N)},\frac{a+1}{2^k}\right)\frac{(a+1)^N}{2^{kN}N!}. \tag{3.56}$$

(ii) case of $u\geq\frac{j}{n}$, iff $\frac{u}{2^k}\geq\frac{j}{2^k n}$, iff $x\geq x-\frac{u}{2^k}+\frac{j}{2^k n}$.
We observe that

$$|\varepsilon(x,u,j)|=$$

$$\left|\int_{\left(x-\frac{u}{2^k}\right)+\frac{j}{2^k n}}^{x}\left(f^{(N)}(t)-f^{(N)}(x)\right)\frac{\left(t-\left[\left(x-\frac{u}{2^k}\right)+\frac{j}{2^k n}\right]\right)^{N-1}}{(N-1)!}dt\right|\leq \tag{3.57}$$

$$\int_{\left(x-\frac{u}{2^k}\right)+\frac{j}{2^k n}}^{x}\left|f^{(N)}(t)-f^{(N)}(x)\right|\frac{\left(t-\left[\left(x-\frac{u}{2^k}\right)+\frac{j}{2^k n}\right]\right)^{N-1}}{(N-1)!}dt\leq$$

$$\int_{\left(x-\frac{u}{2^k}\right)+\frac{j}{2^k n}}^{x}\omega_1\left(f^{(N)},|t-x|\right)\frac{\left(t-\left[\left(x-\frac{u}{2^k}\right)+\frac{j}{2^k n}\right]\right)^{N-1}}{(N-1)!}dt\leq$$

$$\omega_1\left(f^{(N)},\frac{u}{2^k}-\frac{j}{2^k n}\right)\int_{\left(x-\frac{u}{2^k}\right)+\frac{j}{2^k n}}^{x}\frac{\left(t-\left[\left(x-\frac{u}{2^k}\right)+\frac{j}{2^k n}\right]\right)^{N-1}}{(N-1)!}dt\leq$$

$$\omega_1\left(f^{(N)},\frac{a+1}{2^k}\right)\frac{\left(\frac{u}{2^k}-\frac{j}{2^k n}\right)^N}{N!}\leq\omega_1\left(f^{(N)},\frac{a+1}{2^k}\right)\frac{(a+1)^N}{2^{kN}N!}. \tag{3.58}$$

So when $u\geq\frac{j}{n}$, we proved that

$$|\varepsilon(x,u,j)|\leq\omega_1\left(f^{(N)},\frac{a+1}{2^k}\right)\frac{(a+1)^N}{2^{kN}N!}. \tag{3.59}$$

Therefore it always holds

$$\left|\varepsilon\left(x,u,j\right)\right| \leq \omega_1\left(f^{(N)},\frac{a+1}{2^k}\right)\frac{(a+1)^N}{2^{kN}N!}. \tag{3.60}$$

Consequently we derive

$$\sum_{j=0}^{n} w_j \left|\varepsilon\left(x,u,j\right)\right| \leq \omega_1\left(f^{(N)},\frac{a+1}{2^k}\right)\frac{(a+1)^N}{2^{kN}N!} := \psi. \tag{3.61}$$

By (3.52) we find

$$\left|\sum_{j=0}^{n} w_j f\left(\left(x-\frac{u}{2^k}\right)+\frac{j}{2^k n}\right) - f(x)\right| \leq \sum_{i=1}^{N}\frac{\left|f^{(i)}(x)\right|}{i!}\frac{(a+1)^i}{2^{ki}} + \psi. \tag{3.62}$$

Therefore we get

$$\left|\left(\Gamma_k(f)\right)(x) - f(x)\right| =$$

$$\left|(C)\int_{-a}^{a}\left[\sum_{j=0}^{n} w_j f\left(\left(x-\frac{u}{2^k}\right)+\frac{j}{2^k n}\right)\right]\varphi(u)\,d\mu(u) - (C)\int_{-a}^{a} f(x)\,\varphi(u)\,d\mu(u)\right| \overset{(3.9)}{\leq} \tag{3.63}$$

$$(C)\int_{-a}^{a}\left|\sum_{j=0}^{n} w_j f\left(\left(x-\frac{u}{2^k}\right)+\frac{j}{2^k n}\right) - f(x)\right|\varphi(u)\,d\mu(u) \overset{(3.62)}{\leq}$$

$$\left[\sum_{i=1}^{N}\frac{\left|f^{(i)}(x)\right|}{i!}\frac{(a+1)^i}{2^{ki}} + \psi\right]\int_{-a}^{a}\varphi(u)\,d\mu(u) =$$

$$\left[\sum_{i=1}^{N}\frac{\left|f^{(i)}(x)\right|}{i!}\frac{(a+1)^i}{2^{ki}} + \psi\right]\cdot 1 =$$

$$\sum_{i=1}^{N}\frac{\left|f^{(i)}(x)\right|}{i!}\frac{(a+1)^i}{2^{ki}} + \frac{(a+1)^N}{2^{kN}N!}\omega_1\left(f^{(N)},\frac{a+1}{2^k}\right), \tag{3.64}$$

proving the claim. ■

We finish with

Corollary 3.15 *Let $f \in C^N(\mathbb{R}, \mathbb{R}_+)$, $N \geq 1$, $f^{(i)}(x) = 0$, $i = 1, ..., N$. Then*
(i)

$$|(B_k(f))(x) - f(x)| \leq \frac{a^N}{2^{kN} N!} \omega_1\left(f^{(N)}, \frac{a}{2^k}\right), \tag{3.65}$$

and
(ii)

$$|(\Gamma_k(f))(x) - f(x)| \leq \frac{(a+1)^N}{N! 2^{kN}} \omega_1\left(f^{(N)}, \frac{a+1}{2^k}\right), \tag{3.66}$$

$\forall\, k \in \mathbb{Z}$, $\forall\, x \in \mathbb{R}$.

Proof By (3.36) and (3.50). ■

Corollary 3.16 *Let $f \in C^1(\mathbb{R}, \mathbb{R}_+)$, $f'(x) = 0$. Then*
(i)

$$|(B_k(f))(x) - f(x)| \leq \frac{a}{2^k} \omega_1\left(f', \frac{a}{2^k}\right), \tag{3.67}$$

and
(ii)

$$|(\Gamma_k(f))(x) - f(x)| \leq \left(\frac{a+1}{2^k}\right) \omega_1\left(f', \frac{a+1}{2^k}\right), \tag{3.68}$$

$\forall\, k \in \mathbb{Z}$, $\forall\, x \in \mathbb{R}$.

Proof By (3.48) and (3.51). ■

In inequalities (3.65)–(3.68) observe the high speed of convergence and approximation.

3.5 Appendix

Let $f \in C_U(\mathbb{R}, \mathbb{R}_+)$, and the positive sublinear Choquet operator

$$(K(f))(x) := (C) \int_{-a}^{a} f(x+u) \varphi(u) \, d\mu(u), \; \forall\, x \in \mathbb{R}. \tag{3.69}$$

We observe the following (for any $x, y \in \mathbb{R}$):

$$|(K(f))(x) - (K(f))(y)| =$$

$$\left|(C) \int_{-a}^{a} f(x+u) \varphi(u) \, d\mu(u) - (C) \int_{-a}^{a} f(y+u) \varphi(u) \, d\mu(u)\right| \overset{(3.9)}{\leq}$$

$$(C) \int_{-a}^{a} |f(x+u) - f(y+u)| \varphi(u)\, d\mu(u) \leq$$

$$\omega_1(f, |x-y|) \left((C) \int_{-a}^{a} \varphi(u)\, d\mu(u) \right) \overset{(3.3)}{=} \omega_1(f, |x-y|) \cdot 1 = \omega_1(f, |x-y|). \tag{3.70}$$

Therefore it holds the global smoothness preservation property:

$$\omega_1(K(f), \delta) \leq \omega_1(f, \delta), \ \forall\, \delta > 0. \tag{3.71}$$

References

1. G.A. Anastassiou, High order approximation by univariate shift-invariant integral operators, in *Nonlinear Analysis and Applications, 2 Volumes*, vol. I, ed. by R. Agarwal, D. O'Regan (Kluwer, Dordrecht, 2003), pp. 141–164
2. G.A. Anastassiou, *Intelligent Mathematics: Computational Analysis* (Springer, Heidelberg, 2011)
3. G.A. Anastassiou, Approximation by shift invariant univariate sublinear-Choquet operators. Indian J. Math. (2018). Accepted
4. G.A. Anastassiou, H.H. Gonska, On some shift invariant integral operators, univariate case. Ann. Polon. Math. **LXI**(3), 225–243 (1995)
5. G.A. Anastassiou, S. Gal, *Approximation Theory* (Birkhauser, Boston, 2000)
6. G. Choquet, Theory of capacities. Ann. Inst. Fourier (Grenoble) **5**, 131–295 (1954)
7. D. Denneberg, *Non-additive Measure and Integral* (Kluwer, Dordrecht, 1994)
8. S. Gal, uniform and pointwise quantitative approximation by Kantorovich–Choquet type integral operators with respect to monotone and submodular set functions. Mediterr. J. Math. **14**(5), 12 pp. (2017). Art. 205
9. Z. Wang, G.J. Klir, *Generalized Measure Theory* (Springer, New York, 2009)

Chapter 4
Approximation with Rates by Shift Invariant Multivariate Sublinear-Choquet Operators

A very general multivariate positive sublinear Choquet integral type operator is given through a convolution-like iteration of another multivariate general positive sublinear operator with a multivariate scaling type function. For it sufficient conditions are given for shift invariance, preservation of global smoothness, convergence to the unit with rates. Furthermore, two examples of very general multivariate specialized operators are presented fulfilling all the above properties, the higher order of multivariate approximation of these operators is also studied. It follows [3].

4.1 Introduction

Let X, Y be function spaces of functions from $\mathbb{R}^d$ into $\mathbb{R}_+$, $d \in \mathbb{N}$. Let $L_N : X \to Y$, $N \in \mathbb{N}$, be a sequence of operators with the following properties:

(i) (positive homogeneous)

$$L_N(\alpha f) = \alpha L_N(f), \forall\, \alpha \geq 0, \forall\, f \in X.$$

(ii) (Monotonicity) if $f, g \in X$ satisfy $f \leq g$, then $L_N(f) \leq L_N(g)$, $\forall\, N \in \mathbb{N}$, and

(iii) (Subadditivity)

$$L_N(f+g) \leq L_N(f) + L_N(g), \forall\, f, g \in X.$$

We call L_N multivariate positive sublinear operators.

In this chapter we deal with sequences of Choquet multivariate positive sublinear operators that are constructed, with the help of Choquet integral. Our functions spaces are continuous functions from $\mathbb{R}^d$ into $\mathbb{R}_+$, $d \in \mathbb{N}$. The sequence of operators is generated by a basic multivariate operator via dilated translations of convolution type using the Choquet integral. We prove that our multivariate operators possess

G. A. Anastassiou, *Ordinary and Fractional Approximation by Non-additive Integrals: Choquet, Shilkret and Sugeno Integral Approximators*, Studies in Systems, Decision and Control 190, https://doi.org/10.1007/978-3-030-04287-5_4

the following properties: of shift invariance of global smoothness preservation, of convergence to the unit operator with rates. Then we apply our results to two specific families of such multivariate Choquet type operators.

We continue with the higher order quantitative approximation study of these multivariate specific operators.

Earlier similar studies have been done by the author, see [5], Chapters 10–17, and [2], Chapters 16, 17. These serve as motivation and inspiration to our present work.

4.2 Background

Next we present briefly about the Choquet integral.

We make

Definition 4.1 Consider $\Omega \neq \varnothing$ and let $\mathcal{C}$ be a σ-algebra of subsets in Ω.

(i) (see, e.g., [8], p. 63) The set function $\mu : \mathcal{C} \to [0, +\infty]$ is called a monotone set function (or capacity) if $\mu(\varnothing) = 0$ and $\mu(A) \leq \mu(B)$ for all $A, B \in \mathcal{C}$, with $A \subset B$. Also, μ is called submodular if

$$\mu(A \cup B) + \mu(A \cap B) \leq \mu(A) + \mu(B), \text{ for all } A, B \in \mathcal{C}.$$

μ is called bounded if $\mu(\Omega) < +\infty$ and normalized if $\mu(\Omega) = 1$.

(ii) (see, e.g., [8], p. 233, or [6]) If μ is a monotone set function on $\mathcal{C}$ and if $f : \Omega \to \mathbb{R}$ is $\mathcal{C}$-measurable (that is, for any Borel subset $B \subset \mathbb{R}$ it follows $f^{-1}(B) \in \mathcal{C}$), then for any $A \in \mathcal{C}$, the Choquet integral is defined by

$$(C)\int_A f d\mu = \int_0^{+\infty} \mu\left(F_\beta(f) \cap A\right) d\beta + \int_{-\infty}^0 \left[\mu\left(F_\beta(f) \cap A\right) - \mu(A)\right] d\beta,$$

where we used the notation $F_\beta(f) = \{\omega \in \Omega : f(\omega) \geq \beta\}$. Notice that if $f \geq 0$ on A, then in the above formula we get $\int_{-\infty}^0 = 0$.

The integrals on the right-hand side are the usual Riemann integral.

The function f will be called Choquet integrable on A if $(C)\int_A f d\mu \in \mathbb{R}$.

Next we list some well known properties of the Choquet integral.

Remark 4.2 If $\mu : \mathcal{C} \to [0, +\infty]$ is a monotone set function, then the following properties hold:

(i) For all $a \geq 0$ we have $(C)\int_A af d\mu = a \cdot (C)\int_A f d\mu$ (if $f \geq 0$ then see, e.g., [8], Theorem 11.2, (5), p. 228 and if f is arbitrary sign, then see, e.g., [7], p. 64, Proposition 5.1, (ii)).

(ii) For all $c \in \mathbb{R}$ and f of arbitrary sign, we have (see, e.g., [8], pp. 232–233, or [7], p. 65) $(C)\int_A (f + c) d\mu = (C)\int_A f d\mu + c \cdot \mu(A)$.

If μ is submodular too, then for all f, g of arbitrary sign and lower bounded, we have (see, e.g., [7], p. 75, Theorem 6.3)

$$(C)\int_A (f+g)\,d\mu \leq (C)\int_A f\,d\mu + (C)\int_A g\,d\mu.$$

(iii) If $f \leq g$ on A then $(C)\int_A f\,d\mu \leq (C)\int_A g\,d\mu$ (see, e.g., [8], p. 228, Theorem 11.2, (3) if $f, g \geq 0$ and p. 232 if f, g are of arbitrary sign).

(iv) Let $f \geq 0$. If $A \subset B$ then $(C)\int_A f\,d\mu \leq (C)\int_B f\,d\mu$. In addition, if μ is finitely subadditive, then

$$(C)\int_{A\cup B} f\,d\mu \leq (C)\int_A f\,d\mu + (C)\int_B f\,d\mu.$$

(v) It is immediate that $(C)\int_A 1 \cdot d\mu(t) = \mu(A)$.

(vi) The formula $\mu(A) = \gamma(M(A))$, where $\gamma : [0,1] \to [0,1]$ is an increasing and concave function, with $\gamma(0) = 0$, $\gamma(1) = 1$ and M is a probability measure (or only finitely additive) on a σ-algebra on Ω (that is, $M(\emptyset) = 0$, $M(\Omega) = 1$ and M is countably additive), gives simple examples of normalized, monotone and submodular set functions (see, e.g., [7], pp. 16–17, Example 2.1). Such of set functions μ are also called distorsions of countably normalized, additive measures (or distorted measures). For a simple example, we can take $\gamma(t) = \frac{2t}{1+t}$, $\gamma(t) = \sqrt{t}$.

If the above γ function is increasing, concave and satisfies only $\gamma(0) = 0$, then for any bounded Borel measure m, $\mu(A) = \gamma(m(A))$ gives a simple example of bounded, monotone and submodular set function.

(vii) If μ is a countably additive bounded measure, then the Choquet integral $(C)\int_A f\,d\mu$ reduces to the usual Lebesgue type integral (see, e.g., [7], p. 62, or [8], p. 226).

(viii) If $f \geq 0$, then $(C)\int_A f\,d\mu \geq 0$.

4.3 Multivariate Theory

This section is motivated and inspired by [4], and Chapter 11, [5]. In a way it generalizes that material.

Let $C_U(\mathbb{R}^d, \mathbb{R})$, $d \in \mathbb{N}$, the space of uniformly continuous functions from $\mathbb{R}^d$ into $\mathbb{R}$. Let the multivariate first modulus of continuity

$$\omega_1(f,\delta) := \sup_{\substack{x,y\in\mathbb{R}^d \\ \|x-y\|\leq\delta}} |f(x) - f(y)|, \quad \delta > 0,$$

where $f \in C_U(\mathbb{R}^d, \mathbb{R})$, and $\|\cdot\|$ an arbitrary norm in $\mathbb{R}^d$. In [4, 5], pp. 297–298, we proved that $\omega_1(f,\delta)$ is finite for any $\delta > 0$.

We denote by $\omega_{1,\infty}$ the multivariate modulus of continuity with respect to max-norm $\|\cdot\|_\infty$.

Here we deal with the functions spaces $C_U\left(\mathbb{R}^d, \mathbb{R}_+\right)$, $C\left(\mathbb{R}^d, \mathbb{R}_+\right)$ (continuous functions from $\mathbb{R}^d$ into $\mathbb{R}_+$).

Let $\{l_k\}_{k\in\mathbb{Z}}$ be a sequence of multivariate positive sublinear operators that map $C_U\left(\mathbb{R}^d, \mathbb{R}_+\right)$ into $C\left(\mathbb{R}^d, \mathbb{R}_+\right)$ with the property:

$$(l_k(f))(x) := \left(l_0\left(f\left(2^{-k}\cdot\right)\right)\right)(x), \ \forall\, x \in \mathbb{R}^d, \forall\, f \in C_U\left(\mathbb{R}^d, \mathbb{R}_+\right). \tag{4.1}$$

For fixed $a > 0$ we assume that

$$\sup_{\substack{u,y\in\mathbb{R}^d \\ \|u-y\|_\infty \le a}} |(l_0(f))(u) - f(y)| \le \omega_{1,\infty}\left(f, \frac{ma+n}{2^r}\right), \tag{4.2}$$

is true for all $f \in C_U\left(\mathbb{R}^d, \mathbb{R}_+\right)$; where $m \in \mathbb{N}$, $n \in \mathbb{Z}_+$, $r \in \mathbb{Z}$.

Let $\mathcal{L}^d$ be the Lebesgue σ-algebra on $\mathbb{R}^d$, and the set function $\mu : \mathcal{L}^d \to [0, +\infty]$, which is assumed to be monotone and submodular.

Let $\varphi : \mathbb{R}^d \to \mathbb{R}_+$ which is Lebesgue measurable, such that

$$(C)\int_{[-a,a]^d} \varphi(u)\, d\mu(u) = 1. \tag{4.3}$$

Example 4.3 Let $[c_1, c_2] \subset \mathbb{R}$ such that $\left[-\frac{1}{2}, \frac{1}{2}\right] \subset [c_1, c_2]$. Let the characteristic function

$$\varphi(x) := \chi_{[c_1,c_2]}(x) = \begin{cases} 1, x \in [c_1, c_2]^d \\ 0, \text{ elsewhere,} \end{cases} \tag{4.4}$$

and λ is the Lebesgue measure on $\mathbb{R}^d$, where $a = \frac{1}{2}$. Here we take $\mu := \sqrt{\lambda}$, which is a monotone and submodular set function (see Remark 4.2 (vi)).

We notice that

$$(C)\int_{\left[-\frac{1}{2},\frac{1}{2}\right]^d} \chi_{[c_1,c_2]^d}(u)\, d\sqrt{\lambda(u)} =$$

$$(C)\int_{\left[-\frac{1}{2},\frac{1}{2}\right]^d} 1 d\sqrt{\lambda(u)} = \sqrt{\lambda\left(\left[-\frac{1}{2}, \frac{1}{2}\right]^d\right)} = 1, \tag{4.5}$$

fulfilling (4.3).

We define the multivariate positive sublinear-Choquet operators

$$(\mathcal{L}_0(f))(x) := (C)\int_{[-a,a]^d} (l_0 f)(x-u)\, \varphi(u)\, d\mu(u), \tag{4.6}$$

and

$$(\mathcal{L}_k(f))(x) := \left(\mathcal{L}_0\left(f\left(2^{-k}\cdot\right)\right)\right)\left(2^k x\right), \ \forall\, k \in \mathbb{Z}, \forall\, x \in \mathbb{R}^d. \tag{4.7}$$

Therefore it holds

$$(\mathcal{L}_k(f))(x) = (C)\int_{[-a,a]^d} \left(l_0\left(f\left(2^{-k}\cdot\right)\right)\right)\left(2^k x - u\right)\varphi(u)\,d\mu(u) = \tag{4.8}$$

$$(C)\int_{[-a,a]^d} (l_k(f))\left(2^k x - u\right)\varphi(u)\,d\mu(u),$$

$\forall\, x \in \mathbb{R}^d$, $\forall\, k \in \mathbb{Z}$.

Indeed here we have

$$(\mathcal{L}_k(f))(x) \leq (C)\int_{[-a,a]^d} \left\|l_k(f)\left(2^k x - \cdot\right)\right\|_{\infty,[-a,a]^d}\varphi(u)\,d\mu(u) =$$

$$\left\|l_k(f)\left(2^k x - \cdot\right)\right\|_{\infty,[-a,a]^d}\left((C)\int_{[-a,a]^d}\varphi(u)\,d\mu(u)\right) \overset{(4.3)}{=} \tag{4.9}$$

$$\left\|l_k(f)\left(2^k x - \cdot\right)\right\|_{\infty,[-a,a]^d} < +\infty.$$

Hence $(\mathcal{L}_k(f))(x) \in \mathbb{R}_+$ is well-defined, $\forall\, x \in \mathbb{R}^d$, $\forall\, k \in \mathbb{Z}$.

Let $f, g \in \mathcal{M}\left(\mathbb{R}^d, \mathbb{R}_+\right)$ (Lebesgue measurable functions) where $X \in A$, $A \subset \mathbb{R}^d$ is a Lebesgue measurable set.

We observe that

$$f(x) = f(x) - g(x) + g(x) \leq |f(x) - g(x)| + g(x),$$

hence

$$(C)\int_A f(x)\,d\mu(x) \leq (C)\int_A |f(x) - g(x)|\,d\mu(x) + (C)\int_A g(x)\,d\mu(x),$$

and

$$(C)\int_A f(x)\,d\mu(x) - (C)\int_A g(x)\,d\mu(x) \leq (C)\int_A |f(x) - g(x)|\,d\mu(x).$$

Similarly, it holds

$$(C)\int_A g(x)\,d\mu(x) - (C)\int_A f(x)\,d\mu(x) \leq (C)\int_A |f(x) - g(x)|\,d\mu(x).$$

Therefore we derive

$$\left|(C)\int_A f(x)\,d\mu(x) - (C)\int_A g(x)\,d\mu(x)\right| \leq (C)\int_A |f(x) - g(x)|\,d\mu(x). \tag{4.10}$$

We need

Definition 4.4 Let $f_\alpha(\cdot) := f(\cdot + \alpha)$, $\alpha \in \mathbb{R}^d$, and Φ be an operator. If $\Phi(f_\alpha) = (\Phi f)_\alpha$, then Φ is called a shift invariant operator.

We give

Theorem 4.5 *Assume that*

$$\left(l_0\left(f\left(2^{-k}\cdot+\alpha\right)\right)\right)\left(2^k u\right) = \left(l_0\left(f\left(2^{-k}\cdot\right)\right)\right)\left(2^k(u+\alpha)\right), \tag{4.11}$$

for all $k \in \mathbb{Z}$, $\alpha \in \mathbb{R}^d$ *fixed, all* $u \in \mathbb{R}^d$ *and any* $f \in C_U\left(\mathbb{R}^d, \mathbb{R}_+\right)$. *Then* $\mathcal{L}_k$ *is a shift invariant operator for all* $k \in \mathbb{Z}$.

Proof We have that

$$\left(\mathcal{L}_k\left(f(\cdot+\alpha)\right)\right)(x) = \left(\mathcal{L}_k\left(f_\alpha\right)\right)(x) \overset{(4.8)}{=}$$

$$(C)\int_{[-a,a]^d}\left(l_0\left(f_\alpha\left(2^{-k}\cdot\right)\right)\right)\left(2^k x - u\right)\varphi(u)\,d\mu(u) =$$

$$(C)\int_{[-a,a]^d}\left(l_0\left(f\left(2^{-k}\cdot+\alpha\right)\right)\right)\left(2^k x - u\right)\varphi(u)\,d\mu(u) =$$

$$(C)\int_{[-a,a]^d}\left(l_0\left(f\left(2^{-k}\cdot+\alpha\right)\right)\right)\left(2^k\left(x-2^{-k}u\right)\right)\varphi(u)\,d\mu(u) \overset{(4.11)}{=} \tag{4.12}$$

$$(C)\int_{[-a,a]^d}\left(l_0\left(f\left(2^{-k}\cdot\right)\right)\right)\left(2^k\left(x-2^{-k}u+\alpha\right)\right)\varphi(u)\,d\mu(u) =$$

$$(C)\int_{[-a,a]^d}\left(l_0\left(f\left(2^{-k}\cdot\right)\right)\right)\left(2^k(x+\alpha)-u\right)\varphi(u)\,d\mu(u) \overset{(4.8)}{=} \left(\mathcal{L}_k(f)\right)(x+\alpha),$$

that is

$$\mathcal{L}_k\left(f_\alpha\right) = \left(\mathcal{L}_k(f)\right)_\alpha, \tag{4.13}$$

proving the claim. ■

It follows the global smoothness of the operators $\mathcal{L}_k$.

Theorem 4.6 *For any* $f \in C_U\left(\mathbb{R}^d, \mathbb{R}_+\right)$ *assume that, for all* $u \in \mathbb{R}^d$,

$$\left|\left(l_0(f)\right)(x-u) - \left(l_0(f)\right)(y-u)\right| \leq \omega_1\left(f, \|x-y\|\right), \tag{4.14}$$

for any $x, y \in \mathbb{R}^d$, *where* $\|\cdot\|$ *is an arbitrary norm in* $\mathbb{R}^d$. *Then*

$$\omega_1\left(\mathcal{L}_k f, \delta\right) \leq \omega_1(f, \delta), \ \forall\, \delta > 0. \tag{4.15}$$

Proof We observe that

$$|(\mathcal{L}_0(f))(x) - (\mathcal{L}_0(f))(y)| =$$

$$\left|(C)\int_{[-a,a]^d} (l_0 f)(x-u)\varphi(u)\,d\mu(u) - (C)\int_{[-a,a]^d} (l_0 f)(y-u)\varphi(u)\,d\mu(u)\right| \overset{(4.10)}{\leq} \tag{4.16}$$

$$(C)\int_{[-a,a]^d} |(l_0 f)(x-u) - (l_0 f)(y-u)|\,\varphi(u)\,d\mu(u) \overset{(4.14)}{\leq}$$

$$\omega_1(f, \|x-y\|)\left((C)\int_{[-a,a]^d} \varphi(u)\,d\mu(u)\right) \overset{(4.3)}{=} \omega_1(f, \|x-y\|).$$

So that

$$|(\mathcal{L}_0(f))(x) - (\mathcal{L}_0(f))(y)| \leq \omega_1(f, \|x-y\|). \tag{4.17}$$

From (4.7), (4.17) we get

$$|(\mathcal{L}_k(f))(x) - (\mathcal{L}_k(f))(y)| \overset{(4.7)}{=}$$

$$\left|\left(\mathcal{L}_0\left(f\left(2^{-k}\cdot\right)\right)\right)\left(2^k x\right) - \left(\mathcal{L}_0\left(f\left(2^{-k}\cdot\right)\right)\right)\left(2^k y\right)\right| \leq \tag{4.18}$$

$$\omega_1\left(f\left(2^{-k}\cdot\right), 2^k\|x-y\|\right) = \omega_1(f, \|x-y\|),$$

i.e. global smoothness for $\mathcal{L}_k$ has been proved. ■

The convergence of $\mathcal{L}_k$ to the unit operator, as $k \to +\infty$, k with rates follows:

Theorem 4.7 *For $f \in C_U\left(\mathbb{R}^d, \mathbb{R}_+\right)$, under the assumption (4.2), we have*

$$|(\mathcal{L}_k(f))(x) - f(x)| \leq \omega_{1,\infty}\left(f, \frac{ma+n}{2^{k+r}}\right), \tag{4.19}$$

where $m \in \mathbb{N}$, $n \in \mathbb{Z}_+$, $k, r \in \mathbb{Z}$.

Proof We notice that

$$|(\mathcal{L}_k(f))(x) - f(x)| \overset{(4.7)}{=} \left|\left(\mathcal{L}_0\left(f\left(2^{-k}\cdot\right)\right)\right)\left(2^k x\right) - f(x)\right| \overset{(4.8)}{=}$$

$$\left|(C)\int_{[-a,a]^d} \left(l_0\left(f\left(2^{-k}\cdot\right)\right)\right)\left(2^k x - u\right)\varphi(u)\,d\mu(u) - f(x)\right| \overset{(4.3)}{=}$$

$$\left|(C)\int_{[-a,a]^d} \left(l_0\left(f\left(2^{-k}\cdot\right)\right)\right)\left(2^k x - u\right)\varphi(u)\,d\mu(u) - (C)\int_{[-a,a]^d} f(x)\varphi(u)\,d\mu(u)\right|$$

$$\overset{(4.10)}{\leq} (C) \int_{[-a,a]^d} \left| \left(l_0 \left(f \left(2^{-k} \cdot \right) \right) \right) \left(2^k x - u \right) - f(x) \right| \varphi(u) \, d\mu(u) = \tag{4.20}$$

$$(C) \int_{[-a,a]^d} \left| \left(l_0 \left(f \left(2^{-k} \cdot \right) \right) \right) \left(2^k x - u \right) - f \left(2^{-k} \cdot \right) \left(2^k x \right) \right| \varphi(u) \, d\mu(u) \overset{(4.2)}{\leq}$$

(here $\left\| \left(2^k x - u \right) - 2^k x \right\|_{\infty} = \|u\|_{\infty} \leq a$)

$$\omega_{1,\infty} \left(f \left(2^{-k} \cdot \right), \frac{ma + n}{2^r} \right) \left((C) \int_{[-a,a]^d} \varphi(u) \, d\mu(u) \right) \overset{(4.3)}{=}$$

$$\omega_{1,\infty} \left(f \left(2^{-k} \cdot \right), \frac{ma + n}{2^r} \right) \cdot 1 = \omega_{1,\infty} \left(f, \frac{ma + n}{2^{k+r}} \right),$$

proving the claim. ■

We give some applications.

For each $k \in \mathbb{Z}$, we define

(i)

$$(B_k f)(x) := (C) \int_{[-a,a]^d} f \left(x - \frac{u}{2^k} \right) \varphi(u) \, d\mu(u), \ \forall \, x \in \mathbb{R}^d, \tag{4.21}$$

i.e., here

$$\begin{gathered} \left(l_k (f) \right)(u) = f \left(\tfrac{u}{2^k} \right), \\ \text{and} \\ \left(l_0 (f) \right)(u) = f(u), \end{gathered} \tag{4.22}$$

are continuous in $u \in \mathbb{R}^d$.

Also for $k \in \mathbb{Z}$, we define

(ii)

$$\left(\Gamma_k (f) \right)(x) := (C) \int_{[-a,a]^d} \gamma_k^f \left(2^k x - u \right) \varphi(u) \, d\mu(u), \ \forall \, x \in \mathbb{R}^d, \tag{4.23}$$

where

$$\gamma_k^f(u) := \sum_{j_1=0}^{n_1} \ldots \sum_{j_d=0}^{n_d} w_{j_1,\ldots,j_d} f \left(\frac{u_1}{2^k} + \frac{j_1}{2^k n_1}, \ldots, \frac{u_d}{2^k} + \frac{j_d}{2^k n_d} \right), \tag{4.24}$$

$(n_1, \ldots, n_d) \in \mathbb{N}^d$, $w_{j_1,\ldots,j_d} \geq 0$, $\sum_{j_1=0}^{n_1} \ldots \sum_{j_d=0}^{n_d} w_{j_1,\ldots,j_d} = 1$.

That is, here

$$\left(l_k (f) \right)(u) = \gamma_k^f(u) \tag{4.25}$$

is continuous in $u \in \mathbb{R}^d$, where $u = (u_1, ..., u_d)$.

Notice that

$$(l_0(f))(u) = \gamma_0^f(u) = \sum_{j_1=0}^{n_1} ... \sum_{j_d=0}^{n_d} w_{j_1,...,j_d} \cdot f\left(u_1 + \frac{j_1}{n_1}, ..., u_d + \frac{j_d}{n_d}\right) \quad (4.26)$$

is continuous in $u \in \mathbb{R}^d$.

Setting $x = (x_1, ..., x_d) \in \mathbb{R}^d$, $n = (n_1, ..., n_d) \in \mathbb{N}^d$, $j = (j_1, ..., j_d) \in \mathbb{Z}_+^d$, $w_j = w_{j_1,...,j_d} \geq 0$, and $\sum_{j=0}^{n} = \sum_{j_1=0}^{n_1} ... \sum_{j_d=0}^{n_d}$. That is $\sum_{j=0}^{n} w_j = 1$.

So we can write ($k \in \mathbb{Z}$)

$$(l_k(f))(u) = \gamma_k^f(u) = \sum_{j=0}^{n} w_j f\left(\frac{u}{2^k} + \frac{j}{2^k n}\right), \quad (4.27)$$

and

$$(l_0(f))(u) = \gamma_0^f(u) = \sum_{j=0}^{n} w_j f\left(u + \frac{j}{n}\right). \quad (4.28)$$

Indeed we have

$$(\Gamma_k(f))(x) = (C)\int_{[-a,a]^d} \left[\sum_{j=0}^{n} w_j f\left(\left(x - \frac{u}{2^k}\right) + \frac{j}{2^k n}\right)\right] \varphi(u)\, d\mu(u). \quad (4.29)$$

Clearly here we get

$$\begin{gathered} (B_k(f))(x) = \left(B_0\left(f\left(2^{-k}\cdot\right)\right)\right)\left(2^k x\right), \\ \text{and} \\ (\Gamma_k(f))(x) = \left(\Gamma_0\left(f\left(2^{-k}\cdot\right)\right)\right)\left(2^k x\right), \end{gathered} \quad (4.30)$$

$\forall\, k \in \mathbb{Z}$, $\forall\, x \in \mathbb{R}^d$.

We give

Proposition 4.8 *B_k, Γ_k are shift invariant operators.*

Proof (i) For B_k operators: Here $l_0 f = f$. Hence

$$\left(l_0\left(f\left(2^{-k}\cdot + \alpha\right)\right)\right)\left(2^k u\right) = f\left(2^{-k}2^k u + \alpha\right) = f(u + \alpha) = \quad (4.31)$$

$$\left(l_0\left(f\left(2^{-k}\cdot\right)\right)\right)\left(2^k(u + \alpha)\right).$$

(ii) For Γ_k operators:

$$(l_0(f))(u) = \sum_{j=0}^{n} w_j f\left(u + \frac{j}{n}\right).$$

Hence

$$\left(l_0\left(f\left(2^{-k}\cdot+\alpha\right)\right)\right)\left(2^k u\right) = \sum_{j=0}^{n} w_j f\left(2^{-k}\left(2^k u + \frac{j}{n}\right) + \alpha\right) =$$

$$\sum_{j=0}^{n} w_j f\left(2^{-k}\left(2^k(u+\alpha) + \frac{j}{n}\right)\right) = \left(l_0\left(f\left(2^{-k}\cdot\right)\right)\right)\left(2^k(u+\alpha)\right), \tag{4.32}$$

proving the claim. ∎

Next we show that the operators B_k, Γ_k possess the property of global smoothness preservation.

Theorem 4.9 *For all* $f \in C_U\left(\mathbb{R}^d, \mathbb{R}_+\right)$ *and all* $\delta > 0$ *we have*

$$\begin{gathered} \omega_1(B_k f, \delta) \leq \omega_1(f, \delta), \\ \textit{and} \\ \omega_1(\Gamma_k f, \delta) \leq \omega_1(f, \delta). \end{gathered} \tag{4.33}$$

Proof (i) For B_k operators: Here $l_0 f = f$, therefore

$$|(l_0(f))(x-u) - (l_0(f))(y-u)| = |f(x-u) - f(y-u)| \leq \omega_1(f, \|x-y\|). \tag{4.34}$$

(ii) For Γ_k operators: We observe that

$$|(l_0(f))(x-u) - (l_0(f))(y-u)| = \left|\gamma_0^f(x-u) - \gamma_0^f(y-u)\right| =$$

$$\left|\sum_{j=0}^{n} w_j\left(f\left(x-u+\frac{j}{n}\right) - f\left(y-u+\frac{j}{n}\right)\right)\right| \leq$$

$$\sum_{j=0}^{n} w_j\left|f\left(x-u+\frac{j}{n}\right) - f\left(y-u+\frac{j}{n}\right)\right| \leq$$

$$\omega_1(f, \|x-y\|)\left(\sum_{j=0}^{n} w_j\right) = \omega_1(f, \|x-y\|), \tag{4.35}$$

proving the claim. ∎

The operators B_k, Γ_k, $k \in \mathbb{Z}$, converge to the unit operator, as $k \to +\infty$, with rates presented next.

Theorem 4.10 *For $k \in \mathbb{Z}$, $f \in C_U\left(\mathbb{R}^d, \mathbb{R}_+\right)$, it holds*

$$\begin{gathered} \left|\left(B_k(f)\right)(x) - f(x)\right| \leq \omega_{1,\infty}\left(f, \tfrac{a}{2^k}\right), \\ \text{and} \\ \left|\left(\Gamma_k(f)\right)(x) - f(x)\right| \leq \omega_{1,\infty}\left(f, \tfrac{a+1}{2^k}\right). \end{gathered} \tag{4.36}$$

Proof (i) For B_k operators: Here $\left(l_0(f)\right)(u) = f(u)$ and

$$\sup_{\substack{u,y \in \mathbb{R}^d \\ \|u-y\|_\infty \leq a}} \left|\left(l_0(f)\right)(u) - f(y)\right| = \sup_{\substack{u,y \in \mathbb{R}^d \\ \|u-y\|_\infty \leq a}} \left|f(u) - f(y)\right| = \omega_{1,\infty}(f, a), \tag{4.37}$$

and we use Theorem 4.7.

(ii) For Γ_k operators: Here we see that

$$\begin{gathered} \sup_{\substack{u,y \in \mathbb{R}^d \\ \|u-y\|_\infty \leq a}} \left|\left(l_0(f)\right)(u) - f(y)\right| = \sup_{\substack{u,y \in \mathbb{R}^d \\ \|u-y\|_\infty \leq a}} \left|\sum_{j=0}^{n} w_j f\left(u + \frac{j}{n}\right) - f(y)\right| \leq \\ \sup_{\substack{u,y \in \mathbb{R}^d \\ \|u-y\|_\infty \leq a}} \sum_{j=0}^{n} w_j \left|f\left(u + \frac{j}{n}\right) - f(y)\right| \leq \\ \sup_{\substack{u,y \in \mathbb{R}^d \\ \|u-y\|_\infty \leq a}} \sum_{j=0}^{n} w_j \omega_{1,\infty}\left(f, \left\|u + \frac{j}{n} - y\right\|_\infty\right) \leq \\ \sup_{\substack{u,y \in \mathbb{R}^d \\ \|u-y\|_\infty \leq a}} \sum_{j=0}^{n} w_j \omega_{1,\infty}\left(f, \left\|\frac{j}{n}\right\|_\infty + \|u-y\|_\infty\right) \leq \\ \left(\sum_{j=0}^{n} w_j\right) \omega_{1,\infty}(f, 1+\alpha) = \omega_{1,\infty}(f, \alpha+1), \end{gathered} \tag{4.38}$$

where $\frac{j}{n} = \left(\frac{j_1}{n_1}, \ldots, \frac{j_d}{n_d}\right)$.

By (4.19) we are done. ∎

4.4 Higher Order Multivariate Approximation

Here all are as in Sect. 4.3. See also our earlier work [1, 2], Chapter 17.

We present

Theorem 4.11 *Let* $f \in C^N(\mathbb{R}^d, \mathbb{R}_+)$, $N, d \in \mathbb{N}$. *Consider the multivariate Choquet-sublinear operators*

$$(B_k f)(x) = (C)\int_{[-a,a]^d} f\left(x - \frac{u}{2^k}\right)\varphi(u)\,d\mu(u),$$

$\forall\, x \in \mathbb{R}^d$, $\forall\, k \in \mathbb{Z}$. *Here we further assume that all partials* $f_{\overline{\alpha}}$, $|\overline{\alpha}| = N$, *are uniformly continuous or bounded and continuous on* $\mathbb{R}^d$. *Then*

$$|(B_k f)(x) - f(x)| \leq \sum_{j=1}^{N} \frac{a^j}{j!2^{kj}}\left(\left(\sum_{i=1}^{d}\left|\frac{\partial}{\partial x_i}\right|\right)^j f(x)\right)$$

$$+ \frac{a^N d^N}{N!2^{kN}} \max_{\overline{\alpha}:|\overline{\alpha}|=N} \omega_{1,\infty}\left(f_{\overline{\alpha}}, \frac{a}{2^k}\right), \tag{4.39}$$

$\forall\, k \in \mathbb{Z}$, $\forall\, x \in \mathbb{R}^d$.

Remark 4.12 (i) Since the Nth order partials $f_{\overline{\alpha}}$ are uniformly continuous or bounded and continuous, as $k \to +\infty$, we get that $(B_k f)(x) \to f(x)$, pointwise with rates.

(ii) When $N = 1$, inequality (4.39) becomes

$$|(B_k f)(x) - f(x)| \leq \frac{a}{2^k}\left\{\left(\sum_{i=1}^{d}\left|\frac{\partial f(x)}{\partial x_i}\right|\right) + d \max_{i\in\{1,\ldots,d\}} \omega_{1,\infty}\left(\frac{\partial f}{\partial x_i}, \frac{a}{2^k}\right)\right\}, \tag{4.40}$$

$\forall\, k \in \mathbb{Z}$, $\forall\, x \in \mathbb{R}^d$.

(iii) When $f_{\overline{\alpha}}(x) = 0$, $\overline{\alpha} : |\overline{\alpha}| = \rho$, $1 \leq \rho \leq N$. Then (4.39) becomes

$$|(B_k f)(x) - f(x)| \leq \frac{a^N d^N}{N!2^{kN}} \max_{\overline{\alpha}:|\overline{\alpha}|=N} \omega_{1,\infty}\left(f_{\overline{\alpha}}, \frac{a}{2^k}\right), \ \forall\, k \in \mathbb{N}. \tag{4.41}$$

(iv) When $\frac{\partial f(x)}{\partial x_i} = 0$, $i = 1, \ldots, d$, then (4.40) becomes

$$|(B_k f)(x) - f(x)| \leq \frac{a}{2^k} d \max_{i\in\{1,\ldots,d\}} \omega_{1,\infty}\left(\frac{\partial f}{\partial x_i}, \frac{a}{2^k}\right), \tag{4.42}$$

$\forall\, k \in \mathbb{Z}$.

In (4.41), (4.42), we observe the high speed of convergence.

Proof of Theorem 4.11

Set

$$g_{\frac{u}{2^k}}(t) := f\left(x - t\frac{u}{2^k}\right), \text{ all } 0 \le t \le 1. \tag{4.43}$$

We notice that

$$g_{\frac{u}{2^k}}(0) = f(x), \text{ and } g_{\frac{u}{2^k}}(1) = f\left(x - \frac{u}{2^k}\right). \tag{4.44}$$

Then, for $j = 1, 2, ..., N$, we get that

$$g_{\frac{u}{2^k}}^{(j)}(t) = \left(\left(\sum_{i=1}^{d}\left(-\frac{u_i}{2^k}\right)\frac{\partial}{\partial x_i}\right)^j f\right)\left(x - t\frac{u}{2^k}\right). \tag{4.45}$$

By Taylor's formula we find

$$f\left(x - \frac{u}{2^k}\right) - f(x) = g_{\frac{u}{2^k}}(1) - g_{\frac{u}{2^k}}(0) = \sum_{i=1}^{N}\frac{g_{\frac{u}{2^k}}^{(j)}(0)}{j!} + R_N\left(x - \frac{u}{2^k}, 0\right), \tag{4.46}$$

where

$$R_N\left(x - \frac{u}{2^k}, 0\right) := \int_0^1\left(\int_0^{t_1} \dots \left(\int_0^{t_{N-1}}\left(g_{\frac{u}{2^k}}^{(N)}(t_N) - g_{\frac{u}{2^k}}^{(N)}(0)\right)dt_N\right)\dots\right)dt_1. \tag{4.47}$$

Here we consider $u = (u_1, ..., u_d) \in \mathbb{R}^d$, with $|u_i| \le a$, $i = 1, ..., d$.

Clearly here it holds

$$\left|g_{\frac{u}{2^k}}^{(j)}(0)\right| \le \left(\frac{u}{2^k}\right)^j\left(\left(\sum_{i=1}^{d}\left|\frac{\partial}{\partial x_i}\right|\right)^j f(x)\right), \quad j = 1, ..., N. \tag{4.48}$$

Next we estimate $\left|R_N\left(x - \frac{u}{2^k}, 0\right)\right|$, $0 \le t_N \le 1$. We observe that

$$\left|g_{\frac{u}{2^k}}^{(N)}(t_N) - g_{\frac{u}{2^k}}^{(N)}(0)\right| =$$

$$\left|\left\{\left(\sum_{i=1}^{d}\left(-\frac{u_i}{2^k}\right)\frac{\partial}{\partial x_i}\right)^N f\right\}\left(x - t\frac{u}{2^k}\right) - \left\{\left(\sum_{i=1}^{d}\left(-\frac{u_i}{2^k}\right)\frac{\partial}{\partial x_i}\right)^N f\right\}(x)\right| \le$$

$$\frac{a^N d^N}{2^{kN}} \max_{\overline{\alpha}:|\overline{\alpha}|=N} \omega_{1,\infty}\left(f_{\overline{\alpha}}, \frac{a}{2^k}\right). \tag{4.49}$$

Therefore we get

$$\left|R_N\left(x-\frac{u}{2^k},0\right)\right| \leq \int_0^1\left(\int_0^{t_1}\ldots\left(\int_0^{t_{N-1}}\left|g_{\frac{u}{2^k}}^{(N)}(t_N)-g_{\frac{u}{2^k}}^{(N)}(0)\right|dt_N\right)\ldots\right)dt_1 \overset{(4.49)}{\leq}$$

$$\int_0^1\left(\int_0^{t_1}\ldots\left(\int_0^{t_{N-1}}\frac{a^Nd^N}{2^{kN}}\max_{\overline{\alpha}:|\overline{\alpha}|=N}\omega_{1,\infty}\left(f_{\overline{\alpha}},\frac{a}{2^k}\right)dt_N\right)\ldots\right)dt_1 = \tag{4.50}$$

$$\frac{a^Nd^N}{N!2^{kN}}\max_{\overline{\alpha}:|\overline{\alpha}|=N}\omega_{1,\infty}\left(f_{\overline{\alpha}},\frac{a}{2^k}\right).$$

That is

$$\left|R_N\left(x-\frac{u}{2^k},0\right)\right| \leq \frac{a^Nd^N}{N!2^{kN}}\max_{\overline{\alpha}:|\overline{\alpha}|=N}\omega_{1,\infty}\left(f_{\overline{\alpha}},\frac{a}{2^k}\right) := \Psi. \tag{4.51}$$

By (4.46), (4.48) and (4.51), we get

$$\left|f\left(x-\frac{u}{2^k}\right)-f(x)\right| \leq \sum_{j=1}^N\frac{1}{j!}\left(\frac{a}{2^k}\right)^j\left(\left(\sum_{i=1}^d\left|\frac{\partial}{\partial x_i}\right|\right)^j f(x)\right)+\Psi. \tag{4.52}$$

We further observe

$$|(B_kf)(x)-f(x)| \overset{(4.3)}{=}$$

$$\left|(C)\int_{[-a,a]^d}f\left(x-\frac{u}{2^k}\right)\varphi(u)\,d\mu(u)-(C)\int_{[-a,a]^d}f(x)\varphi(u)\,d\mu(u)\right| \overset{(4.10)}{\leq}$$

$$(C)\int_{[-a,a]^d}\left|f\left(x-\frac{u}{2^k}\right)-f(x)\right|\varphi(u)\,d\mu(u) \overset{(4.52)}{\leq}$$

$$\left\{\sum_{j=1}^N\frac{1}{j!}\left(\frac{a}{2^k}\right)^j\left(\left(\sum_{i=1}^d\left|\frac{\partial}{\partial x_i}\right|\right)^j f(x)\right)+\Psi\right\}\left((C)\int_{[-a,a]^d}\varphi(u)\,d\mu(u)\right) \overset{(4.3)}{=}$$

$$\sum_{j=1}^N\frac{a^j}{j!2^{kj}}\left(\left(\sum_{i=1}^d\left|\frac{\partial}{\partial x_i}\right|\right)^j f(x)\right)+\frac{a^Nd^N}{N!2^{kN}}\max_{\overline{\alpha}:|\overline{\alpha}|=N}\omega_{1,\infty}\left(f_{\overline{\alpha}},\frac{a}{2^k}\right), \tag{4.53}$$

proving the claim. ∎

We also give

Theorem 4.13 *Let* $f \in C^N\left(\mathbb{R}^d,\mathbb{R}_+\right)$, $N,d\in\mathbb{N}$. *Consider the multivariate Choquet-sublinear operators*

$$(\Gamma_k(f))(x) = (C)\int_{[-a,a]^d}\left[\sum_{j=0}^n w_j f\left(\left(x-\frac{u}{2^k}\right)+\frac{j}{2^kn}\right)\right]\varphi(u)\,d\mu(u),$$

$\forall\, x \in \mathbb{R}^d$, $\forall\, k \in \mathbb{Z}$. Here we further assume that all partials $f_{\overline{\alpha}}$, $|\overline{\alpha}| = N$, are uniformly continuous or bounded and continuous on $\mathbb{R}^d$. Then

$$|(\Gamma_k f)(x) - f(x)| \leq \sum_{j=1}^{N} \frac{(a+1)^j}{j! 2^{kj}} \left(\left(\sum_{i=1}^{d} \left| \frac{\partial}{\partial x_i} \right| \right)^j f(x) \right) + \tag{4.54}$$

$$\frac{(a+1)^N d^N}{N! 2^{kN}} \max_{\overline{\alpha}: |\overline{\alpha}| = N} \omega_{1,\infty} \left(f_{\overline{\alpha}}, \frac{a+1}{2^k} \right),$$

$\forall\, x \in \mathbb{R}^d$, $\forall\, k \in \mathbb{Z}$.

Remark 4.14 (i) Since the Nth order partials $f_{\overline{\alpha}}$ are uniformly continuous or bounded and continuous and $k \to +\infty$, we get that $(\Gamma_k(f))(x) \to f(x)$ pointwise with rates.

(ii) When $N = 1$, inequality (4.54) becomes

$$|(\Gamma_k(f))(x) - f(x)| \leq$$

$$\left(\frac{a+1}{2^k} \right) \left\{ \left(\sum_{i=1}^{d} \left| \frac{\partial f(x)}{\partial x_i} \right| \right) + d \max_{i \in \{1,\ldots,d\}} \omega_{1,\infty} \left(\frac{\partial f}{\partial x_i}, \frac{a+1}{2^k} \right) \right\}, \tag{4.55}$$

$\forall\, k \in \mathbb{Z}$, $\forall\, x \in \mathbb{R}^d$.

(iii) When $f_{\overline{\alpha}}(x) = 0$, $\overline{\alpha} : |\overline{\alpha}| = \rho$, $1 \leq \rho \leq N$. Then (4.54) becomes

$$|(\Gamma_k(f))(x) - f(x)| \leq \frac{(a+1)^N d^N}{N! 2^{kN}} \max_{\overline{\alpha}: |\overline{\alpha}| = N} \omega_{1,\infty} \left(f_{\overline{\alpha}}, \frac{a+1}{2^k} \right), \ \forall\, k \in \mathbb{Z}. \tag{4.56}$$

(iv) When $\frac{\partial f(x)}{\partial x_i} = 0$, $i = 1, \ldots, d$, then (4.55) becomes

$$|(\Gamma_k(f))(x) - f(x)| \leq \left(\frac{a+1}{2^k} \right) d \max_{i \in \{1,\ldots,d\}} \omega_{1,\infty} \left(\frac{\partial f}{\partial x_i}, \frac{a+1}{2^k} \right), \tag{4.57}$$

$\forall\, k \in \mathbb{Z}$.

In (4.56), (4.57), we observe the high speed of convergence.

Proof of Theorem 4.13

Set

$$g_{\frac{u}{2^k} + \frac{j}{2^k n}}(t) := f\left(x + t \left(-\frac{u}{2^k} + \frac{j}{2^k n} \right) \right), \text{ all } 0 \leq t \leq 1. \tag{4.58}$$

We notice that

$$g_{\frac{u}{2^k} + \frac{j}{2^k n}}(0) = f(x), \text{ and } g_{\frac{u}{2^k} + \frac{j}{2^k n}}(1) = f\left(\left(x - \frac{u}{2^k} \right) + \frac{j}{2^k n} \right). \tag{4.59}$$

Then, for $j^* = 1, 2, ..., N$, we get that

$$g^{(j^*)}_{\frac{u}{2^k}+\frac{j}{2^k n}}(t) = \left\{\left(\sum_{i=1}^{d}\left(-\frac{u_i}{2^k}+\frac{j_i}{2^k n_i}\right)\frac{\partial}{\partial x_i}\right)^{j^*} f\right\}\left(x + t\left(-\frac{u}{2^k}+\frac{j}{2^k n}\right)\right). \tag{4.60}$$

By Taylor's formula we find

$$f\left(\left(x-\frac{u}{2^k}\right)+\frac{j}{2^k n}\right) - f(x) = g_{\frac{u}{2^k}+\frac{j}{2^k n}}(1) - g_{\frac{u}{2^k}+\frac{j}{2^k n}}(0) = \tag{4.61}$$

$$\sum_{j^*=1}^{N}\frac{g^{(j^*)}_{\frac{u}{2^k}+\frac{j}{2^k n}}(0)}{j^*!} + R_N\left(\left(x-\frac{u}{2^k}\right)+\frac{j}{2^k n}, 0\right),$$

where

$$R_N\left(\left(x-\frac{u}{2^k}\right)+\frac{j}{2^k n}, 0\right) :=$$

$$\int_0^1\left(\int_0^{t_1}\cdots\left(\int_0^{t_{N-1}}\left(g^{(N)}_{\frac{u}{2^k}+\frac{j}{2^k n}}(t_N) - g^{(N)}_{\frac{u}{2^k}+\frac{j}{2^k n}}(0)\right)dt_N\right)\cdots\right)dt_1. \tag{4.62}$$

Here we consider $u = (u_1, ..., u_d) \in \mathbb{R}^d$, with $|u_i| \le a$, $i = 1, ..., d$.

Clearly here it holds

$$\left|g^{(j^*)}_{\frac{u}{2^k}+\frac{j}{2^k n}}(0)\right| \le \left(\frac{a+1}{2^k}\right)^{j^*}\left(\left(\sum_{i=1}^{d}\left|\frac{\partial}{\partial x_i}\right|\right)^{j^*} f(x)\right), \quad j^* = 1, ..., N. \tag{4.63}$$

Next we estimate $\left|R_N\left(\left(x-\frac{u}{2^k}\right)+\frac{j}{2^k n}, 0\right)\right|$, $0 \le t_N \le 1$. We observe that

$$\left|g^{(N)}_{\frac{u}{2^k}+\frac{j}{2^k n}}(t_N) - g^{(N)}_{\frac{u}{2^k}+\frac{j}{2^k n}}(0)\right| =$$

$$\left|\left\{\left(\sum_{i=1}^{d}\left(-\frac{u_i}{2^k}+\frac{j_i}{2^k n_i}\right)\frac{\partial}{\partial x_i}\right)^{N} f\right\}\left(x + t_N\left(-\frac{u}{2^k}+\frac{j}{2^k n}\right)\right)\right.$$

$$\left. - \left\{\left(\sum_{i=1}^{d}\left(-\frac{u_i}{2^k}+\frac{j_i}{2^k n_i}\right)\frac{\partial}{\partial x_i}\right)^{N} f\right\}(x)\right| \le$$

$$\frac{(a+1)^N d^N}{2^{kN}} \max_{\overline{\alpha}:|\overline{\alpha}|=N} \omega_{1,\infty}\left(f_{\overline{\alpha}}, \frac{a+1}{2^k}\right). \tag{4.64}$$

Therefore we get

$$\left| R_N\left(\left(x-\frac{u}{2^k}\right)+\frac{j}{2^k n},0\right)\right| \le$$

$$\int_0^1\left(\int_0^{t_1}\dots\left(\int_0^{t_{N-1}}\left|g^{(N)}_{\frac{u}{2^k}+\frac{j}{2^k n}}(t_N)-g^{(N)}_{\frac{u}{2^k}+\frac{j}{2^k n}}(0)\right|dt_N\right)\dots\right)dt_1 \overset{(4.64)}{\le}$$

$$\int_0^1\left(\int_0^{t_1}\dots\left(\int_0^{t_{N-1}}\frac{(a+1)^N d^N}{2^{kN}}\max_{\overline{\alpha}:|\overline{\alpha}|=N}\omega_{1,\infty}\left(f_{\overline{\alpha}},\frac{a+1}{2^k}\right)dt_N\right)\dots\right)dt_1 =$$

$$\frac{(a+1)^N d^N}{N!2^{kN}}\max_{\overline{\alpha}:|\overline{\alpha}|=N}\omega_{1,\infty}\left(f_{\overline{\alpha}},\frac{a+1}{2^k}\right). \tag{4.65}$$

That is

$$\left| R_N\left(\left(x-\frac{u}{2^k}\right)+\frac{j}{2^k n},0\right)\right| \le \frac{(a+1)^N d^N}{N!2^{kN}}\max_{\overline{\alpha}:|\overline{\alpha}|=N}\omega_{1,\infty}\left(f_{\overline{\alpha}},\frac{a+1}{2^k}\right) := \theta. \tag{4.66}$$

By (4.61) we get

$$\sum_{j=0}^{n} w_j f\left(\left(x-\frac{u}{2^k}\right)+\frac{j}{2^k n}\right)-f(x)=$$

$$\sum_{j^*=1}^{N}\frac{1}{j^*!}\sum_{j=0}^{n} w_j\left(g^{(j^*)}_{\frac{u}{2^k}+\frac{j}{2^k n}}(0)\right)+\sum_{j=0}^{n} w_j R_N\left(\left(x-\frac{u}{2^k}\right)+\frac{j}{2^k n},0\right). \tag{4.67}$$

By (4.63) and (4.66), (4.67) we derive

$$\left|\sum_{j=0}^{n} w_j f\left(\left(x-\frac{u}{2^k}\right)+\frac{j}{2^k n}\right)-f(x)\right| \le$$

$$\sum_{j^*=1}^{N}\frac{1}{j^*!}\left(\frac{a+1}{2^k}\right)^{j^*}\left(\left(\sum_{i=1}^{d}\left|\frac{\partial}{\partial x_i}\right|\right)^{j^*} f(x)\right)+\theta, \tag{4.68}$$

above we applied that $\sum_{j=0}^{n} w_j = 1$.

We further observe that

$$|(\Gamma_k(f))(x)-f(x)| \overset{(4.3)}{=}$$

$$\left|(C)\int_{[-a,a]^d}\left[\sum_{j=0}^{n} w_j f\left(\left(x-\frac{u}{2^k}\right)+\frac{j}{2^k n}\right)\right]\varphi(u)\,d\mu(u)\right.$$

$$\left.-(C)\int_{[-a,a]^d} f(x)\,\varphi(u)\,d\mu(u)\right| \overset{(4.10)}{\leq}$$

$$(C)\int_{[-a,a]^d}\left|\sum_{j=0}^{n} w_j f\left(\left(x-\frac{u}{2^k}\right)+\frac{j}{2^k n}\right)-f(x)\right|\varphi(u)\,d\mu(u) \overset{(4.68)}{\leq} \tag{4.69}$$

$$\left\{\sum_{j^*=1}^{N}\frac{1}{j^*!}\left(\frac{a+1}{2^k}\right)^{j^*}\left(\left(\sum_{i=1}^{d}\left|\frac{\partial}{\partial x_i}\right|\right)^{j^*} f(x)\right)+\theta\right\}\left((C)\int_{[-a,a]^d}\varphi(u)\,d\mu(u)\right) \overset{(4.3)}{=}$$

$$\sum_{j^*=1}^{N}\frac{(a+1)^{j^*}}{j^*!2^{kj^*}}\left(\left(\sum_{i=1}^{d}\left|\frac{\partial}{\partial x_i}\right|\right)^{j^*} f(x)\right)+\frac{(a+1)^N d^N}{N!2^{kN}}\max_{\overline{\alpha}:|\overline{\alpha}|=N}\omega_{1,\infty}\left(f_{\overline{\alpha}},\frac{a+1}{2^k}\right),$$

proving the claim. ■

4.5 Appendix

Let $f \in C_U\left(\mathbb{R}^d, \mathbb{R}_+\right)$, and the multivariate positive sublinear Choquet operator

$$(K(f))(x) := (C)\int_{[-a,a]^d} f(x+u)\,\varphi(u)\,d\mu(u),\ \forall\, x \in \mathbb{R}. \tag{4.70}$$

We observe the following (for any $x, y \in \mathbb{R}^d$):

$$|(K(f))(x)-(K(f))(y)| =$$

$$\left|(C)\int_{[-a,a]^d} f(x+u)\,\varphi(u)\,d\mu(u)-(C)\int_{[-a,a]^d} f(y+u)\,\varphi(u)\,d\mu(u)\right| \overset{(4.10)}{\leq}$$

$$(C)\int_{[-a,a]^d}|f(x+u)-f(y+u)|\,\varphi(u)\,d\mu(u) \leq$$

$$\omega_1(f,\|x-y\|)\left((C)\int_{[-a,a]^d}\varphi(u)\,d\mu(u)\right) \overset{(4.3)}{=} \omega_1(f,\|x-y\|)\cdot 1 = \omega_1(f,\|x-y\|). \tag{4.71}$$

Therefore it holds the multivariate global smoothness preservation property:

$$\omega_1\left(K\left(f\right),\delta\right) \leq \omega_1\left(f,\delta\right),\ \forall\ \delta > 0. \tag{4.72}$$

References

1. G.A. Anastassiou, High order approximation by multivariate shift-invariant convolution type operators, in *Computers and Mathematics with Applications*, ed. by G.A. Anastassiou. Special issue on Computational Methods in Analysis, vol. 48 (2004), pp. 1245–1261
2. G.A. Anastassiou, *Intelligent Mathematics: Computational Analysis* (Springer, Heidelberg, 2011)
3. G.A. Anastassiou, Quantitative approximation by shift invariant multivariate sublinear-Choquet operators. J. Appl. Anal. (2018). Accepted
4. G.A. Anastassiou, H.H. Gonska, On some shift invariant integral operators, multivariate case, in *Proceedings of International Conference on Approximation, Probability, and Related Fields, U.C.S.B., Santa Barbara, CA, 20–22 May 1993*, ed. by G.A. Anastassiou, S.T. Rachev (Plenum Press, New York, 1993), pp. 41–64
5. G.A. Anastassiou, S. Gal, *Approximation Theory* (Birkhauser, Boston, 2000)
6. G. Choquet, Theory of capacities. Ann. Inst. Fourier (Grenoble) **5**, 131–295 (1954)
7. D. Denneberg, *Non-additive Measure and Integral* (Kluwer, Dordrecht, 1994)
8. Z. Wang, G.J. Klir, *Generalized Measure Theory* (Springer, New York, 2009)

Chapter 5
Hardy Type Inequalities for Choquet Integrals

Here we present Hardy type integral inequalities for Choquet integrals. These are very general inequalities involving convex and increasing functions. Initially we collect a rich machinery of results about Choquet integrals needed next, and we prove also results of their own merit such as, Choquet–Hölder's inequalities for more than two functions and a multivariate Choquet–Fubini's theorem. The main proving tool here is the property of comonotonicity of functions. We finish with independent estimates on left and right Riemann–Liouville–Choquet fractional integrals. It follows [2].

5.1 Introduction

To motivate the work in this chapter we mention the Riemann–Liouville fractional integrals, see [10]. Let $[a, b]$, $(-\infty < a < b < \infty)$ be a finite interval on the real axis $\mathbb{R}$. The left and right Riemann–Liouville fractional integrals $I_{a+}^{\alpha} f$ and $I_{b-}^{\alpha} f$ (respectively) of order $\alpha > 0$ are defined by

$$\left(I_{a+}^{\alpha} f\right)(x) = \frac{1}{\Gamma(\alpha)} \int_a^x f(t)(x-t)^{\alpha-1} dt, \ (x > a),$$

$$\left(I_{b-}^{\alpha} f\right)(x) = \frac{1}{\Gamma(\alpha)} \int_x^b f(t)(t-x)^{\alpha-1} dt, \ (x < b),$$

where Γ is the Gamma function.

We mention a basic property of the operators $I_{a+}^{\alpha} f$ and $I_{b-}^{\alpha} f$ of order $\alpha > 0$, see also [12]: It holds that the fractional integral operators $I_{a+}^{\alpha} f$ and $I_{b-}^{\alpha} f$ are bounded in $L_p(a, b)$, $1 \leq p \leq \infty$, that is

$$\left\| I_{a+}^{\alpha} f \right\|_p \leq K \|f\|_p, \quad \left\| I_{b-}^{\alpha} f \right\|_p \leq K \|f\|_p,$$

G. A. Anastassiou, *Ordinary and Fractional Approximation by Non-additive Integrals: Choquet, Shilkret and Sugeno Integral Approximators*, Studies in Systems, Decision and Control 190, https://doi.org/10.1007/978-3-030-04287-5_5

where

$$K = \frac{(b-a)^{\alpha}}{\alpha\Gamma(\alpha)}.$$

The first inequality that is the result involving the left-sided fractional integral, was proved by H.G. Hardy in one of his first papers, see [8]. He did not write down the constant, but the calculation of the constant was hidden inside his proof.

General Hardy inequalities of the above type were derived also in [1, 9]. We continue this kind of research for Choquet integrals based on the comonotonicity property of functions and convexity. We derive a wide range of Choquet integral inequalities of Hardy type.

5.2 Background

In this section we give some definitions and basic properties of Choquet integral essential for this work.

Definition 5.1 ([14]) Let X be a non-empty set, $\mathcal{F}$ be a σ-algebra of subsets of X and $\mu : \mathcal{F} \to [0, \infty]$ be a nonnegative real-valued set function, μ is said to be a fuzzy measure iff:

(1) $\mu(\varnothing) = 0$,

(2) for any $A, B \in \mathcal{F}$, $A \subseteq B$ implies $\mu(A) \leq \mu(B)$ (monotonicity),

(3) for $\{A_n\} \subseteq \mathcal{F}$, $A_1 \subseteq A_2 \subseteq ... \subseteq A_n \subseteq ...$, implies $\lim_{n\to\infty} \mu(A_n) = \mu\left(\cup_{n=1}^{\infty} A_n\right)$ (continuity from below)

(4) for $\{A_n\} \subseteq \mathcal{F}$, $A_1 \supseteq A_2 \supseteq ... \supseteq A_n \supseteq ...$, $\mu(A_1) < \infty$, implies $\lim_{n\to\infty} \mu(A_n) = \mu\left(\cap_{n=1}^{\infty} A_n\right)$ (continuity from above).

If μ is a fuzzy measure from $\mathcal{F}$ to $[0, 1]$ with $\mu(X) = 1$, μ is called a regular fuzzy measure. If μ is a fuzzy measure, $(X, \mathcal{F}, \mu)$ is called a fuzzy measure space and $(X, \mathcal{F})$ is a fuzzy measurable space. Clearly μ is not necessarily an additive measure. Let F be the set of all real-valued nonnegative measurable functions defined on X.

Definition 5.2 ([11]) Let $(X, \mathcal{F}, \mu)$ be a fuzzy measure space, μ is said to be submodular (supermodular) if

$$\mu(A \cap B) + \mu(A \cup B) \leq (\geq)\, \mu(A) + \mu(B), \ \forall\, A, B \subseteq \mathcal{F}. \tag{5.1}$$

Definition 5.3 ([5]) Let $f, g \in F$, f and g are said to be comonotonic iff $f(x) < f(x')$ implies $g(x) \leq g(x')$, $\forall\, x, x' \in X$.

Definition 5.4 ([6, 15]) Let $(X, \mathcal{F}, \mu)$ be a fuzzy measure space, $f \in F$ and $A \in \mathcal{F}$. The Choquet integral of f with respect to μ on A is defined by

$$(C)\int_A f d\mu = \int_0^{\infty} \mu(A \cap \{x | f(x) \geq \alpha\})\, d\alpha. \tag{5.2}$$

If $(C)\int_X f d\mu < \infty$, we call f (C)-integrable, $L_1(\mu)$ is the set of all (C)-integrable function.

Clearly $(C)\int_X f d\mu < \infty$, implies $(C)\int_A f d\mu < \infty$.

Theorem 5.5 ([16]) *Let $(X, \mathcal{F}, \mu)$ be a fuzzy measurable space, $\{f_1, f_2, f\} \subset F$, $A, B \in \mathcal{F}$ and $c \geq 0$ constant. Then,*

(1) if $\mu(A) = 0$, then $(C)\int_A f d\mu = 0$,
(2) $(C)\int_A c d\mu = c\mu(A)$,
(3) if $f_1 \leq f_2$, then

$$(C)\int_A f_1 d\mu \leq (C)\int_A f_2 d\mu, \tag{5.3}$$

(4) if $A \subset B$, then $(C)\int_A f d\mu \leq (C)\int_B f d\mu$,
(5) $(C)\int_A (f + c) d\mu = (C)\int_A f d\mu + c\mu(A)$,
(6) $(C)\int_A c f d\mu = c\left((C)\int_A f d\mu\right)$.

Theorem 5.6 ([6]) *Let $(X, \mathcal{F}, \mu)$ be a fuzzy measure space and $f, g \in F$. Then*

(1) if f, g are comonotonic, then for any $A \in \mathcal{F}$,

$$(C)\int_A (f + g) d\mu = (C)\int_A f d\mu + (C)\int_A g d\mu, \tag{5.4}$$

(2) if μ is submodular, then for any $A \in \mathcal{F}$,

$$(C)\int_A (f + g) d\mu \leq (C)\int_A f d\mu + (C)\int_A g d\mu. \tag{5.5}$$

The Jensen's inequality for Choquet integrals follows:

Theorem 5.7 ([18]) *Let $(X, \mathcal{F}, \mu)$ be a fuzzy measure space and $f \in L_1(\mu)$. If μ is a regular fuzzy measure and $\Phi : [0, \infty) \to [0, \infty)$ is a convex function, then*

$$\Phi\left((C)\int_X f d\mu\right) \leq (C)\int_X \Phi(f) d\mu. \tag{5.6}$$

Corollary 5.8 ([18]) *Let $(X, \mathcal{F}, \mu)$ be a fuzzy measure space and $f \in L_1(\mu)$. If μ is a regular fuzzy measure, then*

$$\left((C)\int_X f d\mu\right)^p \leq (C)\int_X f^p d\mu, \tag{5.7}$$

for any $1 < p < \infty$.

Theorem 5.9 ([18]) *(Hölder's inequality) Let $(X, \mathcal{F}, \mu)$ be a fuzzy measure space and $f, g \in F$. If μ is a submodular fuzzy measure and $1 < p, q < \infty$ with $\frac{1}{p} + \frac{1}{q} = 1$, then*

$$(C)\int_X fg\,d\mu \le \left((C)\int_X f^p d\mu\right)^{\frac{1}{p}} \left((C)\int_X g^q d\mu\right)^{\frac{1}{q}}. \tag{5.8}$$

Theorem 5.10 ([18]) *(Minkowski inequality) Let $(X, \mathcal{F}, \mu)$ be a fuzzy measure space and $f, g \in F$. If μ is a submodular fuzzy measure and $1 \le p < \infty$, then*

$$\left((C)\int_X (f+g)^p\,d\mu\right)^{\frac{1}{p}} \le \left((C)\int_X f^p d\mu\right)^{\frac{1}{p}} + \left((C)\int_X g^p d\mu\right)^{\frac{1}{p}}. \tag{5.9}$$

We give

Theorem 5.11 (Hölder's inequality for three functions) *Let $(X, \mathcal{F}, \mu)$ be a fuzzy measure space and $f_1, f_2, f_3 \in F$. If μ is a submodular fuzzy measure and $1 < p_1 \le p_2 \le p_3 < \infty$ with $\frac{1}{p_1} + \frac{1}{p_2} + \frac{1}{p_3} = 1$, then*

$$(C)\int_X f_1 f_2 f_3\,d\mu \le \left((C)\int_X f_1^{p_1} d\mu\right)^{\frac{1}{p_1}} \left((C)\int_X f_2^{p_2} d\mu\right)^{\frac{1}{p_2}} \left((C)\int_X f_3^{p_3} d\mu\right)^{\frac{1}{p_3}}. \tag{5.10}$$

Proof Let $p = \frac{p_3}{p_3 - 1} > 1$ and $q = p_3$. Notice that $\frac{1}{p} + \frac{1}{q} = 1$.

We apply (5.8) as follows

$$(C)\int_X f_1 f_2 f_3\,d\mu \le \left((C)\int_X (f_1 f_2)^p\,d\mu\right)^{\frac{1}{p}} \left((C)\int_X f_3^{p_3} d\mu\right)^{\frac{1}{p_3}}. \tag{5.11}$$

We see that

$$\frac{p}{p_1} + \frac{p}{p_2} = p\left(\frac{1}{p_1} + \frac{1}{p_2}\right) = p\left(1 - \frac{1}{p_3}\right) = p\left(\frac{p_3 - 1}{p_3}\right) = 1. \tag{5.12}$$

Clearly it holds $\frac{p_1}{p}, \frac{p_2}{p} > 1$.

Therefore we get

$$(C)\int_X f_1^p f_2^p\,d\mu \overset{(5.8)}{\le} \left((C)\int_X f_1^{p\frac{p_1}{p}} d\mu\right)^{\frac{p}{p_1}} \left((C)\int_X f_2^{p\frac{p_2}{p}} d\mu\right)^{\frac{p}{p_2}} = \tag{5.13}$$

$$\left((C)\int_X f_1^{p_1} d\mu\right)^{\frac{p}{p_1}} \left((C)\int_X f_2^{p_2} d\mu\right)^{\frac{p}{p_2}}.$$

That is

$$\left((C)\int_X (f_1 f_2)^p\,d\mu\right)^{\frac{1}{p}} \le \left((C)\int_X f_1^{p_1} d\mu\right)^{\frac{1}{p_1}} \left((C)\int_X f_2^{p_2} d\mu\right)^{\frac{1}{p_2}}. \tag{5.14}$$

Combining (5.11) and (5.14), we produce (5.10). ■

In general we have

Theorem 5.12 (Hölder's inequality for n functions) *Let $(X, \mathcal{F}, \mu)$ be a fuzzy measure space and $f_i \in F$, $i = 1, ..., n \in \mathbb{N}$. If μ is a submodular fuzzy measure and $1 < p_1 \leq p_2 \leq ... \leq p_n < \infty$ with $\sum_{i=1}^{n} \frac{1}{p_i} = 1$, then*

$$(C) \int_X \prod_{i=1}^{n} f_i d\mu \leq \prod_{i=1}^{n} \left((C) \int_X f_i^{p_i} d\mu \right)^{\frac{1}{p_i}}. \tag{5.15}$$

Proof By induction. ■

Remark 5.13 Let $\mathcal{A}$ be a σ-algebra, and let $\{A_k\}_{k \in \mathbb{N}} \subseteq \mathcal{A}$ be a family of pairwise disjoint sets. Here $\mathcal{P}$ is a probability measure on $(X, \mathcal{A})$ with only the finite additivity property valid: i.e.,

$$P\left(\cup_{k=1}^{n} A_k\right) = \sum_{k=1}^{n} P(A_k), \ \forall\, n \in \mathbb{N}.$$

We observe that

$$P\left(\cup_{k=1}^{\infty} A_k\right) = \lim_{n\to\infty} P\left(\cup_{k=1}^{n} A_k\right) = \lim_{n\to\infty} \sum_{k=1}^{n} P(A_k) = \sum_{k=1}^{\infty} P(A_k). \tag{5.16}$$

That is, the countable additivity property holds, hence P is a usual probability measure.

Notice that a σ-algebra on X is also an algebra of subsets of X.

Definition 5.14 ([4, 7]) For every space Ω and algebra $\mathcal{A}$ of subsets of Ω a set-function $\sigma : \mathcal{A} \to \mathbb{R}$ is called a (normalized) capacity if it satisfies the following:

(i)

$$\sigma(\varnothing) = 0, \ \sigma(\Omega) = 1, \tag{5.17}$$

(ii) $\forall\, A, B \in \mathcal{A} : A \subseteq B \Rightarrow \sigma(A) \leq \sigma(B)$.

From (i) and (ii) we get that the range of σ is contained in $[0, 1]$.

In general the Choquet integral is defined as follows:

Definition 5.15 ([4, 13]) Let $(\Omega, \mathcal{A})$ be an algebra and $f : \Omega \to \mathbb{R}$ is a bounded $\mathcal{A}$-measurable function and σ is any (normalized) capacity on Ω we define the Choquet integral of f with respect to σ to be the number

$$(C) \int_\Omega f(\omega)\, d\sigma(\omega) = \int_0^\infty \sigma(\{\omega \in \Omega : f(\omega) \geq \alpha\})\, d\alpha + \tag{5.18}$$

$$\int_{-\infty}^{0} \left[\sigma \left(\omega \in \Omega : f\left(\omega\right) \geq \alpha\}\right) - 1\right] d\alpha,$$

where the integrals are taken in the sense of Riemann.

A (normalized) capacity σ is called probability [7] iff

$$\forall\, A, B \in \mathcal{A} : \sigma\left(A \cup B\right) + \sigma\left(A \cap B\right) = \sigma\left(A\right) + \sigma\left(B\right). \tag{5.19}$$

Notice that since the integrands are monotone, the Choquet integral always exists, and if σ is a probability it collapses to a usual Lebesgue integral.

Definition 5.16 ([7]) Let $f, g : \Omega \to \mathbb{R}$ be two bounded $\mathcal{A}$-measurable functions. We say that f and g are comonotonic, if for every $\omega, \omega' \in \Omega$,

$$\left(f\left(\omega\right) - f\left(\omega'\right)\right)\left(g\left(\omega\right) - g\left(\omega'\right)\right) \geq 0. \tag{5.20}$$

A class of functions $\mathcal{F}^*$ is said to be comonotonic if for every $f, g \in \mathcal{F}^*$, f and g are comonotonic.

Proposition 5.17 ([7]) *If σ and λ are (normalized) capacities on the algebra $(\Omega, \mathcal{A})$, and $f, g : \Omega \to \mathbb{R}$ are bounded $\mathcal{A}$-measurable functions then:*

(i)

$$(C)\int_{\Omega} 1_A d\sigma = \sigma\left(A\right), \forall\, A \in \mathcal{A}, \tag{5.21}$$

where 1_A is the characteristic function on A,

(ii) (positive homogeneity)

$$(C)\int_{\Omega} pf d\sigma = p\left((C)\int_{\Omega} f d\sigma\right), \text{ for every } p \geq 0, \tag{5.22}$$

(iii) (monotonicity) $f \geq g$ implies

$$(C)\int_{\Omega} f d\sigma \geq (C)\int_{\Omega} g d\sigma, \tag{5.23}$$

(iv)

$$(C)\int_{\Omega} (f + p)\, d\sigma = (C)\int_{\Omega} f d\sigma + p, \forall\, p \in \mathbb{R}, \tag{5.24}$$

(v) (comonotonic additivity) If f, g are comonotonic then

$$(C)\int_{\Omega} (f + g)\, d\sigma = (C)\int_{\Omega} f d\sigma + (C)\int_{\Omega} g d\sigma. \tag{5.25}$$

We need the very important

Lemma 5.18 ([7]) *Let $(\Omega, \mathcal{A})$ be an algebra. Suppose that $\mathcal{F}^*$ is a comonotonic class of bounded and $\mathcal{A}$-measurable functions from Ω into $\mathbb{R}$ and σ is a (normalized) capacity on $(\Omega, \mathcal{A})$. Then there exists a probability measure P on $(\Omega, \mathcal{A})$ such that for every $f \in \mathcal{F}^*$*

$$\int_\Omega f d\sigma = \int_\Omega f dP. \tag{5.26}$$

Here $\int_\Omega f dP$ is a standard integral of Lebesgue type.

Based on Remark 5.13, Lemma 5.18 is still valid in case that $(\Omega, \mathcal{A})$ is a σ-algebra.

Definition 5.19 ([7]) Let X, Y be two sets and $Z = X \times Y$. Let $f : Z \to \mathbb{R}$. We say that f has comonotonic x-sections if for every $x, x' \in X$, $f(x, \cdot) : Y \to \mathbb{R}$, and $f\left(x', \cdot\right) : Y \to \mathbb{R}$ are comonotonic functions. Comonotonicity of y-sections is similarly defined. We call f slice-comonotonic if it has both comonotonic x-sections and y-sections.

Remark 5.20 Notice that Definitions 5.14–5.16 and Proposition 5.17, are still valid when $(\Omega, \mathcal{A})$ is a σ-algebra.

Next we mention Fubini's theorem for Choquet integrals.

Theorem 5.21 ([3]) *Let (Ω_1, Σ_1), (Ω_2, Σ_2) be σ-algebras. Let u_i, $i = 1, 2$ be submodular (or supermodular) regular fuzzy measures on Ω_i, respectively. Let $\Omega = \Omega_1 \times \Omega_2$ be endowed with the product σ-algebra $\Sigma = \Sigma_1 \otimes \Sigma_2$. Let $f : \Omega_1 \times \Omega_2 \to \mathbb{R}$ be a slice-comonotonic bounded Σ-measurable mapping, then:*

(1) $f(\cdot, \omega_2)$ is Σ_1-measurable and $\omega_2 \in \Omega_2 \to (C)\int_{\Omega_1} f(s, \omega_2)\, du_1(s)$ is bounded and Σ_2-measurable,

$f(\omega_1, \cdot)$ is Σ_2-measurable and $\omega_1 \in \Omega_1 \to (C)\int_{\Omega_2} f(\omega_1, t)\, du_2(t)$ is bounded and Σ_1-measurable,

(2) the iterated integrals $(C)\int_{\Omega_2}\int_{\Omega_1} f du_1 du_2$, $(C)\int_{\Omega_1}\int_{\Omega_2} f du_2 du_1$ exist and are equal:

$$(C)\int_{\Omega_2}\left((C)\int_{\Omega_1} f(\omega_1, \omega_2)\, du_1\right) du_2 = (C)\int_{\Omega_1}\left((C)\int_{\Omega_2} f(\omega_1, \omega_2)\, du_2\right) du_1. \tag{5.27}$$

We give

Definition 5.22 Let $f : \prod_{i=1}^{n} \Omega_i \to \mathbb{R}$, $n \in \mathbb{N}$. If the i-sections $f(x_1, ..., x_{i-1}, \cdot, x_{i+1}, ..., x_n)$ and $f\left(x_1', ..., x_{i-1}', \cdot, x_{i+1}', ..., x_n'\right)$ are comonotonic functions, for all $i = 1, ..., n$; where the vectors $(x_1, ..., x_{i-1}, x_{i+1}, ..., x_n)$, $\left(x_1', ..., x_{i-1}', x_{i+1}', ..., x_n'\right) \in \prod_{\substack{j=1 \\ j \neq i}}^{n-1} \Omega_j$ are different, for all $i = 1, 2, ..., n$, we call f slice n comonotonic function.

We denote by θ a permutation of the set $\{1, 2, ..., n\}$ into itself, $n \in \mathbb{N}$. There are $n!$ permutations.

In [3] is mentioned that Theorem 5.21 can be generalized for n spaces. Next we state in brief Fubini's theorem for n Choquet iterated integrals.

Theorem 5.23 *Let (Ω_i, Σ_i) be σ-algebras, $i = 1, 2, ..., n \in \mathbb{N}$. Let $u_i, i = 1, 2, ..., n$ be submodular (or supermodular) regular fuzzy measures on Ω_i, respectively. Let $\Omega = \prod_{i=1}^{n} \Omega_i$ be endowed with the product σ-algebra $\Sigma = \otimes_{i=1}^{n} \Sigma_i$. Let $f : \prod_{i=1}^{n} \Omega_i \rightarrow \mathbb{R}$ be a slice-comonotonic bounded Σ-measurable mapping, then*

$$(C) \int_{\Omega_n} \int_{\Omega_{n-1}} ... \int_{\Omega_1} f du_1 du_2 ... du_n =$$

$$(C) \int_{\Omega_{\theta(n)}} \int_{\Omega_{\theta(n-1)}} ... \int_{\Omega_{\theta(1)}} f du_{\theta(1)} du_{\theta(2)} ... du_{\theta(n)}, \tag{5.28}$$

for any permutation θ on the set $\{1, ..., n\}$. All the iterated Choquet integrals in (5.28) exist and are equal.

Proof By induction, (5.23) and using Theorem 5.21. ■

Remark 5.24 If μ is a countably additive bounded measure, then the Choquet integral $(C) \int_A f d\mu$ reduces to the usual Lebesgue type integral (see, e.g. [6], p. 62, or [17], p. 226), above it is $A \subseteq \Omega$.

5.3 Main Results

This section is motivated by [9].

Let the fuzzy measure spaces $(\Omega_1, \Sigma_1, \mu_1)$ and $(\Omega_2, \Sigma_2, \mu_2)$, where μ_1, μ_2 are regular fuzzy measures, furthermore μ_1, μ_2 are assumed to be submodular.

Let $k : \Omega_1 \times \Omega_2 \rightarrow \mathbb{R}_+$ which is a bounded measurable function and $k(x, y)$ is $slice-$comonotonic and belongs to a comonotonic class F_1^* $as\ a\ function\ of\ y$.

Consider the function

$$K(x) = (C) \int_{\Omega_2} k(x, y) d\mu_2(y), \quad x \in \Omega_1, \tag{5.29}$$

and assume that $K(x) > 0$.

Notice that K is bounded.

Denote by $W(k)$ the class of functions $g : \Omega_1 \rightarrow \mathbb{R}_+$, such that

$$g(x) = (C) \int_{\Omega_2} k(x, y) f(y) d\mu_2(y), \tag{5.30}$$

where $f : \Omega_2 \rightarrow \mathbb{R}_+$ is a bounded measurable function, such that $k(x, y) f(y)$ is $slice-$comonotonic and belongs to a comonotonic class F_2^* $as\ a\ function\ of\ y$.

Notice that g is also bounded.
We give

Theorem 5.25 *Let u be a nonnegative measurable function on Ω_1. Assume that $\frac{u(x)}{K(x)}$ is bounded on Ω_1. Define υ on Ω_2 by*

$$\upsilon(y) = (C)\int_{\Omega_1} \frac{u(x)}{K(x)} k(x, y)\, d\mu_1(x), \tag{5.31}$$

which is bounded. Let $\Phi : \mathbb{R}_+ \to \mathbb{R}_+$ be a convex and increasing function, such that $k(x, y)\,\Phi(f(y))$ is x-section comonotonic with comonotonic class F_3^*. Assume here that $\left(F_1^* \cup F_2^* \cup F_3^*\right) \subseteq F^*$, where F^* is one comonotonic class of functions on Ω_2. Assume further that $u(x)(K(x))^{-1} k(x, y)\,\Phi(f(y))$ is slice-comonotonic. Then

$$(C)\int_{\Omega_1} u(x)\,\Phi\left(\frac{g(x)}{K(x)}\right) d\mu_1(x) \le (C)\int_{\Omega_2} \upsilon(y)\,\Phi(f(y))\, d\mu_2(y), \tag{5.32}$$

holds for all $g \in W(k)$, with f as in (5.30).

Proof We observe that

$$(C)\int_{\Omega_1} u(x)\,\Phi\left(\frac{g(x)}{K(x)}\right) d\mu_1(x) =$$

$$(C)\int_{\Omega_1} u(x)\,\Phi\left(\frac{1}{K(x)}(C)\int_{\Omega_2} k(x, y) f(y)\, d\mu_2(y)\right) d\mu_1(x) = \tag{5.33}$$

(next we use Lemma 5.18, where P is a probability measure on Ω_2)

$$(C)\int_{\Omega_1} u(x)\,\Phi\left(\frac{1}{K(x)}(C)\int_{\Omega_2} k(x, y) f(y)\, dP(y)\right) d\mu_1(x) \le$$

(we can also write $K(x) = \int_{\Omega_2} k(x, y)\, dP(y)$, hence by classic Jensen's inequality)

$$(C)\int_{\Omega_1} u(x)(K(x))^{-1}\left((C)\int_{\Omega_2} k(x, y)\,\Phi(f(y))\, dP(y)\right) d\mu_1(x) = \tag{5.34}$$

(again by Lemma 5.18)

$$(C)\int_{\Omega_1} u(x)(K(x))^{-1}\left((C)\int_{\Omega_2} k(x, y)\,\Phi(f(y))\, d\mu_2(y)\right) d\mu_1(x) =$$

$$(C)\int_{\Omega_1}\left((C)\int_{\Omega_2} u(x)(K(x))^{-1} k(x, y)\,\Phi(f(y))\, d\mu_2(y)\right) d\mu_1(x) =$$

(since the functions $\Phi(f(y))$ and $u(x)(K(x))^{-1}k(x,y)\Phi(f(y))$ are bounded and the second one is slice-comonotonic, we can apply Fubini's Theorem 5.21)

$$(C)\int_{\Omega_2}\left((C)\int_{\Omega_1}u(x)(K(x))^{-1}k(x,y)\Phi(f(y))\,d\mu_1(x)\right)d\mu_2(y)=$$

$$(C)\int_{\Omega_2}\Phi(f(y))\left((C)\int_{\Omega_1}u(x)(K(x))^{-1}k(x,y)\,d\mu_1(x)\right)d\mu_2(y)\overset{(5.31)}{=} \tag{5.35}$$

$$(C)\int_{\Omega_2}\Phi(f(y))\,\upsilon(y)\,d\mu_2(y),$$

proving the claim. ∎

We also give

Corollary 5.26 *All as in Theorem 5.25, with $\Phi=$ identity mapping. Then*

$$(C)\int_{\Omega_1}\frac{u(x)}{K(x)}g(x)\,d\mu_1(x)\le(C)\int_{\Omega_2}\upsilon(y)f(y)\,d\mu_2(y), \tag{5.36}$$

holds for all $g\in W(k)$, with f as in (5.30).

Corollary 5.27 *All as in Theorem 5.25, with $\Phi(x)=x^p$, $\forall\,x\in\mathbb{R}_+$, $p>1$. Then*

$$(C)\int_{\Omega_1}\frac{u(x)}{K^p(x)}g^p(x)\,d\mu_1(x)\le(C)\int_{\Omega_2}\upsilon(y)f^p(y)\,d\mu_2(y), \tag{5.37}$$

holds for all $g\in W(k)$, with f as in (5.30).

Corollary 5.28 *All as in Theorem 5.25, with $\Phi(x)=e^x$, $\forall\,x\in\mathbb{R}_+$. Then*

$$(C)\int_{\Omega_1}u(x)\,e^{\frac{g(x)}{K(x)}}d\mu_1(x)\le(C)\int_{\Omega_2}\upsilon(y)\,e^{f(y)}d\mu_2(y), \tag{5.38}$$

holds for all $g\in W(k)$, with f as in (5.30).

Corollary 5.29 *All as in Theorem 5.25, with $\Phi=$ identity mapping and $u(x)=K(x)$. Then*

$$(C)\int_{\Omega_1}g(x)\,d\mu_1(x)\le(C)\int_{\Omega_2}\upsilon(y)f(y)\,d\mu_2(y), \tag{5.39}$$

holds for all $g\in W(k)$, with f as in (5.30). Here $\upsilon(y)=(C)\int_{\Omega_1}k(x,y)\,d\mu_1(x)$ is bounded.

Corollary 5.30 *All as in Theorem 5.25, with $\Phi(x)=x^p$, $\forall\,x\in\mathbb{R}_+$, $p>1$, and $u(x)=K^p(x)$. Then*

$$(C)\int_{\Omega_1} g^p(x)\,d\mu_1(x) \le (C)\int_{\Omega_2} \upsilon(y)\, f^p(y)\,d\mu_2(y), \tag{5.40}$$

holds for all $g \in W(k)$, with f as in (5.30). Here

$$\upsilon(y) = (C)\int_{\Omega_1} K^{p-1}(x)\,k(x,y)\,d\mu_1(x) \ \textit{is bounded}. \tag{5.41}$$

Remark 5.31 (*on Corollary* 5.30) Let us assume that $k(x,y) \le M$, $M > 0$, $\forall (x,y) \in \Omega_1 \times \Omega_2$, then $K(x) \le M$. And from (5.41), $\upsilon(y) \le M^p$.

Consequently, from (5.40), it holds

$$(C)\int_{\Omega_1} g^p(x)\,d\mu_1(x) \le M^p\left((C)\int_{\Omega_2} f^p(y)\,d\mu_2(y)\right), \tag{5.42}$$

and even better written

$$\left((C)\int_{\Omega_1} g^p(x)\,d\mu_1(x)\right)^{\frac{1}{p}} \le M\left((C)\int_{\Omega_2} f^p(y)\,d\mu_2(y)\right)^{\frac{1}{p}}. \tag{5.43}$$

Next we rewrite the result of (5.43) in detail.

Theorem 5.32 *Assume that $k(x,y) \le M$, $M > 0$, $\forall (x,y) \in \Omega_1 \times \Omega_2$, and let $p > 1$. Define*

$$\upsilon(y) = (C)\int_{\Omega_1} K^{p-1}(x)\,k(x,y)\,d\mu_1(x), \tag{5.44}$$

which is bounded. Here $k(x,y)(f(y))^p$ is x-section comonotonic with comonotonic class F_3^*. Assume that $\left(F_1^* \cup F_2^* \cup F_3^*\right) \subseteq F^*$, where F^* one comonotonic class on Ω_2. Assume further that $(K(x))^{p-1}\,k(x,y)(f(y))^p$ is slice-comonotonic. Then

$$\left((C)\int_{\Omega_1} g^p(x)\,d\mu_1(x)\right)^{\frac{1}{p}} \le M\left((C)\int_{\Omega_2} f^p(y)\,d\mu_2(y)\right)^{\frac{1}{p}}, \tag{5.45}$$

holds for all $g \in W(k)$, with f as in (5.30).

Remark 5.33 Assume that $k(x,y) \le M$, $M > 0$, $\forall (x,y) \in \Omega_1 \times \Omega_2$. Hence directly by (5.30) we get

$$g(x) \le M\left((C)\int_{\Omega_2} f(y)\,d\mu_2(y)\right), \ \forall\, x \in \Omega_1.$$

Therefore

$$\int_{\Omega_1} g(x)\,d\mu_1(x) \le M\left((C)\int_{\Omega_2} f(y)\,d\mu_2(y)\right), \tag{5.46}$$

holds for all $g \in W(k)$, with f as in (5.30).

Theorem 5.34 *Define υ on Ω_2 by $\upsilon(y) = (C)\int_{\Omega_1} k(x, y)\, d\mu_1(x)$, which is bounded. Let $p > 1$. Here $k(x, y)(f(y))^p$ is slice–comonotonic and belongs to a comonotonic class F_3^* as a function of y. Assume that $\left(F_1^* \cup F_2^* \cup F_3^*\right) \subseteq F^*$, where F^* one comonotonic class on Ω_2. Then*

$$(C)\int_{\Omega_1} (K(x))^{1-p} g^p(x)\, d\mu_1(x) \leq (C)\int_{\Omega_2} \upsilon(y) f^p(y)\, d\mu_2(y), \tag{5.47}$$

holds for all $g \in W(k)$, with f as in (5.30).

Proof By Theorem 5.25, take $f(x) = x^p$, $x \geq 0$, $p > 1$, and $u(x) = K(x)$. ■

Corollary 5.35 *All as in Theorem 5.34. Then*

$$\left((C)\int_{\Omega_1} g^p(x)\, d\mu_1(x)\right)^{\frac{1}{p}} \leq M\left((C)\int_{\Omega_2} f^p(y)\, d\mu_2(y)\right)^{\frac{1}{p}}. \tag{5.48}$$

holds for all $g \in W(k)$, with f as in (5.30). Here $k(x, y) \leq M$, $M > 0$, $\forall (x, y) \in \Omega_1 \times \Omega_2$.

Proof Since $p > 1$, $1 - p < 0$. Hence the left hand side of (5.47) is greater equal to $M^{1-p}\left((C)\int_{\Omega_1} g^p(x)\, d\mu_1(x)\right)$, by $K(x) \leq M$ and $(K(x))^{1-p} \geq M^{1-p}$. And the right hand side of (5.47) is less equal to $M\left((C)\int_{\Omega_2} f^p(y)\, d\mu_2(y)\right)$, by $\upsilon(y) \leq M$. Therefore

$$M^{1-p}\left((C)\int_{\Omega_1} g^p(x)\, d\mu_1(x)\right) \leq M\left((C)\int_{\Omega_2} f^p(y)\, d\mu_2(y)\right), \tag{5.49}$$

proving the claim. ■

5.4 Appendix

Here $\mathcal{B}$ stands for the Borel σ-algebra on $[a, b]$.

Let the fuzzy measure spaces $([a, b], \mathcal{B}, \mu_1)$ and $([a, b], \mathcal{B}, \mu_2)$, where $[a, b] \subset \mathbb{R}$ and μ_1, μ_2 are bounded fuzzy measures with μ_2 submodular. Let $p, q > 1$ such that $\frac{1}{p} + \frac{1}{q} = 1$. Let $f : [a, b] \to \mathbb{R}_+$ which is bounded and $\mathcal{B}$-measurable.

We define the left and right Riemann–Liouville–Choquet fractional integrals of order $\alpha > 1$ (respectively):

$$\left(I_{a+}^{\alpha} f\right)(x) = \frac{1}{\Gamma(\alpha)} (C)\int_a^x (x - t)^{\alpha - 1} f(t)\, d\mu_2(t), \tag{5.50}$$

and

$$\left(I_{b-}^{\alpha} f\right)(x) = \frac{1}{\Gamma(\alpha)} (C) \int_{x}^{b} (t-x)^{\alpha-1} f(t)\, d\mu_2(t), \tag{5.51}$$

$\forall\, x \in [a,b]$, where Γ is the gamma function.

We assume that $\left(I_{a+}^{\alpha} f\right)$ and $\left(I_{b-}^{\alpha} f\right)$ are $\mathcal{B}$-measurable functions. Clearly $I_{a+}^{\alpha} f$, $I_{b-}^{\alpha} f$ are nonnegative and bounded over $[a,b]$.

Remark 5.36 By Theorem 5.9 we obtain

$$\left(I_{a+}^{\alpha} f\right)(x) \leq \frac{1}{\Gamma(\alpha)} \left((C) \int_{a}^{x} (x-t)^{p(\alpha-1)} d\mu_2(t)\right)^{\frac{1}{p}} \left((C) \int_{a}^{x} f^{q}(t)\, d\mu_2(t)\right)^{\frac{1}{q}} \leq \tag{5.52}$$

$$\frac{1}{\Gamma(\alpha)} \left((b-a)^{p(\alpha-1)} \mu_2([a,b])\right)^{\frac{1}{p}} \left((C) \int_{a}^{b} f^{q}(t)\, d\mu_2(t)\right)^{\frac{1}{q}}.$$

Hence it holds

$$\left(\left(I_{a+}^{\alpha} f\right)(x)\right)^{p} \leq \frac{1}{(\Gamma(\alpha))^{p}} (b-a)^{p(\alpha-1)} \mu_2([a,b]) \left((C) \int_{a}^{b} f^{q}(t)\, d\mu_2(t)\right)^{\frac{p}{q}}, \tag{5.53}$$

$\forall\, x \in [a,b]$.

Therefore

$$(C) \int_{a}^{b} \left(\left(I_{a+}^{\alpha} f\right)(x)\right)^{p} d\mu_1(x) \leq$$

$$\frac{\mu_1([a,b])}{(\Gamma(\alpha))^{p}} (b-a)^{p(\alpha-1)} \mu_2([a,b]) \left((C) \int_{a}^{b} f^{q}(t)\, d\mu_2(t)\right)^{\frac{p}{q}}. \tag{5.54}$$

We have proved that

$$\left((C) \int_{a}^{b} \left(\left(I_{a+}^{\alpha} f\right)(x)\right)^{p} d\mu_1(x)\right)^{\frac{1}{p}} \leq$$

$$\frac{\left(\mu_1([a,b])\, \mu_2([a,b])\right)^{\frac{1}{p}} (b-a)^{(\alpha-1)}}{\Gamma(\alpha)} \left((C) \int_{a}^{b} f^{q}(t)\, d\mu_2(t)\right)^{\frac{1}{q}}. \tag{5.55}$$

Similarly, we have

$$\left(I_{b-}^{\alpha} f\right)(x) \overset{(5.8)}{\leq} \frac{1}{\Gamma(\alpha)} \left((C) \int_{x}^{b} (t-x)^{p(\alpha-1)} d\mu_2(t)\right)^{\frac{1}{p}} \left((C) \int_{x}^{b} f^{q}(t)\, d\mu_2(t)\right)^{\frac{1}{q}} \tag{5.56}$$

$$\leq \frac{1}{\Gamma(\alpha)} \left((b-a)^{p(\alpha-1)} \mu_2([a,b])\right)^{\frac{1}{p}} \left((C)\int_a^b f^q(t)\, d\mu_2(t)\right)^{\frac{1}{q}}.$$

As before we obtain

$$\left((C)\int_a^b \left(\left(I_{b-}^{\alpha} f\right)(x)\right)^p d\mu_1(x)\right)^{\frac{1}{p}} \leq$$

$$\frac{\left(\mu_1([a,b])\, \mu_2([a,b])\right)^{\frac{1}{p}} (b-a)^{(\alpha-1)}}{\Gamma(\alpha)} \left((C)\int_a^b f^q(t)\, d\mu_2(t)\right)^{\frac{1}{q}}. \tag{5.57}$$

We have proved

Theorem 5.37 *Here $\alpha > 1$ and the rest are as in this section. It holds*

$$\max\left\{\left((C)\int_a^b \left(\left(I_{a+}^{\alpha} f\right)(x)\right)^p d\mu_1(x)\right)^{\frac{1}{p}}, \left((C)\int_a^b \left(\left(I_{b-}^{\alpha} f\right)(x)\right)^p d\mu_1(x)\right)^{\frac{1}{p}}\right\}$$

$$\leq \frac{\left(\mu_1([a,b])\, \mu_2([a,b])\right)^{\frac{1}{p}} (b-a)^{(\alpha-1)}}{\Gamma(\alpha)} \left((C)\int_a^b f^q(t)\, d\mu_2(t)\right)^{\frac{1}{q}}. \tag{5.58}$$

Remark 5.38 From (5.52) we get

$$\left(I_{a+}^{\alpha} f\right)(x) \leq \frac{1}{\Gamma(\alpha)} \left((x-a)^{p(\alpha-1)} \mu_2([a,x])\right)^{\frac{1}{p}} \left((C)\int_a^b f^q(t)\, d\mu_2(t)\right)^{\frac{1}{q}}, \tag{5.59}$$

and from (5.56) we derive (by exchanging the roles of p and q)

$$\left(I_{b-}^{\alpha} f\right)(x) \leq \frac{1}{\Gamma(\alpha)} \left((b-x)^{q(\alpha-1)} \mu_2([x,b])\right)^{\frac{1}{q}} \left((C)\int_a^b f^p(t)\, d\mu_2(t)\right)^{\frac{1}{p}}. \tag{5.60}$$

Therefore by multiplying (5.59), (5.60) we get

$$\left(I_{a+}^{\alpha} f\right)(x) \left(I_{b-}^{\alpha} f\right)(x) \leq \frac{1}{(\Gamma(\alpha))^2} \left((x-a)^{p(\alpha-1)} \mu_2([a,x])\right)^{\frac{1}{p}} \cdot \tag{5.61}$$

$$\left((b-x)^{q(\alpha-1)} \mu_2([x,b])\right)^{\frac{1}{q}} \left((C)\int_a^b f^q(t)\, d\mu_2(t)\right)^{\frac{1}{q}} \left((C)\int_a^b f^p(t)\, d\mu_2(t)\right)^{\frac{1}{p}}$$

(using Young's inequality for $a, b \geq 0$, $a^{\frac{1}{p}} b^{\frac{1}{q}} \leq \frac{a}{p} + \frac{b}{q}$)

$$\leq \frac{1}{(\Gamma(\alpha))^2}\left(\frac{(x-a)^{p(\alpha-1)}\mu_2([a,x])}{p}+\frac{(b-x)^{q(\alpha-1)}\mu_2([x,b])}{q}\right)\cdot$$

$$\left((C)\int_a^b f^q(t)\,d\mu_2(t)\right)^{\frac{1}{q}}\left((C)\int_a^b f^p(t)\,d\mu_2(t)\right)^{\frac{1}{p}}. \tag{5.62}$$

We have that

$$\frac{\left(I_{a+}^{\alpha}f\right)(x)\left(I_{b-}^{\alpha}f\right)(x)}{\left[\frac{(x-a)^{p(\alpha-1)}\mu_2([a,x])}{p}+\frac{(b-x)^{q(\alpha-1)}\mu_2([x,b])}{q}\right]}\leq$$

$$\frac{1}{(\Gamma(\alpha))^2}\left((C)\int_a^b f^q(t)\,d\mu_2(t)\right)^{\frac{1}{q}}\left((C)\int_a^b f^p(t)\,d\mu_2(t)\right)^{\frac{1}{p}}. \tag{5.63}$$

Notice that the denominator of left hand side of (5.63) is never zero.

Integrating (5.63) with respect to x we obtain:

Theorem 5.39 *Here $\alpha > 1$ and the rest are as in this section. It holds*

$$(C)\int_a^b \frac{\left(I_{a+}^{\alpha}f\right)(x)\left(I_{b-}^{\alpha}f\right)(x)\,d\mu_1(x)}{\left[\frac{(x-a)^{p(\alpha-1)}\mu_2([a,x])}{p}+\frac{(b-x)^{q(\alpha-1)}\mu_2([x,b])}{q}\right]}\leq$$

$$\frac{\mu_1([a,b])}{(\Gamma(\alpha))^2}\left((C)\int_a^b f^p(t)\,d\mu_2(t)\right)^{\frac{1}{p}}\left((C)\int_a^b f^q(t)\,d\mu_2(t)\right)^{\frac{1}{q}}. \tag{5.64}$$

Inequality (5.64) is a Hilbert–Pachpatte type inequality for Choquet fractional integrals.

References

1. G. Anastassiou, *Intelligent Comparisons: Analytic Inequalities* (Springer, Heidelberg, 2016)
2. G.A. Anastassiou, Hardy type inequalities for Choquet integrals. J. Comput. Anal. Appl. (2018). Accepted
3. A. Chateauneuf, J.P. Lefort, Some Fubini theorems on product sigma-algebras for non-additive measures. Int. J. Approx. Reason **48**(3), 686–696 (2008)
4. G. Choquet, Theory of capacities. Ann. Inst. Fourier **5**, 131–295 (1953)
5. L.M. de Campos, M.J. Bolanos, Characterization and comparison of Sugeno and Choquet integrals. Fuzzy Sets Syst. **52**, 61–67 (1992)
6. D. Denneberg, *Nonadditive Measure and Integral* (Kluwer Academic, Dordrecht, 1994)
7. P. Ghirardato, On independence for non-additive measures, with a Fubini theorem. J. Econ. Theory **73**, 261–291 (1997)
8. H.G. Hardy, Notes on some points in the integral calculus. Messenger Math. **47**(10), 145–150 (1918)

9. S. Iqbal, K. Krulic, J. Pecaric, On an inequality of G. Hardy. J. Inequalities Appl. **2010**, 23 (2010). Article ID 264347
10. A.A. Kilbas, H.M. Srivastava, J.J. Trujillo, *Theory and Applications of Fractional Differential Equations*. North-Holland Mathematics Studies, vol. 204 (Elsevier, New York, 2006)
11. E. Pap, *Null-Additive Set Functions* (Kluwer Academic, Dordrecht, 1995)
12. S.G. Samko, A.A. Kilbas, O.I. Marichev, *Fractional Integral and Derivatives: Theory and Applications* (Gordon and Breach Science, Yverdon, 1993)
13. D. Schmeidler, Subjective probability and expected utility without additivity. Econometrica **57**, 571–587 (1989)
14. Z. Wang, G. Klir, *Fuzzy Measure Theory* (Plenum, New York, 1992)
15. Z. Wang, G.J. Klir, W. Wang, Monotone set functions defined by Choquet integrals. Fuzzy Sets Syst. **81**, 241–250 (1996)
16. Z. Wang, Convergence theorems for sequences of Choquet integrals. Int. J. Gen. Syst. **26**, 133–143 (1997)
17. Z. Wang, G.J. Klir, *Generalized Measure Theory* (Springer, New York, 2009)
18. R.-S. Wang, Some inequalities and convergence theorems for Choquet integrals. J. Appl. Math. Comput. **35**, 305–321 (2011)

Chapter 6
Quantitative Approximation by Choquet Integrals

Here we present the quantitative approximation of positive sublinear operators to the unit operator. These are given a precise Choquet integral interpretation. Initially we start with the study of the rate of the convergence of the well-known Bernstein–Kantorovich–Choquet and Bernstein–Durrweyer–Choquet polynomial Choquet-integral operators. Then we study the very general comonotonic positive sublinear operators based on the representation theorem of Schmeidler [9]. We finish with the approximation by the very general direct Choquet-integral form positive sublinear operators. All approximations are given via inequalities involving the modulus of continuity of the approximated function or its higher order derivative. It follows [3].

6.1 Introduction

G. Choquet [4], introduced the capacities and his integral. Initially these were applied to statistical mechanics and potential theory, and they gave rise to the study of non-additive measure theory. Slowly but steady these ideas of Choquet started to attract economists especially after the very important work of Shapley [11] in the study of cooperative games. Capacities and Choquet integrals became main stream to Decision theorists since 1989 when D. Schmeidler [10] was the first to use them in a model of choice with non-additive beliefs. The expected utility results are strengthened by the use of Choquet capacities instead of probability measures.

In now days Choquet integral has wide applications, among others, to decision making under risk and uncertainty, in finance, in economics, in portofolio problems and in insurance.

Our motivation also comes from the foundations of Bayesian decision theory and subjective probability.

G. A. Anastassiou, *Ordinary and Fractional Approximation by Non-additive Integrals: Choquet, Shilkret and Sugeno Integral Approximators*, Studies in Systems, Decision and Control 190, https://doi.org/10.1007/978-3-030-04287-5_6

Because of the paramount importance of Choquet integral, we decided to research the related positive sublinear operators approximation, part of it is exhibited in this work.

6.2 Background - I

Next we present briefly about the Choquet integral, see also [8].

We make

Definition 6.1 Consider $\Omega \neq \varnothing$ and let $\mathcal{C}$ be a σ-algebra of subsets in Ω.

(i) (see, e.g., [12], p. 63) The set function $\mu : \mathcal{C} \to [0, +\infty]$ is called a monotone set function (or capacity) if $\mu(\varnothing) = 0$ and $\mu(A) \leq \mu(B)$ for all $A, B \in \mathcal{C}$, with $A \subset B$. Also, μ is called submodular if

$$\mu(A \cup B) + \mu(A \cap B) \leq \mu(A) + \mu(B), \text{ for all } A, B \in \mathcal{C}.$$

μ is called bounded if $\mu(\Omega) < +\infty$ and normalized if $\mu(\Omega) = 1$.

(ii) (see, e.g., [12], p. 233, or [4]) If μ is a monotone set function on $\mathcal{C}$ and if $f : \Omega \to \mathbb{R}$ is $\mathcal{C}$-measurable (that is, for any Borel subset $B \subset \mathbb{R}$ it follows $f^{-1}(B) \in \mathcal{C}$), then for any $A \in \mathcal{C}$, the Choquet integral is defined by

$$(C)\int_A f d\mu = \int_0^{+\infty} \mu\left(F_\beta(f) \cap A\right) d\beta + \int_{-\infty}^0 \left[\mu\left(F_\beta(f) \cap A\right) - \mu(A)\right] d\beta,$$

where we used the notation $F_\beta(f) = \{\omega \in \Omega : f(\omega) \geq \beta\}$. Notice that if $f \geq 0$ on A, then in the above formula we get $\int_{-\infty}^0 = 0$.

The integrals on the right-hand side are the usual Riemann integral.

The function f will be called Choquet integrable on A if $(C)\int_A f d\mu \in \mathbb{R}$.

Next we list some well known properties of the Choquet integral.

Remark 6.2 If $\mu : \mathcal{C} \to [0, +\infty]$ is a monotone set function, then the following properties hold:

(i) For all $a \geq 0$ we have $(C)\int_A af d\mu = a \cdot (C)\int_A f d\mu$ (if $f \geq 0$ then see, e.g., [12], Theorem 11.2, (5), p. 228 and if f is arbitrary sign, then see, e.g., [5], p. 64, Proposition 5.1, (ii)).

(ii) For all $c \in \mathbb{R}$ and f of arbitrary sign, we have (see, e.g., [12], pp. 232-233, or [5], p. 65) $(C)\int_A (f + c) d\mu = (C)\int_A f d\mu + c \cdot \mu(A)$.

If μ is submodular too, then for all f, g of arbitrary sign and lower bounded, we have (see, e.g., [5], p. 75, Theorem 6.3)

$$(C)\int_A (f + g) d\mu \leq (C)\int_A f d\mu + (C)\int_A g d\mu.$$

(iii) If $f \leq g$ on A then $(C)\int_A f d\mu \leq (C)\int_A g d\mu$ (see, e.g., [12], p. 228, Theorem 11.2, (3) if $f, g \geq 0$ and p. 232 if f, g are of arbitrary sign).

(iv) Let $f \geq 0$. If $A \subset B$ then $(C)\int_A f d\mu \leq (C)\int_B f d\mu$. In addition, if μ is finitely subadditive, then

$$(C)\int_{A\cup B} f d\mu \leq (C)\int_A f d\mu + (C)\int_B f d\mu.$$

(v) It is immediate that $(C)\int_A 1 \cdot d\mu(t) = \mu(A)$.

(vi) The formula $\mu(A) = \gamma(M(A))$, where $\gamma : [0, 1] \to [0, 1]$ is an increasing and concave function, with $\gamma(0) = 0$, $\gamma(1) = 1$ and M is a probability measure (or only finitely additive) on a σ-algebra on Ω (that is, $M(\emptyset) = 0$, $M(\Omega) = 1$ and M is countably additive), gives simple examples of normalized, monotone and submodular set functions (see, e.g., [5], pp. 16-17, Example 2.1). Such of set functions μ are also called distorsions of countably normalized, additive measures (or distorted measures). For a simple example, we can take $\gamma(t) = \frac{2t}{1+t}$, $\gamma(t) = \sqrt{t}$.

If the above γ function is increasing, concave and satisfies only $\gamma(0) = 0$, then for any bounded Borel measure m, $\mu(A) = \gamma(m(A))$ gives a simple example of bounded, monotone and submodular set function.

(vii) If μ is a countably additive bounded measure, then the Choquet integral $(C)\int_A f d\mu$ reduces to the usual Lebesgue type integral (see, e.g., [5], p. 62, or [12], p. 226).

(viii) If $f \geq 0$, then $(C)\int_A f d\mu \geq 0$.

(ix) Let $\mu = \sqrt{M}$, where M is the Lebesgue measure on $[0, +\infty)$, then μ is a monotone and submodular set function, furthermore μ is strictly positive, see [7].

(x) If $\Omega = \mathbb{R}^N$, $N \in \mathbb{N}$, we call μ strictly positive if $\mu(A) > 0$, for any open subset $A \subseteq \mathbb{R}^N$.

We need some possibility theory:

Definition 6.3 ([6]) For the $\Omega \neq \emptyset$, the power set $\mathcal{P}(\Omega)$ denotes the family of all subsets of Ω.

(i) A function $\lambda : \Omega \to [0, 1]$ with the property $\sup\{\lambda(s) : s \in \Omega\} = 1$, is called possibility distribution on Ω.

(ii) $P : \mathcal{P}(\Omega) \to [0, 1]$ is called possibility measure, if it satisfies $P(\emptyset) = 0$, $P(\Omega) = 1$, and $P(\cup_{i \in I} A_i) = \sup\{P(A_i) : i \in I\}$ for all $A_i \subset \Omega$, and any I, an at most countable family of indices. Note that if $A, B \subset \Omega$, $A \subset B$, then the last property implies $P(A) \leq P(B)$ and that $P(A \cup B) \leq P(A) + P(B)$.

Any possibility distribution λ on Ω, induces the possibility measure $P_\lambda : \mathcal{P}(\Omega) \to [0, 1]$, $P_\lambda(A) = \sup\{\lambda(s) : s \in A\}$, $A \subset \Omega$. Also, if $f : \Omega \to \mathbb{R}_+$, then the possibilistic integral of f on $A \subset \Omega$ with respect to P_λ is defined by $(Pos)\int_A f dP_\lambda = \sup\{f(t)\lambda(t) : t \in A\}$ (see [6], chapter 1).

Note that any possibility measure μ is normalized, monotone and submodular. From $\mu(A \cup B) = \max\{\mu(A), \mu(B)\}$ we get monotonicity, and from $\mu(A \cap B) \leq \min\{\mu(A), \mu(B)\}$ we derive the submodularity.

6.3 Background - II

We need

Definition 6.4 ([1, 2]) Let $I \subset \mathbb{R}$ be a bounded or unbounded interval, $n \in \mathbb{N}$, and

$$CB_{+}^{n}(I) = \left\{ f : I \to \mathbb{R}_{+} : f^{(i)} \text{ is continuous and bounded on } I, \text{ for both } i = 0, n \right\}. \tag{6.1}$$

We define for

$$f \in CB_{+}(I) = \{ f : I \to \mathbb{R}_{+} : f \text{ is continuous and bounded on } I \}, \tag{6.2}$$

the first modulus of continuity

$$\omega_1(f, \delta) = \sup_{\substack{x, y \in I: \\ |x-y| \leq \delta}} |f(x) - f(y)|, \tag{6.3}$$

where $0 < \delta \leq \text{diameter}(I)$.

Call $C_{+}(I) = \{ f : I \to \mathbb{R}_{+} : f \text{ is continuous on } I \}$.

Let $L_N : C_{+}(I) \to CB_{+}(I)$, $n, N \in \mathbb{N}$ be a sequence of operators satisfying the following properties:

(i) (positive homogeneous)

$$L_N(\alpha f) = \alpha L_N(f), \forall \alpha \geq 0, f \in C_{+}(I), \tag{6.4}$$

(ii) (Monotonicity)

$$\text{if } f, g \in C_{+}(I) \text{ satisfy } f \leq g, \text{ then } L_N(f) \leq L_N(g), \ \forall N \in \mathbb{N}, \tag{6.5}$$

and

(iii) (Subadditivity)

$$L_N(f + g) \leq L_N(f) + L_N(g), \ \forall f, g \in C_{+}(I). \tag{6.6}$$

We call L_N positive sublinear operators.

In particular we will study the restrictions $L_N|_{CB_{+}^{n}(I)} : CB_{+}^{n}(I) \to CB_{+}(I)$.

We will be using the following results:

Theorem 6.5 ([2]) *Let $(L_N)_{N \in \mathbb{N}}$ be a sequence of positive sublinear operators from $C_{+}(I)$ into $CB_{+}(I)$, and $f \in CB_{+}^{n}(I)$, where $n \in \mathbb{N}$ and $I \subset \mathbb{R}$ a bounded or unbounded interval. Assume $L_N(1) = 1$, $\forall N \in \mathbb{N}$, and $f^{(i)}(x) = 0$, $i = 1, \ldots, n$, for a fixed $x \in I$, and $\delta > 0$. Then*

$$|L_N(f)(x) - f(x)| \leq \frac{\omega_1\left(f^{(n)}, \delta\right)}{n!}\left[L_N\left(|\cdot - x|^n\right)(x) + \frac{L_N\left(|\cdot - x|^{n+1}\right)(x)}{(n+1)\delta}\right], \tag{6.7}$$

$\forall\, N \in \mathbb{N}$.

Theorem 6.6 ([1]) *Let $(L_N)_{N\in\mathbb{N}}$ be a sequence of positive sublinear operators from $C_+(I)$ into $CB_+(I)$, and $f \in CB_+^n(I)$, where $n \in \mathbb{N}$ and $I \subset \mathbb{R}$ a bounded or unbounded interval. Assume $L_N(1) = 1$, $\forall\, N \in \mathbb{N}$, and $f^{(i)}(x) = 0$, $i = 1, \ldots, n$, for a fixed $x \in I$. Furthermore assume that $L_N\left(|\cdot - x|^{n+1}\right)(x) > 0$, $\forall\, N \in \mathbb{N}$.*

Then

$$|L_N(f)(x) - f(x)| \leq \frac{\omega_1\left(f^{(n)}, \left(L_N\left(|\cdot - x|^{n+1}\right)(x)\right)^{\frac{1}{n+1}}\right)}{n!}\cdot$$

$$\left[L_N\left(|\cdot - x|^n\right)(x) + \frac{\left(L_N\left(|\cdot - x|^{n+1}\right)(x)\right)^{\frac{n}{n+1}}}{(n+1)}\right], \ \forall\, N \in \mathbb{N}. \tag{6.8}$$

We give ($n = 1$ case)

Corollary 6.7 ([1]) *Let $(L_N)_{N\in\mathbb{N}}$ be a sequence of positive sublinear operators from $C_+(I)$ into $CB_+(I)$, and $f \in CB_+^1(I)$, and $I \subset \mathbb{R}$ a bounded or unbounded interval. Assume $L_N(1) = 1$, $\forall\, N \in \mathbb{N}$, and $f'(x) = 0$, for a fixed $x \in I$. Furthermore assume that $L_N\left((\cdot - x)^2\right)(x) > 0$, $\forall\, N \in \mathbb{N}$.*

Then

$$|L_N(f)(x) - f(x)| \leq \omega_1\left(f', \sqrt{\left(L_N\left((\cdot - x)^2\right)(x)\right)}\right)\cdot$$

$$\left[L_N\left(|\cdot - x|\right)(x) + \frac{\sqrt{\left(L_N\left((\cdot - x)^2\right)(x)\right)}}{2}\right], \ \forall\, N \in \mathbb{N}. \tag{6.9}$$

Remark 6.8 ([1]) (i) To Theorem 6.6: Assuming $f^{(n)}$ is uniformly continuous on I, and $L_N\left(|\cdot - x|^{n+1}\right)(x) \to 0$, by [1], we get that $(L_N(f))(x) \to f(x)$, as $N \to \infty$.

(ii) to Corollary 6.7: Assuming f' is uniformly continuous on I, and $L_N\left((\cdot - x)^2\right)(x) \to 0$, by [1], we get that $(L_N(f))(x) \to f(x)$, as $N \to \infty$.

(iii) The right hand sides of (6.8), (6.9) are finite.

We also give the basic result ($n = 0$ case).

Theorem 6.9 ([1]) *Let $(L_N)_{N\in\mathbb{N}}$ be a sequence of positive sublinear operators from $C_+(I)$ into $CB_+(I)$, and $f \in CB_+(I)$, where $I \subset \mathbb{R}$ a bounded or unbounded interval. Assume that $L_N(|\cdot - x|)(x) > 0$, for some fixed $x \in I$, $\forall\, N \in \mathbb{N}$. Then*

(1)

$$|L_N(f)(x) - f(x)| \leq f(x)\,|L_N(1)(x) - 1| +$$

$$[L_N(1)(x)+1]\,\omega_1(f, L_N(|\cdot - x|)(x)), \ \forall N \in \mathbb{N}, \tag{6.10}$$

(2) when $L_N(1) = 1$, *we get*

$$|L_N(f)(x) - f(x)| \leq 2\omega_1(f, L_N(|\cdot - x|)(x)), \ \forall N \in \mathbb{N}. \tag{6.11}$$

Remark 6.10 ([1]) (To Theorem 6.9) Here $x \in I$ is fixed.

(i) Assume $L_N(1)(x) \to 1$, as $N \to \infty$, and $L_N(|\cdot - x|)(x) \to 0$, as $N \to \infty$, given that f is uniformly continuous we get that $L_n(f)(x) \to f(x)$, as $N \to \infty$ (use of (6.10)). Notice here that $L_N(1)(x)$ is bounded.

(ii) Assume that $L_N(1) = 1$, and $L_N(|\cdot - x|)(x) \to 0$, as $N \to \infty$, and f is uniformly continuous on I, then $L_n(f)(x) \to f(x)$, as $N \to \infty$ (use of (6.11)).

(iii) The right hand sides of (6.10) and (6.11) are finite.

We mention

Definition 6.11 ([7]) Let $I = [0,1]$, $\mathcal{B}_I$ the σ-algebra of all Borel measurable subsets of I, $(\Gamma_{N,x})_{N\in\mathbb{N}, x\in I}$ will be the collection of the family $\Gamma_{N,x} = \{\mu_{N,k,x}\}_{k=0}^N$, of monotone, submodular and strictly positive set functions $\mu_{N,k,x}$ on $\mathcal{B}_I$.

Let $f : [0,1] \to \mathbb{R}_+$ be a B_I-measurable function which is bounded, and call $p_{N,k}(x) = \binom{N}{k} x^k (1-x)^{N-k}$, for any $x \in [0,1]$.

The Bernstein–Kantorovich–Choquet operators are defined by the formula

$$K_{N,\Gamma_{N,x}}(f)(x) = \sum_{k=0}^N p_{N,k}(x) \frac{(C)\int_{\frac{k}{(N+1)}}^{\frac{(k+1)}{(N+1)}} f(t)\, d\mu_{N,k,x}(t)}{\mu_{N,k,x}\left(\left[\frac{k}{(N+1)}, \frac{(k+1)}{(N+1)}\right]\right)}, \ \forall x \in [0,1]. \tag{6.12}$$

If $\mu_{N,k,x} = \mu$, for all N, x, k, we will denote $K_{N,\Gamma_{N,x}}(f) := K_{N,\mu}(f)$.

Theorem 6.12 ([7]) *Suppose that* $\mu_{N,k,x} = \mu := \sqrt{M}$, *for all* N, k *and* x, *where* M *is the Lebesgue measure on* $[0,1]$. *Then*

$$\left|K_{N,\mu}(f)(x) - f(x)\right| \leq 2\omega_1\left(f, \frac{\sqrt{x(1-x)}}{\sqrt{N}} + \frac{1}{N}\right), \tag{6.13}$$

$\forall N \in \mathbb{N}$, $x \in [0,1]$, $f \in C_+([0,1])$, *above* ω_1 *is over* $[0,1]$.

Remark 6.13 By [7] we have that

$$K_{N,\mu}(|\cdot - x|)(x) \leq \frac{\sqrt{x(1-x)}}{\sqrt{N}} + \frac{1}{N}, \ \forall N \in \mathbb{N}. \tag{6.14}$$

Let $m \in \mathbb{N} - \{1\}$, notice that $|\cdot - x|^{m-1} \leq 1$, therefore

$$|\cdot - x|^m = |\cdot - x|\,|\cdot - x|^{m-1} \leq |\cdot - x|,$$

hence

$$K_{N,\mu}\left(|\cdot - x|^m\right)(x) \leq K_{N,\mu}\left(|\cdot - x|\right)(x),$$

that is

$$K_{N,\mu}\left(|\cdot - x|^m\right)(x) \leq \frac{\sqrt{x(1-x)}}{\sqrt{N}} + \frac{1}{N}, \ \forall x \in [0,1], \ N, m \in \mathbb{N}. \tag{6.15}$$

Notice that $K_{N,\mu}(1) = 1, \forall N \in \mathbb{N}$.

Clearly $K_{N,\mu}$ operators are positive sublinear operators from $C_+([0,1])$ into itself.

We mention

Definition 6.14 ([8]) Here we consider measures of possibility. Denoting $p_{N,k}(x) = \binom{N}{k} x^k (1-x)^{N-k}$, let us defined

$$\lambda_{N,k}(t) := \frac{p_{N,k}(t)}{k^k N^{-N} (N-k)^{N-k} \binom{N}{k}} = \frac{t^k (1-t)^{N-k}}{k^k N^{-N} (N-k)^{N-k}}, \ k = 0, ..., N. \tag{6.16}$$

By convention we assume that $0^0 = 1$, so that the cases $k = 0$, and $k = N$ make sense. By considering the root $\frac{k}{N}$ of $p'_{N,k}(x)$, it is clear that

$$\max\{p_{N,k}(t) : t \in [0,1]\} = k^k N^{-N} (N-k)^{N-k} \binom{N}{k},$$

which implies that each $\lambda_{N,k}$ is a possibility distribution on $[0,1]$.

Denoting by $P_{\lambda_{N,k}}$ the possibility measure induced by $\lambda_{N,k}$ and $\Gamma_{n,x} := \Gamma_N := \{P_{\lambda_{N,k}}\}_{k=0}^N$ (that is Γ_N is independent of x), we define the nonlinear Bernstein–Durrmeyer–Choquet polynomial operators with respect to the set functions in Γ_N given by the formula

$$D_{N,\Gamma_N}(f)(x) := \sum_{k=0}^{N} p_{N,k}(x) \frac{(C)\int_0^1 f(t)\, t^k (1-t)^{N-k} dP_{\lambda_{N,k}}(t)}{(C)\int_0^1 t^k (1-t)^{N-k} dP_{\lambda_{N,k}}(t)}, \tag{6.17}$$

$\forall x \in [0,1], N \in \mathbb{N}, f \in C_+([0,1])$.

Remark 6.15 Above $P_{\lambda_{N,k}}$ is bounded, monotone, submodular and strictly positive, $N \in \mathbb{N}, k = 0, 1, ..., N$. Notice that $D_{N,\Gamma_N}(1) = 1, \forall N \in \mathbb{N}$.

Clearly D_{N,Γ_N} operators are positive sublinear operators mapping $C_+([0,1])$ into itself.

We mention

Theorem 6.16 ([8]) *For every* $f \in C_+([0,1])$, $x \in [0,1]$ *and* $N \in \mathbb{N} - \{1\}$, *we have*

$$\left|D_{N,\Gamma_N}(f)(x) - f(x)\right| \leq 2\omega_1\left(f, \frac{\left(1+\sqrt{2}\right)\sqrt{x(1-x)} + \sqrt{2}\sqrt{x}}{\sqrt{N}} + \frac{1}{N}\right), \tag{6.18}$$

where ω_1 *is on* $[0,1]$.

Remark 6.17 By [8] we have that

$$D_{N,\Gamma_N}(|\cdot - x|)(x) \leq \frac{\left(1+\sqrt{2}\right)\sqrt{x(1-x)} + \sqrt{2}\sqrt{x}}{\sqrt{N}} + \frac{1}{N}, \ \forall\, N \in \mathbb{N} - \{1\}. \tag{6.19}$$

Let $m \in \mathbb{N} - \{1\}$, notice that $|\cdot - x|^{m-1} \leq 1$, therefore

$$|\cdot - x|^m = |\cdot - x|\,|\cdot - x|^{m-1} \leq |\cdot - x|,$$

hence

$$D_{N,\Gamma_N}\left(|\cdot - x|^m\right)(x) \leq D_{N,\Gamma_N}(|\cdot - x|)(x),$$

that is

$$D_{N,\Gamma_N}\left(|\cdot - x|^m\right)(x) \leq \frac{\left(1+\sqrt{2}\right)\sqrt{x(1-x)} + \sqrt{2}\sqrt{x}}{\sqrt{N}} + \frac{1}{N}, \tag{6.20}$$

$\forall\, N \in \mathbb{N} - \{1\}, \forall\, m \in \mathbb{N}, \forall\, x \in [0,1]$.

We make

Remark 6.18 When $x \in [0,1]$, then the $\max(x(1-x)) = \frac{1}{4}$, at $x = \frac{1}{2}$. Therefore it holds

$$\frac{\sqrt{x(1-x)}}{\sqrt{N}} + \frac{1}{N} \leq \frac{1}{2\sqrt{N}} + \frac{1}{N}, \tag{6.21}$$

$\forall\, x \in [0,1], \forall\, N \in \mathbb{N}$.

Similarly, it holds

$$\frac{\left(1+\sqrt{2}\right)\sqrt{x(1-x)} + \sqrt{2}\sqrt{x}}{\sqrt{N}} + \frac{1}{N} \leq \frac{1+3\sqrt{2}}{2\sqrt{N}} + \frac{1}{N}, \tag{6.22}$$

$\forall\, x \in [0,1], \forall\, N \in \mathbb{N} - \{1\}$.

Corollary 6.19 (To Theorem 6.12) *It holds*

$$\left\| K_{N,\mu}(f) - f \right\|_{\infty} \leq 2\omega_1 \left(f, \frac{1}{2\sqrt{N}} + \frac{1}{N} \right), \tag{6.23}$$

$\forall\, N \in \mathbb{N},\ f \in C_+([0,1])$.

Corollary 6.20 (To Theorem 6.16) *It holds*

$$\left\| D_{N,\Gamma_N}(f) - f \right\|_{\infty} \leq 2\omega_1 \left(f, \frac{1+3\sqrt{2}}{2\sqrt{N}} + \frac{1}{N} \right), \tag{6.24}$$

$\forall\, N \in \mathbb{N} - \{1\},\ f \in C_+([0,1])$.

6.4 Main Results

Here we start by applying Theorem 6.5 to the Bernstein–Kantorovich–Choquet operators $K_{N,\mu}$, where $\mu := \sqrt{M}$, with M the Lebesgue measure on $[0,1]$. More precisely here it is

$$K_{N,\mu}(f)(x) = \sum_{k=0}^{N} p_{N,k}(x) \frac{(C)\int_{\frac{k}{(N+1)}}^{\frac{(k+1)}{(N+1)}} f(t)\, d\mu(t)}{\mu\left(\left[\frac{k}{(N+1)}, \frac{(k+1)}{(N+1)}\right]\right)}, \tag{6.25}$$

$\forall\, x \in [0,1],\ \forall\, N \in \mathbb{N},\ f \in C_+([0,1])$.

Next, we also apply Theorem 6.5 to Bernstein–Durrmeyer–Choquet operators D_{N,Γ_N}, see (6.17).

Denote by $C_+^n([a,b]) = \{f : [a,b] \to \mathbb{R}_+ \text{ and } f \in C^n([a,b]), n \in \mathbb{N}\}$.

We present

Theorem 6.21 *Let $f \in C_+^n([0,1])$, $n \in \mathbb{N}$, such that $f^{(i)}(x) = 0$, $i = 1, ..., n$, for a fixed $x \in [0,1]$. Then*

$$\left| \left(K_{N,\mu}(f) \right)(x) - f(x) \right| \leq \frac{\omega_1 \left(f^{(n)}, \left(\sqrt{\frac{x(1-x)}{N}} + \frac{1}{N} \right)^{\frac{1}{n+1}} \right)}{n!} \tag{6.26}$$

$$\left[\left(\sqrt{\frac{x(1-x)}{N}} + \frac{1}{N} \right) + \frac{1}{(n+1)} \left[\sqrt{\frac{x(1-x)}{N}} + \frac{1}{N} \right]^{\frac{n}{n+1}} \right] \leq$$

$$\frac{\omega_1\left(f^{(n)},\left(\frac{1}{2\sqrt{N}}+\frac{1}{N}\right)^{\frac{1}{n+1}}\right)}{n!}\left[\left(\frac{1}{2\sqrt{N}}+\frac{1}{N}\right)+\frac{1}{(n+1)}\left[\frac{1}{2\sqrt{N}}+\frac{1}{N}\right]^{\frac{n}{n+1}}\right], \tag{6.27}$$

$\forall\, N \in \mathbb{N}$.

We get that $\lim_{N\to+\infty} K_{N,\mu}(f)(x) = f(x)$.

Proof By Theorem 6.5 and inequality (6.7) we have ($\delta > 0$):

$$\left|K_{N,\mu}(f)(x) - f(x)\right| \leq$$

$$\frac{\omega_1\left(f^{(n)},\delta\right)}{n!}\left[K_{N,\mu}\left(|\cdot - x|^n\right)(x) + \frac{K_{N,\mu}\left(|\cdot - x|^{n+1}\right)(x)}{(n+1)\,\delta}\right] \overset{(6.15)}{\leq}$$

$$\frac{\omega_1\left(f^{(n)},\delta\right)}{n!}\left[\left(\sqrt{\frac{x(1-x)}{N}}+\frac{1}{N}\right)+\frac{\left(\sqrt{\frac{x(1-x)}{N}}+\frac{1}{N}\right)}{(n+1)\,\delta}\right]$$

(choose $\delta := \left(\sqrt{\frac{x(1-x)}{N}}+\frac{1}{N}\right)^{\frac{1}{n+1}} > 0$, then $\delta^{n+1} = \sqrt{\frac{x(1-x)}{N}}+\frac{1}{N}$, and $\delta^n = \left(\sqrt{\frac{x(1-x)}{N}}+\frac{1}{N}\right)^{\frac{n}{n+1}}$)

$$= \frac{\omega_1\left(f^{(n)},\left(\sqrt{\frac{x(1-x)}{N}}+\frac{1}{N}\right)^{\frac{1}{n+1}}\right)}{n!}\cdot \tag{6.28}$$

$$\left[\left(\sqrt{\frac{x(1-x)}{N}}+\frac{1}{N}\right)+\frac{1}{(n+1)}\left[\sqrt{\frac{x(1-x)}{N}}+\frac{1}{N}\right]^{\frac{n}{n+1}}\right],$$

along with (6.21), we have proved inequalities (6.26), (6.27). ■

We also give

Theorem 6.22 *Let* $f \in C_+^n([0,1])$, $n \in \mathbb{N}$, *such that* $f^{(i)}(x) = 0$, $i = 1, ..., n$, *for a fixed* $x \in [0,1]$. *Then*

$$\left|\left(D_{N,\Gamma_N}(f)\right)(x) - f(x)\right| \leq \frac{\omega_1\left(f^{(n)},\left(\frac{(1+\sqrt{2})\sqrt{x(1-x)}+\sqrt{2}\sqrt{x}}{\sqrt{N}}+\frac{1}{N}\right)^{\frac{1}{n+1}}\right)}{n!} \tag{6.29}$$

$$\left[\left(\frac{\left(1+\sqrt{2}\right)\sqrt{x\left(1-x\right)}+\sqrt{2}\sqrt{x}}{\sqrt{N}}+\frac{1}{N}\right)+\right.$$

$$\left.\frac{1}{(n+1)}\left[\frac{\left(1+\sqrt{2}\right)\sqrt{x\left(1-x\right)}+\sqrt{2}\sqrt{x}}{\sqrt{N}}+\frac{1}{N}\right]^{\frac{n}{n+1}}\right]\leq$$

$$\frac{\omega_1\left(f^{(n)},\left(\frac{1+3\sqrt{2}}{2\sqrt{N}}+\frac{1}{N}\right)^{\frac{1}{n+1}}\right)}{n!}\cdot$$

$$\left[\left(\frac{1+3\sqrt{2}}{2\sqrt{N}}+\frac{1}{N}\right)+\frac{1}{(n+1)}\left[\frac{1+3\sqrt{2}}{2\sqrt{N}}+\frac{1}{N}\right]^{\frac{n}{n+1}}\right], \tag{6.30}$$

$\forall\, N\in\mathbb{N}-\{1\}$.

We notice that $\lim_{N\to+\infty} D_{N,\Gamma_N}(f)(x)=f(x)$.

Proof By Theorem 6.5 and inequality (6.7) we get ($\delta>0$):

$$\left|D_{N,\Gamma_N}(f)(x)-f(x)\right|\leq$$

$$\frac{\omega_1\left(f^{(n)},\delta\right)}{n!}\left[D_{N,\Gamma_N}\left(\left|\cdot-x\right|^n\right)(x)+\frac{D_{N,\Gamma_N}\left(\left|\cdot-x\right|^{n+1}\right)(x)}{(n+1)\,\delta}\right]\overset{(6.20)}{\leq}$$

$$\frac{\omega_1\left(f^{(n)},\delta\right)}{n!}\left[\left(\frac{\left(1+\sqrt{2}\right)\sqrt{x\left(1-x\right)}+\sqrt{2}\sqrt{x}}{\sqrt{N}}+\frac{1}{N}\right)+\right. \tag{6.31}$$

$$\left.\frac{1}{(n+1)\,\delta}\left(\frac{\left(1+\sqrt{2}\right)\sqrt{x\left(1-x\right)}+\sqrt{2}\sqrt{x}}{\sqrt{N}}+\frac{1}{N}\right)\right]$$

(choose $\delta:=\left(\frac{(1+\sqrt{2})\sqrt{x(1-x)}+\sqrt{2}\sqrt{x}}{\sqrt{N}}+\frac{1}{N}\right)^{\frac{1}{n+1}}>0$, then
$\delta^{n+1}=\frac{(1+\sqrt{2})\sqrt{x(1-x)}+\sqrt{2}\sqrt{x}}{\sqrt{N}}+\frac{1}{N}$, and $\delta^{n}=\left(\frac{(1+\sqrt{2})\sqrt{x(1-x)}+\sqrt{2}\sqrt{x}}{\sqrt{N}}+\frac{1}{N}\right)^{\frac{n}{n+1}}$)

$$=\frac{\omega_1\left(f^{(n)},\left(\frac{(1+\sqrt{2})\sqrt{x(1-x)}+\sqrt{2}\sqrt{x}}{\sqrt{N}}+\frac{1}{N}\right)^{\frac{1}{n+1}}\right)}{n!}\cdot \tag{6.32}$$

$$\left[\left(\frac{\left(1+\sqrt{2}\right)\sqrt{x\left(1-x\right)}+\sqrt{2}\sqrt{x}}{\sqrt{N}} + \frac{1}{N} \right) + \right.$$

$$\left. \frac{1}{(n+1)} \left[\frac{\left(1+\sqrt{2}\right)\sqrt{x\left(1-x\right)}+\sqrt{2}\sqrt{x}}{\sqrt{N}} + \frac{1}{N} \right]^{\frac{n}{n+1}} \right],$$

along with (6.22), proving the claims. ■

We need

Definition 6.23 Let Ω be a set, and let $f, g : \Omega \to \mathbb{R}$ be bounded functions. We say that f and g are comonotonic, if for every $\omega, \omega' \in \Omega$,

$$\left(f\left(\omega\right) - f\left(\omega'\right)\right)\left(g\left(\omega\right) - g\left(\omega'\right)\right) \geq 0. \tag{6.33}$$

We also need the famous Schmeidler's Representation Theorem (Schmeidler 1986)

Theorem 6.24 ([9]) *Denote with $\mathcal{L}_\infty(\mathcal{A})$ the vector space of $\mathcal{A}$-measurable bounded real valued functions on Ω, where $\mathcal{A} \subset 2^\Omega$ is a σ -algebra. Given a real functional $\Gamma : \mathcal{L}_\infty(\mathcal{A}) \to \mathbb{R}$, assume that for $f, g \in \mathcal{L}_\infty(\mathcal{A})$:*
(i) $\Gamma(cf) = c\Gamma(f)$, $\forall\, c > 0$,
(ii) $f \leq g$, implies $\Gamma(f) \leq \Gamma(g)$,
and
(iii) $\Gamma(f+g) = \Gamma(f) + \Gamma(g)$, for any comonotonic f, g.
Then $\gamma(A) := \Gamma(1_A)$, $\forall\, A \in \mathcal{A}$, defines a finite monotone set function on $\mathcal{A}$, and Γ is the Choquet integral with respect to γ, i.e.

$$\Gamma(f) = (C)\int_\Omega f(t)\, d\gamma(t), \;\; \forall\, f \in \mathcal{L}_\infty(\mathcal{A}). \tag{6.34}$$

Above 1_A denotes the characteristic function on A.

Next we give a nice interpretation of Theorems 6.5 and 6.6, Corollary 6.7 and Theorem 6.9(2) involving Choquet integrals and based on Theorem 6.24.

We make

Remark 6.25 Consider here $[a, b] \subset \mathbb{R}$, $\mathcal{B} = \mathcal{B}([a, b])$ is the Borel σ -algebra on $[a, b]$, and $\mathcal{L}_\infty(\mathcal{B})$ is the vector space of $\mathcal{B}$-measurable bounded real valued functions on $[a, b]$. Let $(L_N)_{N\in\mathbb{N}}$ be a sequence of positive sublinear operators from $\mathcal{L}_\infty(\mathcal{B})$ into $C_+([a, b])$, and $x \in [a, b]$. That is here L_N fulfills the positive homogeneity, monotonicity and subadditivity properties, see (6.4)–(6.6).

Assume $L_N(1) = 1$, $\forall$ $N \in \mathbb{N}$. Clearly here $\mathcal{L}_\infty(\mathcal{B}) \supset C_+([a, b]) = CB_+([a, b])$. In particular we treat $L_N|_{C_+([a,b])}$, just denoted for simplicity by L_N, $\forall\, N \in \mathbb{N}$.

It is clear that $L_N(\cdot)(x) : \mathcal{L}_\infty(\mathcal{B}) \to \mathbb{R}$ is a functional, $\forall\, N \in \mathbb{N}$. It has the properties:

(i)

$$L_N(cf)(x) = cL_N(f)(x), \ \forall\, c > 0, \ \forall\, f \in \mathcal{L}_\infty(\mathcal{B}), \tag{6.35}$$

(ii)

$$f \le g, \text{ implies } L_N(f)(x) \le L_N(g)(x), \quad \text{where } f, g \in \mathcal{L}_\infty(\mathcal{B}), \tag{6.36}$$

and

(iii)

$$L_N(f+g)(x) \le L_N(f)(x) + L_N(g)(x), \ \forall\, f, g \in \mathcal{L}_\infty(\mathcal{B}). \tag{6.37}$$

For comonotonic $f, g \in \mathcal{L}_\infty(\mathcal{B})$, we further assume that

$$L_N(f+g)(x) = L_N(f)(x) + L_N(g)(x). \tag{6.38}$$

By Theorem 6.24 we get that:

$$\gamma_{N,x}(A) := L_N(1_A)(x), \forall\, A \in \mathcal{B}, \forall\, N \in \mathbb{N},$$

defines a finite monotone set function on $\mathcal{B}$, and

$$L_N(f)(x) = (C)\int_a^b f(t)\, d\gamma_{N,x}(t), \tag{6.39}$$

$\forall\, f \in \mathcal{L}_\infty(\mathcal{B})$, $\forall\, N \in \mathbb{N}$.

In particular (6.39) is valid for any $f \in C_+([a,b])$. Furthermore $\gamma_{N,x}$ is normalized, that is $\gamma_{N,x}([a,b]) = 1$, $\forall\, N \in \mathbb{N}$.

Remind that

$$C_+^n([a,b]) = \{f : [a,b] \to \mathbb{R}_+, \ f \in C^n([a,b])\}, n \in \mathbb{N}.$$

We give

Theorem 6.26 *Let $(L_N)_{N\in\mathbb{N}}$ be a sequence of positive sublinear operators from $\mathcal{L}_\infty(\mathcal{B})$ into $C_+([a,b])$, and $f \in C_+^n([a,b])$, $n \in \mathbb{N}$. Assume $L_N(1) = 1$, $\forall\, N \in \mathbb{N}$, and $f^{(i)}(x) = 0$, $i = 1, \ldots, n$, for a fixed $x \in [a,b]$, and $\delta > 0$. Also assume that $L_N(\cdot)(x)$ is comonotonic (see (6.38)), $\forall\, N \in \mathbb{N}$. Then*

$$|L_N(f)(x) - f(x)| \le \frac{\omega_1\left(f^{(n)}, \delta\right)}{n!}.$$

$$\left[(C)\int_a^b |t-x|^n \, d\gamma_{N,x}(t) + \frac{1}{(n+1)\,\delta}(C)\int_a^b |t-x|^{n+1} \, d\gamma_{N,x}(t)\right], \tag{6.40}$$

$\forall\, N \in \mathbb{N}$.

Theorem 6.27 *Same assumptions as in Theorem 6.26. Furthermore assume that* $(C)\int_a^b |t-x|^{n+1} \, d\gamma_{N,x}(t) > 0$, $\forall\, N \in \mathbb{N}$. *Then*

$$|L_N(f)(x) - f(x)| \le \frac{\omega_1\left(f^{(n)}, \left((C)\int_a^b |t-x|^{n+1} \, d\gamma_{N,x}(t)\right)^{\frac{1}{n+1}}\right)}{n!} \cdot$$

$$\left[(C)\int_a^b |t-x|^n \, d\gamma_{N,x}(t) + \frac{1}{(n+1)}\left((C)\int_a^b |t-x|^{n+1} \, d\gamma_{N,x}(t)\right)^{\frac{n}{n+1}}\right], \tag{6.41}$$

$\forall\, N \in \mathbb{N}$.

We give ($n = 1$ case)

Corollary 6.28 *Let* $(L_N)_{N\in\mathbb{N}}$ *be a sequence of positive sublinear operators from* $\mathcal{L}_\infty(\mathcal{B})$ *into* $C_+([a,b])$, *and* $f \in C_+^1([a,b])$. *Assume* $L_N(1) = 1$, $\forall\, N \in \mathbb{N}$, *and* $f'(x) = 0$, *for a fixed* $x \in [a,b]$. *Also assume that* $L_N(\cdot)(x)$ *is comonotonic and that* $(C)\int_a^b (t-x)^2 \, d\gamma_{N,x}(t) > 0$, $\forall\, N \in \mathbb{N}$. *Then*

$$|L_N(f)(x) - f(x)| \le \omega_1\left(f', \left((C)\int_a^b (t-x)^2 \, d\gamma_{N,x}(t)\right)^{\frac{1}{2}}\right) \cdot$$

$$\left[(C)\int_a^b |t-x| \, d\gamma_{N,x}(t) + \frac{1}{2}\left((C)\int_a^b (t-x)^2 \, d\gamma_{N,x}(t)\right)^{\frac{1}{2}}\right], \tag{6.42}$$

$\forall\, N \in \mathbb{N}$.

Remark 6.29 (i) To Theorem 6.27: Assuming $(C)\int_a^b |t-x|^{n+1} \, d\gamma_{N,x}(t) \to 0$, we get that $L_N(f)(x) \to f(x)$, as $N \to \infty$.

(ii) To Corollary 6.28: Assuming that $(C)\int_a^b (t-x)^2 \, d\gamma_{N,x}(t) \to 0$, then $L_N(f)(x) \to f(x)$, as $N \to \infty$.

(iii) The right hand sides of (6.41), (6.42) are finite.

We also cover the case $n = 0$, next.

Theorem 6.30 *Let* $(L_N)_{N\in\mathbb{N}}$ *be a sequence of positive sublinear operators from* $\mathcal{L}_\infty(\mathcal{B})$ *into* $C_+([a,b])$, *and* $f \in C_+([a,b])$, *along with* $L_N(1) = 1$, $\forall\, N \in \mathbb{N}$; *and fix* $x \in [a,b]$. *Assume that* $L_N(\cdot)(x)$ *is comonotonic, and* $(C)\int_a^b |t-x| \, d\gamma_{N,x}(t) > 0$, $\forall\, N \in \mathbb{N}$. *Then*

$$|L_N(f)(x) - f(x)| \leq 2\omega_1 \left(f, \left((C) \int_a^b |t - x| \, d\gamma_{N,x}(t) \right) \right), \tag{6.43}$$

$\forall\, N \in \mathbb{N}$.

If $(C) \int_a^b |t - x| \, d\gamma_{N,x}(t) \to 0$, *then* $L_N(f)(x) \to f(x)$, *as* $N \to \infty$.

The right hand side of (6.43) is finite.

We make

Remark 6.31 Consider again $[a, b] \subset \mathbb{R}$, $\mathcal{B} = \mathcal{B}([a, b])$ the Borel σ -algebra on $[a, b]$. For each $N \in \mathbb{N}$ and each $x \in [a, b]$ consider the monotone set functions $\mu_{N,x}; \mathcal{B} \to \mathbb{R}_+$. We assume that all $\mu_{N,x}$ are normalized, that is $\mu_{N,x}([a, b]) = 1$, and submodular. Here we consider the operators $T_N : C_+([a, b]) \to C_+([a, b])$ given by the formula

$$T_N(f)(x) = (C) \int_a^b f(t) \, d\mu_{N,x}(t), \tag{6.44}$$

$\forall\, N \in \mathbb{N}, \forall\, x \in [a, b]$.

Infact here $\mu_{N,x}$ are chosen so that $T_N(C_+([a, b])) \subseteq C_+([a, b])$.

We notice here that hold:

(i)

$$T_N(\alpha f)(x) = \alpha T_N(f)(x), \; \forall\, \alpha \geq 0, \tag{6.45}$$

(ii)

$$f \leq g, \text{ implies } T_N(f)(x) \leq T_N(g)(x), \tag{6.46}$$

and

(iii)

$$T_N(f + g)(x) \leq T_N(f)(x) + T_N(g)(x), \tag{6.47}$$

$\forall\, N \in \mathbb{N}, \forall\, x \in [a, b], \forall\, f, g \in C_+([a, b])$.

Clearly T_N are positive sublinear operators, compare to (6.4)–(6.6). We also have that $T_N(1) = 1$, $\forall\, N \in \mathbb{N}$.

Next we apply Theorems 6.5–6.6, Corollary 6.7 and Theorem 6.9(2) to T_N operators.

Theorem 6.32 *Let* $f \in C_+^n([a, b])$, $n \in \mathbb{N}$. *Assume* $f^{(i)}(x) = 0$, $i = 1, \ldots, n$, *for a fixed* $x \in [a, b]$, *and* $\delta > 0$. *Then*

$$|T_N(f)(x) - f(x)| \leq \frac{\omega_1\left(f^{(n)}, \delta\right)}{n!} \cdot$$

$$\left[(C) \int_a^b |t - x|^n \, d\mu_{N,x}(t) + \frac{1}{(n+1)\delta} (C) \int_a^b |t - x|^{n+1} \, d\mu_{N,x}(t) \right], \tag{6.48}$$

$\forall\, N \in \mathbb{N}$.

We continue with

Theorem 6.33 *All as in Theorem 6.32. Additionally assume that* $(C)\int_a^b |t-x|^{n+1} d\mu_{N,x}(t) > 0$, $\forall N \in \mathbb{N}$. *Then*

$$|T_N(f)(x) - f(x)| \le \frac{\omega_1\left(f^{(n)}, \left((C)\int_a^b |t-x|^{n+1} d\mu_{N,x}(t)\right)^{\frac{1}{n+1}}\right)}{n!}.$$

$$\left[(C)\int_a^b |t-x|^n d\mu_{N,x}(t) + \frac{1}{(n+1)}\left((C)\int_a^b |t-x|^{n+1} d\mu_{N,x}(t)\right)^{\frac{n}{n+1}}\right], \tag{6.49}$$

$\forall N \in \mathbb{N}$.

We give ($n = 1$ case)

Corollary 6.34 *Let* $f \in C_+^1([a,b])$. *Assume* $f'(x) = 0$, *for a fixed* $x \in [a,b]$, *and* $(C)\int_a^b (t-x)^2 d\mu_{N,x}(t) > 0$, $\forall N \in \mathbb{N}$. *Then*

$$|T_N(f)(x) - f(x)| \le \omega_1\left(f', \left((C)\int_a^b (t-x)^2 d\mu_{N,x}(t)\right)^{\frac{1}{2}}\right).$$

$$\left[(C)\int_a^b |t-x| d\mu_{N,x}(t) + \frac{1}{2}\left((C)\int_a^b (t-x)^2 d\mu_{N,x}(t)\right)^{\frac{1}{2}}\right], \tag{6.50}$$

$\forall N \in \mathbb{N}$.

Remark 6.35 (i) To Theorem 6.33: Assuming $(C)\int_a^b |t-x|^{n+1} d\mu_{N,x}(t) \to 0$, we get that $T_N(f)(x) \to f(x)$, as $N \to \infty$.

(ii) To Corollary 6.34: Assuming that $(C)\int_a^b (t-x)^2 d\mu_{N,x}(t) \to 0$, we get that $T_N(f)(x) \to f(x)$, as $N \to \infty$.

(iii) The right hand sides of (6.49), (6.50) are finite.

We finish with the $n = 0$ case.

Theorem 6.36 *Let* $f \in C_+([a,b])$. *Assume* $(C)\int_a^b |t-x| d\mu_{N,x}(t) > 0$, *for some fixed* $x \in [a,b]$, $\forall N \in \mathbb{N}$. *Then*

$$|T_N(f)(x) - f(x)| \le 2\omega_1\left(f, \left((C)\int_a^b |t-x| d\mu_{N,x}(t)\right)\right), \tag{6.51}$$

$\forall N \in \mathbb{N}$.

If $(C)\int_a^b |t-x| d\mu_{N,x}(t) \to 0$, *then* $T_N(f)(x) \to f(x)$, *as* $N \to \infty$.

The right hand side of (6.51) is finite.

References

1. G. Anastassiou, Approximation by sublinear operators. Acta Mathematica Universitatis Comenianae (2018). Accepted for publication
2. G. Anastassiou, Approximation by max-product operators. Fasc. Mat. **60**(1), 5–28 (2018)
3. G. Anastassiou, Approximation by Choquet integrals (2018). Submitted
4. G. Choquet, Theory of capacities. Ann. Inst. Fourier (Grenoble) **5**, 131–295 (1954)
5. D. Denneberg, *Non-Additive Measure and Integral* (Kluwer, Dordrecht, 1994)
6. D. Dubois, H. Prade, *Possibility Theory* (Plenum Press, New York, 1988)
7. S. Gal, Uniform and pointwise quantitative approximation by Kantorovich-Choquet type integral operators with respect to monotone and submodular set functions. Mediterr. J. Math. **14**(5), 12 (2017). Art. 205
8. S. Gal, S. Trifa, Quantitative estimates in uniform and pointwise approximation by Bernstein-Durrmeyer-Choquet operators. Carpathian J. Math. **33**(1), 49–58 (2017)
9. D. Schmeidler, Integral representation without additivity. Proc. Am. Math. Soc. **97**, 255–261 (1986)
10. D. Schmeidler, Subjective probability and expected utility without additivity. Econometrica **57**, 571–587 (1989)
11. L.S. Shapley, A value for n-person games, in *Contributions to the Theory of Games*, ed. by H.W. Kuhn, A.W. Tucker. Annals of Mathematical Studies, vol. 28, (Princeton University, Princeton, 1953), pp. 307–317
12. Z. Wang, G.J. Klir, *Generalized Measure Theory* (Springer, New York, 2009)

Chapter 7
Conformable Fractional Approximation by Choquet Integrals

Here we present the conformable fractional quantitative approximation of positive sublinear operators to the unit operator. These are given a precise Choquet integral interpretation. Initially we start with the study of the conformable fractional rate of the convergence of the well-known Bernstein–Kantorovich–Choquet and Bernstein–Durrweyer–Choquet polynomial Choquet-integral operators. Then we study in the fractional sense the very general comonotonic positive sublinear operators based on the representation theorem of Schmeidler [12]. We continue with the conformable fractional approximation by the very general direct Choquet-integral form positive sublinear operators. The case of convexity is also studied thoroughly and the estimates become much simpler. All approximations are given via inequalities involving the modulus of continuity of the approximated function's higher order conformable fractional derivative. It follows [5].

7.1 Introduction

G. Choquet [6], introduced the capacities and his integral. Initially these were applied to statistical mechanics and potential theory, and they gave rise to the study of non-additive measure theory. Slowly but steady these ideas of Choquet started to attract economists especially after the very important work of Shapely [14] in the study of cooperative games. Capacities and Choquet integrals became main stream to Decision theorists since 1989 when D. Schmeidler [13] was the first to use them in an axiomatic model of choice with non-additive beliefs. The expected utility results are strengthened by the use of Choquet capacities instead of probability measures.

G. A. Anastassiou, *Ordinary and Fractional Approximation by Non-additive Integrals: Choquet, Shilkret and Sugeno Integral Approximators*, Studies in Systems, Decision and Control 190, https://doi.org/10.1007/978-3-030-04287-5_7

In now days Choquet integral has wide applications, among others, to decision making under risk and uncertainty, in finance, in economics, in portofolio problems and in insurance.

Our motivation also comes from the foundations of Bayesian decision theory and subjective probability.

Because of the paramount importance of Choquet integral, we decided to research the related positive sublinear operators approximation, part of it is exhibited in this work in the conformable fractional sense.

7.2 Background - I

Next we present briefly about the Choquet integral, see also [10].

We make

Definition 7.1 Consider $\Omega \neq \varnothing$ and let $\mathcal{C}$ be a σ-algebra of subsets in Ω.

(i) (see, e.g., [15], p. 63) The set function $\mu : \mathcal{C} \to [0, +\infty]$ is called a monotone set function (or capacity) if $\mu(\varnothing) = 0$ and $\mu(A) \leq \mu(B)$ for all $A, B \in \mathcal{C}$, with $A \subset B$. Also, μ is called submodular if

$$\mu(A \cup B) + \mu(A \cap B) \leq \mu(A) + \mu(B), \text{ for all } A, B \in \mathcal{C}.$$

μ is called bounded if $\mu(\Omega) < +\infty$ and normalized if $\mu(\Omega) = 1$.

(ii) (see, e.g., [15], p. 233, or [6]) If μ is a monotone set function on $\mathcal{C}$ and if $f : \Omega \to \mathbb{R}$ is $\mathcal{C}$-measurable (that is, for any Borel subset $B \subset \mathbb{R}$ it follows $f^{-1}(B) \in \mathcal{C}$), then for any $A \in \mathcal{C}$, the Choquet integral is defined by

$$(C)\int_A f d\mu = \int_0^{+\infty} \mu\left(F_\beta(f) \cap A\right) d\beta + \int_{-\infty}^0 \left[\mu\left(F_\beta(f) \cap A\right) - \mu(A)\right] d\beta,$$

where we used the notation $F_\beta(f) = \{\omega \in \Omega : f(\omega) \geq \beta\}$. Notice that if $f \geq 0$ on A, then in the above formula we get $\int_{-\infty}^0 = 0$.

The integrals on the right-hand side are the usual Riemann integral.

The function f will be called Choquet integrable on A if $(C)\int_A f d\mu \in \mathbb{R}$.

Next we list some well known properties of the Choquet integral.

Remark 7.2 If $\mu : \mathcal{C} \to [0, +\infty]$ is a monotone set function, then the following properties hold:

(i) For all $a \geq 0$ we have $(C)\int_A af d\mu = a \cdot (C)\int_A f d\mu$ (if $f \geq 0$ then see, e.g., [15], Theorem 11.2, (5), p. 228 and if f is arbitrary sign, then see, e.g., [7], p. 64, Proposition 5.1, (ii)).

(ii) For all $c \in \mathbb{R}$ and f of arbitrary sign, we have (see, e.g., [15], pp. 232–233, or [7], p. 65) $(C)\int_A (f + c) d\mu = (C)\int_A f d\mu + c \cdot \mu(A)$.

If μ is submodular too, then for all f, g of arbitrary sign and lower bounded, we have (see, e.g., [7], p. 75, Theorem 6.3)

$$(C)\int_A (f+g)\,d\mu \le (C)\int_A f d\mu + (C)\int_A g d\mu.$$

(iii) If $f \le g$ on A then $(C)\int_A f d\mu \le (C)\int_A g d\mu$ (see, e.g., [15], p. 228, Theorem 11.2, (3) if $f, g \ge 0$ and p. 232 if f, g are of arbitrary sign).

(iv) Let $f \ge 0$. If $A \subset B$ then $(C)\int_A f d\mu \le (C)\int_B f d\mu$. In addition, if μ is finitely subadditive, then

$$(C)\int_{A\cup B} f d\mu \le (C)\int_A f d\mu + (C)\int_B f d\mu.$$

(v) It is immediate that $(C)\int_A 1 \cdot d\mu(t) = \mu(A)$.

(vi) The formula $\mu(A) = \gamma(M(A))$, where $\gamma : [0, 1] \to [0, 1]$ is an increasing and concave function, with $\gamma(0) = 0$, $\gamma(1) = 1$ and M is a probability measure (or only finitely additive) on a σ-algebra on Ω (that is, $M(\varnothing) = 0$, $M(\Omega) = 1$ and M is countably additive), gives simple examples of normalized, monotone and submodular set functions (see, e.g., [7], pp. 16–17, Example 2.1). Such of set functions μ are also called distorsions of countably normalized, additive measures (or distorted measures). For a simple example, we can take $\gamma(t) = \frac{2t}{1+t}$, $\gamma(t) = \sqrt{t}$.

If the above γ function is increasing, concave and satisfies only $\gamma(0) = 0$, then for any bounded Borel measure m, $\mu(A) = \gamma(m(A))$ gives a simple example of bounded, monotone and submodular set function.

(vii) If μ is a countably additive bounded measure, then the Choquet integral $(C)\int_A f d\mu$ reduces to the usual Lebesgue type integral (see, e.g., [7], p. 62, or [15], p. 226).

(viii) If $f \ge 0$, then $(C)\int_A f d\mu \ge 0$.

(ix) Let $\mu = \sqrt{M}$, where M is the Lebesgue measure on $[0, +\infty)$, then μ is a monotone and submodular set function, furthermore μ is strictly positive, see [9].

(x) If $\Omega = \mathbb{R}^N$, $N \in \mathbb{N}$, we call μ strictly positive if $\mu(A) > 0$, for any open subset $A \subseteq \mathbb{R}^N$.

We need some possibility theory:

Definition 7.3 ([8]) For the $\Omega \ne \varnothing$, the power set $\mathcal{P}(\Omega)$ denotes the family of all subsets of Ω.

(i) A function $\lambda : \Omega \to [0, 1]$ with the property $\sup\{\lambda(s) : s \in \Omega\} = 1$, is called possibility distribution on Ω.

(ii) $P : \mathcal{P}(\Omega) \to [0, 1]$ is called possibility measure, if it satisfies $P(\varnothing) = 0$, $P(\Omega) = 1$, and $P(\cup_{i\in I} A_i) = \sup\{P(A_i) : i \in I\}$ for all $A_i \subset \Omega$, and any I, an at most countable family of indices. Note that if $A, B \subset \Omega$, $A \subset B$, then the last property implies $P(A) \le P(B)$ and that $P(A \cup B) \le P(A) + P(B)$.

Any possibility distribution λ on Ω, induces the possibility measure $P_\lambda : \mathcal{P}(\Omega) \to [0, 1]$, $P_\lambda(A) = \sup\{\lambda(s) : s \in A\}$, $A \subset \Omega$. Also, if $f : \Omega \to \mathbb{R}_+$, then the possibilistic integral of f on $A \subset \Omega$ with respect to P_λ is defined by $(Pos)\int_A f dP_\lambda = \sup\{f(t)\lambda(t) : t \in A\}$ (see [8], Chapter 1).

Note that any possibility measure μ is normalized, monotone and submodular. From $\mu(A \cup B) = \max\{\mu(A), \mu(B)\}$ we get monotonicity, and from $\mu(A \cap B) \leq \min\{\mu(A), \mu(B)\}$ we derive the submodularity.

7.3 Background - II

We make

Definition 7.4 ([3]) Let $f : [a, b] \subseteq [0, \infty) \to \mathbb{R}$ and $\alpha \in (0, 1]$. We say that f is an α-fractional continuous function, iff $\forall\, \varepsilon > 0\, \exists\, \delta > 0$: for any $x, y \in [a, b]$ such that $|x^\alpha - y^\alpha| \leq \delta$ we get that $|f(x) - f(y)| \leq \varepsilon$.

We mention

Theorem 7.5 ([3]) *Over $[a, b] \subseteq [0, \infty)$, $\alpha \in [0, 1]$, an α-fractional continuous function is a uniformly continuous function and vice versa, a uniformly continuous function is an α-fractional continuous function.*

We need

Definition 7.6 ([3]) Let $[a, b] \subseteq [0, \infty)$, $\alpha \in [0, 1]$. We define the α-fractional modulus of continuity:

$$\omega_1^\alpha(f, \delta) := \sup_{\substack{x, y \in [a,b]: \\ |x^\alpha - y^\alpha| \leq \delta}} |f(x) - f(y)|, \; \delta > 0. \tag{7.1}$$

Properties ([3]):

(1) $\omega_1^\alpha(f, 0) = 0$.

(2) $\omega_1^\alpha(f, \delta) \to 0$ as $\delta \downarrow 0$, iff f is in the set of all α-fractional continuous functions, denoted as $f \in C_\alpha([a, b], \mathbb{R}) (= C([a, b], \mathbb{R}))$.

(3) ω_1^α is ≥ 0 and non-decreasing on $\mathbb{R}_+$.

(4) ω_1^α is subadditive:

$$\omega_1^\alpha(f, t_1 + t_2) \leq \omega_1^\alpha(f, t_1) + \omega_1^\alpha(f, t_2). \tag{7.2}$$

(5) ω_1^α is continuous on $\mathbb{R}_+$.

(6) Clearly it holds

$$\omega_1^\alpha(f, t_1 + \ldots + t_n) \leq \omega_1^\alpha(f, t_1) + \ldots + \omega_1^\alpha(f, t_n), \tag{7.3}$$

for $t = t_1 = \ldots = t_n$, we obtain

$$\omega_1^\alpha(f, nt) = n\omega_1^\alpha(f, t). \tag{7.4}$$

(7) Let $\lambda \geq 0$, $\lambda \notin \mathbb{N}$, we get

$$\omega_1^{\alpha}(f, \lambda t) \leq (\lambda + 1)\, \omega_1^{\alpha}(f, t). \tag{7.5}$$

We notice that $\omega_1^{\alpha}(f, \delta)$ is finite when f is uniformly continuous on $[a, b] \subseteq [0, \infty)$.

We need

Definition 7.7 ([1, 11]) Let $f : [0, \infty) \to \mathbb{R}$. The conformable α-fractional derivative for $\alpha \in (0, 1]$ is given by

$$D_{\alpha} f(t) := \lim_{\varepsilon \to 0} \frac{f\left(t + \varepsilon t^{1-\alpha}\right) - f(t)}{\varepsilon}, \tag{7.6}$$

$$D_{\alpha} f(0) = \lim_{t \to 0+} D_{\alpha} f(t).$$

If f is differentiable, then

$$D_{\alpha} f(t) = t^{1-\alpha} f'(t), \tag{7.7}$$

where f' is the usual derivative.

We define $D_{\alpha}^{n} f = D_{\alpha}^{n-1}(D_{\alpha} f)$, $D_{\alpha}^{0} f = f$.

If $f : [0, \infty) \to \mathbb{R}$ is α-differentiable at $t_0 > 0$, $\alpha \in (0, 1]$, then f is continuous at t_0, see [11].

We need

Definition 7.8 ([3]) Here $C_+([a, b]) := \{f : [a, b] \subseteq [0, \infty) \to \mathbb{R}_+$, continuous functions$\}$. Let $L_N : C_+([a, b]) \to C_+([a, b])$, operators, $\forall\, N \in \mathbb{N}$, such that

(i)

$$L_N(\alpha f) = \alpha L_N(f), \ \forall \alpha \geq 0, \forall f \in C_+([a, b]), \tag{7.8}$$

(ii) if $f, g \in C_+([a, b]) : f \leq g$, then

$$L_N(f) \leq L_N(g), \ \forall N \in \mathbb{N}, \tag{7.9}$$

(iii)

$$L_N(f + g) \leq L_N(f) + L_N(g), \ \forall\, f, g \in C_+([a, b]). \tag{7.10}$$

We call $\{L_N\}_{N \in \mathbb{N}}$ positive sublinear operators.

We need

Theorem 7.9 ([3]) *Let $\alpha \in (0, 1]$, $[a, b] \subseteq [0, \infty)$. Suppose f is α-conformable fractional differentiable on $[a, b]$. $D_{\alpha} f$ is continuous on $[a, b]$. Let an $x \in [a, b]$ such that $D_{\alpha} f(x) = 0$, and L_N from $C_+([a, b])$ into itself, positive sublinear operators.*

Assume that $L_N(1) = 1$ *and* $L_N\left(|\cdot - x|^{\alpha+1}\right)(x), L_N\left((\cdot - x)^{2(\alpha+1)}\right)(x) > 0, \forall N \in \mathbb{N}$.

Then

$$|L_N(f)(x) - f(x)| \leq \frac{\omega_1^{\alpha}\left(D_{\alpha} f, \left(L_N\left((\cdot - x)^{2(\alpha+1)}\right)(x)\right)^{\frac{\alpha}{2(\alpha+1)}}\right)}{\alpha} \cdot$$

$$\left[\left(L_N\left(|\cdot - x|^{\alpha+1}\right)(x)\right)^{\frac{\alpha}{\alpha+1}} + \frac{1}{2}\left(L_N\left((\cdot - x)^{2(\alpha+1)}\right)(x)\right)^{\frac{\alpha}{2(\alpha+1)}}\right], \forall N \in \mathbb{N}. \quad (7.11)$$

We make

Remark 7.10 ([3]) By [3], we get that

$$L_N\left(|\cdot - x|^{\alpha+1}\right)(x) \leq \left(L_N\left((\cdot - x)^{2(\alpha+1)}\right)(x)\right)^{\frac{1}{2}}. \quad (7.12)$$

As $N \to +\infty$, by (7.11) and (7.12), and $L_N\left((\cdot - x)^{2(\alpha+1)}\right)(x) \to 0$, we obtain that $L_N(f)(x) \to f(x)$.

We need

Theorem 7.11 ([3]) *Let* $\alpha \in (0, 1]$, $n \in \mathbb{N}$. *Suppose* f *is* n *times conformable* α*-fractional differentiable on* $[a, b] \subseteq [0, \infty)$*, and* $D_{\alpha}^n f$ *is continuous on* $[a, b]$*. For a fixed* $x \in [a, b]$ *we have* $D_{\alpha}^k f(x) = 0$, $k = 1, ..., n$*. Let positive sublinear operators* $\{L_N\}_{N \in \mathbb{N}}$ *from* $C_+([a, b])$ *into itself, such that* $L_N(1) = 1$*, and* $L_N\left(|\cdot - x|^{n(\alpha+1)}\right)(x), L_N\left(|\cdot - x|^{(n+1)(\alpha+1)}\right)(x) > 0, \forall N \in \mathbb{N}$*. Then*

$$|L_N(f)(x) - f(x)| \leq \frac{\omega_1^{\alpha}\left(D_{\alpha}^n f, \left(L_N\left(|\cdot - x|^{(n+1)(\alpha+1)}\right)(x)\right)^{\frac{\alpha}{(n+1)(\alpha+1)}}\right)}{\alpha^n n!} \cdot \quad (7.13)$$

$$\left[\left(L_N\left(|\cdot - x|^{n(\alpha+1)}\right)(x)\right)^{\frac{\alpha}{\alpha+1}} + \frac{1}{(n+1)}\left(L_N\left(|\cdot - x|^{(n+1)(\alpha+1)}\right)(x)\right)^{\frac{n\alpha}{(n+1)(\alpha+1)}}\right],$$

$\forall N \in \mathbb{N}$.

We make

Remark 7.12 ([3]) By [3], we get that

$$L_N\left(|\cdot - x|^{n(\alpha+1)}\right)(x) \leq \left(L_N\left(|\cdot - x|^{(n+1)(\alpha+1)}\right)(x)\right)^{\frac{n}{n+1}}. \quad (7.14)$$

As $N \to +\infty$, by (7.13), (7.14), and $L_N\left(|\cdot - x|^{(n+1)(\alpha+1)}\right)(x) \to 0$, we derive that $L_N(f)(x) \to f(x)$.

We also need

Definition 7.13 Let $f \in C([a,b])$. We define the usual first modulus of continuity of f as:

$$\omega_1(f,\delta) := \sup_{\substack{x,y\in[a,b]:\\ |x-y|\leq\delta}} |f(x)-f(y)|, \ \delta>0. \tag{7.15}$$

We need

Theorem 7.14 ([4]) *Let $\alpha \in (0,1]$ and $n \in \mathbb{N}$. Suppose $f \in C_+([a,b])$ is n times conformable α-fractional differentiable on $[a,b] \subseteq [0,\infty)$, and $x \in (a,b)$, and $D_\alpha^n f$ is continuous on $[a,b]$. Let $0 < h \leq \min(x-a, b-x)$ and assume $\left|D_\alpha^n f\right|$ is convex over $[a,b]$. Furthermore assume that $D_\alpha^k f(x) = 0$, $k = 1, \ldots, n$. Let $\{L_N\}_{N\in\mathbb{N}}$ from $C_+([a,b])$ into itself, positive sublinear operators such that: $L_N(1) = 1$, $\forall N \in \mathbb{N}$. Then*

$$|L_N(f)(x) - f(x)| \leq \left(\frac{\omega_1\left(D_\alpha^n f, h\right) b^{1-\alpha}}{(n+1)!\alpha^{n+1} h}\right) L_N\left(|\cdot - x|^{(n+1)\alpha}\right)(x), \ \forall N \in \mathbb{N}. \tag{7.16}$$

We have

Theorem 7.15 ([4]) *All as in Theorem 7.14. Additionally assume that $L_N\left(|\cdot - x|^{(n+1)(\alpha+1)}\right)(x) > 0$, $\forall N \in \mathbb{N}$. Then*

$$|L_N(f)(x) - f(x)| \leq \left(\frac{\omega_1\left(D_\alpha^n f, h\right) b^{1-\alpha}}{(n+1)!\alpha^{n+1} h}\right) \left(L_N\left(|\cdot - x|^{(n+1)(\alpha+1)}\right)(x)\right)^{\frac{\alpha}{\alpha+1}}, \tag{7.17}$$

$\forall N \in \mathbb{N}$.

An application of Theorem 7.15 follows:

Theorem 7.16 ([4]) *Let $\{L_N\}_{N\in\mathbb{N}}$ from $C_+([a,b])$ into itself, positive sublinear operators: $L_N(1) = 1$, $\forall N \in \mathbb{N}$. Also $x \in (a,b)$ and $L_N\left(|\cdot - x|^{(n+1)(\alpha+1)}\right)(x) > 0$, $\forall N \in \mathbb{N}$. Here $\alpha \in (0,1]$ and $n \in \mathbb{N}$. Suppose $f \in C_+([a,b])$ is n times conformable α-fractional differentiable on $[a,b] \subseteq [0,\infty)$, and $D_\alpha^n f$ is continuous on $[a,b]$. Assume here that $0 < \left(L_N\left(|\cdot - x|^{(n+1)(\alpha+1)}\right)(x)\right)^{\frac{\alpha}{\alpha+1}} \leq \min(x-a, b-x)$, $\forall N \in \mathbb{N} : N \geq N^* \in \mathbb{N}$, and assume $\left|D_\alpha^n f\right|$ is convex over $[a,b]$. Furthermore assume that $D_\alpha^k f(x) = 0$, $k = 1, \ldots, n$. Then*

$$|L_N(f)(x) - f(x)| \leq \frac{b^{1-\alpha}\omega_1\left(D_\alpha^n f, \left(L_N\left(|\cdot - x|^{(n+1)(\alpha+1)}\right)(x)\right)^{\frac{\alpha}{\alpha+1}}\right)}{(n+1)!\alpha^{n+1}}, \tag{7.18}$$

$\forall N \in \mathbb{N} : N \geq N^* \in \mathbb{N}$.

If $L_N\left(|\cdot - x|^{(n+1)(\alpha+1)}\right)(x) \to 0$, then $L_N(f)(x) \to f(x)$, as $N \to +\infty$.

An application of Theorem 7.14 follows:

Theorem 7.17 ([4]) *Let $\{L_N\}_{N\in\mathbb{N}}$ from $C_+([a,b])$ into itself, positive sublinear operators: $L_N(1)=1$, $\forall\, N\in\mathbb{N}$. Also $L_N\left(|\cdot-x|^{(n+1)\alpha}\right)(x)>0$, $\forall\, N\in\mathbb{N}$. Here $\alpha\in(0,1]$, $n\in\mathbb{N}$ and $x\in(a,b)$; $[a,b]\subseteq[0,\infty)$. Suppose $f\in C_+([a,b])$ is n times conformable α-fractional differentiable on $[a,b]$, and $D_\alpha^n f$ is continuous on $[a,b]$. Let $0<L_N\left(|\cdot-x|^{(n+1)\alpha}\right)(x)\leq\min(x-a,b-x)$, $\forall\, N\geq N^*$; $N, N^*\in\mathbb{N}$, and assume $\left|D_\alpha^n f\right|$ is convex over $[a,b]$. Furthermore assume that $D_\alpha^k f(x)=0$, $k=1,...,n$. Then*

$$|L_N(f)(x)-f(x)|\leq\frac{b^{1-\alpha}\omega_1\left(D_\alpha^n f, L_N\left(|\cdot-x|^{(n+1)\alpha}\right)(x)\right)}{(n+1)!\alpha^{n+1}}, \tag{7.19}$$

$\forall\, N\geq N^*$, where $N, N^*\in\mathbb{N}$.

If $L_N\left(|\cdot-x|^{(n+1)\alpha}\right)(x)\to 0$, then $L_N(f)(x)\to f(x)$, as $N\to+\infty$.

7.4 Background - III

We mention

Definition 7.18 ([9]) Let $I=[0,1]$, $\mathcal{B}_I$ the σ-algebra of all Borel measurable subsets of I, $\left(\Gamma_{N,x}\right)_{N\in\mathbb{N},x\in I}$ will be the collection of the family $\Gamma_{N,x}=\{\mu_{N,k,x}\}_{k=0}^N$, of monotone, submodular and strictly positive set functions $\mu_{N,k,x}$ on $\mathcal{B}_I$.

Let $f:[0,1]\to\mathbb{R}_+$ be a B_I-measurable function which is bounded, and call $p_{N,k}(x)=\binom{N}{k}x^k(1-x)^{N-k}$, for any $x\in[0,1]$.

The Bernstein–Kantorovich–Choquet operators are defined by the formula

$$K_{N,\Gamma_{N,x}}(f)(x)=\sum_{k=0}^N p_{N,k}(x)\frac{(C)\int_{\frac{k}{(N+1)}}^{\frac{(k+1)}{(N+1)}}f(t)\,d\mu_{N,k,x}(t)}{\mu_{N,k,x}\left(\left[\frac{k}{(N+1)},\frac{(k+1)}{(N+1)}\right]\right)},\ \forall\, x\in[0,1]. \tag{7.20}$$

If $\mu_{N,k,x}=\mu$, for all N, x, k, we will denote $K_{N,\Gamma_{N,x}}(f):=K_{N,\mu}(f)$.

Theorem 7.19 ([9]) *Suppose that $\mu_{N,k,x}=\mu:=\sqrt{M}$, for all N, k and x, where M is the Lebesgue measure on $[0,1]$. Then*

$$\left|K_{N,\mu}(f)(x)-f(x)\right|\leq 2\omega_1\left(f,\frac{\sqrt{x(1-x)}}{\sqrt{N}}+\frac{1}{N}\right), \tag{7.21}$$

$\forall\, N\in\mathbb{N}$, $x\in[0,1]$, $f\in C_+([0,1])$, above ω_1 is over $[0,1]$.

Remark 7.20 By [9] we have that

$$K_{N,\mu}(|\cdot-x|)(x)\leq\frac{\sqrt{x(1-x)}}{\sqrt{N}}+\frac{1}{N},\ \forall\, N\in\mathbb{N}. \tag{7.22}$$

Let $m > 1$, notice that $|\cdot - x|^{m-1} \leq 1$, therefore

$$|\cdot - x|^m = |\cdot - x| \, |\cdot - x|^{m-1} \leq |\cdot - x| \, ,$$

hence

$$K_{N,\mu} \left(|\cdot - x|^m \right) (x) \leq K_{N,\mu} \left(|\cdot - x| \right) (x) \, ,$$

that is

$$K_{N,\mu} \left(|\cdot - x|^m \right) (x) \leq \frac{\sqrt{x(1-x)}}{\sqrt{N}} + \frac{1}{N}, \; \forall \, x \in [0,1] \, , \; N \in \mathbb{N}, \; m \geq 1. \quad (7.23)$$

Notice that $K_{N,\mu}(1) = 1, \forall \, N \in \mathbb{N}$.

Clearly $K_{N,\mu}$ operators are positive sublinear operators from $C_+([0,1])$ into itself.

We mention

Definition 7.21 ([10]) Here we consider measures of possibility. Denoting $p_{N,k}(x) = \binom{N}{k} x^k (1-x)^{N-k}$, let us defined

$$\lambda_{N,k}(t) := \frac{p_{N,k}(t)}{k^k N^{-N} (N-k)^{N-k} \binom{N}{k}} = \frac{t^k (1-t)^{N-k}}{k^k N^{-N} (N-k)^{N-k}}, \; k = 0, ..., N. \tag{7.24}$$

By convention we assume that $0^0 = 1$, so that the cases $k = 0$, and $k = N$ make sense. By considering the root $\frac{k}{N}$ of $p'_{N,k}(x)$, it is clear that

$$\max\{p_{N,k}(t) : t \in [0,1]\} = k^k N^{-N} (N-k)^{N-k} \binom{N}{k},$$

which implies that each $\lambda_{N,k}$ is a possibility distribution on $[0,1]$.

Denoting by $P_{\lambda_{N,k}}$ the possibility measure induced by $\lambda_{N,k}$ and $\Gamma_{n,x} := \Gamma_N := \{P_{\lambda_{N,k}}\}_{k=0}^N$ (that is Γ_N is independent of x), we define the nonlinear Bernstein–Durrmeyer–Choquet polynomial operators with respect to the set functions in Γ_N given by the formula

$$D_{N,\Gamma_N}(f)(x) := \sum_{k=0}^{N} p_{N,k}(x) \frac{(C) \int_0^1 f(t) \, t^k (1-t)^{N-k} \, dP_{\lambda_{N,k}}(t)}{(C) \int_0^1 t^k (1-t)^{N-k} \, dP_{\lambda_{N,k}}(t)}, \tag{7.25}$$

$\forall \, x \in [0,1]$, $N \in \mathbb{N}$, $f \in C_+([0,1])$.

Remark 7.22 Above $P_{\lambda_{N,k}}$ is bounded, monotone, submodular and strictly positive, $N \in \mathbb{N}$, $k = 0, 1, ..., N$. Notice that $D_{N,\Gamma_N}(1) = 1, \forall \, N \in \mathbb{N}$.

Clearly D_{N,Γ_N} operators are positive sublinear operators mapping $C_+([0,1])$ into itself.

We mention

Theorem 7.23 ([10]) *For every* $f \in C_+([0,1])$, $x \in [0,1]$ *and* $N \in \mathbb{N} - \{1\}$, *we have*

$$\left|D_{N,\Gamma_N}(f)(x) - f(x)\right| \leq 2\omega_1\left(f, \frac{\left(1+\sqrt{2}\right)\sqrt{x(1-x)} + \sqrt{2}\sqrt{x}}{\sqrt{N}} + \frac{1}{N}\right), \tag{7.26}$$

where ω_1 *is on* $[0,1]$.

Remark 7.24 By [10] we have that

$$D_{N,\Gamma_N}(|\cdot - x|)(x) \leq \frac{\left(1+\sqrt{2}\right)\sqrt{x(1-x)} + \sqrt{2}\sqrt{x}}{\sqrt{N}} + \frac{1}{N}, \ \forall\, N \in \mathbb{N} - \{1\}. \tag{7.27}$$

Let $m > 1$, notice that $|\cdot - x|^{m-1} \leq 1$, therefore

$$|\cdot - x|^m = |\cdot - x|\,|\cdot - x|^{m-1} \leq |\cdot - x|,$$

hence

$$D_{N,\Gamma_N}\left(|\cdot - x|^m\right)(x) \leq D_{N,\Gamma_N}(|\cdot - x|)(x),$$

that is

$$D_{N,\Gamma_N}\left(|\cdot - x|^m\right)(x) \leq \frac{\left(1+\sqrt{2}\right)\sqrt{x(1-x)} + \sqrt{2}\sqrt{x}}{\sqrt{N}} + \frac{1}{N}, \tag{7.28}$$

$\forall\, N \in \mathbb{N} - \{1\}$, $\forall\, x \in [0,1]$, $m \geq 1$.

We make

Remark 7.25 When $x \in [0,1]$, then the $\max(x(1-x)) = \frac{1}{4}$, at $x = \frac{1}{2}$. Therefore it holds

$$\frac{\sqrt{x(1-x)}}{\sqrt{N}} + \frac{1}{N} \leq \frac{1}{2\sqrt{N}} + \frac{1}{N}, \tag{7.29}$$

$\forall x \in [0,1]$, $\forall N \in \mathbb{N}$.

Similarly, it holds

$$\frac{\left(1+\sqrt{2}\right)\sqrt{x(1-x)} + \sqrt{2}\sqrt{x}}{\sqrt{N}} + \frac{1}{N} \leq \frac{1+3\sqrt{2}}{2\sqrt{N}} + \frac{1}{N}, \tag{7.30}$$

$\forall\, x \in [0,1]$, $\forall\, N \in \mathbb{N} - \{1\}$.

Corollary 7.26 (to Theorem 7.19) *It holds*

$$\left\| K_{N,\mu}(f) - f \right\|_{\infty} \leq 2\omega_1 \left(f, \frac{1}{2\sqrt{N}} + \frac{1}{N} \right), \tag{7.31}$$

$\forall\, N \in \mathbb{N},\ f \in C_+([0,1])$.

Corollary 7.27 (to Theorem 7.23) *It holds*

$$\left\| D_{N,\Gamma_N}(f) - f \right\|_{\infty} \leq 2\omega_1 \left(f, \frac{1+3\sqrt{2}}{2\sqrt{N}} + \frac{1}{N} \right), \tag{7.32}$$

$\forall\, N \in \mathbb{N} - \{1\},\ f \in C_+([0,1])$.

7.5 Main Results

Here first we apply some of the main theorems mentioned in Sect. 7.3 to the Bernstein–Kantorovich–Choquet operators $K_{N,\mu}$, where $\mu := \sqrt{M}$, with M the Lebesgue measure on $[0, 1]$. More precisely here it is

$$K_{N,\mu}(f)(x) = \sum_{k=0}^{N} p_{N,k}(x) \frac{(C)\int_{\frac{k}{(N+1)}}^{\frac{(k+1)}{(N+1)}} f(t)\, d\mu(t)}{\mu\left(\left[\frac{k}{(N+1)}, \frac{(k+1)}{(N+1)}\right]\right)}, \tag{7.33}$$

$\forall\, x \in [0,1],\ \forall\, N \in \mathbb{N},\ f \in C_+([0,1])$.

It follows applications to Bernstein–Durrmeyer–Choquet operators D_{N,Γ_N}, see (7.25).

In particular we need (a variation of Theorem 7.11):

Theorem 7.28 ([3]) *Let $\alpha \in (0,1]$ and $n \in \mathbb{N} : n\alpha \geq 1$. That is $\frac{1}{n} \leq \alpha \leq 1$. Suppose f is n times conformable α-fractional differentiable on $[a,b] \subseteq [0,\infty)$, and $D_\alpha^n f$ is continuous on $[a,b]$. For a fixed $x \in [a,b]$ we have $D_\alpha^k f(x) = 0$, $k = 1, ..., n$. Let positive sublinear operators $\{L_N\}_{N \in \mathbb{N}}$ from $C_+([a,b])$ into itself, such that $L_N(1) = 1$, $\forall\, N \in \mathbb{N}$, and $\delta > 0$. Then*

$$|L_N(f)(x) - f(x)| \leq \frac{\omega_1^\alpha \left(D_\alpha^n f, \delta\right)}{\alpha^n n!} \cdot$$

$$\left[L_N\left(|\cdot - x|^{n\alpha}\right)(x) + \frac{1}{(n+1)\delta} L_N\left(|\cdot - x|^{(n+1)\alpha}\right)(x) \right], \tag{7.34}$$

$\forall\, N \in \mathbb{N}$.

We present

Theorem 7.29 *Let $\alpha \in (0,1]$ and $n \in \mathbb{N} : n\alpha \geq 1$. Suppose f is n times conformable α-fractional differentiable on $[0,1]$, and $D_{\alpha}^{n} f$ is continuous on $[0,1]$. For a fixed $x \in [0,1]$ we have $D_{\alpha}^{k} f(x) = 0$, $k = 1, ..., n$. Then*

$$\left|K_{N,\mu}(f)(x) - f(x)\right| \leq \frac{\omega_1^{\alpha}\left(D_{\alpha}^{n} f, \left(\sqrt{\frac{x(1-x)}{N}} + \frac{1}{N}\right)^{\frac{1}{n+1}}\right)}{\alpha^n n!} \cdot$$

$$\left[\left(\sqrt{\frac{x(1-x)}{N}} + \frac{1}{N}\right) + \frac{1}{(n+1)}\left(\sqrt{\frac{x(1-x)}{N}} + \frac{1}{N}\right)^{\frac{n}{n+1}}\right] \leq$$

$$\frac{\omega_1^{\alpha}\left(D_{\alpha}^{n} f, \left(\frac{1}{2\sqrt{N}} + \frac{1}{N}\right)^{\frac{1}{n+1}}\right)}{\alpha^n n!}\left[\left(\frac{1}{2\sqrt{N}} + \frac{1}{N}\right) + \frac{1}{(n+1)}\left(\frac{1}{2\sqrt{N}} + \frac{1}{N}\right)^{\frac{n}{n+1}}\right], \tag{7.35}$$

$\forall N \in \mathbb{N}$.

Notice that $\lim_{N \to \infty} K_{N,\mu}(f)(x) = f(x)$.

Proof By (7.34) we have

$$\left|K_{N,\mu}(f)(x) - f(x)\right| \leq \frac{\omega_1^{\alpha}\left(D_{\alpha}^{n} f, \delta\right)}{\alpha^n n!} \cdot$$

$$\left[K_{N,\mu}\left(|\cdot - x|^{n\alpha}\right)(x) + \frac{1}{(n+1)\delta} K_{N,\mu}\left(|\cdot - x|^{(n+1)\alpha}\right)(x)\right] \overset{(7.23)}{\leq}$$

$$\frac{\omega_1^{\alpha}\left(D_{\alpha}^{n} f, \delta\right)}{\alpha^n n!}\left[\left(\sqrt{\frac{x(1-x)}{N}} + \frac{1}{N}\right) + \frac{1}{(n+1)\delta}\left(\sqrt{\frac{x(1-x)}{N}} + \frac{1}{N}\right)\right] \tag{7.36}$$

(choose $\delta := \left(\sqrt{\frac{x(1-x)}{N}} + \frac{1}{N}\right)^{\frac{1}{n+1}} > 0$, then $\delta^{n+1} = \sqrt{\frac{x(1-x)}{N}} + \frac{1}{N}$, and $\delta^n = \left(\sqrt{\frac{x(1-x)}{N}} + \frac{1}{N}\right)^{\frac{n}{n+1}}$)

$$= \frac{\omega_1^{\alpha}\left(D_{\alpha}^{n} f, \left(\sqrt{\frac{x(1-x)}{N}} + \frac{1}{N}\right)^{\frac{1}{n+1}}\right)}{\alpha^n n!} \cdot$$

$$\left[\left(\sqrt{\frac{x(1-x)}{N}} + \frac{1}{N}\right) + \frac{1}{(n+1)}\left(\sqrt{\frac{x(1-x)}{N}} + \frac{1}{N}\right)^{\frac{n}{n+1}}\right] \overset{(7.29)}{\leq}$$

$$\frac{\omega_1^{\alpha}\left(D_{\alpha}^{n}f, \left(\frac{1}{2\sqrt{N}}+\frac{1}{N}\right)^{\frac{1}{n+1}}\right)}{\alpha^n n!}\left[\left(\frac{1}{2\sqrt{N}}+\frac{1}{N}\right)+\frac{1}{(n+1)}\left(\frac{1}{2\sqrt{N}}+\frac{1}{N}\right)^{\frac{n}{n+1}}\right], \tag{7.37}$$

proving the claim. ■

We continue with

Theorem 7.30 *All as in Theorem 7.29. Then*

$$\left|\left(D_{N,\Gamma_N}(f)\right)(x)-f(x)\right| \leq \frac{\omega_1^{\alpha}\left(D_{\alpha}^{n}f, \left(\frac{(1+\sqrt{2})\sqrt{x(1-x)}+\sqrt{2x}}{\sqrt{N}}+\frac{1}{N}\right)^{\frac{1}{n+1}}\right)}{\alpha^n n!}$$

$$\left[\left(\frac{\left(1+\sqrt{2}\right)\sqrt{x(1-x)}+\sqrt{2x}}{\sqrt{N}}+\frac{1}{N}\right)+\right.$$

$$\left.\frac{1}{(n+1)}\left[\frac{\left(1+\sqrt{2}\right)\sqrt{x(1-x)}+\sqrt{2x}}{\sqrt{N}}+\frac{1}{N}\right]^{\frac{n}{n+1}}\right] \leq$$

$$\frac{\omega_1^{\alpha}\left(D_{\alpha}^{n}f, \left(\frac{1+3\sqrt{2}}{2\sqrt{N}}+\frac{1}{N}\right)^{\frac{1}{n+1}}\right)}{\alpha^n n!}\cdot \tag{7.38}$$

$$\left[\left(\frac{1+3\sqrt{2}}{2\sqrt{N}}+\frac{1}{N}\right)+\frac{1}{(n+1)}\left[\frac{1+3\sqrt{2}}{2\sqrt{N}}+\frac{1}{N}\right]^{\frac{n}{n+1}}\right],$$

$\forall\, N \in \mathbb{N}-\{1\}$.

Notice that $\lim_{N\to+\infty} D_{N,\Gamma_N}(f)(x) = f(x)$.

Proof By (7.34) we have

$$\left|D_{N,\Gamma_N}(f)(x)-f(x)\right| \leq \frac{\omega_1^{\alpha}\left(D_{\alpha}^{n}f,\delta\right)}{\alpha^n n!}\cdot$$

$$\left[D_{N,\Gamma_N}\left(|\cdot-x|^{n\alpha}\right)(x)+\frac{1}{(n+1)\delta}D_{N,\Gamma_N}\left(|\cdot-x|^{(n+1)\alpha}\right)(x)\right] \overset{(7.28)}{\leq}$$

$$\frac{\omega_1^{\alpha}\left(D_{\alpha}^{n}f,\delta\right)}{\alpha^n n!}\left[\left(\frac{\left(1+\sqrt{2}\right)\sqrt{x(1-x)}+\sqrt{2x}}{\sqrt{N}}+\frac{1}{N}\right)+\right.$$

$$\frac{1}{(n+1)\,\delta}\left(\frac{\left(1+\sqrt{2}\right)\sqrt{x(1-x)}+\sqrt{2x}}{\sqrt{N}}+\frac{1}{N}\right)\Bigg] \tag{7.39}$$

(choose $\delta := \left(\frac{(1+\sqrt{2})\sqrt{x(1-x)}+\sqrt{2x}}{\sqrt{N}}+\frac{1}{N}\right)^{\frac{1}{n+1}} > 0$, then
$\delta^{n+1} = \frac{(1+\sqrt{2})\sqrt{x(1-x)}+\sqrt{2x}}{\sqrt{N}}+\frac{1}{N}$, and $\delta^{n} = \left(\frac{(1+\sqrt{2})\sqrt{x(1-x)}+\sqrt{2x}}{\sqrt{N}}+\frac{1}{N}\right)^{\frac{n}{n+1}}$)

$$= \frac{\omega_1^{\alpha}\left(D_{\alpha}^{n}f, \left(\frac{(1+\sqrt{2})\sqrt{x(1-x)}+\sqrt{2x}}{\sqrt{N}}+\frac{1}{N}\right)^{\frac{1}{n+1}}\right)}{\alpha^{n}n!}\cdot$$

$$\left[\left(\frac{\left(1+\sqrt{2}\right)\sqrt{x(1-x)}+\sqrt{2x}}{\sqrt{N}}+\frac{1}{N}\right)+\right.$$

$$\left.\frac{1}{(n+1)}\left[\frac{\left(1+\sqrt{2}\right)\sqrt{x(1-x)}+\sqrt{2x}}{\sqrt{N}}+\frac{1}{N}\right]^{\frac{n}{n+1}}\right] \overset{(7.30)}{\leq} \tag{7.40}$$

$$\frac{\omega_1^{\alpha}\left(D_{\alpha}^{n}f, \left(\frac{1+3\sqrt{2}}{2\sqrt{N}}+\frac{1}{N}\right)^{\frac{1}{n+1}}\right)}{\alpha^{n}n!}\cdot$$

$$\left[\left(\frac{1+3\sqrt{2}}{2\sqrt{N}}+\frac{1}{N}\right)+\frac{1}{(n+1)}\left[\frac{1+3\sqrt{2}}{2\sqrt{N}}+\frac{1}{N}\right]^{\frac{n}{n+1}}\right],$$

$\forall\, N \in \mathbb{N}-\{1\}$, proving the claim. ∎

Next we apply Theorem 7.14.

We give

Theorem 7.31 *Let $\alpha \in (0,1]$ and $n \in \mathbb{N}$ such that $(n+1)\,\alpha \geq 1$, that is $\frac{1}{n+1} \leq \alpha \leq 1$. Suppose $f \in C_{+}([0,1])$ is n times conformable α-fractional differentiable on $[0,1]$, and $x \in (0,1)$, and $D_{\alpha}^{n}f$ is continuous on $[0,1]$. Let $N^{*} \in \mathbb{N}$ such that $\frac{1}{2\sqrt{N^{*}}}+\frac{1}{N^{*}} \leq \min(x, 1-x)$ and assume $\left|D_{\alpha}^{n}f\right|$ is convex over $[0,1]$. Furthermore assume that $D_{\alpha}^{k}f(x)=0$, $k=1,\ldots,n$. Then*

$$\left|\left(K_{N,\mu}(f)\right)(x)-f(x)\right| \leq \frac{\omega_1\left(D_{\alpha}^{n}f, \frac{1}{2\sqrt{N}}+\frac{1}{N}\right)}{(n+1)!\alpha^{n+1}}, \tag{7.41}$$

$\forall\, N \geq N^{*}$, $N \in \mathbb{N}$.

It holds $\lim_{N \to +\infty} K_{N,\mu}(f)(x) = f(x)$.

Proof By (7.16) we get

$$\left|K_{N,\mu}(f)(x) - f(x)\right| \leq \frac{\omega_1\left(D_\alpha^n f, h\right)}{(n+1)!\alpha^{n+1} h} K_{N,\mu}\left(|\cdot - x|^{(n+1)\alpha}\right)(x) \overset{(7.23)}{\leq}$$

$$\frac{\omega_1\left(D_\alpha^n f, h\right)}{(n+1)!\alpha^{n+1} h}\left(\sqrt{\frac{x(1-x)}{N}} + \frac{1}{N}\right) \overset{(7.29)}{\leq}$$

$$\frac{\omega_1\left(D_\alpha^n f, h\right)}{(n+1)!\alpha^{n+1} h}\left(\frac{1}{2\sqrt{N}} + \frac{1}{N}\right) = \tag{7.42}$$

(setting $h := \frac{1}{2\sqrt{N}} + \frac{1}{N} > 0$)

$$\frac{\omega_1\left(D_\alpha^n f, \frac{1}{2\sqrt{N}} + \frac{1}{N}\right)}{(n+1)!\alpha^{n+1}},$$

proving the claim. ■

We continue with

Theorem 7.32 *Let* $x \in (0,1)$ *and* $N^* \in \mathbb{N} - \{1\} : \frac{(1+3\sqrt{2})}{2\sqrt{N^*}} + \frac{1}{N^*} \leq \min(x, 1-x)$. *The rest as in Theorem 7.31. Then*

$$\left|\left(D_{N,\Gamma_N}(f)\right)(x) - f(x)\right| \leq \frac{\omega_1\left(D_\alpha^n f, \frac{(1+3\sqrt{2})}{2\sqrt{N}} + \frac{1}{N}\right)}{(n+1)!\alpha^{n+1}}, \tag{7.43}$$

$\forall\, N \geq N^*$, $N \in \mathbb{N} - \{1\}$.

It holds $\lim_{N \to +\infty} D_{N,\Gamma_N}(f)(x) = f(x)$.

Proof We use Theorem 7.14:

By (7.16) we get

$$\left|D_{N,\Gamma_N}(f)(x) - f(x)\right| \leq \frac{\omega_1\left(D_\alpha^n f, h\right)}{(n+1)!\alpha^{n+1} h} D_{N,\Gamma_N}\left(|\cdot - x|^{(n+1)\alpha}\right)(x) \overset{(7.28)}{\leq}$$

$$\frac{\omega_1\left(D_\alpha^n f, h\right)}{(n+1)!\alpha^{n+1} h}\left(\frac{\left(1+\sqrt{2}\right)\sqrt{x(1-x)} + \sqrt{2x}}{\sqrt{N}} + \frac{1}{N}\right) \overset{(7.30)}{\leq}$$

$$\frac{\omega_1\left(D_\alpha^n f, h\right)}{(n+1)!\alpha^{n+1} h}\left(\frac{\left(1+3\sqrt{2}\right)}{2\sqrt{N}} + \frac{1}{N}\right) \tag{7.44}$$

(setting $h := \frac{(1+3\sqrt{2})}{2\sqrt{N}} + \frac{1}{N} > 0$)

$$= \frac{\omega_1\left(D_\alpha^n f, \frac{(1+3\sqrt{2})}{2\sqrt{N}} + \frac{1}{N}\right)}{(n+1)!\alpha^{n+1}},$$

proving the claim. ■

We need

Definition 7.33 Let Ω be a set, and let $f, g : \Omega \to \mathbb{R}$ be bounded functions. We say that f and g are comonotonic, if for every $\omega, \omega' \in \Omega$,

$$\left(f(\omega) - f\left(\omega'\right)\right)\left(g(\omega) - g\left(\omega'\right)\right) \geq 0. \tag{7.45}$$

We also need the famous Schmeidler's Representation Theorem (Schmeidler 1986)

Theorem 7.34 ([12]) *Denote with $\mathcal{L}_\infty(\mathcal{A})$ the vector space of $\mathcal{A}$-measurable bounded real valued functions on Ω, where $\mathcal{A} \subset 2^\Omega$ is a σ -algebra. Given a real functional $\Gamma : \mathcal{L}_\infty(\mathcal{A}) \to \mathbb{R}$, assume that for $f, g \in \mathcal{L}_\infty(\mathcal{A})$:*

(i) $\Gamma(cf) = c\Gamma(f)$, $\forall\, c > 0$,

(ii) $f \leq g$, implies $\Gamma(f) \leq \Gamma(g)$,

and

(iii) $\Gamma(f+g) = \Gamma(f) + \Gamma(g)$, for any comonotonic f, g.

Then $\gamma(A) := \Gamma(1_A)$, $\forall\, A \in \mathcal{A}$, defines a finite monotone set function on $\mathcal{A}$, and Γ is the Choquet integral with respect to γ, i.e.

$$\Gamma(f) = (C)\int_\Omega f(t)\, d\gamma(t), \;\; \forall\, f \in \mathcal{L}_\infty(\mathcal{A}). \tag{7.46}$$

Above 1_A denotes the characteristic function on A.

Next we give nice interpretations of Theorems 7.9, 7.11, 7.16 and 7.17 involving Choquet integrals and based on Theorem 7.34.

We make

Remark 7.35 Consider here $[a, b] \subset \mathbb{R}_+$, $\mathcal{B} = \mathcal{B}([a, b])$ is the Borel σ-algebra on $[a, b]$, and $\mathcal{L}_\infty(\mathcal{B})$ is the vector space of $\mathcal{B}$-measurable bounded real valued functions on $[a, b]$. Let $(L_N)_{N\in\mathbb{N}}$ be a sequence of positive sublinear operators from $\mathcal{L}_\infty(\mathcal{B})$ into $C_+([a, b])$, and $x \in [a, b]$. That is here L_N fulfills the positive homogeneity, monotonicity and subadditivity properties, see (7.8)–(7.10).

Assume $L_N(1) = 1, \forall\, N \in \mathbb{N}$. Clearly here $\mathcal{L}_\infty(\mathcal{B}) \supset C_+([a, b])$, where $[a, b] \subset [0, \infty)$. In particular we treat $L_N|_{C_+([a,b])}$, just denoted for simplicity by L_N, $\forall\, N \in \mathbb{N}$.

It is clear that $L_N(\cdot)(x) : \mathcal{L}_\infty(\mathcal{B}) \to \mathbb{R}$ is a functional, $\forall\, N \in \mathbb{N}$. It has the properties:

(i)

$$L_N(cf)(x) = cL_N(f)(x), \ \forall\, c > 0, \ \forall\, f \in \mathcal{L}_\infty(\mathcal{B}), \tag{7.47}$$

(ii)

$$f \leq g, \text{ implies } L_N(f)(x) \leq L_N(g)(x), \quad \text{where } f, g \in \mathcal{L}_\infty(\mathcal{B}), \tag{7.48}$$

and

(iii)

$$L_N(f+g)(x) \leq L_N(f)(x) + L_N(g)(x), \ \forall\, f, g \in \mathcal{L}_\infty(\mathcal{B}). \tag{7.49}$$

For comonotonic $f, g \in \mathcal{L}_\infty(\mathcal{B})$, we further assume that

$$L_N(f+g)(x) = L_N(f)(x) + L_N(g)(x). \tag{7.50}$$

In that case L_N is called comonotonic.

By Theorem 7.34 we get that:

$$\gamma_{N,x}(A) := L_N(1_A)(x), \ \forall\, A \in \mathcal{B}, \forall\, N \in \mathbb{N}, \tag{7.51}$$

defines a finite monotone set function on $\mathcal{B}$, and

$$L_N(f)(x) = (C)\int_a^b f(t)\, d\gamma_{N,x}(t), \tag{7.52}$$

$\forall\, f \in \mathcal{L}_\infty(\mathcal{B}), \forall\, N \in \mathbb{N}$.

In particular (7.52) is valid for any $f \in C_+([a,b])$. Furthermore $\gamma_{N,x}$ is normalized, that is $\gamma_{N,x}([a,b]) = 1, \forall\, N \in \mathbb{N}$.

We give

Theorem 7.36 *Let $\alpha \in (0,1]$, $[a,b] \subseteq [0,\infty)$. Suppose f is $\mathbb{R}_+$ valued and is α-conformable fractional differentiable on $[a,b]$, with $D_\alpha f$ being continuous on $[a,b]$. Let $x \in [a,b]$ such that $D_\alpha f(x) = 0$, and $(L_N)_{N\in\mathbb{N}}$ be a sequence of positive sublinear comonotonic operators from $\mathcal{L}_\infty(\mathcal{B})$ into $C_+([a,b])$. We assume that $L_N(1) = 1$, and $(C)\int_a^b |t-x|^{\alpha+1}\, d\gamma_{N,x}(t) > 0$, $(C)\int_a^b (t-x)^{2(\alpha+1)}$ $d\gamma_{N,x}(t) > 0$, $\forall\, N \in \mathbb{N}$. Then*

$$|L_N(f)(x) - f(x)| \leq \frac{\omega_1^\alpha\left(D_\alpha f, \left((C)\int_a^b (t-x)^{2(\alpha+1)}\, d\gamma_{N,x}(t)\right)^{\frac{\alpha}{2(\alpha+1)}}\right)}{\alpha}.$$

$$\left[\left((C)\int_a^b |t-x|^{\alpha+1}\,d\gamma_{N,x}(t)\right)^{\frac{\alpha}{\alpha+1}} + \frac{1}{2}\left((C)\int_a^b (t-x)^{2(\alpha+1)}\,d\gamma_{N,x}(t)\right)^{\frac{\alpha}{2(\alpha+1)}}\right], \tag{7.53}$$

$\forall\, N \in \mathbb{N}$.

As $(C)\int_a^b (t-x)^{2(\alpha+1)}\,d\gamma_{N,x}(t) \to 0$, $N \to \infty$, *we get that* $\lim_{N\to+\infty} L_N(f)(x) = f(x)$.

Proof By Theorems 7.9 and 7.34. ∎

Theorem 7.37 *Let* $\alpha \in (0,1]$, $n \in \mathbb{N}$. *Suppose* f *is* $\mathbb{R}_+$ *valued and is* n *times conformable* α*-fractional differentiable on* $[a,b] \subseteq [0,\infty)$, *and* $D_\alpha^n f$ *is continuous on* $[a,b]$. *For a fixed* $x \in [a,b]$ *we have* $D_\alpha^k f(x) = 0$, $k = 1, \ldots, n$. *Let positive sublinear comonotonic operators* $\{L_N\}_{N\in\mathbb{N}}$ *from* $\mathcal{L}_\infty(\mathcal{B})$ *into* $C_+([a,b])$, *such that* $(C)\int_a^b |t-x|^{n(\alpha+1)}\,d\gamma_{N,x}(t)$, $(C)\int_a^b |t-x|^{(n+1)(\alpha+1)}\,d\gamma_{N,x}(t) > 0$, $\forall\, N \in \mathbb{N}$. *Then*

$$|L_N(f)(x) - f(x)| \le \frac{\omega_1^\alpha\left(D_\alpha^n f, \left((C)\int_a^b |t-x|^{(n+1)(\alpha+1)}\,d\gamma_{N,x}(t)\right)^{\frac{\alpha}{(n+1)(\alpha+1)}}\right)}{\alpha^n n!}\cdot \tag{7.54}$$

$$\left[\left((C)\int_a^b |t-x|^{n(\alpha+1)}\,d\gamma_{N,x}(t)\right)^{\frac{\alpha}{\alpha+1}} + \right.$$

$$\left.\frac{1}{(n+1)}\left((C)\int_a^b |t-x|^{(n+1)(\alpha+1)}\,d\gamma_{N,x}(t)\right)^{\frac{n\alpha}{(n+1)(\alpha+1)}}\right],$$

$\forall\, N \in \mathbb{N}$.

As $(C)\int_a^b |t-x|^{(n+1)(\alpha+1)}\,d\gamma_{N,x}(t) \to 0$, *when* $N \to \infty$, *we get that* $\lim_{N\to+\infty} L_N(f)(x) = f(x)$.

Proof By Theorems 7.11 and 7.34. ∎

We continue with

Theorem 7.38 *Let* $\{L_N\}_{N\in\mathbb{N}}$ *from* $\mathcal{L}_\infty(\mathcal{B})$ *into* $C_+([a,b])$ *positive sublinear comonotonic operators, such that* $L_N(1) = 1$, $\forall\, N \in \mathbb{N}$. *Additionally assume that* $(C)\int_a^b |t-x|^{(n+1)(\alpha+1)}\,d\gamma_{N,x}(t) > 0$, $\forall\ N \in \mathbb{N}$; $x \in (a,b)$. *Here* $\alpha \in (0,1]$, *and* $n \in \mathbb{N}$. *Suppose* $f \in C_+([a,b])$ *is* n *times conformable* α*-fractional differentiable on* $[a,b] \subseteq [0,\infty)$, *and* $D_\alpha^n f$ *is continuous on* $[a,b]$. *Assume here* $0 < \left((C)\int_a^b |t-x|^{(n+1)(\alpha+1)}\,d\gamma_{N,x}(t)\right)^{\frac{\alpha}{\alpha+1}} \le \min(x-a, b-x)$, $\forall\, N \in \mathbb{N} : N \ge N^* \in \mathbb{N}$, *and assume* $\left|D_\alpha^n f\right|$ *is convex over* $[a,b]$. *Furthermore assume that* $D_\alpha^k f(x) = 0$, $k = 1, \ldots, n$. *Then*

$$|L_N(f)(x) - f(x)| \leq \frac{b^{1-\alpha}\omega_1\left(D_\alpha^n f, \left((C)\int_a^b |t-x|^{(n+1)(\alpha+1)} d\gamma_{N,x}(t)\right)^{\frac{\alpha}{\alpha+1}}\right)}{(n+1)!\alpha^{n+1}}, \tag{7.55}$$

$\forall N \geq N^*$; $N, N^* \in \mathbb{N}$.

If $(C)\int_a^b |t-x|^{(n+1)(\alpha+1)} d\gamma_{N,x}(t) \to 0$, *then* $L_N(f)(x) \to f(x)$ *as* $N \to \infty$.

Proof By Theorems 7.16 and 7.34. ∎

Theorem 7.39 *Let* $\{L_N\}_{N\in\mathbb{N}}$ *from* $\mathcal{L}_\infty(\mathcal{B})$ *into* $C_+([a,b])$ *positive sublinear comonotonic operators, such that* $L_N(1) = 1$, $\forall N \in \mathbb{N}$. *Additionally assume that* $(C)\int_a^b |t-x|^{(n+1)\alpha} d\gamma_{N,x}(t) > 0$, $\forall N \in \mathbb{N}$; $x \in (a,b)$. *Here* $\alpha \in (0,1]$, *and* $n \in \mathbb{N}$. *Suppose* $f \in C_+([a,b])$ *is* n *times conformable* α *-fractional differentiable on* $[a,b] \subseteq [0,\infty)$, *and* $D_\alpha^n f$ *is continuous on* $[a,b]$. *Assume here* $0 < \left((C)\int_a^b |t-x|^{(n+1)\alpha} d\gamma_{N,x}(t)\right) \leq \min(x-a, b-x)$, $\forall N \in \mathbb{N} : N \geq N^* \in \mathbb{N}$, *and assume* $\left|D_\alpha^n f\right|$ *is convex over* $[a,b]$. *Furthermore assume that* $D_\alpha^k f(x) = 0$, $k = 1, ..., n$. *Then*

$$|L_N(f)(x) - f(x)| \leq \frac{b^{1-\alpha}\omega_1\left(D_\alpha^n f, (C)\int_a^b |t-x|^{(n+1)\alpha} d\gamma_{N,x}(t)\right)}{(n+1)!\alpha^{n+1}}, \tag{7.56}$$

$\forall N \geq N^*$, *where* $N, N^* \in \mathbb{N}$.

If $(C)\int_a^b |t-x|^{(n+1)\alpha} d\gamma_{N,x}(t) \to 0$, *then* $L_N(f)(x) \to f(x)$ *as* $N \to \infty$.

Proof By Theorems 7.17 and 7.34. ∎

We make

Remark 7.40 Consider again $[a,b] \subset \mathbb{R}_+$, $\mathcal{B} = \mathcal{B}([a,b])$ the Borel σ algebra on $[a,b]$. For each $N \in \mathbb{N}$ and each $x \in [a,b]$ consider the monotone set functions $\mu_{N,x}$; $\mathcal{B} \to \mathbb{R}_+$. We assume that all $\mu_{N,x}$ are normalized, that is $\mu_{N,x}([a,b]) = 1$, and submodular. Here we consider the operators $T_N : C_+([a,b]) \to C_+([a,b])$ given by the formula

$$T_N(f)(x) = (C)\int_a^b f(t)\, d\mu_{N,x}(t), \tag{7.57}$$

$\forall N \in \mathbb{N}$, $\forall x \in [a,b]$.

In fact here $\mu_{N,x}$ are chosen so that $T_N(C_+([a,b])) \subseteq C_+([a,b])$.

We notice here that hold:

(i)

$$T_N(\alpha f)(x) = \alpha T_N(f)(x), \ \forall \alpha \geq 0, \tag{7.58}$$

(ii)

$$f \leq g, \text{ implies } T_N(f)(x) \leq T_N(g)(x), \tag{7.59}$$

and

(iii)
$$T_N(f+g)(x) \leq T_N(f)(x) + T_N(g)(x), \tag{7.60}$$

$\forall N \in \mathbb{N}, \forall x \in [a,b], \forall f, g \in C_+([a,b])$.

Clearly T_N are positive sublinear operators, compare to (7.8)–(7.10). We also have that $T_N(1) = 1, \forall N \in \mathbb{N}$.

We give

Theorem 7.41 *Let $\alpha \in (0,1]$, $[a,b] \subseteq [0,\infty)$. Suppose f is α-conformable fractional differentiable on $[a,b]$. $D_\alpha f$ is continuous on $[a,b]$. Let an $x \in [a,b]$ such that $D_\alpha f(x) = 0$. Assume $(C)\int_a^b |t-x|^{\alpha+1} d\mu_{N,x}(t)$, $(C)\int_a^b (t-x)^{2(\alpha+1)} d\mu_{N,x}(t) > 0$, $\forall N \in \mathbb{N}$. Then*

$$|T_N(f)(x) - f(x)| \leq \frac{\omega_1^\alpha\left(D_\alpha f, \left((C)\int_a^b (t-x)^{2(\alpha+1)} d\mu_{N,x}(t)\right)^{\frac{\alpha}{2(\alpha+1)}}\right)}{\alpha}.$$

$$\left[\left((C)\int_a^b |t-x|^{\alpha+1} d\mu_{N,x}(t)\right)^{\frac{\alpha}{\alpha+1}} + \frac{1}{2}\left((C)\int_a^b (t-x)^{2(\alpha+1)} d\mu_{N,x}(t)\right)^{\frac{\alpha}{2(\alpha+1)}}\right], \tag{7.61}$$

$\forall N \in \mathbb{N}$.

As $N \to \infty$, and $(C)\int_a^b (t-x)^{2(\alpha+1)} d\mu_{N,x}(t) \to 0$, we obtain $T_N(f)(x) \to f(x)$.

Proof By Theorem 7.9. ■

Theorem 7.42 *Let $\alpha \in (0,1]$, $n \in \mathbb{N}$. Suppose f is n times conformable α-fractional differentiable on $[a,b] \subseteq [0,\infty)$ and takes values on $\mathbb{R}_+$. $D_\alpha^n f$ is continuous on $[a,b]$. For a fixed $x \in [a,b]$ we have $D_\alpha^k f(x) = 0$, $k = 1,...,n$. Assume that $(C)\int_a^b |t-x|^{n(\alpha+1)} d\mu_{N,x}(t)$, $(C)\int_a^b |t-x|^{(n+1)(\alpha+1)} d\mu_{N,x}(t) > 0$, $\forall\ N \in \mathbb{N}$. Then*

$$|T_N(f)(x) - f(x)| \leq \frac{\omega_1^\alpha\left(D_\alpha^n f, \left((C)\int_a^b |t-x|^{(n+1)(\alpha+1)} d\mu_{N,x}(t)\right)^{\frac{\alpha}{(n+1)(\alpha+1)}}\right)}{\alpha^n n!}.$$

$$\left[\left((C)\int_a^b |t-x|^{n(\alpha+1)} d\mu_{N,x}(t)\right)^{\frac{\alpha}{\alpha+1}} + \right.$$

$$\left.\frac{1}{(n+1)}\left((C)\int_a^b |t-x|^{(n+1)(\alpha+1)} d\mu_{N,x}(t)\right)^{\frac{n\alpha}{(n+1)(\alpha+1)}}\right], \tag{7.62}$$

$\forall N \in \mathbb{N}$.

As $N \to \infty$, and $(C)\int_a^b |t-x|^{(n+1)(\alpha+1)} d\mu_{N,x}(t) \to 0$, we get $T_N(f)(x) \to f(x)$.

Proof By Theorem 7.11. ∎

We continue with

Theorem 7.43 *Assume $(C)\int_a^b |t-x|^{(n+1)(\alpha+1)} d\mu_{N,x}(t) > 0, \forall N \in \mathbb{N}; x \in (a,b)$. Here $\alpha \in (0,1]$, and $n \in \mathbb{N}$. Suppose $f \in C_+([a,b])$ is n times conformable α-fractional differentiable on $[a,b] \subseteq [0,\infty)$, and $D_\alpha^n f$ is continuous on $[a,b]$. Assume here that $0 < \left((C)\int_a^b |t-x|^{(n+1)(\alpha+1)} d\mu_{N,x}(t)\right)^{\frac{\alpha}{\alpha+1}} \leq \min(x-a, b-x)$, $\forall N \in \mathbb{N} : N \geq N^* \in \mathbb{N}$, and assume $\left|D_\alpha^n f\right|$ is convex over $[a,b]$. Furthermore assume that $D_\alpha^k f(x) = 0$, $k = 1, ..., n$. Then*

$$|T_N(f)(x) - f(x)| \leq \frac{b^{1-\alpha}\omega_1\left(D_\alpha^n f, \left((C)\int_a^b |t-x|^{(n+1)(\alpha+1)} d\mu_{N,x}(t)\right)^{\frac{\alpha}{\alpha+1}}\right)}{(n+1)!\alpha^{n+1}}, \tag{7.63}$$

$\forall N \in \mathbb{N} : N \geq N^* \in \mathbb{N}$.

If $(C)\int_a^b |t-x|^{(n+1)(\alpha+1)} d\mu_{N,x}(t) \to 0$, then $T_N(f)(x) \to f(x)$ as $N \to \infty$.

Proof By Theorem 7.16. ∎

Theorem 7.44 *Assume $(C)\int_a^b |t-x|^{(n+1)\alpha} d\mu_{N,x}(t) > 0$, $\forall N \in \mathbb{N}$. Here $\alpha \in (0,1]$, $n \in \mathbb{N}$ and $x \in (a,b)$; $[a,b] \subseteq [0,\infty)$. Suppose $f \in C_+([a,b])$ is n times conformable α-fractional differentiable on $[a,b]$, and $D_\alpha^n f$ is continuous on $[a,b]$. Let $0 < (C)\int_a^b |t-x|^{(n+1)\alpha} d\mu_{N,x}(t) \leq \min(x-a, b-x)$, $\forall N \geq N^*$; $N, N^* \in \mathbb{N}$, and assume $\left|D_\alpha^n f\right|$ is convex over $[a,b]$. Furthermore assume that $D_\alpha^k f(x) = 0$, $k = 1, ..., n$. Then*

$$|T_N(f)(x) - f(x)| \leq \frac{b^{1-\alpha}\omega_1\left(D_\alpha^n f, (C)\int_a^b |t-x|^{(n+1)\alpha} d\mu_{N,x}(t)\right)}{(n+1)!\alpha^{n+1}}, \tag{7.64}$$

$\forall N \geq N^*$, where $N, N^* \in \mathbb{N}$.

If $(C)\int_a^b |t-x|^{(n+1)\alpha} d\mu_{N,x}(t) \to 0$, then $T_N(f)(x) \to f(x)$ as $N \to \infty$.

Proof By Theorem 7.17. ∎

References

1. M. Abu Hammad, R. Khalil, Abel's formula and Wronskian for conformable fractional differential equations. Int. J. Differ. Equ. Appl. **13**(3), 177–183 (2014)
2. G. Anastassiou, *Moments in Probability and Approximation Theory*. Pitman Research Notes in Mathematics Series (Longman Group UK, New York, 1993)

3. G. Anastassiou, Conformable fractional approximation by max-product operators. Studia Mathematica Babes Bolyai **63**(1), 3–22 (2018)
4. G. Anastassiou, Conformable fractional approximations by max-product operators using convexity. Arab. J. Math. (2018). Accepted for publication
5. G.A. Anastassiou, Conformable fractional approximation by Choquet integrals. J. Comput. Anal. Appl. (2018). Accepted
6. G. Choquet, Theory of capacities. Ann. Inst. Fourier (Grenoble) **5**, 131–295 (1954)
7. D. Denneberg, *Non-additive Measure and Integral* (Kluwer, Dordrecht, 1994)
8. D. Dubois, H. Prade, *Possibility Theory* (Plenum Press, New York, 1988)
9. S. Gal, Uniform and pointwise quantitative approximation by Kantorovich–Choquet type integral operators with respect to monotone and submodular set functions. Mediterr. J. Math. **14**(5), 12 pp. (2017). Art. 205
10. S. Gal, S. Trifa, Quantitative estimates in uniform and pointwise approximation by Bernstein–Durrmeyer–Choquet operators. Carpathian J. Math. **33**(1), 49–58 (2017)
11. R. Khalil, M. Al Horani, A. Yousef, M. Sababheh, A new definition of fractional derivative. J. Comput. Appl. Math. **264**, 65–70 (2014)
12. D. Schmeidler, Integral representation without additivity. Proc. Am. Math. Soc. **97**, 255–261 (1986)
13. D. Schmeidler, Subjective probability and expected utility without additivity. Econometrica **57**, 571–587 (1989)
14. L.S. Shapley, A Value for n-person games, in *Contributions to the Theory of Games*, ed. by H.W. Kuhn, A.W. Tucker. Annals of Mathematical Studies, vol. 28 (Princeton University Press, Princeton, 1953), pp. 307–317
15. Z. Wang, G.J. Klir, *Generalized Measure Theory* (Springer, New York, 2009)

Chapter 8
Multivariate and Convex Quantitative Approximation by Choquet Integrals

Here we consider the quantitative approximation of positive sublinear operators to the unit operator. These are given a precise Choquet integral interpretation. Initially we start with the study of the rate of the convergence of the well-known Bernstein–Kantorovich–Choquet and Bernstein–Durrweyer–Choquet polynomial Choquet-integral operators. We introduce also their multivariate analogs. Then we study the very general comonotonic positive sublinear operators based on the representation theorem of Schmeidler [10]. We finish with the approximation by the very general direct Choquet-integral form positive sublinear operators. All approximations are given via inequalities involving the modulus of continuity of the approximated function or its higher order derivative. We derive univariate and multivariate results without or with convexity assumptions. In the latter case estimates become very elegant and brief. It follows [4].

8.1 Introduction

G. Choquet [5], introduced the capacities and his integral. Initially these were applied to statistical mechanics and potential theory, and they gave rise to the study of non-additive measure theory. Slowly but steady these ideas of Choquet started to attract economists especially after the very important work of Shapley [12] in the study of cooperative games. Capacities and Choquet integrals became main stream to Decision theorists since 1989 when D. Schmeidler [11] was the first to use them in an axiomatic model of choice with non-additive beliefs. The expected utility results are strengthened by the use of Choquet capacities instead of probability measures.

In now days Choquet integral has wide applications, among others, to decision making under risk and uncertainty, in finance, in economics, in portofolio problems and in insurance.

Our motivation also comes from the foundations of Bayesian decision theory and subjective probability.

G. A. Anastassiou, *Ordinary and Fractional Approximation by Non-additive Integrals: Choquet, Shilkret and Sugeno Integral Approximators*, Studies in Systems, Decision and Control 190, https://doi.org/10.1007/978-3-030-04287-5_8

Because of the paramount importance of Choquet integral, we decided to research the related positive sublinear operators approximation, part of it is exhibited in this work in the multivariate and convex sense.

8.2 Background - I

Next we present briefly about the Choquet integral, see also [9].

We make

Definition 8.1 Consider $\Omega \neq \varnothing$ and let $\mathcal{C}$ be a σ-algebra of subsets in Ω.

(i) (see, e.g., [13], p. 63) The set function $\mu : \mathcal{C} \to [0, +\infty]$ is called a monotone set function (or capacity) if $\mu(\varnothing) = 0$ and $\mu(A) \leq \mu(B)$ for all $A, B \in \mathcal{C}$, with $A \subset B$. Also, μ is called submodular if

$$\mu(A \cup B) + \mu(A \cap B) \leq \mu(A) + \mu(B), \text{ for all } A, B \in \mathcal{C}.$$

μ is called bounded if $\mu(\Omega) < +\infty$ and normalized if $\mu(\Omega) = 1$.

(ii) (see, e.g., [13], p. 233, or [5]) If μ is a monotone set function on $\mathcal{C}$ and if $f : \Omega \to \mathbb{R}$ is $\mathcal{C}$-measurable (that is, for any Borel subset $B \subset \mathbb{R}$ it follows $f^{-1}(B) \in \mathcal{C}$), then for any $A \in \mathcal{C}$, the Choquet integral is defined by

$$(C)\int_A f d\mu = \int_0^{+\infty} \mu\left(F_\beta(f) \cap A\right) d\beta + \int_{-\infty}^0 \left[\mu\left(F_\beta(f) \cap A\right) - \mu(A)\right] d\beta,$$

where we used the notation $F_\beta(f) = \{\omega \in \Omega : f(\omega) \geq \beta\}$. Notice that if $f \geq 0$ on A, then in the above formula we get $\int_{-\infty}^0 = 0$.

The integrals on the right-hand side are the usual Riemann integral.

The function f will be called Choquet integrable on A if $(C)\int_A f d\mu \in \mathbb{R}$.

Next we list some well known properties of the Choquet integral.

Remark 8.2 If $\mu : \mathcal{C} \to [0, +\infty]$ is a monotone set function, then the following properties hold:

(i) For all $a \geq 0$ we have $(C)\int_A af d\mu = a \cdot (C)\int_A f d\mu$ (if $f \geq 0$ then see, e.g., [13], Theorem 11.2, (5), p. 228 and if f is arbitrary sign, then see, e.g., [6], p. 64, Proposition 5.1, (ii)).

(ii) For all $c \in \mathbb{R}$ and f of arbitrary sign, we have (see, e.g., [13], pp. 232–233, or [6], p. 65) $(C)\int_A (f + c) d\mu = (C)\int_A f d\mu + c \cdot \mu(A)$.

If μ is submodular too, then for all f, g of arbitrary sign and lower bounded, we have (see, e.g., [6], p. 75, Theorem 6.3)

$$(C)\int_A (f + g) d\mu \leq (C)\int_A f d\mu + (C)\int_A g d\mu.$$

(iii) If $f \leq g$ on A then $(C)\int_A f d\mu \leq (C)\int_A g d\mu$ (see, e.g., [13], p. 228, Theorem 11.2, (3) if $f, g \geq 0$ and p. 232 if f, g are of arbitrary sign).

(iv) Let $f \geq 0$. If $A \subset B$ then $(C)\int_A f d\mu \leq (C)\int_B f d\mu$. In addition, if μ is finitely subadditive, then

$$(C)\int_{A\cup B} f d\mu \leq (C)\int_A f d\mu + (C)\int_B f d\mu.$$

(v) It is immediate that $(C)\int_A 1 \cdot d\mu(t) = \mu(A)$.

(vi) The formula $\mu(A) = \gamma(M(A))$, where $\gamma : [0, 1] \to [0, 1]$ is an increasing and concave function, with $\gamma(0) = 0$, $\gamma(1) = 1$ and M is a probability measure (or only finitely additive) on a σ-algebra on Ω (that is, $M(\emptyset) = 0$, $M(\Omega) = 1$ and M is countably additive), gives simple examples of normalized, monotone and submodular set functions (see, e.g., [6], pp. 16–17, Example 2.1). Such of set functions μ are also called distorsions of countably normalized, additive measures (or distorted measures). For a simple example, we can take $\gamma(t) = \frac{2t}{1+t}$, $\gamma(t) = \sqrt{t}$.

If the above γ function is increasing, concave and satisfies only $\gamma(0) = 0$, then for any bounded Borel measure m, $\mu(A) = \gamma(m(A))$ gives a simple example of bounded, monotone and submodular set function.

(vii) If μ is a countably additive bounded measure, then the Choquet integral $(C)\int_A f d\mu$ reduces to the usual Lebesgue type integral (see, e.g., [6], p. 62, or [13], p. 226).

(viii) If $f \geq 0$, then $(C)\int_A f d\mu \geq 0$.

(ix) Let $\mu = \sqrt{M}$, where M is the Lebesgue measure on $[0, +\infty)$, then μ is a monotone and submodular set function, furthermore μ is strictly positive, see [8].

(x) If $\Omega = \mathbb{R}^N$, $N \in \mathbb{N}$, we call μ strictly positive if $\mu(A) > 0$, for any open subset $A \subseteq \mathbb{R}^N$.

We need some possibility theory:

Definition 8.3 ([7]) For the $\Omega \neq \emptyset$, the power set $\mathcal{P}(\Omega)$ denotes the family of all subsets of Ω.

(i) A function $\lambda : \Omega \to [0, 1]$ with the property $\sup\{\lambda(s) : s \in \Omega\} = 1$, is called possibility distribution on Ω.

(ii) $P : \mathcal{P}(\Omega) \to [0, 1]$ is called possibility measure, if it satisfies $P(\emptyset) = 0$, $P(\Omega) = 1$, and $P(\cup_{i\in I} A_i) = \sup\{P(A_i) : i \in I\}$ for all $A_i \subset \Omega$, and any I, an at most countable family of indices. Note that if $A, B \subset \Omega$, $A \subset B$, then the last property implies $P(A) \leq P(B)$ and that $P(A \cup B) \leq P(A) + P(B)$.

Any possibility distribution λ on Ω, induces the possibility measure $P_\lambda : \mathcal{P}(\Omega) \to [0, 1]$, $P_\lambda(A) = \sup\{\lambda(s) : s \in A\}$, $A \subset \Omega$. Also, if $f : \Omega \to \mathbb{R}_+$, then the possibilistic integral of f on $A \subset \Omega$ with respect to P_λ is defined by $(Pos)\int_A f dP_\lambda = \sup\{f(t)\lambda(t) : t \in A\}$ (see [7], chapter 1).

Note that any possibility measure μ is normalized, monotone and submodular. From $\mu(A \cup B) = \max\{\mu(A), \mu(B)\}$ we get monotonicity, and from $\mu(A \cap B) \leq \min\{\mu(A), \mu(B)\}$ we derive the submodularity.

8.3 Background - II

Here is about univariate convex approximation. Let $f \in C([a,b])$, we define the first modulus of continuity of f:

$$\omega_1(f,h) := \sup_{\substack{x,y\in[a,b]\\ |x-y|\le h}} |f(x)-f(y)|. \tag{8.1}$$

We mention

Definition 8.4 ([3]) Call $C_+([a,b]) := \{f : [a,b] \to \mathbb{R}_+ \text{ and continuous }\}$. Let L_N from $C_+([a,b])$ into $C_+([a,b])$ be a sequence of operators satisfying the following properties:
(i) (positive homogeneous)

$$L_N(\alpha f) = \alpha L_N(f),\ \forall \alpha \ge 0,\ \forall f \in C_+([a,b]), \tag{8.2}$$

(ii) (Monotonicity)
if $f, g \in C_+([a,b])$ satisfy $f \le g$, then

$$L_N(f) \le L_N(g),\ \forall N \in \mathbb{N}, \tag{8.3}$$

(iii) (Subadditivity)

$$L_N(f+g) \le L_N(f) + L_N(g),\ \forall f, g \in C_+([a,b]). \tag{8.4}$$

We call L_N positive sublinear operators.

We need:

Theorem 8.5 ([3]) *Let $f \in C_+([a,b])$, $x \in (a,b)$, $0 < h \le \min(x-a, b-x)$, and $|f(\cdot) - f(x)|$ is a convex function over $[a,b]$. Let $\{L_N\}_{N\in\mathbb{N}}$ positive sublinear operators from $C_+([a,b])$ into itself, such that $L_N(1) = 1$, $\forall N \in \mathbb{N}$. Then*

$$|L_N(f)(x) - f(x)| \le \frac{\omega_1(f,h)}{h} L_N(|\cdot - x|)(x),\ \forall N \in \mathbb{N}. \tag{8.5}$$

Theorem 8.6 ([3]) *Let $f \in C^n([a,b], \mathbb{R}_+)$, $n \in \mathbb{N}$, $x \in (a,b)$, $0 < h \le \min(x-a, b-x)$, and $\left|f^{(n)}\right|$ is convex over $[a,b]$. Assume that $f^{(k)}(x) = 0$, $k = 1, ..., n$. Let $\{L_N\}_{N\in\mathbb{N}}$ positive sublinear operators from $C_+([a,b])$ into itself, such that $L_N(1) = 1$, $\forall N \in \mathbb{N}$. Then*

$$|L_N(f)(x) - f(x)| \le \frac{\omega_1\left(f^{(n)}, h\right)}{h(n+1)!} L_N\left(|\cdot - x|^{n+1}\right)(x),\ \forall N \in \mathbb{N}. \tag{8.6}$$

We also need:

Theorem 8.7 ([3]) *Let* $f \in C_+([a,b])$, $x \in (a,b)$, $0 < L_N(|\cdot - x|)(x) \leq \min(x-a, b-x)$, $\forall N \in \mathbb{N}$, *and* $|f(\cdot) - f(x)|$ *is a convex function over* $[a,b]$. *Here* L_N *are positive sublinear operators from* $C_+([a,b])$ *into itself, such that* $L_N(1) = 1$, $\forall N \in \mathbb{N}$. *Then*

$$|L_N(f)(x) - f(x)| \leq \omega_1(f, L_N(|\cdot - x|)(x)), \ \forall N \in \mathbb{N}. \tag{8.7}$$

If $L_N(|\cdot - x|)(x) \to 0$, *then* $L_N(f)(x) \to f(x)$, *as* $N \to +\infty$.

Theorem 8.8 ([3]) *Let* $f \in C^n([a,b], \mathbb{R}_+)$, $n \in \mathbb{N}$, $x \in (a,b)$, $0 < L_N(|\cdot - x|^{n+1})(x) \leq \min(x-a, b-x)$, $\forall N \in \mathbb{N}$, *and* $|f^{(n)}|$ *is convex over* $[a,b]$. *Assume that* $f^{(k)}(x) = 0$, $k = 1, ..., n$. *Here* $\{L_N\}_{N\in\mathbb{N}}$ *are positive sublinear operators from* $C_+([a,b])$ *into itself, such that* $L_N(1) = 1$, $\forall N \in \mathbb{N}$. *Then*

$$|L_N(f)(x) - f(x)| \leq \frac{\omega_1\left(f^{(n)}, L_N\left(|\cdot - x|^{n+1}\right)(x)\right)}{(n+1)!}, \ \forall N \in \mathbb{N}. \tag{8.8}$$

If $L_N\left(|\cdot - x|^{n+1}\right)(x) \to 0$, *then* $L_N(f)(x) \to f(x)$, *as* $N \to +\infty$.

Next we combine both Theorems 8.5 and 8.6:

Theorem 8.9 ([3]) *Let* $f \in C^n([a,b], \mathbb{R}_+)$, $n \in \mathbb{Z}_+$, $x \in (a,b)$, $0 < h \leq \min(x-a, b-x)$, *and* $|f^{(n)}|$ *is convex over* $[a,b]$. *Assume that* $f^{(k)}(x) = 0$, $k = 1, ..., n$. *Let* $\{L_N\}_{N\in\mathbb{N}}$ *positive sublinear operators from* $C_+([a,b])$ *into itself, such that* $L_N(1) = 1$, $\forall N \in \mathbb{N}$. *Then*

$$|L_N(f)(x) - f(x)| \leq \frac{\omega_1\left(f^{(n)}, h\right)}{h(n+1)!} L_N\left(|\cdot - x|^{n+1}\right)(x), \ \forall N \in \mathbb{N}; n \in \mathbb{Z}_+. \tag{8.9}$$

The initial conditions $f^{(k)}(x) = 0$, $k = 1, ..., n$, *are void when* $n = 0$, *in that case we require* $|f(\cdot) - f(x)|$ *to be convex over* $[a,b]$.

8.4 Background - III

Here is about multivariate approximation.

Let Q be a compact and convex subset of $\mathbb{R}^k$, $k \in \mathbb{N} - \{1\}$ and let $x_0 := (x_{01}, ..., x_{0k}) \in Q$ be fixed. Let $f \in C^n(Q)$ and suppose that each nth partial derivative $f_\alpha = \frac{\partial^\alpha f}{\partial x^\alpha}$, where $\alpha := (\alpha_1, ..., \alpha_k)$, $\alpha_i \in \mathbb{Z}^+$, $i = 1, ..., k$, and $|\alpha| := \sum_{i=1}^k \alpha_i = n$, has relative to Q and the l_1-norm $\|\cdot\|$, a modulus of continuity $\omega_1(f_\alpha, h) \leq w$, where h and w are fixed positive numbers. Here

$$\omega_1(f_\alpha, h) := \sup_{\substack{x,y \in Q \\ \|x-y\|_{l_1} \leq h}} |f_\alpha(x) - f_\alpha(y)|. \tag{8.10}$$

We need

Definition 8.10 ([1]) Let Q be a compact and convex subset of $\mathbb{R}^k$, $k \in \mathbb{N} - \{1\}$. Here we denote

$$C_+(Q) = \{f : Q \to \mathbb{R}_+ \text{ and continuous}\}.$$

Let $L_N : C_+(Q) \to C_+(Q)$, $N \in \mathbb{N}$, be a sequence of operators satisfying the following properties:
(i) (positive homogeneous)

$$L_N(\alpha f) = \alpha L_N(f), \ \forall \alpha \geq 0, f \in C_+(Q); \tag{8.11}$$

(ii) (monotonicity)
if $f, g \in C_+(Q)$ satisfy $f \leq g$, then

$$L_N(f) \leq L_N(g), \ \forall N \in \mathbb{N}, \tag{8.12}$$

and
(iii) (subadditivity)

$$L_N(f+g) \leq L_N(f) + L_N(g), \ \forall f, g \in C_+(Q). \tag{8.13}$$

We call L_N positive sublinear operators.

We need

Theorem 8.11 ([1]) *Let Q be a compact and convex subset of $\mathbb{R}^k$, $k \in \mathbb{N} - \{1\}$ and let $x \in Q$ be fixed. Let $f \in C^n(Q, \mathbb{R}_+)$, $n \in \mathbb{N}$, $h > 0$. We assume that $f_\alpha(x) = 0$, for all $\alpha : |\alpha| = 1, ..., n$. Let $\{L_N\}_{N \in \mathbb{N}}$ positive sublinear operators mapping $C_+(Q)$ into itself, such that $L_N(1) = 1$. Then*

$$|L_N(f)(x) - f(x)| \leq \left(\max_{\alpha:|\alpha|=n} \omega_1(f_\alpha, h)\right) \cdot$$

$$\left(\frac{k^n}{(n+1)!h}\left(\sum_{i=1}^k L_N\left(|t_i - x_i|^{n+1}\right)(x)\right) + \frac{k^{n-1}}{2n!}\left(\sum_{i=1}^k L_N\left(|t_i - x_i|^n\right)(x)\right)\right.$$

$$\left.+\frac{hk^{n-2}}{8(n-1)!}\left(\sum_{i=1}^k L_N\left(|t_i - x_i|^{n-1}\right)(x)\right)\right), \ \forall N \in \mathbb{N}. \tag{8.14}$$

We will also use

Corollary 8.12 ([1]) (to Theorem 8.11, case of $n = 1$) *Let Q be a compact and convex subset of $\mathbb{R}^k$, $k \in \mathbb{N} - \{1\}$ and let $x \in Q$. Let $f \in C^1(Q, \mathbb{R}_+)$, $h > 0$. We assume that $\frac{\partial f(x)}{\partial x_i} = 0$, for $i = 1, ..., k$. Let $\{L_N\}_{N \in \mathbb{N}}$ be positive sublinear operators from $C_+(Q)$ into $C_+(Q)$: $L_N(1) = 1$, $\forall N \in \mathbb{N}$. Then*

$$|L_N(f)(x) - f(x)| \leq \left(\max_{i=1,\dots,k} \omega_1\left(\frac{\partial f}{\partial x_i}, h\right)\right) \cdot$$

$$\left[\frac{k}{2h}\left(\sum_{i=1}^{k} L_N\left((t_i - x_i)^2\right)(x)\right) + \frac{1}{2}\left(\sum_{i=1}^{k} L_N(|t_i - x_i|)(x)\right) + \frac{h}{8}\right], \quad (8.15)$$

$\forall N \in \mathbb{N}$.

We need

Theorem 8.13 ([1]) *Let Q with $\|\cdot\|$ the l_1-norm, be a compact and convex subset of $\mathbb{R}^k$, $k \in \mathbb{N} - \{1\}$, and $f \in C_+(Q)$; $h > 0$. We denote $\omega_1(f, h) := \sup_{\substack{x,y \in Q: \\ \|x-y\| \leq h}} |f(x) - f(y)|$, the modulus of continuity of f. Let $\{L_N\}_{N \in \mathbb{N}}$ be positive sublinear operators from $C_+(Q)$ into itself such that $L_N(1) = 1$, $\forall N \in \mathbb{N}$. Then*

$$|L_N(f)(x) - f(x)| \leq \omega_1(f, h)\left(1 + \frac{1}{h} L_N(\|t - x\|)(x)\right) \leq$$

$$\omega_1(f, h)\left(1 + \frac{1}{h}\left(\sum_{i=1}^{k} L_N(|t_i - x_i|)(x)\right)\right), \quad (8.16)$$

$\forall N \in \mathbb{N}$, $\forall x \in Q$, where $x := (x_1, \dots, x_k)$; $t = (t_1, \dots, t_k) \in Q$.

Theorem 8.14 ([1]) *Let Q with $\|\cdot\|$ the l_1-norm, be a compact and convex subset of $\mathbb{R}^k$, $k \in \mathbb{N} - \{1\}$, and $f \in C_+(Q)$. Let $\{L_N\}_{N \in \mathbb{N}}$ be positive sublinear operators from $C_+(Q)$ into itself, such that $L_N(1) = 1$, $\forall N \in \mathbb{N}$. We assume further that $L_N(\|t - x\|)(x) > 0$, $\forall N \in \mathbb{N}$. Then*

$$|L_N(f)(x) - f(x)| \leq 2\omega_1(f, L_N(\|t - x\|)(x)), \quad (8.17)$$

$\forall N \in \mathbb{N}$, $x = (x_1, \dots, x_k) \in Q$; $t = (t_1, \dots, t_k) \in Q$.

If $L_N(\|t - x\|)(x) \to 0$, then $L_N(f)(x) \to f(x)$, as $N \to +\infty$.

We also mention

Theorem 8.15 ([1]) *Let $(Q, \|\cdot\|)$, where $\|\cdot\|$ is the l_1-norm, be a compact and convex subset of $\mathbb{R}^k$, $k \in \mathbb{N} - \{1\}$, and let $x \in Q$ ($x = (x_1, \dots, x_k)$) be fixed. Let $f \in C^n(Q, \mathbb{R}_+)$, $n \in \mathbb{N}$, $h > 0$. We assume that $f_\alpha(x) = 0$, for all $\alpha : |\alpha| = 1, \dots, n$. Let $\{L_N\}_{N \in \mathbb{N}}$ be positive sublinear operators from $C_+(Q)$ into $C_+(Q)$, such that $L_N(1) = 1$, $\forall N \in \mathbb{N}$. Then*

$$|L_N(f)(x) - f(x)| \leq \left(\max_{\alpha:|\alpha|=n} \omega_1(f_\alpha, h)\right) \cdot \quad (8.18)$$

$$\left[\frac{L_N\left(\|\cdot - x\|^{n+1}\right)(x)}{(n+1)!h} + \frac{L_N\left(\|\cdot - x\|^{n}\right)(x)}{2n!} + \frac{h}{8(n-1)!} L_N\left(\|\cdot - x\|^{n-1}\right)(x)\right],$$

$\forall\, N \in \mathbb{N}$.

It follows

Theorem 8.16 ([1]) *All as in Theorem 8.15. Additionally assume that* $L_N\left(\|\cdot - x\|^{n+1}\right)(x) > 0$, $\forall\, N \in \mathbb{N}$. *Then*

$$|L_N(f)(x) - f(x)| \le \frac{1}{2n!}\left(3 + \frac{n}{4(n+1)}\right)\cdot$$

$$\left(\max_{\alpha:|\alpha|=n} \omega_1\left(f_\alpha, \frac{1}{(n+1)}\left(L_N\left(\|\cdot - x\|^{n+1}\right)(x)\right)^{\frac{1}{n+1}}\right)\right)\left(L_N\left(\|\cdot - x\|^{n+1}\right)(x)\right)^{\frac{n}{n+1}}, \tag{8.19}$$

$\forall\, N \in \mathbb{N}$, $x = (x_1, ..., x_k) \in Q$.

If $L_N\left(\|\cdot - x\|^{n+1}\right)(x) \to 0$, *then* $L_N(f)(x) \to f(x)$, *as* $N \to +\infty$.

An application for $n = 1$ follows:

Corollary 8.17 ([1]) *Let* $(Q, \|\cdot\|)$, *where* $\|\cdot\|$ *is the* l_1*-norm, be a compact and convex subset of* $\mathbb{R}^k$, $k \in \mathbb{N} - \{1\}$, *and let* $x \in Q$ ($x = (x_1, ..., x_k)$) *be fixed. Let* $f \in C^1(Q, \mathbb{R}_+)$. *We assume that* $\frac{\partial f}{\partial x_i}(x) = 0$, $i = 1, ..., k$. *Let* $\{L_N\}_{N \in \mathbb{N}}$ *be positive sublinear operators from* $C_+(Q)$ *into* $C_+(Q)$, *such that* $L_N(1) = 1$, $\forall\, N \in \mathbb{N}$. *Assume that* $L_N\left(\|\cdot - x\|^2\right)(x) > 0$, $\forall\, N \in \mathbb{N}$. *Then*

$$|L_N(f)(x) - f(x)| \le \frac{25}{16}\left(\max_{i=1,...,k} \omega_1\left(\frac{\partial f}{\partial x_i}, \frac{1}{2}\left(L_N\left(\|\cdot - x\|^2\right)(x)\right)^{\frac{1}{2}}\right)\right)\cdot$$

$$\left(L_N\left(\|\cdot - x\|^2\right)(x)\right)^{\frac{1}{2}}, \tag{8.20}$$

$\forall\, N \in \mathbb{N}$.

If $L_N\left(\|\cdot - x\|^2\right)(x) \to 0$, *then* $L_N(f)(x) \to f(x)$, *as* $N \to +\infty$.

8.5 Background - IV

Here is about multivariate convex approximation.

Let Q be a compact and convex subset of $\mathbb{R}^k$, $k \in \mathbb{N} - \{1\}$ and let $x_0 := (x_{01}, ..., x_{0k}) \in Q^o$ be fixed. Let $f \in C^n(Q)$ and suppose that each nth partial derivative $f_\alpha = \frac{\partial^\alpha f}{\partial x^\alpha}$, where $\alpha := (\alpha_1, ..., \alpha_k)$, $\alpha_i \in \mathbb{Z}^+$, $i = 1, ..., k$, and $|\alpha| := \sum_{i=1}^k \alpha_i = n$, has relative to Q and the l_1-norm $\|\cdot\|$, a modulus of continuity $\omega_1(f_\alpha, h) \le w$, where h and w are fixed positive numbers. Here again

$$\omega_1(f_\alpha, h) := \sup_{\substack{x,y \in Q \\ \|x-y\|_{l_1} \le h}} |f_\alpha(x) - f_\alpha(y)|. \tag{8.21}$$

We would assume that the ball $B(x_0, h) \subset Q$. We also would assume that $f_\alpha(x_0) = 0$, all $\alpha : |\alpha| = 1, ..., n$, and that $|f_\alpha(x)|$ is convex in x, all $\alpha : |\alpha| = n$.

We need

Theorem 8.18 ([2]) *Let Q with the l_1-norm $\|\cdot\|$, be a compact and convex subset of $\mathbb{R}^k$, $k \ge 1$, let $x = (x_1, ..., x_k) \in Q^o$ be fixed. Let $f \in C^n(Q, \mathbb{R}_+)$, $n \in \mathbb{N}$ and suppose that each nth partial derivative $f_\alpha = \frac{\partial^\alpha f}{\partial x^\alpha}$, where $\alpha = (\alpha_1, ..., \alpha_k)$, $\alpha_i \ge 0$, $i = 1, ..., k$, and $|\alpha| = \sum_{i=1}^k \alpha_i = n$, has relative to Q and the l_1-norm a modulus of continuity $\omega_1(f_\alpha, h)$, and that each $|f_\alpha(t)|$ is a convex function of $t \in Q$, all $\alpha : |\alpha| = n$. Assume further that $f_\alpha(x) = 0$, for all $\alpha : |\alpha| = 1, ..., n$; and $h > 0$ such that the ball in $\mathbb{R}^k : B(x, h) \subset Q$. Let $\{L_N\}_{N\in\mathbb{N}}$ be positive sublinear operators mapping $C_+(Q)$ into itself, such that $L_N(1) = 1$. Then*

$$|L_N(f)(x) - f(x)| \le \frac{\left(\max_{\alpha:|\alpha|=n} \omega_1(f_\alpha, h)\right) k^n}{h(n+1)!} \left(\sum_{i=1}^k L_N\left(|t_i - x_i|^{n+1}\right)(x)\right), \tag{8.22}$$

$\forall\, N \in \mathbb{N}$.

Theorem 8.19 ([2]) *Let Q with the l_1-norm$\|\cdot\|$, be a compact and convex subset of $\mathbb{R}^k$, $k \in \mathbb{N} - \{1\}$, and $f \in C_+(Q) : |f(t) - f(x)|$ is a convex function in $t := (t_1, ..., t_k) \in Q$ for a fixed $x = (x_1, ..., x_k) \in Q^o$. We denote $\omega_1(f, h) := \sup_{\substack{x,y \in Q: \\ \|x-y\| \le h}} |f(x) - f(y)|$, $h > 0$, the first modulus of continuity of f. We assume that the ball in $\mathbb{R}^k : B(x, h) \subset Q$. Let $\{L_N\}_{N\in\mathbb{N}}$ be positive sublinear operators from $C_+(Q)$ into $C_+(Q)$, $L_N(1) = 1$, $\forall\, N \in \mathbb{N}$. Then*

$$|L_N(f)(x) - f(x)| \le \frac{\omega_1(f, h)}{h} L_N(\|t - x\|)(x) \le$$

$$\frac{\omega_1(f, h)}{h} \left(\sum_{i=1}^k L_N(|t_i - x_i|)(x)\right), \tag{8.23}$$

$\forall\, N \in \mathbb{N}$.

We mention

Theorem 8.20 ([2]) *Let Q with the l_1-norm $\|\cdot\|$, be a compact and convex subset of $\mathbb{R}^k$, $k \in \mathbb{N} - \{1\}$, and $f \in C_+(Q) : |f(t) - f(x)|$ is a convex function in $t \in Q$ for a fixed $x \in Q^o$. The first modulus of continuity $\omega_1(f, \cdot)$ is with respect to l_1-norm. Let $\{L_N\}_{N\in\mathbb{N}}$ be positive sublinear operators from $C_+(Q)$ into itself*

: $L_N(1) = 1$, $\forall\ N \in \mathbb{N}$. *We assume that* $L_N(\|t-x\|)(x) > 0$ *and the ball in* $\mathbb{R}^k$: $B(x, L_N(\|t-x\|)(x)) \subset Q$, $\forall\ N \in \mathbb{N}$. *Then*

$$|L_N(f)(x) - f(x)| \leq \omega_1(f, L_N(\|t-x\|)(x)),\ \forall\ N \in \mathbb{N}. \tag{8.24}$$

If $L_N(\|t-x\|)(x) \to 0$, *then* $L_N(f)(x) \to f(x)$, *as* $N \to +\infty$.

We also need

Theorem 8.21 ([2]) *All as in Theorem 8.18. Then*

$$|L_N(f)(x) - f(x)| \leq \frac{\left(\max_{\alpha:|\alpha|=n} \omega_1(f_\alpha, h)\right)}{h(n+1)!}\left(L_N\left(\|t-x\|^{n+1}\right)\right)(x), \tag{8.25}$$

$\forall\ N \in \mathbb{N}$.

Theorem 8.22 ([2]) *Let* Q *with the* l_1*-norm* $\|\cdot\|$, *be a compact and convex subset of* $\mathbb{R}^k$, $k \geq 1$, *let* $x \in Q^o$ *be fixed. Let* $f \in C^n(Q, \mathbb{R}_+)$, $n \in \mathbb{N}$, *and* f_α *with* $\alpha : |\alpha| = n$, *has a first modulus of continuity* $\omega_1(f_\alpha, \cdot)$ *relative to* Q *with respect to* l_1*-norm. Each* $|f_\alpha|$ *is a convex function over* Q, *all* $\alpha : |\alpha| = n$. *Assume further that* $f_\alpha(x) = 0$, *all* $\alpha : |\alpha| = 1, \ldots, n$. *Let* $\{L_N\}_{N\in\mathbb{N}}$ *be positive sublinear operators from* $C_+(Q)$ *into* $C_+(Q)$: $L_N(1) = 1$, $\forall\ N \in \mathbb{N}$. *We further assume that* $L_N\left(\|t-x\|^{n+1}\right)(x) > 0$ *and the ball in* $\mathbb{R}^k$: $B\left(x, L_N\left(\|t-x\|^{n+1}\right)(x)\right) \subset Q$, $\forall\ N \in \mathbb{N}$. *Then*

$$|L_N(f)(x) - f(x)| \leq \frac{\left(\max_{\alpha:|\alpha|=n} \omega_1\left(f_\alpha, \left(L_N\left(\|t-x\|^{n+1}\right)\right)(x)\right)\right)}{(n+1)!},\ \forall\ N \in \mathbb{N}. \tag{8.26}$$

It holds, as $L_N\left(\|t-x\|^{n+1}\right)(x) \to 0$, *then* $L_N(f)(x) \to f(x)$, *when* $N \to +\infty$.

8.6 Background - V

Here we describe two well-known specific positive sublinear operators using the Choquet integral.

We mention

Definition 8.23 ([8]) Let $I = [0, 1]$, $\mathcal{B}_I$ the σ-algebra of all Borel measurable subsets of I, $\left(\Gamma_{N,x}\right)_{N\in\mathbb{N},x\in I}$ will be the collection of the family $\Gamma_{N,x} = \{\mu_{N,k,x}\}_{k=0}^N$, of monotone, submodular and strictly positive set functions $\mu_{N,k,x}$ on $\mathcal{B}_I$.

Let $f : [0, 1] \to \mathbb{R}_+$ be a B_I-measurable function which is bounded, and call $p_{N,k}(x) = \binom{N}{k} x^k (1-x)^{N-k}$, for any $x \in [0, 1]$.

The Bernstein–Kantorovich–Choquet operators are defined by the formula

$$K_{N,\Gamma_{N,x}}(f)(x) = \sum_{k=0}^{N} p_{N,k}(x) \frac{(C)\int_{\frac{k}{(N+1)}}^{\frac{(k+1)}{(N+1)}} f(t)\, d\mu_{N,k,x}(t)}{\mu_{N,k,x}\left(\left[\frac{k}{(N+1)}, \frac{(k+1)}{(N+1)}\right]\right)}, \; \forall\, x \in [0,1]. \quad (8.27)$$

If $\mu_{N,k,x} = \mu$, for all N, x, k, we will denote $K_{N,\Gamma_{N,x}}(f) := K_{N,\mu}(f)$.

In particular we study the Bernstein–Kantorovich–Choquet operators $K_{N,\mu}$, where $\mu := \sqrt{M}$, with M the Lebesgue measure on [0, 1]. We have that

$$K_{N,\mu}(f)(x) = \sum_{k=0}^{N} p_{N,k}(x) \frac{(C)\int_{\frac{k}{(N+1)}}^{\frac{(k+1)}{(N+1)}} f(t)\, d\mu(t)}{\mu\left(\left[\frac{k}{(N+1)}, \frac{(k+1)}{(N+1)}\right]\right)}, \quad (8.28)$$

$\forall\, x \in [0,1]$, $\forall\, N \in \mathbb{N}$, $f \in C_+([0,1])$.

We mention

Theorem 8.24 ([8]) *Suppose that* $\mu_{N,k,x} = \mu := \sqrt{M}$, *for all* N, k *and* x, *where* M *is the Lebesgue measure on* $[0,1]$. *Then*

$$\left|K_{N,\mu}(f)(x) - f(x)\right| \leq 2\omega_1\left(f, \frac{\sqrt{x(1-x)}}{\sqrt{N}} + \frac{1}{N}\right), \quad (8.29)$$

$\forall\, N \in \mathbb{N}$, $x \in [0,1]$, $f \in C_+([0,1])$, *above* ω_1 *is over* $[0,1]$.

Remark 8.25 By [8] we have that

$$K_{N,\mu}(|\cdot - x|)(x) \leq \frac{\sqrt{x(1-x)}}{\sqrt{N}} + \frac{1}{N}, \; \forall\, N \in \mathbb{N}. \quad (8.30)$$

Let $m > 1$, notice that $|\cdot - x|^{m-1} \leq 1$, therefore

$$|\cdot - x|^m = |\cdot - x|\,|\cdot - x|^{m-1} \leq |\cdot - x|,$$

hence

$$K_{N,\mu}\left(|\cdot - x|^m\right)(x) \leq K_{N,\mu}(|\cdot - x|)(x),$$

that is

$$K_{N,\mu}\left(|\cdot - x|^m\right)(x) \leq \frac{\sqrt{x(1-x)}}{\sqrt{N}} + \frac{1}{N}, \; \forall\, x \in [0,1], \; N \in \mathbb{N}, \; m \geq 1. \quad (8.31)$$

Notice that $K_{N,\mu}(1) = 1$, $\forall\, N \in \mathbb{N}$.

Clearly $K_{N,\mu}$ operators are positive sublinear operators from $C_+([0,1])$ into itself.

We mention

Definition 8.26 ([9]) Here we consider measures of possibility. Denoting $p_{N,k}(x) = \binom{N}{k} x^k (1-x)^{N-k}$, let us defined

$$\lambda_{N,k}(t) := \frac{p_{N,k}(t)}{k^k N^{-N} (N-k)^{N-k} \binom{N}{k}} = \frac{t^k (1-t)^{N-k}}{k^k N^{-N} (N-k)^{N-k}}, \ k = 0, \ldots, N. \tag{8.32}$$

By convention we assume that $0^0 = 1$, so that the cases $k = 0$, and $k = N$ make sense. By considering the root $\frac{k}{N}$ of $p'_{N,k}(x)$, it is clear that

$$\max\{p_{N,k}(t) : t \in [0,1]\} = k^k N^{-N} (N-k)^{N-k} \binom{N}{k}, \tag{8.33}$$

which implies that each $\lambda_{N,k}$ is a possibility distribution on $[0,1]$.

Denoting by $P_{\lambda_{N,k}}$ the possibility measure induced by $\lambda_{N,k}$ and $\Gamma_{N,x} := \Gamma_N := \{P_{\lambda_{N,k}}\}_{k=0}^N$ (that is Γ_N is independent of x), we define the nonlinear Bernstein–Durrmeyer–Choquet polynomial operators with respect to the set functions in Γ_N given by the formula

$$D_{N,\Gamma_N}(f)(x) := \sum_{k=0}^{N} p_{N,k}(x) \frac{(C)\int_0^1 f(t)\, t^k (1-t)^{N-k} dP_{\lambda_{N,k}}(t)}{(C)\int_0^1 t^k (1-t)^{N-k} dP_{\lambda_{N,k}}(t)}, \tag{8.34}$$

$\forall\, x \in [0,1]$, $N \in \mathbb{N}$, $f \in C_+([0,1])$.

Remark 8.27 Above $P_{\lambda_{N,k}}$ is bounded, monotone, submodular and strictly positive, $N \in \mathbb{N}$, $k = 0, 1, \ldots, N$. Notice that $D_{N,\Gamma_N}(1) = 1$, $\forall\, N \in \mathbb{N}$.

Clearly D_{N,Γ_N} operators are positive sublinear operators mapping $C_+([0,1])$ into itself.

We mention

Theorem 8.28 ([9]) *For every $f \in C_+([0,1])$, $x \in [0,1]$ and $N \in \mathbb{N} - \{1\}$, we have*

$$\left|D_{N,\Gamma_N}(f)(x) - f(x)\right| \leq 2\omega_1\left(f, \frac{\left(1+\sqrt{2}\right)\sqrt{x(1-x)} + \sqrt{2}\sqrt{x}}{\sqrt{N}} + \frac{1}{N}\right), \tag{8.35}$$

where ω_1 is on $[0,1]$.

Remark 8.29 By [9] we have that

$$D_{N,\Gamma_N}(|\cdot - x|)(x) \leq \frac{\left(1+\sqrt{2}\right)\sqrt{x(1-x)} + \sqrt{2}\sqrt{x}}{\sqrt{N}} + \frac{1}{N}, \ \forall N \in \mathbb{N} - \{1\}. \tag{8.36}$$

Let $m > 1$, notice that $|\cdot - x|^{m-1} \leq 1$, therefore

$$|\cdot - x|^m = |\cdot - x| \, |\cdot - x|^{m-1} \leq |\cdot - x|,$$

hence

$$D_{N,\Gamma_N}\left(|\cdot - x|^m\right)(x) \leq D_{N,\Gamma_N}(|\cdot - x|)(x),$$

that is

$$D_{N,\Gamma_N}\left(|\cdot - x|^m\right)(x) \leq \frac{\left(1+\sqrt{2}\right)\sqrt{x(1-x)} + \sqrt{2}\sqrt{x}}{\sqrt{N}} + \frac{1}{N}, \tag{8.37}$$

$\forall N \in \mathbb{N} - \{1\}$, $\forall x \in [0, 1]$, $m \geq 1$.

We make

Remark 8.30 When $x \in [0, 1]$, then the max $(x(1-x)) = \frac{1}{4}$, at $x = \frac{1}{2}$. Therefore it holds

$$\frac{\sqrt{x(1-x)}}{\sqrt{N}} + \frac{1}{N} \leq \frac{1}{2\sqrt{N}} + \frac{1}{N}, \tag{8.38}$$

$\forall x \in [0, 1]$, $\forall N \in \mathbb{N}$.

Similarly, it holds

$$\frac{\left(1+\sqrt{2}\right)\sqrt{x(1-x)} + \sqrt{2}\sqrt{x}}{\sqrt{N}} + \frac{1}{N} \leq \frac{1+3\sqrt{2}}{2\sqrt{N}} + \frac{1}{N}, \tag{8.39}$$

$\forall x \in [0, 1]$, $\forall N \in \mathbb{N} - \{1\}$.

Corollary 8.31 (to Theorem 8.24) *It holds*

$$\left\| K_{N,\mu}(f) - f \right\|_\infty \leq 2\omega_1\left(f, \frac{1}{2\sqrt{N}} + \frac{1}{N}\right), \tag{8.40}$$

$\forall N \in \mathbb{N}$, $f \in C_+([0, 1])$.

Corollary 8.32 (to Theorem 8.28) *It holds*

$$\left\| D_{N,\Gamma_N}(f) - f \right\|_\infty \leq 2\omega_1\left(f, \frac{1+3\sqrt{2}}{2\sqrt{N}} + \frac{1}{N}\right), \tag{8.41}$$

$\forall N \in \mathbb{N} - \{1\}$, $f \in C_+([0, 1])$.

8.7 Background - VI

Here we describe two new specific multivariate positive sublinear operators using the integrated Choquet integral.

We make

Remark 8.33 (I) Let $f \in C_{+}([a,b] \times [c,d])$ be a Lipschitz function such that

$$|f(x_1, y_1) - f(x_2, y_2)| \leq K(|x_1 - x_2| + |y_1 - y_2|), \tag{8.42}$$

where $K > 0$, $\forall (x_1, y_1), (x_2, y_2) \in [a,b] \times [c,d]$.

Consider

$$F(y) := (C) \int_a^b f(x, y) \, d\mu(x), \tag{8.43}$$

where μ is an $\mathbb{R}_{+}$-valued monotone set function which is submodular.

We will prove that $F \in C_{+}([c,d])$.

Let $(y_n)_{n \in \mathbb{N}} \in [c,d] : y_n \to y \in [c,d]$, as $n \to \infty$.

Indeed we have

$$f(x, y_n) = f(x, y_n) - f(x, y) + f(x, y) \leq |f(x, y_n) - f(x, y)| + f(x, y), \tag{8.44}$$

hence

$$(C) \int_a^b f(x, y_n) \, d\mu(x) \leq (C) \int_a^b |f(x, y_n) - f(x, y)| \, d\mu(x) + (C) \int_a^b f(x, y) \, d\mu(x), \tag{8.45}$$

that is

$$(C) \int_a^b f(x, y_n) \, d\mu(x) - (C) \int_a^b f(x, y) \, d\mu(x) \leq (C) \int_a^b |f(x, y_n) - f(x, y)| \, d\mu(x). \tag{8.46}$$

Similarly we have

$$f(x, y) = f(x, y) - f(x, y_n) + f(x, y_n) \leq |f(x, y_n) - f(x, y)| + f(x, y_n), \tag{8.47}$$

hence

$$(C) \int_a^b f(x, y) \, d\mu(x) \leq (C) \int_a^b |f(x, y_n) - f(x, y)| \, d\mu(x) + (C) \int_a^b f(x, y_n) \, d\mu(x), \tag{8.48}$$

that is

$$(C) \int_a^b f(x, y) \, d\mu(x) - (C) \int_a^b f(x, y_n) \, d\mu(x) \leq (C) \int_a^b |f(x, y_n) - f(x, y)| \, d\mu(x). \tag{8.49}$$

We have proved that

$$\left|(C)\int_a^b f(x,y_n)\,d\mu(x)-(C)\int_a^b f(x,y)\,d\mu(x)\right|\leq (C)\int_a^b |f(x,y_n)-f(x,y)|\,d\mu(x). \tag{8.50}$$

Because of (8.42) we get:

$$(C)\int_a^b |f(x,y_n)-f(x,y)|\,d\mu(x)\leq$$

$$K\left((C)\int_a^b |y_n-y|\,d\mu(x)\right)=K\mu([a,b])\,|y_n-y|\to 0, \tag{8.51}$$

as $n\to\infty$.

Together with (8.50), proving the continuity of F.

So assuming in general that F is continuous is not strange at all.

(II) Let $\Omega\neq\varnothing$ and a σ-algebra $\mathcal{A}$ on it. Let $A,B\in\mathcal{A}$, and $f:\Omega\to\mathbb{R}_+$ is $\mathcal{A}$-measurable. Here μ_1,μ_2 are $\mathbb{R}_+$-valued monotone set functions. We assume that f is a Choquet μ_1-integrable function. Then

$$(C)\int_A\left((C)\int_B f(t_1)\,d\mu_1(t_1)\right)d\mu_2(t_2)=\mu_2(A)\left((C)\int_B f(t_1)\,d\mu_1(t_1)\right). \tag{8.52}$$

We give the following definition

Definition 8.34 Here $\mu=\sqrt{M}$, where M is the Lebesgue measure on $[0,1]$. We consider $f\in C_+\left([0,1]^d\right)$, $d\in\mathbb{N}-\{1\}$; $\overrightarrow{N}:=(N_1,\ldots,N_d)$, with $N_i\in\mathbb{N}$, $i=1,\ldots,d$; $\overrightarrow{x}:=(x_1,\ldots,x_d)\in[0,1]^d$. For any $\prod_{i=1}^{d}[a_i,b_i]\subset[0,1]^d$, we define and assume that

$$F_1(x_1,x_2,\ldots,x_{d-1}):=(C)\int_{a_d}^{b_d} f(x_1,\ldots,x_d)\,d\mu(x_d)\in C_+\left([0,1]^{d-1}\right), \tag{8.53}$$

$$F_2(x_1,\ldots,x_{d-2}):=(C)\int_{a_{d-1}}^{b_{d-1}} F_1(x_1,\ldots,x_{d-1})\,d\mu(x_{d-1})\in C_+\left([0,1]^{d-2}\right), \tag{8.54}$$

$$\vdots$$

$$F_{d-1}(x_1):=(C)\int_{a_2}^{b_2} F_{d-2}(x_1,x_2)\,d\mu(x_2)\in C_+([0,1]). \tag{8.55}$$

So that the iterated Choquet integral

$$(C)\int_{a_1}^{b_1}\left((C)\int_{a_2}^{b_2}\left(...\left((C)\int_{a_d}^{b_d} f(x_1,x_2,...,x_d)\,d\mu(x_d)\right)d\mu(x_{d-1})\right)...\right)d\mu(x_1) \tag{8.56}$$

$$= (C)\int_{a_1}^{b_1} F_{d-1}(x_1)\,d\mu(x_1) \geq 0,$$

and it makes sense.

Denote by

$$p_{N_i,k_i}(x_i) := \binom{N_i}{k_i} x_i^{k_i}(1-x_i)^{N_i-k_i}, \text{ for any } x_i \in [0,1],\ i=1,...,d. \tag{8.57}$$

We define the multivariate Bernstein–Knatorovich–Choquet operators:

$$K_{\vec{N},\mu}(f)(\vec{x}) := \sum_{k_1=0}^{N_1}\sum_{k_2=0}^{N_2}...\sum_{k_d=0}^{N_d} p_{N_1,k_1}(x_1)\,p_{N_2,k_2}(x_2)...p_{N_d,k_d}(x_d)\cdot$$

$$\frac{(C)\int_{\frac{k_1}{(N_1+1)}}^{\left(\frac{k_1+1}{N_1+1}\right)}...(C)\int_{\frac{k_d}{N_d+1}}^{\left(\frac{k_d+1}{N_d+1}\right)} f(t_1,...,t_d)\,d\mu(t_d)...d\mu(t_1)}{\prod_{i=1}^{d}\mu\left(\left[\frac{k_i}{N_i+1},\frac{k_i+1}{N_i+1}\right]\right)}. \tag{8.58}$$

We make

Remark 8.35 (*to Definition* 8.34) (I) We notice the following:

$$(C)\int_{a_1}^{b_1}\left((C)\int_{a_2}^{b_2}\left(...\left((C)\int_{a_d}^{b_d} g(x_\lambda)\,d\mu(x_d)\right)d\mu(x_{d-1})\right)...\right)d\mu(x_1) \tag{8.59}$$

$$= \left((C)\int_{a_\lambda}^{b_\lambda} g(x_\lambda)\,d\mu(x_\lambda)\right)\prod_{\substack{i=1\\ i\neq\lambda}}^{d}\mu([a_i,b_i]);$$

above it is $g(x_\lambda) \in C_+([0,1])$, $\lambda \in \{1,...,d\}$, $x_\lambda \in [0,1]$.

By (8.59) we get that

$$K_{\vec{N},\mu}(g)(\vec{x}) = K_{N_\lambda,\mu}(g)(x_\lambda),\ \ \lambda \in \{1,...,d\}, \tag{8.60}$$

where $K_{N_\lambda,\mu}$ as in (8.28).

(II) Assume that $f \geq 0$ is a Lipschitz function, i.e.

$$|f(x_1, ..., x_d) - f(y_1, ..., y_d)| \leq K \left\| \overrightarrow{x} - \overrightarrow{y} \right\|_{l_1} \tag{8.61}$$

$$= K \left(\sum_{i=1}^{d} |x_i - y_i| \right), \text{ where } K > 0,$$

where $\overrightarrow{y} := (y_1, ..., y_d) \in [0, 1]^d$. Infact (8.61) implies continuity of f, i.e. $f \in C_+ \left([0, 1]^d\right)$.

Then it holds (see (8.53))

$$|F_1(x_1, x_2, ..., x_{d-1}) - F_1(y_1, y_2, ..., y_{d-1})| =$$

$$\left|(C) \int_{a_d}^{b_d} f(x_1, ..., x_{d-1}, x_d) \, d\mu(x_d) - (C) \int_{a_d}^{b_d} f(y_1, ..., y_{d-1}, x_d) \, d\mu(x_d)\right| \leq$$

(acting as in (8.50))

$$(C) \int_{a_d}^{b_d} |f(x_1, .., x_{d-1}, x_d) - f(y_1, ..., y_{d-1}, x_d)| \, d\mu(x_d) \leq$$

$$K\mu([a_d, b_d]) \left(\sum_{i=1}^{d-1} |x_i - y_i| \right). \tag{8.62}$$

Therefore, $F_1(x_1, ..., x_{d-1})$, see (8.53), is Lipschitz and hence continuous in $C_+ \left([0, 1]^{d-1}\right)$. Similarly acting (see (8.54)):

$$|F_2(x_1, x_2, ..., x_{d-2}) - F_2(y_1, y_2, ..., y_{d-2})| =$$

$$\left|(C) \int_{a_{d-1}}^{b_{d-1}} F_1(x_1, ..., x_{d-2}, x_{d-1}) \, d\mu(x_{d-1}) - (C) \int_{a_{d-1}}^{b_{d-1}} F_1(y_1, ..., y_{d-2}, x_{d-1}) \, d\mu(x_{d-1})\right| \leq$$

$$(C) \int_{a_{d-1}}^{b_{d-1}} |F_1(x_1, .., x_{d-2}, x_{d-1}) - F_1(y_1, ..., y_{d-2}, x_{d-1})| \, d\mu(x_{d-1}) \leq$$

$$K\mu([a_d, b_d]) \, \mu\left(\left[a_{d-1}, b_{d-1}\right]\right) \left(\sum_{i=1}^{d-2} |x_i - y_i| \right). \tag{8.63}$$

Therefore, $F_2(x_1, ..., x_{d-2})$, see (8.54), is Lipschitz and thus continuous in $C_+ \left([0, 1]^{d-2}\right)$. By finite induction, we find that $F_{d-1}(x_1)$, see (8.55), is Lipschitz and thus continuous in $C_+([0, 1])$.

So, if $f \geq 0$ is Lipschitz function with respect to l_1 norm, we can drop the assumptions (8.53)–(8.55), as they are implied by the latter. Clearly then, the iterated Choquet integral (8.56) and $K_{\overrightarrow{N},\mu}(f)\left(\overrightarrow{x}\right)$, see (8.58), make sense.

(III) Consider $f, g \geq 0$ that are Lipschitz functions with respect to l_1-norm. We easily derive that

(i)

$$K_{\overrightarrow{N},\mu}(\alpha f) = \alpha K_{\overrightarrow{N},\mu}(f), \ \forall \alpha \geq 0, \tag{8.64}$$

(ii) if $f \leq g$, then

$$K_{\overrightarrow{N},\mu}(f) \leq K_{\overrightarrow{N},\mu}(g), \tag{8.65}$$

and

(iii)

$$K_{\overrightarrow{N},\mu}(f+g) \leq K_{\overrightarrow{N},\mu}(f) + K_{\overrightarrow{N},\mu}(g), \tag{8.66}$$

by the sum of two Lipschitz functions is a Lipschitz function.

Furthermore it holds $K_{\overrightarrow{N},\mu}(1) = 1$.

Denote by

$$Lip_+\left([0,1]^d\right) := \{f : [0,1]^d \to \mathbb{R}_+ : f \text{ is a Lipschitz function}\}. \tag{8.67}$$

Of course it holds $Lip_+\left([0,1]^d\right) \subset C_+\left([0,1]^d\right)$.

Conclusion: $K_{\overrightarrow{N},\mu}$ is a positive sublinear operator on $Lip_+\left([0,1]^d\right)$ with $K_{\overrightarrow{N},\mu}(1) = 1$, which maps $Lip_+\left([0,1]^d\right)$ into $C_+\left([0,1]^d\right)$.

We give

Definition 8.36 Here we define ($N_i \in \mathbb{N}$)

$$\lambda_{N_i,k_i}(t_i) := \frac{p_{N_i,k_i}(t_i)}{k_i^{k_i} N_i^{-N_i} (N_i - k_i)^{N_i - k_i} \binom{N_i}{k_i}}, \tag{8.68}$$

$k_i = 0, 1, ..., N_i$; $i = 1, ..., d \in \mathbb{N} - \{1\}$; $t_i \in [0,1]$. Each λ_{N_i,k_i} is a possibility distribution on $[0,1]$, $i = 1, ..., d$, see (8.33). Denote by $P_{\lambda_{N_i,k_i}}$ the possibility measure induced by λ_{N_i,k_i}, $i = 1, ..., d$, and by $\Gamma_{N_i,x_i} := \Gamma_{N_i} := \{P_{\lambda_{N_i,k_i}}\}_{k_i=0}^{N_i}$ (that is Γ_{N_i} is independent of x_i), $\overrightarrow{N} := (N_1, ..., N_d)$, $\overrightarrow{x} := (x_1, ..., x_d) \in [0,1]^d$.

Let $f \in Lip_+\left([0,1]^d\right)$. We define the multivariate nonlinear Bernstein–Durremeyer–Choquet polynomial operators with respect to the set functions in $\otimes_{i=1}^d \Gamma_{N_i}$ given by the formula:

$$D_{\overrightarrow{N},\Gamma_{\overrightarrow{N}}}(f)\left(\overrightarrow{x}\right) = \sum_{k_1=0}^{N_1} \ldots \sum_{k_d=0}^{N_d} p_{N_1,k_1}(x_1) \ldots p_{N_d,k_d}(x_d) \cdot$$

$$\frac{(C)\int_0^1 ...(C)\int_0^1 f(t_1,...,t_d)\left(\prod_{i=1}^{d} t_i^{k_i}(1-t_i)^{N_i-k_i}\right)\left(\prod_{i=d}^{1} dP_{\lambda_{N_i,k_i}}(t_i)\right)}{(C)\int_0^1 ...(C)\int_0^1 \left(\prod_{i=1}^{d} t_i^{k_i}(1-t_i)^{N_i-k_i}\right)\left(\prod_{i=d}^{1} dP_{\lambda_{N_i,k_i}}(t_i)\right)}. \tag{8.69}$$

One can rewrite the last as

$$D_{\overrightarrow{N},\Gamma_{\overrightarrow{N}}}(f)\left(\overrightarrow{x}\right) = \sum_{k_1=0}^{N_1}...\sum_{k_d=0}^{N_d} p_{N_1,k_1}(x_1)...p_{N_d,k_d}(x_d)\cdot$$

$$\frac{(C)\int_0^1 ...(C)\int_0^1 f(t_1,...,t_d)\left(\prod_{i=1}^{d} t_i^{k_i}(1-t_i)^{N_i-k_i}\right)\left(\prod_{i=d}^{1} dP_{\lambda_{N_i,k_i}}(t_i)\right)}{\prod_{i=1}^{d}(C)\int_0^1 t_i^{k_i}(1-t_i)^{N_i-k_i}\, dP_{\lambda_{N_i,k_i}}(t_i)}. \tag{8.70}$$

We make

Remark 8.37 Above $P_{\lambda_{N_i,k_i}}$, $i = 1,...,d$, is bounded, monotone, submodular and strictly positive.

Let $g(x_r) \in Lip_+([0,1])$, $r \in \{1,...,d\}$, then it holds

$$D_{\overrightarrow{N},\Gamma_{\overrightarrow{N}}}(g)\left(\overrightarrow{x}\right) = D_{N_r,\Gamma_r}(g)(x_r), \tag{8.71}$$

where $D_{N_r,\Gamma_r}(g)(x_r)$ is as in (8.34).

Notice that $D_{\overrightarrow{N},\Gamma_{\overrightarrow{N}}}(1) = 1$, and $D_{\overrightarrow{N},\Gamma_{\overrightarrow{N}}}$ maps $Lip_+\left([0,1]^d\right)$ into $C_+\left([0,1]^d\right)$.

For $f, g \in Lip_+\left([0,1]^d\right)$ we have

$$D_{\overrightarrow{N},\Gamma_{\overrightarrow{N}}}(\alpha f) = \alpha D_{\overrightarrow{N},\Gamma_{\overrightarrow{N}}}(f),\ \forall \alpha \geq 0, \tag{8.72}$$

if $f \leq g$, then

$$D_{\overrightarrow{N},\Gamma_{\overrightarrow{N}}}(f) \leq D_{\overrightarrow{N},\Gamma_{\overrightarrow{N}}}(g), \tag{8.73}$$

and

$$D_{\overrightarrow{N},\Gamma_{\overrightarrow{N}}}(f+g) \leq D_{\overrightarrow{N},\Gamma_{\overrightarrow{N}}}(f) + D_{\overrightarrow{N},\Gamma_{\overrightarrow{N}}}(g). \tag{8.74}$$

Hence $D_{\overrightarrow{N},\Gamma_{\overrightarrow{N}}}$ is a positive sublinear operator on $Lip_+\left([0,1]^d\right)$.

Remark 8.38 Let $f \in Lip_+\left([0,1]^d\right)$ with a Lipschitz constant $K > 0$.

One has

$$(C)\int_0^1 f(t_1,..,t_d)\left(\prod_{i=1}^{d} t_i^{k_i}(1-t_i)^{N_i-k_i}\right) dP_{\lambda_{N_d,k_d}}(t_d) = \tag{8.75}$$

$$\left(\prod_{i=1}^{d-1} t_i^{k_i} (1-t_i)^{N_i-k_i}\right) (C)\int_0^1 f(t_1,...,t_d)\, t_d^{k_d} (1-t_d)^{N_d-k_d}\, dP_{\lambda_{N_d,k_d}}(t_d).$$

Call

$$G_1(t_1,t_2,...,t_{d-1}) := (C)\int_0^1 f(t_1,..,t_d)\, t_d^{k_d} (1-t_d)^{N_d-k_d}\, dP_{\lambda_{N_d,k_d}}(t_d).$$

We observe that

$$\left|G_1(t_1,t_2,...,t_{d-1}) - G_1\left(t_1',t_2',...,t_{d-1}'\right)\right| =$$

$$\left|(C)\int_0^1 f(t_1,..,t_d)\, t_d^{k_d} (1-t_d)^{N_d-k_d}\, dP_{\lambda_{N_d,k_d}}(t_d) - \right.$$

$$\left.(C)\int_0^1 f\left(t_1',..,t_{d-1}',t_d\right) t_d^{k_d} (1-t_d)^{N_d-k_d}\, dP_{\lambda_{N_d,k_d}}(t_d)\right| \leq \tag{8.76}$$

$$(C)\int_0^1 \left|f(t_1,..,t_d) - f\left(t_1',..,t_{d-1}',t_d\right)\right| t_d^{k_d} (1-t_d)^{N_d-k_d}\, dP_{\lambda_{N_d,k_d}}(t_d) \leq$$

$$K\left((C)\int_0^1 t_d^{k_d} (1-t_d)^{N_d-k_d}\, dP_{\lambda_{N_d,k_d}}(t_d)\right)\left(\sum_{i=1}^{d-1} \left|t_i - t_i'\right|\right). \tag{8.77}$$

That is, $G_1 \in Lip_+\left([0,1]^{d-1}\right)$ and $G_1 \in C_+\left([0,1]^{d-1}\right)$.

One can write

$$J := (C)\int_0^1 ...(C)\int_0^1 f(t_1,...,t_d)\left(\prod_{i=1}^{d} t_i^{k_i} (1-t_i)^{N_i-k_i}\right)\left(\prod_{i=d}^{1} dP_{\lambda_{N_i,k_i}}(t_i)\right)$$

$$= (C)\int_0^1 ...(C)\int_0^1 G_1(t_1,t_2...,t_{d-1})\left(\prod_{i=1}^{d-1} t_i^{k_i} (1-t_i)^{N_i-k_i}\right)\left(\prod_{i=d-1}^{1} dP_{\lambda_{N_i,k_i}}(t_i)\right). \tag{8.78}$$

We have that

$$(C)\int_0^1 G_1(t_1,t_2...,t_{d-1})\left(\prod_{i=1}^{d-1} t_i^{k_i} (1-t_i)^{N_i-k_i}\right) dP_{\lambda_{N_{d-1},k_{d-1}}}(t_{d-1}) =$$

$$\left(\prod_{i=1}^{d-2} t_i^{k_i} (1-t_i)^{N_i-k_i}\right) (C)\int_0^1 G_1(t_1,t_2...,t_{d-1})\cdot$$

$$t_{d-1}^{k_{d-1}} (1 - t_{d-1})^{N_{d-1} - k_{d-1}} \, dP_{\lambda_{N_{d-1}, k_{d-1}}} (t_{d-1}), \tag{8.79}$$

and the above integrals exist.

Call

$$G_2 (t_1, t_2, ..., t_{d-2}) :=$$

$$(C) \int_0^1 G_1 (t_1, .., t_{d-1}) \, t_{d-1}^{k_{d-1}} (1 - t_{d-1})^{N_{d-1} - k_{d-1}} \, dP_{\lambda_{N_{d-1}, k_{d-1}}} (t_{d-1}). \tag{8.80}$$

As earlier, we can prove that $G_2 \in Lip_+ \left([0, 1]^{d-2} \right)$ and $G_2 \in C_+ \left([0, 1]^{d-2} \right)$.

So working similarly, by finite induction we can prove that the Choquet iterated integral J makes sense, hence the definition of $D_{\vec{N}, \Gamma_{\vec{N}}}$ also makes sense, see (8.69). Clearly then, all properties of $D_{\vec{N}, \Gamma_{\vec{N}}}$ listed in Remark 8.37 can easily be proved and all mentioned there are valid.

8.8 Background - VII

Here we talk about representations of positive sublinear operators by Choquet integrals.

We need

Definition 8.39 Let Ω be a set, and let $f, g : \Omega \to \mathbb{R}$ be bounded functions. We say that f and g are comonotonic, if for every $\omega, \omega' \in \Omega$,

$$\left(f (\omega) - f \left(\omega' \right) \right) \left(g (\omega) - g \left(\omega' \right) \right) \geq 0. \tag{8.81}$$

We also need the famous Schmeidler's Representation Theorem (Schmeidler 1986)

Theorem 8.40 ([10]) *Denote with $\mathcal{L}_\infty (\mathcal{A})$ the vector space of $\mathcal{A}$-measurable bounded real valued functions on Ω, where $\mathcal{A} \subset 2^\Omega$ is a σ-algebra. Given a real functional $\Gamma : \mathcal{L}_\infty (\mathcal{A}) \to \mathbb{R}$, assume that for $f, g \in \mathcal{L}_\infty (\mathcal{A})$:*

(i) $\Gamma (cf) = c\Gamma (f)$, $\forall \, c > 0$,

(ii) $f \leq g$, implies $\Gamma (f) \leq \Gamma (g)$,

and

(iii) $\Gamma (f + g) = \Gamma (f) + \Gamma (g)$, for any comonotonic f, g.

Then $\gamma (A) := \Gamma (1_A)$, $\forall \, A \in \mathcal{A}$, defines a finite monotone set function on $\mathcal{A}$, and Γ is the Choquet integral with respect to γ, i.e.

$$\Gamma (f) = (C) \int_\Omega f (t) \, d\gamma (t), \; \forall \, f \in \mathcal{L}_\infty (\mathcal{A}). \tag{8.82}$$

Above 1_A denotes the characteristic function on A.

We make

Remark 8.41 Consider here Q a compact and convex subset of $\mathbb{R}^k$, $k \in \mathbb{N} - \{1\}$, $\mathcal{B} = \mathcal{B}(Q)$ is the Borel σ-algebra on Q, and $\mathcal{L}_\infty(\mathcal{B})$ is the vector space of $\mathcal{B}$-measurable bounded real valued functions on Q. Let $(L_N)_{N\in\mathbb{N}}$ be a sequence of positive sublinear operators from $\mathcal{L}_\infty(\mathcal{B})$ into $C_+(Q)$, and $x \in Q$. That is here L_N fulfills the positive homogeneity, monotonicity and subadditivity properties, see (8.11)–(8.13).

Assume $L_N(1) = 1, \forall N \in \mathbb{N}$. Clearly here $\mathcal{L}_\infty(\mathcal{B}) \supset C_+(Q)$. In particular we treat $L_N|_{C_+(Q)}$, just denoted for simplicity by L_N, $\forall N \in \mathbb{N}$.

It is clear that $L_N(\cdot)(x) : \mathcal{L}_\infty(\mathcal{B}) \to \mathbb{R}$ is a functional, $\forall N \in \mathbb{N}$. It has the properties:

(i)

$$L_N(cf)(x) = cL_N(f)(x), \ \forall c > 0, \ \forall f \in \mathcal{L}_\infty(\mathcal{B}), \tag{8.83}$$

(ii)

$$f \leq g, \text{ implies } L_N(f)(x) \leq L_N(g)(x), \quad \text{where } f, g \in \mathcal{L}_\infty(\mathcal{B}), \tag{8.84}$$

and

(iii)

$$L_N(f+g)(x) \leq L_N(f)(x) + L_N(g)(x), \ \forall f, g \in \mathcal{L}_\infty(\mathcal{B}). \tag{8.85}$$

For comonotonic $f, g \in \mathcal{L}_\infty(\mathcal{B})$, we further assume that

$$L_N(f+g)(x) = L_N(f)(x) + L_N(g)(x). \tag{8.86}$$

In that case L_N is called comonotonic.

By Theorem 8.40 we get that:

$$\gamma_{N,x}(A) := L_N(1_A)(x), \ \forall A \in \mathcal{B}, \forall N \in \mathbb{N}, \tag{8.87}$$

defines a finite monotone set function on $\mathcal{B}$, and

$$L_N(f)(x) = (C)\int_Q f(t)\, d\gamma_{N,x}(t), \tag{8.88}$$

$\forall f \in \mathcal{L}_\infty(\mathcal{B}), \forall N \in \mathbb{N}$.

In particular (8.88) is valid for any $f \in C_+(Q)$. Furthermore $\gamma_{N,x}$ is normalized, that is $\gamma_{N,x}(Q) = 1, \forall N \in \mathbb{N}$.

A different kind of general operators follows:

We make

Remark 8.42 Consider again Q a compact and convex subset of $\mathbb{R}^k$, $k \in \mathbb{N}$, $\mathcal{B} = \mathcal{B}(Q)$ the Borel σ-algebra on Q. For each $N \in \mathbb{N}$ and each $x \in Q$ consider the monotone set functions $\mu_{N,x}$; $\mathcal{B} \to \mathbb{R}_+$. We assume that all $\mu_{N,x}$ are normalized, that is $\mu_{N,x}(Q) = 1$, and submodular. Here we consider the operators $T_N : C_+(Q) \to C_+(Q)$ given by the formula

$$T_N(f)(x) = (C)\int_Q f(t)\, d\mu_{N,x}(t), \tag{8.89}$$

$\forall\, N \in \mathbb{N}$, $\forall\, x \in Q$.

Infact here $\mu_{N,x}$ are chosen so that $T_N(C_+(Q)) \subseteq C_+(Q)$.

We notice here that hold:

(i)

$$T_N(\alpha f)(x) = \alpha T_N(f)(x),\ \forall \alpha \geq 0, \tag{8.90}$$

(ii)

$$f \leq g, \text{ implies } T_N(f)(x) \leq T_N(g)(x), \tag{8.91}$$

and

(iii)

$$T_N(f+g)(x) \leq T_N(f)(x) + T_N(g)(x), \tag{8.92}$$

$\forall\, N \in \mathbb{N}$, $\forall\, x \in Q$, $\forall\, f, g \in C_+(Q)$.

Clearly T_N are positive sublinear operators, compare to (8.11)–(8.13). We also have that $T_N(1) = 1$, $\forall\, N \in \mathbb{N}$.

On $\mathbb{R}$, it will be $Q = [a, b]$.

8.9 Main Results - I

We start with estimates on the specific operators mentioned earlier. First univariate results.

We give

Theorem 8.43 *Let* $f \in C_+([0,1])$, $x \in (0,1)$, $\frac{1}{2\sqrt{N^*}} + \frac{1}{N^*} \leq \min(x, 1-x)$, *for some* $N^* \in \mathbb{N}$, *and* $|f(\cdot) - f(x)|$ *is a convex function over* $[0,1]$. *Then*

$$\left|K_{N,\mu}(f)(x) - f(x)\right| \leq \omega_1\left(f, \frac{1}{2\sqrt{N}} + \frac{1}{N}\right), \tag{8.93}$$

$\forall\, N \geq N^*$, $N \in \mathbb{N}$, *where* $K_{N,\mu}$ *is as in (8.28).*

We have $\lim_{N\to\infty} K_{N,\mu}(f)(x) = f(x)$.

Proof By (8.5) we have

$$\left|K_{N,\mu}(f)(x)-f(x)\right| \leq \frac{\omega_1(f,h)}{h} K_{N,\mu}(|\cdot - x|)(x) \overset{(8.30)}{\leq}$$

$$\frac{\omega_1(f,h)}{h}\left(\frac{\sqrt{x(1-x)}}{\sqrt{N}}+\frac{1}{N}\right) \overset{(8.38)}{\leq} \frac{\omega_1(f,h)}{h}\left(\frac{1}{2\sqrt{N}}+\frac{1}{N}\right) \tag{8.94}$$

(choosing $h := \frac{1}{2\sqrt{N}}+\frac{1}{N}>0$)

$$=\omega_1\left(f,\frac{1}{2\sqrt{N}}+\frac{1}{N}\right),$$

proving the claim. ■

Theorem 8.44 *Let* $f \in C^n([0,1],\mathbb{R}_+)$, $n \in \mathbb{N}$, $x \in (0,1)$, $\frac{1}{2\sqrt{N^*}}+\frac{1}{N^*} \leq \min(x,1-x)$, *for some* $N^* \in \mathbb{N}$, *and* $\left|f^{(n)}\right|$ *is convex over* $[0,1]$. *Assume that* $f^{(k)}(x)=0$, $k=1,...,n$. *Then*

$$\left|K_{N,\mu}(f)(x)-f(x)\right| \leq \frac{\omega_1\left(f^{(n)},\frac{1}{2\sqrt{N}}+\frac{1}{N}\right)}{(n+1)!}, \tag{8.95}$$

$\forall\, N \geq N^*$, $N \in \mathbb{N}$, *where* $K_{N,\mu}$ *is as in (8.28).*

We have $\lim_{N\to\infty} K_{N,\mu}(f)(x)=f(x)$.

Proof By (8.6) we have

$$\left|K_{N,\mu}(f)(x)-f(x)\right| \leq \frac{\omega_1\left(f^{(n)},h\right)}{h(n+1)!} K_{N,\mu}\left(|\cdot - x|^{n+1}\right)(x) \overset{(8.31)}{\leq}$$

$$\frac{\omega_1\left(f^{(n)},h\right)}{h(n+1)!}\left(\frac{\sqrt{x(1-x)}}{\sqrt{N}}+\frac{1}{N}\right) \overset{(8.38)}{\leq} \frac{\omega_1\left(f^{(n)},h\right)}{h(n+1)!}\left(\frac{1}{2\sqrt{N}}+\frac{1}{N}\right) \tag{8.96}$$

(choosing $h := \frac{1}{2\sqrt{N}}+\frac{1}{N}>0$)

$$=\frac{\omega_1\left(f^{(n)},\frac{1}{2\sqrt{N}}+\frac{1}{N}\right)}{(n+1)!},$$

proving the claim. ■

We continue with

Theorem 8.45 *Let $f \in C_+([0,1])$, $x \in (0,1)$, $\frac{1+3\sqrt{2}}{2\sqrt{N^*}} + \frac{1}{N^*} \leq \min(x, 1-x)$, for some $N^* \in \mathbb{N} - \{1\}$, and $|f(\cdot) - f(x)|$ is a convex function over $[0,1]$. Then*

$$\left| D_{N,\Gamma_N}(f)(x) - f(x) \right| \leq \omega_1\left(f, \frac{1+3\sqrt{2}}{2\sqrt{N}} + \frac{1}{N} \right), \tag{8.97}$$

$\forall\, N \geq N^*$, $N \in \mathbb{N} - \{1\}$, where D_{N,Γ_N} is as in (8.34).
We have $\lim_{N\to\infty} D_{N,\Gamma_N}(f)(x) = f(x)$.

Proof By (8.5) we get

$$\left| D_{N,\Gamma_N}(f)(x) - f(x) \right| \leq \frac{\omega_1(f,h)}{h} D_{N,\Gamma_N}(|\cdot - x|)(x) \overset{(8.36)}{\leq}$$

$$\frac{\omega_1(f,h)}{h}\left(\frac{\left(1+\sqrt{2}\right)\sqrt{x(1-x)} + \sqrt{2x}}{\sqrt{N}} + \frac{1}{N} \right) \overset{(8.39)}{\leq} \frac{\omega_1(f,h)}{h}\left(\frac{1+3\sqrt{2}}{2\sqrt{N}} + \frac{1}{N} \right) \tag{8.98}$$

(choosing $h := \frac{1+3\sqrt{2}}{2\sqrt{N}} + \frac{1}{N}$)

$$= \omega_1\left(f, \left(\frac{1+3\sqrt{2}}{2\sqrt{N}} + \frac{1}{N} \right) \right),$$

proving the claim. ∎

Theorem 8.46 *Let $f \in C^n([0,1], \mathbb{R}_+)$, $n \in \mathbb{N}$, $x \in (0,1)$, $\left(\frac{1+3\sqrt{2}}{2\sqrt{N^*}} + \frac{1}{N^*} \right) \leq \min(x, 1-x)$, for some $N^* \in \mathbb{N} - \{1\}$, and $\left| f^{(n)} \right|$ is convex over $[0,1]$. Assume that $f^{(k)}(x) = 0$, $k = 1, \ldots, n$. Then*

$$\left| D_{N,\Gamma_N}(f)(x) - f(x) \right| \leq \frac{\omega_1\left(f^{(n)}, \left(\frac{1+3\sqrt{2}}{2\sqrt{N}} + \right) \frac{1}{N} \right)}{(n+1)!}, \tag{8.99}$$

$\forall\, N \geq N^*$, $N \in \mathbb{N} - \{1\}$, where D_{N,Γ_N} is as in (8.34).
We have $\lim_{N\to\infty} D_{N,\Gamma_N}(f)(x) = f(x)$.

Proof By (8.6) we get

$$\left| D_{N,\Gamma_N}(f)(x) - f(x) \right| \leq \frac{\omega_1\left(f^{(n)}, h \right)}{h(n+1)!} D_{N,\Gamma_N}\left(|\cdot - x|^{n+1} \right)(x) \overset{(8.37)}{\leq}$$

$$\frac{\omega_1\left(f^{(n)},h\right)}{h\,(n+1)!}\left(\frac{\left(1+\sqrt{2}\right)\sqrt{x\,(1-x)}+\sqrt{2x}}{\sqrt{N}}+\frac{1}{N}\right)\overset{(8.39)}{\leq}$$

$$\frac{\omega_1\left(f^{(n)},h\right)}{h\,(n+1)!}\left(\frac{1+3\sqrt{2}}{2\sqrt{N}}+\frac{1}{N}\right) \tag{8.100}$$

(choosing $h := \frac{1+3\sqrt{2}}{2\sqrt{N}}+\frac{1}{N}$)

$$=\frac{\omega_1\left(f^{(n)},\left(\frac{1+3\sqrt{2}}{2\sqrt{N}}+\frac{1}{N}\right)\right)}{(n+1)!},$$

proving the claim. ■

We continue with multivariate approximation results for specific operators. Some related general results follow:

Theorem 8.47 *Let* $[0,1]^d$, $d\in\mathbb{N}-\{1\}$, *endowed with* l_1*-norm, and* $f\in Lip_+\left([0,1]^d\right)$. *Then*

$$\left|K_{\overrightarrow{N},\mu}(f)\left(\overrightarrow{x}\right)-f\left(\overrightarrow{x}\right)\right|\leq(d+1)\,K\left(\frac{1}{2\sqrt{N_{\min}}}+\frac{1}{N_{\min}}\right), \tag{8.101}$$

where $K>0$, $\forall\,\overrightarrow{x}\in[0,1]^d$, $\forall\,\overrightarrow{N}\in\mathbb{N}^d$, *where* $N_{\min}:=\min\{N_1,\dots,N_d\}$.

That is

$$\left\|K_{\overrightarrow{N},\mu}(f)-f\right\|_\infty\leq(d+1)\,K\left(\frac{1}{2\sqrt{N_{\min}}}+\frac{1}{N_{\min}}\right). \tag{8.102}$$

It holds that $\lim_{\overrightarrow{N}\to(\infty,\dots,\infty)}K_{\overrightarrow{N},\mu}(f)=f$, *uniformly.*

Proof By (8.16) we have

$$\left|K_{\overrightarrow{N},\mu}(f)\left(\overrightarrow{x}\right)-f\left(\overrightarrow{x}\right)\right|\leq\omega_1(f,h)\left(1+\frac{1}{h}\left(\sum_{i=1}^{d}K_{\overrightarrow{N},\mu}\left(|t_i-x_i|\right)\left(\overrightarrow{x}\right)\right)\right)\overset{(8.60)}{=}$$

$$\omega_1(f,h)\left(1+\frac{1}{h}\left(\sum_{i=1}^{d}K_{N_i,\mu}\left(|t_i-x_i|\right)(x_i)\right)\right)\overset{(\text{by }(8.30),\,(8.38))}{\leq} \tag{8.103}$$

$$\omega_1(f,h)\left(1+\frac{1}{h}\left(\sum_{i=1}^{d}\left(\frac{1}{2\sqrt{N_i}}+\frac{1}{N_i}\right)\right)\right)\leq$$

$$\omega_1(f,h)\left(1+\frac{d}{h}\left(\frac{1}{2\sqrt{N_{\min}}}+\frac{1}{N_{\min}}\right)\right)$$

(choose $h := \left(\frac{1}{2\sqrt{N_{\min}}}+\frac{1}{N_{\min}}\right) > 0$)

$$=\omega_1\left(f,\left(\frac{1}{2\sqrt{N_{\min}}}+\frac{1}{N_{\min}}\right)\right)(d+1)\leq(d+1)K\left(\frac{1}{2\sqrt{N_{\min}}}+\frac{1}{N_{\min}}\right),$$

proving the claim. ■

Theorem 8.48 *Let* $[0,1]^d$, $d\in\mathbb{N}-\{1\}$, *endowed with* l_1*-norm, and* $f\in Lip_+\left([0,1]^d\right)\cap C^1\left([0,1]^d,\mathbb{R}_+\right)$; $\overrightarrow{x}\in[0,1]^d$: $\frac{\partial f(\overrightarrow{x})}{\partial x_i}=0$, *for* $i=1,\ldots,d$. *Then*

$$\left|K_{\overrightarrow{N},\mu}(f)\left(\overrightarrow{x}\right)-f\left(\overrightarrow{x}\right)\right|\leq\frac{1}{2}\left(\max_{i=1,\ldots,d}\omega_1\left(\frac{\partial f}{\partial x_i},\sqrt{\frac{1}{2\sqrt{N_{\min}}}+\frac{1}{N_{\min}}}\right)\right)\cdot$$

$$\left[d^2\left(\sqrt{\frac{1}{2\sqrt{N_{\min}}}+\frac{1}{N_{\min}}}\right)+d\left(\frac{1}{2\sqrt{N_{\min}}}+\frac{1}{N_{\min}}\right)+\frac{1}{4}\sqrt{\frac{1}{2\sqrt{N_{\min}}}+\frac{1}{N_{\min}}}\right],\tag{8.104}$$

$\forall\,\overrightarrow{N}\in\mathbb{N}^d$, *where* $N_{\min}:=\min\{N_1,\ldots,N_d\}$.

Also it holds $\lim_{\overrightarrow{N}\to(\infty,\ldots,\infty)}K_{\overrightarrow{N},\mu}(f)\left(\overrightarrow{x}\right)=f\left(\overrightarrow{x}\right)$.

Proof By (8.15) we have

$$\left|K_{\overrightarrow{N},\mu}(f)\left(\overrightarrow{x}\right)-f\left(\overrightarrow{x}\right)\right|\leq\left(\max_{i=1,\ldots,d}\omega_1\left(\frac{\partial f}{\partial x_i},h\right)\right)\cdot$$

$$\left[\frac{d}{2h}\left(\sum_{i=1}^{d}K_{\overrightarrow{N},\mu}\left((t_i-x_i)^2\right)\left(\overrightarrow{x}\right)\right)+\frac{1}{2}\left(\sum_{i=1}^{d}K_{\overrightarrow{N},\mu}\left(|t_i-x_i|\right)\left(\overrightarrow{x}\right)\right)+\frac{h}{8}\right]\overset{(8.60)}{=}\tag{8.105}$$

$$\left(\max_{i=1,\ldots,d}\omega_1\left(\frac{\partial f}{\partial x_i},h\right)\right)\left[\frac{d}{2h}\left(\sum_{i=1}^{d}K_{N_i,\mu}\left((t_i-x_i)^2\right)(x_i)\right)\right.$$

$$\left.+\frac{1}{2}\left(\sum_{i=1}^{d}K_{N_i,\mu}\left(|t_i-x_i|\right)(x_i)\right)+\frac{h}{8}\right]\overset{\text{(by (8.31), (8.38))}}{\leq}$$

$$\left(\max_{i=1,\ldots,d}\omega_1\left(\frac{\partial f}{\partial x_i},h\right)\right)\left[\frac{d}{2h}\left(\sum_{i=1}^{d}\left(\frac{1}{2\sqrt{N_i}}+\frac{1}{N_i}\right)\right)\right.$$

$$+\frac{1}{2}\left(\sum_{i=1}^{d}\left(\frac{1}{2\sqrt{N_i}}+\frac{1}{N_i}\right)\right)+\frac{h}{8}\right] \leq$$

$$\left(\max_{i=1,\dots,d}\omega_1\left(\frac{\partial f}{\partial x_i},h\right)\right)\left[\frac{d^2}{2h}\left(\frac{1}{2\sqrt{N_{\min}}}+\frac{1}{N_{\min}}\right)\right.$$

$$\left.+\frac{d}{2}\left(\frac{1}{2\sqrt{N_{\min}}}+\frac{1}{N_{\min}}\right)+\frac{h}{8}\right]$$

(choose $h := \left(\frac{1}{2\sqrt{N_{\min}}}+\frac{1}{N_{\min}}\right)^{\frac{1}{2}} > 0$, then $h^2 = \frac{1}{2\sqrt{N_{\min}}}+\frac{1}{N_{\min}}$)

$$= \max_{i=1,\dots,d}\omega_1\left(\frac{\partial f}{\partial x_i},\left(\frac{1}{2\sqrt{N_{\min}}}+\frac{1}{N_{\min}}\right)^{\frac{1}{2}}\right)\cdot \tag{8.106}$$

$$\left[\frac{d^2}{2}\left(\frac{1}{2\sqrt{N_{\min}}}+\frac{1}{N_{\min}}\right)^{\frac{1}{2}}+\frac{d}{2}\left(\frac{1}{2\sqrt{N_{\min}}}+\frac{1}{N_{\min}}\right)+\frac{1}{8}\left(\frac{1}{2\sqrt{N_{\min}}}+\frac{1}{N_{\min}}\right)^{\frac{1}{2}}\right]$$

$$=\frac{1}{2}\left(\max_{i=1,\dots,d}\omega_1\left(\frac{\partial f}{\partial x_i},\sqrt{\frac{1}{2\sqrt{N_{\min}}}+\frac{1}{N_{\min}}}\right)\right)\cdot$$

$$\left[d^2\left(\sqrt{\frac{1}{2\sqrt{N_{\min}}}+\frac{1}{N_{\min}}}\right)+d\left(\frac{1}{2\sqrt{N_{\min}}}+\frac{1}{N_{\min}}\right)+\frac{1}{4}\sqrt{\frac{1}{2\sqrt{N_{\min}}}+\frac{1}{N_{\min}}}\right], \tag{8.107}$$

proving the claim. ■

Theorem 8.49 *Let* $[0,1]^d$, $d \in \mathbb{N}-\{1\}$, *endowed with* l_1*-norm, and* $f \in Lip_+([0,1]^d) \cap C^n([0,1]^d, \mathbb{R}_+)$, $n \in \mathbb{N}-\{1\}$; $\overrightarrow{x} \in [0,1]^d : f_\alpha(\overrightarrow{x}) = 0$, *for all* $\alpha : |\alpha| = 1, \dots, n$. *Then*

$$\left|K_{\overrightarrow{N},\mu}(f)(\overrightarrow{x})-f(\overrightarrow{x})\right| \leq$$

$$\left(\max_{\alpha:|\alpha|=n}\omega_1\left(f_\alpha,\left(\frac{1}{2\sqrt{N_{\min}}}+\frac{1}{N_{\min}}\right)^{\frac{1}{n+1}}\right)\right)\cdot \tag{8.108}$$

$$\left[\frac{d^{n+1}}{(n+1)!}\left(\frac{1}{2\sqrt{N_{\min}}}+\frac{1}{N_{\min}}\right)^{\frac{n}{n+1}}+\frac{d^n}{2n!}\left(\frac{1}{2\sqrt{N_{\min}}}+\frac{1}{N_{\min}}\right)\right.$$

$$+\frac{d^{n-1}}{8(n-1)!}\left(\frac{1}{2\sqrt{N_{\min}}}+\frac{1}{N_{\min}}\right)^{\frac{n+2}{n+1}}\Bigg],$$

$\forall\ \overrightarrow{N}\in\mathbb{N}^d$, *where* $N_{\min}:=\min\{N_1,...,N_d\}$.

We have that $\lim_{\overrightarrow{N}\to(\infty,...,\infty)} K_{\overrightarrow{N},\mu}(f)\left(\overrightarrow{x}\right)=f\left(\overrightarrow{x}\right)$.

Proof By (8.14) we have

$$\left|K_{\overrightarrow{N},\mu}(f)\left(\overrightarrow{x}\right)-f\left(\overrightarrow{x}\right)\right|\leq\left(\max_{\alpha:|\alpha|=n}\omega_1\left(f_\alpha,h\right)\right)\cdot$$

$$\left[\frac{d^n}{(n+1)!h}\left(\sum_{i=1}^{d}K_{\overrightarrow{N},\mu}\left(|t_i-x_i|^{n+1}\right)\left(\overrightarrow{x}\right)\right)+\frac{d^{n-1}}{2n!}\left(\sum_{i=1}^{d}K_{\overrightarrow{N},\mu}\left(|t_i-x_i|^{n}\right)\left(\overrightarrow{x}\right)\right)\right.$$

$$\left.+\frac{hd^{n-2}}{8(n-1)!}\left(\sum_{i=1}^{d}K_{\overrightarrow{N},\mu}\left(|t_i-x_i|^{n-1}\right)\left(\overrightarrow{x}\right)\right)\right]\overset{(8.60)}{=}\tag{8.109}$$

$$\left(\max_{\alpha:|\alpha|=n}\omega_1\left(f_\alpha,h\right)\right)\left[\frac{d^n}{(n+1)!h}\left(\sum_{i=1}^{d}K_{N_i,\mu}\left(|t_i-x_i|^{n+1}\right)(x_i)\right)\right.$$

$$\left.+\frac{d^{n-1}}{2n!}\left(\sum_{i=1}^{d}K_{N_i,\mu}\left(|t_i-x_i|^{n}\right)(x_i)\right)+\frac{hd^{n-2}}{8(n-1)!}\left(\sum_{i=1}^{d}K_{N_i,\mu}\left(|t_i-x_i|^{n-1}\right)(x_i)\right)\right]$$

$$\overset{\text{(by (8.31), (8.38))}}{\leq}\left(\max_{\alpha:|\alpha|=n}\omega_1\left(f_\alpha,h\right)\right)\left[\frac{d^n}{(n+1)!h}\left(\sum_{i=1}^{d}\left(\frac{1}{2\sqrt{N_i}}+\frac{1}{N_i}\right)\right)\right.$$

$$\left.+\frac{d^n}{2n!}\left(\sum_{i=1}^{d}\left(\frac{1}{2\sqrt{N_i}}+\frac{1}{N_i}\right)\right)+\frac{hd^{n-2}}{8(n-1)!}\left(\sum_{i=1}^{d}\left(\frac{1}{2\sqrt{N_i}}+\frac{1}{N_i}\right)\right)\right]\leq\tag{8.110}$$

$$\left(\max_{\alpha:|\alpha|=n}\omega_1\left(f_\alpha,h\right)\right)\left[\frac{d^{n+1}}{(n+1)!h}\left(\frac{1}{2\sqrt{N_{\min}}}+\frac{1}{N_{\min}}\right)\right.$$

$$\left.+\frac{d^n}{2n!}\left(\frac{1}{2\sqrt{N_{\min}}}+\frac{1}{N_{\min}}\right)+\frac{hd^{n-1}}{8(n-1)!}\left(\frac{1}{2\sqrt{N_{\min}}}+\frac{1}{N_{\min}}\right)\right]$$

(choose $h:=\left(\frac{1}{2\sqrt{N_{\min}}}+\frac{1}{N_{\min}}\right)^{\frac{1}{n+1}}>0$, then $h^n=\left(\frac{1}{2\sqrt{N_{\min}}}+\frac{1}{N_{\min}}\right)^{\frac{n}{n+1}}$, and $h^{n+1}=\frac{1}{2\sqrt{N_{\min}}}+\frac{1}{N_{\min}}$)

$$= \left(\max_{\alpha:|\alpha|=n} \omega_1 \left(f_\alpha, \left(\frac{1}{2\sqrt{N_{\min}}} + \frac{1}{N_{\min}} \right)^{\frac{1}{n+1}} \right) \right) \cdot \tag{8.111}$$

$$\left[\frac{d^{n+1}}{(n+1)!} \left(\frac{1}{2\sqrt{N_{\min}}} + \frac{1}{N_{\min}} \right)^{\frac{n}{n+1}} \right.$$

$$\left. + \frac{d^n}{2n!} \left(\frac{1}{2\sqrt{N_{\min}}} + \frac{1}{N_{\min}} \right) + \frac{d^{n-1}}{8(n-1)!} \left(\frac{1}{2\sqrt{N_{\min}}} + \frac{1}{N_{\min}} \right)^{\frac{n+2}{n+1}} \right],$$

proving the claim. ∎

We continue with

Theorem 8.50 *Let* $[0,1]^d$, $d \in \mathbb{N} - \{1\}$, *endowed with* l_1*-norm, and* $f \in Lip_+\left([0,1]^d\right)$. *Then*

$$\left| D_{\vec{N},\Gamma_{\vec{N}}}(f)(\vec{x}) - f(\vec{x}) \right| \leq (d+1) K \left(\frac{1+3\sqrt{2}}{2\sqrt{N_{\min}}} + \frac{1}{N_{\min}} \right), \tag{8.112}$$

where $K > 0$, $\forall \vec{x} \in [0,1]^d$, $\forall \vec{N} \in (\mathbb{N} - \{1\})^d$, *where* $N_{\min} := \min\{N_1, \ldots, N_d\}$.

That is

$$\left\| D_{\vec{N},\Gamma_{\vec{N}}}(f) - f \right\|_\infty \leq (d+1) K \left(\frac{1+3\sqrt{2}}{2\sqrt{N_{\min}}} + \frac{1}{N_{\min}} \right), \tag{8.113}$$

$\forall \vec{N} \in (\mathbb{N} - \{1\})^d$.

It holds that $\lim_{\vec{N} \to (\infty,\ldots,\infty)} D_{\vec{N},\Gamma_{\vec{N}}}(f) = f$, *uniformly.*

Proof By (8.16) we have

$$\left| D_{\vec{N},\Gamma_{\vec{N}}}(f)(\vec{x}) - f(\vec{x}) \right| \leq \omega_1(f,h) \left(1 + \frac{1}{h} \left(\sum_{i=1}^d D_{\vec{N},\Gamma_{\vec{N}}}(|t_i - x_i|)(\vec{x}) \right) \right) \overset{(8.71)}{=}$$

$$\omega_1(f,h) \left(1 + \frac{1}{h} \left(\sum_{i=1}^d D_{N_i,\Gamma_{N_i}}(|t_i - x_i|)(x_i) \right) \right) \overset{\text{(by (8.36), (8.39))}}{\leq}$$

$$\omega_1(f,h) \left(1 + \frac{1}{h} \left(\sum_{i=1}^d \left(\frac{1+3\sqrt{2}}{2\sqrt{N_i}} + \frac{1}{N_i} \right) \right) \right) \leq$$

$$\omega_1(f,h)\left(1+\frac{d}{h}\left(\frac{1+3\sqrt{2}}{2\sqrt{N_{\min}}}+\frac{1}{N_{\min}}\right)\right) \tag{8.114}$$

(choose $h := \left(\frac{1+3\sqrt{2}}{2\sqrt{N_{\min}}}+\frac{1}{N_{\min}}\right) > 0$)

$$= \omega_1\left(f,\left(\frac{1+3\sqrt{2}}{2\sqrt{N_{\min}}}+\frac{1}{N_{\min}}\right)\right)(d+1) \leq (d+1)\,K\left(\frac{1+3\sqrt{2}}{2\sqrt{N_{\min}}}+\frac{1}{N_{\min}}\right),$$

proving the claim. ∎

Theorem 8.51 *Let* $[0,1]^d$, $d \in \mathbb{N}-\{1\}$, *endowed with* l_1*-norm, and* $f \in Lip_+\left([0,1]^d\right) \cap C^1\left([0,1]^d,\mathbb{R}_+\right)$; $\overrightarrow{x} \in [0,1]^d : \frac{\partial f(\overrightarrow{x})}{\partial x_i} = 0$, *for* $i = 1,\ldots,d$. *Then*

$$\left|D_{\overrightarrow{N},\Gamma_{\overrightarrow{N}}}(f)(\overrightarrow{x}) - f(\overrightarrow{x})\right| \leq \frac{1}{2}\left(\max_{i=1,\ldots,d}\omega_1\left(\frac{\partial f}{\partial x_i},\sqrt{\frac{1+3\sqrt{2}}{2\sqrt{N_{\min}}}+\frac{1}{N_{\min}}}\right)\right)\cdot$$

$$\left[d^2\left(\sqrt{\frac{1+3\sqrt{2}}{2\sqrt{N_{\min}}}+\frac{1}{N_{\min}}}\right)+d\left(\frac{1+3\sqrt{2}}{2\sqrt{N_{\min}}}+\frac{1}{N_{\min}}\right)+\frac{1}{4}\sqrt{\frac{1+3\sqrt{2}}{2\sqrt{N_{\min}}}+\frac{1}{N_{\min}}}\right], \tag{8.115}$$

$\forall\, \overrightarrow{N} \in (\mathbb{N}-\{1\})^d$, *where* $N_{\min} := \min\{N_1,\ldots,N_d\}$.

Also it holds $\lim_{\overrightarrow{N}\to(\infty,\ldots,\infty)} D_{\overrightarrow{N},\Gamma_{\overrightarrow{N}}}(f)(\overrightarrow{x}) = f(\overrightarrow{x})$.

Proof By (8.15) we have

$$\left|D_{\overrightarrow{N},\Gamma_{\overrightarrow{N}}}(f)(\overrightarrow{x}) - f(\overrightarrow{x})\right| \leq \left(\max_{i=1,\ldots,d}\omega_1\left(\frac{\partial f}{\partial x_i},h\right)\right)\cdot$$

$$\left[\frac{d}{2h}\left(\sum_{i=1}^{d} D_{\overrightarrow{N},\Gamma_{\overrightarrow{N}}}\left((t_i-x_i)^2\right)(\overrightarrow{x})\right)+\frac{1}{2}\left(\sum_{i=1}^{d} D_{\overrightarrow{N},\Gamma_{\overrightarrow{N}}}\left(|t_i-x_i|\right)(\overrightarrow{x})\right)+\frac{h}{8}\right] \overset{(8.71)}{=}$$

$$\left(\max_{i=1,\ldots,d}\omega_1\left(\frac{\partial f}{\partial x_i},h\right)\right)\left[\frac{d}{2h}\left(\sum_{i=1}^{d} D_{N_i,\Gamma_{N_i}}\left((t_i-x_i)^2\right)(x_i)\right)\right.$$

$$\left.+\frac{1}{2}\left(\sum_{i=1}^{d} D_{N_i,\Gamma_{N_i}}\left(|t_i-x_i|\right)(x_i)\right)+\frac{h}{8}\right] \overset{\text{(by (8.36), (8.39))}}{\leq} \tag{8.116}$$

$$\left(\max_{i=1,\ldots,d}\omega_1\left(\frac{\partial f}{\partial x_i},h\right)\right)\left[\frac{d}{2h}\left(\sum_{i=1}^{d}\left(\frac{1+3\sqrt{2}}{2\sqrt{N_i}}+\frac{1}{N_i}\right)\right)\right.$$

$$+\frac{1}{2}\left(\sum_{i=1}^{d}\left(\frac{1+3\sqrt{2}}{2\sqrt{N_i}}+\frac{1}{N_i}\right)\right)+\frac{h}{8}\Bigg]\leq$$

$$\left(\max_{i=1,\ldots,d}\omega_1\left(\frac{\partial f}{\partial x_i},h\right)\right)\left[\frac{d^2}{2h}\left(\frac{1+3\sqrt{2}}{2\sqrt{N_{\min}}}+\frac{1}{N_{\min}}\right)\right.$$

$$\left.+\frac{d}{2}\left(\frac{1+3\sqrt{2}}{2\sqrt{N_{\min}}}+\frac{1}{N_{\min}}\right)+\frac{h}{8}\right]$$

(choose $h:=\left(\frac{1+3\sqrt{2}}{2\sqrt{N_{\min}}}+\frac{1}{N_{\min}}\right)^{\frac{1}{2}}>0$, then $h^2=\frac{1+3\sqrt{2}}{2\sqrt{N_{\min}}}+\frac{1}{N_{\min}}$)

$$=\max_{i=1,\ldots,d}\omega_1\left(\frac{\partial f}{\partial x_i},\left(\frac{1+3\sqrt{2}}{2\sqrt{N_{\min}}}+\frac{1}{N_{\min}}\right)^{\frac{1}{2}}\right)\cdot \tag{8.117}$$

$$\left[\frac{d^2}{2}\left(\frac{1+3\sqrt{2}}{2\sqrt{N_{\min}}}+\frac{1}{N_{\min}}\right)^{\frac{1}{2}}+\frac{d}{2}\left(\frac{1+3\sqrt{2}}{2\sqrt{N_{\min}}}+\frac{1}{N_{\min}}\right)+\frac{1}{8}\left(\frac{1+3\sqrt{2}}{2\sqrt{N_{\min}}}+\frac{1}{N_{\min}}\right)^{\frac{1}{2}}\right]$$

$$=\frac{1}{2}\left(\max_{i=1,\ldots,d}\omega_1\left(\frac{\partial f}{\partial x_i},\sqrt{\frac{1+3\sqrt{2}}{2\sqrt{N_{\min}}}+\frac{1}{N_{\min}}}\right)\right)\cdot$$

$$\left[d^2\left(\sqrt{\frac{1+3\sqrt{2}}{2\sqrt{N_{\min}}}+\frac{1}{N_{\min}}}\right)+d\left(\frac{1+3\sqrt{2}}{2\sqrt{N_{\min}}}+\frac{1}{N_{\min}}\right)+\frac{1}{4}\sqrt{\frac{1+3\sqrt{2}}{2\sqrt{N_{\min}}}+\frac{1}{N_{\min}}}\right], \tag{8.118}$$

proving the claim. ■

Theorem 8.52 *Let* $[0,1]^d$, $d\in\mathbb{N}-\{1\}$, *endowed with* l_1*-norm, and* $f\in Lip_+\left([0,1]^d\right)\cap C^n\left([0,1]^d,\mathbb{R}_+\right)$, $n\in\mathbb{N}-\{1\}$; $\overrightarrow{x}\in[0,1]^d:f_\alpha\left(\overrightarrow{x}\right)=0$, *for all* $\alpha:|\alpha|=1,\ldots,n$. *Then*

$$\left|D_{\overrightarrow{N},\Gamma_{\overrightarrow{N}}}(f)\left(\overrightarrow{x}\right)-f\left(\overrightarrow{x}\right)\right|\leq$$

$$\left(\max_{\alpha:|\alpha|=n}\omega_1\left(f_\alpha,\left(\frac{1+3\sqrt{2}}{2\sqrt{N_{\min}}}+\frac{1}{N_{\min}}\right)^{\frac{1}{n+1}}\right)\right)\cdot \tag{8.119}$$

$$\left[\frac{d^{n+1}}{(n+1)!}\left(\frac{1+3\sqrt{2}}{2\sqrt{N_{\min}}}+\frac{1}{N_{\min}}\right)^{\frac{n}{n+1}}+\frac{d^n}{2n!}\left(\frac{1+3\sqrt{2}}{2\sqrt{N_{\min}}}+\frac{1}{N_{\min}}\right)\right.$$

$$+\frac{d^{n-1}}{8(n-1)!}\left(\frac{1+3\sqrt{2}}{2\sqrt{N_{\min}}}+\frac{1}{N_{\min}}\right)^{\frac{n+2}{n+1}}\Bigg],$$

$\forall\ \overrightarrow{N} \in (\mathbb{N}-\{1\})^d$, *where* $N_{\min} := \min\{N_1, \ldots, N_d\}$.

We have that $\lim\limits_{\overrightarrow{N}\to(\infty,\ldots,\infty)} D_{\overrightarrow{N},\Gamma_{\overrightarrow{N}}}(f)\left(\overrightarrow{x}\right) = f\left(\overrightarrow{x}\right)$.

Proof By (8.14) we have

$$\left|D_{\overrightarrow{N},\Gamma_{\overrightarrow{N}}}(f)\left(\overrightarrow{x}\right) - f\left(\overrightarrow{x}\right)\right| \leq \left(\max_{\alpha:|\alpha|=n}\omega_1(f_\alpha,h)\right)\cdot$$

$$\Bigg[\frac{d^n}{(n+1)!h}\left(\sum_{i=1}^{d} D_{\overrightarrow{N},\Gamma_{\overrightarrow{N}}}\left(|t_i-x_i|^{n+1}\right)\left(\overrightarrow{x}\right)\right)+\frac{d^{n-1}}{2n!}\left(\sum_{i=1}^{d} D_{\overrightarrow{N},\Gamma_{\overrightarrow{N}}}\left(|t_i-x_i|^{n}\right)\left(\overrightarrow{x}\right)\right)$$

$$+\frac{hd^{n-2}}{8(n-1)!}\left(\sum_{i=1}^{d} D_{\overrightarrow{N},\Gamma_{\overrightarrow{N}}}\left(|t_i-x_i|^{n-1}\right)\left(\overrightarrow{x}\right)\right)\Bigg] \overset{(8.71)}{=}$$

$$\left(\max_{\alpha:|\alpha|=n}\omega_1(f_\alpha,h)\right)\Bigg[\frac{d^n}{(n+1)!h}\left(\sum_{i=1}^{d} D_{N_i,\Gamma_{N_i}}\left(|t_i-x_i|^{n+1}\right)(x_i)\right) \tag{8.120}$$

$$\frac{d^{n-1}}{2n!}\left(\sum_{i=1}^{d} D_{N_i,\Gamma_{N_i}}\left(|t_i-x_i|^{n}\right)(x_i)\right)+$$

$$+\frac{hd^{n-2}}{8(n-1)!}\left(\sum_{i=1}^{d} D_{N_i,\Gamma_{N_i}}\left(|t_i-x_i|^{n-1}\right)(x_i)\right)\Bigg] \overset{(\text{by } (8.50),\ (8.59))}{\leq}$$

$$\left(\max_{\alpha:|\alpha|=n}\omega_1(f_\alpha,h)\right)\Bigg[\frac{d^n}{(n+1)!h}\left(\sum_{i=1}^{d}\left(\frac{1+3\sqrt{2}}{2\sqrt{N_i}}+\frac{1}{N_i}\right)\right)$$

$$+\frac{d^{n-1}}{2n!}\left(\sum_{i=1}^{d}\left(\frac{1+3\sqrt{2}}{2\sqrt{N_i}}+\frac{1}{N_i}\right)\right)+\frac{hd^{n-2}}{8(n-1)!}\left(\sum_{i=1}^{d}\left(\frac{1+3\sqrt{2}}{2\sqrt{N_i}}+\frac{1}{N_i}\right)\right)\Bigg] \leq \tag{8.121}$$

$$\left(\max_{\alpha:|\alpha|=n}\omega_1(f_\alpha,h)\right)\Bigg[\frac{d^{n+1}}{(n+1)!h}\left(\frac{1+3\sqrt{2}}{2\sqrt{N_{\min}}}+\frac{1}{N_{\min}}\right)$$

$$+\frac{d^n}{2n!}\left(\frac{1+3\sqrt{2}}{2\sqrt{N_{\min}}}+\frac{1}{N_{\min}}\right)+\frac{hd^{n-1}}{8(n-1)!}\left(\frac{1+3\sqrt{2}}{2\sqrt{N_{\min}}}+\frac{1}{N_{\min}}\right)\Bigg]$$

(choose $h := \left(\frac{1+3\sqrt{2}}{2\sqrt{N_{\min}}} + \frac{1}{N_{\min}}\right)^{\frac{1}{n+1}} > 0$, then $h^n = \left(\frac{1+3\sqrt{2}}{2\sqrt{N_{\min}}} + \frac{1}{N_{\min}}\right)^{\frac{n}{n+1}}$, and $h^{n+1} = \frac{1+3\sqrt{2}}{2\sqrt{N_{\min}}} + \frac{1}{N_{\min}}$)

$$= \left(\max_{\alpha:|\alpha|=n} \omega_1 \left(f_\alpha, \left(\frac{1+3\sqrt{2}}{2\sqrt{N_{\min}}} + \frac{1}{N_{\min}}\right)^{\frac{1}{n+1}}\right)\right) \cdot$$

$$\left[\frac{d^{n+1}}{(n+1)!}\left(\frac{1+3\sqrt{2}}{2\sqrt{N_{\min}}} + \frac{1}{N_{\min}}\right)^{\frac{n}{n+1}} + \frac{d^n}{2n!}\left(\frac{1+3\sqrt{2}}{2\sqrt{N_{\min}}} + \frac{1}{N_{\min}}\right) \right. \tag{8.122}$$

$$\left. + \frac{d^{n-1}}{8(n-1)!}\left(\frac{1+3\sqrt{2}}{2\sqrt{N_{\min}}} + \frac{1}{N_{\min}}\right)^{\frac{n+2}{n+1}}\right],$$

proving the claim. ■

We continue with convex approximation by specific multivariate positive sublinear operators.

Theorem 8.53 *Let* $[0,1]^d$, $d \in \mathbb{N} - \{1\}$, *endowed with* l_1*-norm, and* $f \in Lip_+\left([0,1]^d\right) : \left|f\left(\overrightarrow{t}\right) - f\left(\overrightarrow{x}\right)\right|$ *is a convex function in* $\overrightarrow{t} = (t_1, ..., t_k) \in [0,1]^d$ *for a fixed* $\overrightarrow{x} = (x_1, ..., x_k) \in (0,1)^d$. *We assume that the ball in* $\mathbb{R}^d : B\left(\overrightarrow{x}, \left(\frac{1}{2\sqrt{N^*_{\min}}} + \frac{1}{N^*_{\min}}\right)\right) \subset [0,1]^d$, *for a sufficiently large* $N^*_{\min} := \min\{N^*_1, ..., N^*_d\}$*; where* $\overrightarrow{N}^* := \left(N^*_1, ..., N^*_d\right) \in \mathbb{N}^d$. *Then*

$$\left|K_{\overrightarrow{N},\mu}(f)\left(\overrightarrow{x}\right) - f\left(\overrightarrow{x}\right)\right| \le dK\left(\frac{1}{2\sqrt{N_{\min}}} + \frac{1}{N_{\min}}\right), \tag{8.123}$$

where $K > 0$, $\forall\, \overrightarrow{N} \in \mathbb{N}^d$, *where* $\overrightarrow{N} := (N_1, ..., N_d)$, *and* $N_{\min} := \min\{N_1, ..., N_d\}$, *with* $N_{\min} \ge N^*_{\min}$.

It holds $\lim_{\overrightarrow{N} \to (\infty,...,\infty)} K_{\overrightarrow{N},\mu}(f)\left(\overrightarrow{x}\right) = f\left(\overrightarrow{x}\right)$.

Proof By (8.23) we have

$$\left|K_{\overrightarrow{N},\mu}(f)\left(\overrightarrow{x}\right) - f\left(\overrightarrow{x}\right)\right| \le \frac{\omega_1(f,h)}{h}\left(\sum_{i=1}^d K_{\overrightarrow{N},\mu}(|t_i - x_i|)\left(\overrightarrow{x}\right)\right) \overset{(8.60)}{=}$$

$$\frac{\omega_1(f,h)}{h}\left(\sum_{i=1}^d K_{N_i,\mu}(|t_i - x_i|)(x_i)\right) \overset{(\text{by } (8.30),\ (8.38))}{\le}$$

$$\frac{\omega_1(f,h)}{h}\left(\sum_{i=1}^{d}\left(\frac{1}{2\sqrt{N_i}}+\frac{1}{N_i}\right)\right)\leq\frac{d\omega_1(f,h)}{h}\left(\frac{1}{2\sqrt{N_{\min}}}+\frac{1}{N_{\min}}\right) \tag{8.124}$$

(setting $h:=\frac{1}{2\sqrt{N_{\min}}}+\frac{1}{N_{\min}}>0$)

$$=d\omega_1\left(f,\left(\frac{1}{2\sqrt{N_{\min}}}+\frac{1}{N_{\min}}\right)\right)\leq dK\left(\frac{1}{2\sqrt{N_{\min}}}+\frac{1}{N_{\min}}\right),$$

proving the claim. ∎

Theorem 8.54 *Let* $[0,1]^d$, $d\in\mathbb{N}-\{1\}$ *with the* l_1*-norm and let* $\overrightarrow{x}=(x_1,...,x_k)\in(0,1)^d$ *be fixed. Let* $f\in Lip_+\left([0,1]^d\right)\cap C^n\left([0,1]^d,\mathbb{R}_+\right)$, $n\in\mathbb{N}$ *and suppose that each nth partial derivative* $f_\alpha=\frac{\partial^\alpha f}{\partial x^\alpha}$, *where* $\alpha=(\alpha_1,...,\alpha_d)$, $\alpha_i\geq 0$, $i=1,...,d$, *and* $|\alpha|=\sum_{i=1}^{d}\alpha_i=n$, *has, relative to* $[0,1]^d$ *and the* l_1*-norm a modulus of continuity* $\omega_1(f_\alpha,\cdot)$, *and that each* $\left|f_\alpha\left(\overrightarrow{t}\right)\right|$ *is a convex function of* $\overrightarrow{t}\in[0,1]^d$, *all* $\alpha:|\alpha|=n$. *Assume further that* $f_\alpha(x)=0$, *for all* $\alpha:|\alpha|=1,...,n$; *and the ball in* $\mathbb{R}^d: B\left(\overrightarrow{x},\left(\frac{1}{2\sqrt{N^*_{\min}}}+\frac{1}{N^*_{\min}}\right)\right)\subset[0,1]^d$, *for a sufficiently large* $N^*_{\min}:=\min\{N^*_1,...,N^*_d\}$; *where* $\overrightarrow{N}^*:=\left(N^*_1,...,N^*_d\right)\in\mathbb{N}^d$. *Then*

$$\left|K_{\overrightarrow{N},\mu}(f)\left(\overrightarrow{x}\right)-f\left(\overrightarrow{x}\right)\right|\leq\frac{d^{n+1}}{(n+1)!}\left(\max_{\alpha:|\alpha|=n}\omega_1\left(f_\alpha,\left(\frac{1}{2\sqrt{N_{\min}}}+\frac{1}{N_{\min}}\right)\right)\right), \tag{8.125}$$

$\forall\,\overrightarrow{N}\in\mathbb{N}^d$, *where* $\overrightarrow{N}:=(N_1,...,N_d)$, *and* $N_{\min}:=\min\{N_1,...,N_d\}$, *with* $N_{\min}\geq N^*_{\min}$.

It holds $\lim_{\overrightarrow{N}\to(\infty,...,\infty)}K_{\overrightarrow{N},\mu}(f)\left(\overrightarrow{x}\right)=f\left(\overrightarrow{x}\right)$

Proof By (8.22) we have

$$\left|K_{\overrightarrow{N},\mu}(f)\left(\overrightarrow{x}\right)-f\left(\overrightarrow{x}\right)\right|\leq$$

$$\frac{\left(\max_{\alpha:|\alpha|=n}\omega_1(f_\alpha,h)\right)d^n}{h(n+1)!}\left(\sum_{i=1}^{d}K_{\overrightarrow{N},\mu}\left(|t_i-x_i|^{n+1}\right)\left(\overrightarrow{x}\right)\right)\overset{(8.60)}{=}$$

$$\frac{\left(\max_{\alpha:|\alpha|=n}\omega_1(f_\alpha,h)\right)d^n}{h(n+1)!}\left(\sum_{i=1}^{d}K_{N_i,\mu}\left(|t_i-x_i|^{n+1}\right)(x_i)\right)\overset{\text{(by (8.30), (8.38))}}{\leq}$$

$$\frac{\left(\max_{\alpha:|\alpha|=n} \omega_1\left(f_\alpha, h\right)\right) d^n}{h\,(n+1)!}\left(\sum_{i=1}^{d}\left(\frac{1}{2\sqrt{N_i}}+\frac{1}{N_i}\right)\right) \leq \tag{8.126}$$

$$\frac{\left(\max_{\alpha:|\alpha|=n} \omega_1\left(f_\alpha, h\right)\right) d^{n+1}}{h\,(n+1)!}\left(\frac{1}{2\sqrt{N_{\min}}}+\frac{1}{N_{\min}}\right)$$

(setting $h := \frac{1}{2\sqrt{N_{\min}}}+\frac{1}{N_{\min}} > 0$)

$$=\frac{\left(\max_{\alpha:|\alpha|=n} \omega_1\left(f_\alpha, \left(\frac{1}{2\sqrt{N_{\min}}}+\frac{1}{N_{\min}}\right)\right)\right) d^{n+1}}{(n+1)!}, \tag{8.127}$$

proving the claim. ■

We continue with

Theorem 8.55 *Let* $[0,1]^d$, $d \in \mathbb{N}-\{1\}$ *endowed with* l_1*-norm, and* $f \in Lip_+\left([0,1]^d\right)$: $\left|f\left(\overrightarrow{t}\right)-f\left(\overrightarrow{x}\right)\right|$ *is a convex function in* $\overrightarrow{t} \in [0,1]^d$ *for a fixed* $\overrightarrow{x} \in (0,1)^d$. *We assume that the ball in* $\mathbb{R}^d$: $B\left(\overrightarrow{x}, \left(\frac{1+3\sqrt{2}}{2\sqrt{N^*_{\min}}}+\frac{1}{N^*_{\min}}\right)\right) \subset [0,1]^d$, *for a sufficiently large* $N^*_{\min} := \min\{N_1^*, ..., N_d^*\}$; *where* $\overrightarrow{N}^* := \left(N_1^*, ..., N_d^*\right) \in (\mathbb{N}-\{1\})^d$. *Then*

$$\left|D_{\overrightarrow{N},\Gamma_{\overrightarrow{N}}}(f)\left(\overrightarrow{x}\right)-f\left(\overrightarrow{x}\right)\right| \leq dK\left(\frac{1+3\sqrt{2}}{2\sqrt{N_{\min}}}+\frac{1}{N_{\min}}\right), \tag{8.128}$$

where $K>0$, $\forall\, \overrightarrow{N} \in (\mathbb{N}-\{1\})^d$, *where* $\overrightarrow{N} := (N_1, ..., N_d)$, *and* $N_{\min} := \min\{N_1, ..., N_d\}$, *with* $N_{\min} \geq N^*_{\min}$.

It holds $\lim_{\overrightarrow{N}\to(\infty,...,\infty)} D_{\overrightarrow{N},\Gamma_{\overrightarrow{N}}}(f)\left(\overrightarrow{x}\right) = f\left(\overrightarrow{x}\right)$.

Proof By (8.23) we have

$$\left|D_{\overrightarrow{N},\Gamma_{\overrightarrow{N}}}(f)\left(\overrightarrow{x}\right)-f\left(\overrightarrow{x}\right)\right| \leq \frac{\omega_1(f,h)}{h}\left(\sum_{i=1}^{d} D_{\overrightarrow{N},\Gamma_{\overrightarrow{N}}}\left(|t_i-x_i|\right)\left(\overrightarrow{x}\right)\right) \overset{(8.71)}{=}$$

$$\frac{\omega_1(f,h)}{h}\left(\sum_{i=1}^{d} D_{N_i,\Gamma_{N_i}}\left(|t_i-x_i|\right)(x_i)\right) \overset{\text{(by (8.36), (8.39))}}{\leq}$$

$$\frac{\omega_1(f,h)}{h}\left(\sum_{i=1}^{d}\left(\frac{1+3\sqrt{2}}{2\sqrt{N_i}}+\frac{1}{N_i}\right)\right)\leq\frac{d\omega_1(f,h)}{h}\left(\frac{1+3\sqrt{2}}{2\sqrt{N_{\min}}}+\frac{1}{N_{\min}}\right) \quad (8.129)$$

(setting $h := \frac{1+3\sqrt{2}}{2\sqrt{N_{\min}}}+\frac{1}{N_{\min}} > 0$)

$$= d\omega_1\left(f,\left(\frac{1+3\sqrt{2}}{2\sqrt{N_{\min}}}+\frac{1}{N_{\min}}\right)\right)\leq dK\left(\frac{1+3\sqrt{2}}{2\sqrt{N_{\min}}}+\frac{1}{N_{\min}}\right),$$

proving the claim. ■

Theorem 8.56 *Let $[0,1]^d$, $d \in \mathbb{N}-\{1\}$ with the l_1-norm and let $\overrightarrow{x} = (x_1, ..., x_k) \in (0,1)^d$ be fixed. Let $f \in Lip_+\left([0,1]^d\right)\cap C^n\left([0,1]^d, \mathbb{R}_+\right)$, $n \in \mathbb{N}$ and suppose that each nth partial derivative $f_\alpha = \frac{\partial^\alpha f}{\partial x^\alpha}$, where $\alpha = (\alpha_1, ..., \alpha_d)$, $\alpha_i \geq 0$, $i = 1, ..., d$, and $|\alpha| = \sum_{i=1}^d \alpha_i = n$, has, relative to $[0,1]^d$ and the l_1-norm a modulus of continuity $\omega_1(f_\alpha, \cdot)$, and that each $\left|f_\alpha\left(\overrightarrow{t}\right)\right|$ is a convex function of $\overrightarrow{t} \in [0,1]^d$, all $\alpha : |\alpha| = n$. Assume further that $f_\alpha(x) = 0$, for all $\alpha : |\alpha| = 1, ..., n$; and the ball in $\mathbb{R}^d$: $B\left(\overrightarrow{x},\left(\frac{1+3\sqrt{2}}{2\sqrt{N^*_{\min}}}+\frac{1}{N^*_{\min}}\right)\right) \subset [0,1]^d$, for a sufficiently large $N^*_{\min} := \min\{N^*_1, ..., N^*_d\}$; where $\overrightarrow{N}^* := \left(N^*_1, ..., N^*_d\right) \in (\mathbb{N}-\{1\})^d$. Then*

$$\left|D_{\overrightarrow{N},\Gamma_{\overrightarrow{N}}}(f)\left(\overrightarrow{x}\right)-f\left(\overrightarrow{x}\right)\right|\leq\frac{d^{n+1}}{(n+1)!}\left(\max_{\alpha:|\alpha|=n}\omega_1\left(f_\alpha,\left(\frac{1+3\sqrt{2}}{2\sqrt{N_{\min}}}+\frac{1}{N_{\min}}\right)\right)\right), \quad (8.130)$$

$\forall\ \overrightarrow{N} \in (\mathbb{N}-\{1\})^d$, where $\overrightarrow{N} := (N_1, ..., N_d)$, and $N_{\min} := \min\{N_1, ..., N_d\}$, with $N_{\min} \geq N^*_{\min}$.

It holds $\lim_{\overrightarrow{N}\to(\infty,...,\infty)} D_{\overrightarrow{N},\Gamma_{\overrightarrow{N}}}(f)\left(\overrightarrow{x}\right) = f\left(\overrightarrow{x}\right)$.

Proof By (8.22) we have

$$\left|D_{\overrightarrow{N},\Gamma_{\overrightarrow{N}}}(f)\left(\overrightarrow{x}\right)-f\left(\overrightarrow{x}\right)\right|\leq$$

$$\frac{\left(\max_{\alpha:|\alpha|=n}\omega_1(f_\alpha,h)\right)d^n}{h(n+1)!}\left(\sum_{i=1}^{d}D_{\overrightarrow{N},\Gamma_{\overrightarrow{N}}}\left(|t_i-x_i|^{n+1}\right)\left(\overrightarrow{x}\right)\right)\overset{(8.71)}{=}$$

$$\frac{\left(\max_{\alpha:|\alpha|=n}\omega_1(f_\alpha,h)\right)d^n}{h(n+1)!}\left(\sum_{i=1}^{d}D_{N_i,\Gamma_{N_i}}\left(|t_i-x_i|^{n+1}\right)(x_i)\right)\overset{\text{(by (8.36), (8.39))}}{\leq}$$

$$\frac{\left(\max\limits_{\alpha:|\alpha|=n}\omega_1\left(f_\alpha,h\right)\right)d^n}{h\left(n+1\right)!}\left(\sum_{i=1}^{d}\left(\frac{1+3\sqrt{2}}{2\sqrt{N_i}}+\frac{1}{N_i}\right)\right)\leq \tag{8.131}$$

$$\frac{\left(\max\limits_{\alpha:|\alpha|=n}\omega_1\left(f_\alpha,h\right)\right)d^{n+1}}{h\left(n+1\right)!}\left(\frac{1+3\sqrt{2}}{2\sqrt{N_{\min}}}+\frac{1}{N_{\min}}\right)$$

(setting $h:=\frac{1+3\sqrt{2}}{2\sqrt{N_{\min}}}+\frac{1}{N_{\min}}>0$)

$$=\frac{d^{n+1}}{\left(n+1\right)!}\left(\max\limits_{\alpha:|\alpha|=n}\omega_1\left(f_\alpha,\left(\frac{1+3\sqrt{2}}{2\sqrt{N_{\min}}}+\frac{1}{N_{\min}}\right)\right)\right), \tag{8.132}$$

proving the claim. ■

8.10 Main Results - II

Here we give interpretations of general comonotonic positive sublinear operator approximation results via Choquet integrals. This is done by the use of Theorem 8.40, and Remark 8.41, we will use similar notations and symbols. We give

Theorem 8.57 *Let $(L_N)_{N\in\mathbb{N}}$ be a sequence of positive sublinear comonotonic operators from $\mathcal{L}_\infty\left(\mathcal{B}\left([a,b]\right)\right)$ into $C_+\left([a,b]\right)$, $L_N\left(1\right)=1$, $\forall\ N\in\mathbb{N}$. Let $f\in C_+\left([a,b]\right)$, $x\in(a,b)$, $0<(C)\int_a^b|t-x|\,d\gamma_{N,x}\left(t\right)\leq\min\left(x-a,b-x\right)$, $\forall\ N\in\mathbb{N}$, and $|f\left(\cdot\right)-f\left(x\right)|$ is a convex function over $[a,b]$. Then*

$$\left|L_N\left(f\right)\left(x\right)-f\left(x\right)\right|\leq\omega_1\left(f,(C)\int_a^b|t-x|\,d\gamma_{N,x}\left(t\right)\right),\ \forall N\in\mathbb{N}. \tag{8.133}$$

If $(C)\int_a^b|t-x|\,d\gamma_{N,x}\left(t\right)\to 0$, then $L_N\left(f\right)\left(x\right)\to f\left(x\right)$, as $N\to+\infty$.

Proof By Theorem 8.7. ■

Theorem 8.58 *Let $(L_N)_{N\in\mathbb{N}}$ be a sequence of positive sublinear comonotonic operators from $\mathcal{L}_\infty\left(\mathcal{B}\left([a,b]\right)\right)$ into $C_+\left([a,b]\right)$, $L_N\left(1\right)=1$, $\forall\ N\in\mathbb{N}$. Let $f\in C^n\left([a,b],\mathbb{R}_+\right)$, $n\in\mathbb{N}$, $x\in(a,b)$, $0<(C)\int_a^b|t-x|^{n+1}\,d\gamma_{N,x}\left(t\right)\leq\min\left(x-a,b-x\right)$, $\forall\ N\in\mathbb{N}$, and $\left|f^{(n)}\right|$ is convex over $[a,b]$. Assume that $f^{(k)}\left(x\right)=0$, $k=1,\ldots,n$. Then*

$$|L_N(f)(x) - f(x)| \leq \frac{\omega_1\left(f^{(n)}, (C)\int_a^b |t-x|^{n+1} d\gamma_{N,x}(t)\right)}{(n+1)!}, \forall N \in \mathbb{N}. \tag{8.134}$$

If $(C)\int_a^b |t-x|^{n+1} d\gamma_{N,x}(t) \to 0$, *then* $L_N(f)(x) \to f(x)$, *as* $N \to +\infty$.

Proof By Theorem 8.8. ■

We continue with

Theorem 8.59 *Let* Q *with* $\|\cdot\|$ *the* l_1*-norm, be a compact and convex subset of* $\mathbb{R}^d$, $d \in \mathbb{N} - \{1\}$, *and* $f \in C_+(Q)$. *Let* $\{L_N\}_{N\in\mathbb{N}}$ *be positive sublinear comonotonic operators from* $\mathcal{L}_\infty(\mathcal{B}(Q))$ *into* $C_+(Q)$, *such that* $L_N(1) = 1$, $\forall N \in \mathbb{N}$. *We assume further that* $(C)\int_Q \left\|\overrightarrow{t} - \overrightarrow{x}\right\| d\gamma_{N,x}(t) > 0$, $\forall N \in \mathbb{N}$. *Then*

$$\left|L_N(f)\left(\overrightarrow{x}\right) - f\left(\overrightarrow{x}\right)\right| \leq 2\omega_1\left(f, (C)\int_Q \left\|\overrightarrow{t} - \overrightarrow{x}\right\| d\gamma_{N,\overrightarrow{x}}\left(\overrightarrow{t}\right)\right), \tag{8.135}$$

$\forall N \in \mathbb{N}$, $\overrightarrow{x} \in Q$; $\forall \overrightarrow{t} \in Q$.

If $(C)\int_Q \left\|\overrightarrow{t} - \overrightarrow{x}\right\| d\gamma_{N,\overrightarrow{x}}\left(\overrightarrow{t}\right) \to 0$, *then* $L_N(f)\left(\overrightarrow{x}\right) \to f\left(\overrightarrow{x}\right)$, *as* $N \to +\infty$.

Proof By Theorem 8.14. ■

Theorem 8.60 *Let* $(Q, \|\cdot\|)$, *where* $\|\cdot\|$ *is the* l_1*-norm, be a compact and convex subset of* $\mathbb{R}^d$, $d \in \mathbb{N} - \{1\}$, *and let* $\overrightarrow{x} \in Q$ *be fixed. Let* $f \in C^n(Q, \mathbb{R}_+)$, $n \in \mathbb{N}$, $h > 0$. *We assume that* $f_\alpha\left(\overrightarrow{x}\right) = 0$, *for all* $\alpha : |\alpha| = 1, ..., n$. *Let* $\{L_N\}_{N\in\mathbb{N}}$ *be positive sublinear comonotonic operators from* $\mathcal{L}_\infty(\mathcal{B}(Q))$ *into* $C_+(Q)$, *such that* $L_N(1) = 1$, $\forall N \in \mathbb{N}$. *Then*

$$\left|L_N(f)\left(\overrightarrow{x}\right) - f\left(\overrightarrow{x}\right)\right| \leq \left(\max_{\alpha:|\alpha|=n} \omega_1(f_\alpha, h)\right) \cdot$$

$$\left[\frac{(C)\int_Q \left\|\overrightarrow{t} - \overrightarrow{x}\right\|^{n+1} d\gamma_{N,\overrightarrow{x}}\left(\overrightarrow{t}\right)}{(n+1)!h} + \frac{(C)\int_Q \left\|\overrightarrow{t} - \overrightarrow{x}\right\|^{n} d\gamma_{N,\overrightarrow{x}}\left(\overrightarrow{t}\right)}{2n!}\right.$$

$$\left. + \frac{h}{8(n-1)!}(C)\int_Q \left\|\overrightarrow{t} - \overrightarrow{x}\right\|^{n-1} d\gamma_{N,\overrightarrow{x}}\left(\overrightarrow{t}\right)\right], \tag{8.136}$$

$\forall N \in \mathbb{N}$.

Proof By Theorem 8.15. ■

Theorem 8.61 *All as in Theorem 8.60. Additionally assume that* $(C)\int_Q \left\|\overrightarrow{t}-\overrightarrow{x}\right\|^{n+1} d\gamma_{N,\overrightarrow{x}}\left(\overrightarrow{t}\right) > 0$, $\forall\, N \in \mathbb{N}$. *Then*

$$\left|L_N(f)\left(\overrightarrow{x}\right) - f\left(\overrightarrow{x}\right)\right| \leq \frac{1}{2n!}\left(3 + \frac{n}{4(n+1)}\right)\cdot$$

$$\left(\max_{\alpha:|\alpha|=n} \omega_1\left(f_\alpha, \frac{1}{(n+1)}\left((C)\int_Q \left\|\overrightarrow{t}-\overrightarrow{x}\right\|^{n+1} d\gamma_{N,\overrightarrow{x}}\left(\overrightarrow{t}\right)\right)^{\frac{1}{n+1}}\right)\right)\cdot$$

$$\left((C)\int_Q \left\|\overrightarrow{t}-\overrightarrow{x}\right\|^{n+1} d\gamma_{N,\overrightarrow{x}}\left(\overrightarrow{t}\right)\right)^{\frac{n}{n+1}}, \tag{8.137}$$

$\forall\, N \in \mathbb{N}$, $\overrightarrow{x} \in Q$.

If $(C)\int_Q \left\|\overrightarrow{t}-\overrightarrow{x}\right\|^{n+1} d\gamma_{N,\overrightarrow{x}}\left(\overrightarrow{t}\right) \to 0$, *then* $L_N(f)(x) \to f(x)$, *as* $N \to +\infty$.

Proof By Theorem 8.16. ■

Corollary 8.62 *Let* $(Q, \|\cdot\|)$, *where* $\|\cdot\|$ *is the* l_1*-norm, be a compact and convex subset of* $\mathbb{R}^d$, $d \in \mathbb{N} - \{1\}$, *and let* $\overrightarrow{x} \in Q$ *be fixed*, $f \in C^1(Q, \mathbb{R}_+)$. *We assume that* $\frac{\partial f}{\partial x_i}\left(\overrightarrow{x}\right) = 0$, $i = 1, \ldots, d$. *Let* $\{L_N\}_{N\in\mathbb{N}}$ *be positive sublinear comonotonic operators from* $\mathcal{L}_\infty(\mathcal{B}(Q))$ *into* $C_+(Q)$, *such that* $L_N(1) = 1$, $\forall\, N \in \mathbb{N}$. *Assume that* $(C)\int_Q \left\|\overrightarrow{t}-\overrightarrow{x}\right\|^{2} d\gamma_{N,\overrightarrow{x}}\left(\overrightarrow{t}\right) > 0$, $\forall\, N \in \mathbb{N}$. *Then*

$$\left|L_N(f)\left(\overrightarrow{x}\right) - f\left(\overrightarrow{x}\right)\right| \leq \frac{25}{16}\cdot$$

$$\left(\max_{i=1,\ldots,d} \omega_1\left(\frac{\partial f}{\partial x_i}, \frac{1}{2}\left((C)\int_Q \left\|\overrightarrow{t}-\overrightarrow{x}\right\|^{2} d\gamma_{N,\overrightarrow{x}}\left(\overrightarrow{t}\right)\right)^{\frac{1}{2}}\right)\right)\cdot$$

$$\left((C)\int_Q \left\|\overrightarrow{t}-\overrightarrow{x}\right\|^{2} d\gamma_{N,\overrightarrow{x}}\left(\overrightarrow{t}\right)\right)^{\frac{1}{2}}, \tag{8.138}$$

$\forall\, N \in \mathbb{N}$.

If $(C)\int_Q \left\|\overrightarrow{t}-\overrightarrow{x}\right\|^{2} d\gamma_{N,\overrightarrow{x}}\left(\overrightarrow{t}\right) \to 0$, *then* $L_N(f)\left(\overrightarrow{x}\right) \to f\left(\overrightarrow{x}\right)$, *as* $N \to +\infty$.

Proof By Corollary 8.17. ■

We continue with

Theorem 8.63 *Let Q with the l_1-norm $\|\cdot\|$, be a compact and convex subset of $\mathbb{R}^d$, $d \in \mathbb{N} - \{1\}$, and $f \in C_+(Q) : |f(t) - f(x)|$ is a convex function in $t \in Q$ for a fixed $x \in Q^o$. The first modulus of continuity $\omega_1(f, \cdot)$ is with respect to l_1-norm. Let $\{L_N\}_{N \in \mathbb{N}}$ be positive sublinear comonotonic operators from $\mathcal{L}_\infty(\mathcal{B}(Q))$ into $C_+(Q)$, $L_N(1) = 1$, $\forall N \in \mathbb{N}$. We assume that $(C)\int_Q \|t - x\| \, d\gamma_{N,x}(t) > 0$ and the ball in $\mathbb{R}^d$: $B\left(x, (C)\int_Q \|t - x\| \, d\gamma_{N,x}(t)\right) \subset Q$, $\forall N \in \mathbb{N}$. Then*

$$|L_N(f)(x) - f(x)| \leq \omega_1\left(f, (C)\int_Q \|t - x\| \, d\gamma_{N,x}(t)\right), \; \forall N \in \mathbb{N}. \quad (8.139)$$

If $(C)\int_Q \|t - x\| \, d\gamma_{N,x}(t) \to 0$, then $L_N(f)(x) \to f(x)$, as $N \to +\infty$.

Proof By Theorem 8.20. ■

Theorem 8.64 *Let Q with the l_1-norm $\|\cdot\|$, be a compact and convex subset of $\mathbb{R}^d$, $d \geq 1$, let $x \in Q^o$ be fixed. Let $f \in C^n(Q, \mathbb{R}_+)$, $n \in \mathbb{N}$, and f_α with $\alpha : |\alpha| = n$, has a first modulus of continuity $\omega_1(f_\alpha, \cdot)$ relative to Q with respect to l_1-norm. Each $|f_\alpha|$ is a convex function over Q, all $\alpha : |\alpha| = n$. Assume further that $f_\alpha(x) = 0$, all $\alpha : |\alpha| = 1, ..., n$. Let $\{L_N\}_{N \in \mathbb{N}}$ be positive sublinear comonotonic operators from $\mathcal{L}_\infty(\mathcal{B}(Q))$ into $C_+(Q) : L_N(1) = 1, \forall N \in \mathbb{N}$. We further assume that $(C)\int_Q \|t - x\|^{n+1} d\gamma_{N,x}(t) > 0$ and the ball in $\mathbb{R}^d$: $B\left(x, (C)\int_Q \|t - x\|^{n+1} d\gamma_{N,x}(t)\right) \subset Q$, $\forall N \in \mathbb{N}$. Then*

$$|L_N(f)(x) - f(x)| \leq \frac{\left(\max_{\alpha:|\alpha|=n} \omega_1\left(f_\alpha, (C)\int_Q \|t - x\|^{n+1} d\gamma_{N,x}(t)\right)\right)}{(n+1)!}, \; \forall N \in \mathbb{N}. \quad (8.140)$$

It holds, as $(C)\int_Q \|t - x\|^{n+1} d\gamma_{N,x}(t) \to 0$, then $L_N(f)(x) \to f(x)$, when $N \to +\infty$.

Proof By Theorem 8.22. ■

8.11 Main Results - III

Here we elaborate on the T_N general positive sublinear operators as described in Remark 8.42.

We give

Theorem 8.65 *Let $f \in C_+([a, b])$, $x \in (a, b)$, $0 < (C)\int_a^b |t - x| \, d\mu_{N,x}(t) \leq \min(x - a, b - x)$, $\forall N \in \mathbb{N}$, and $|f(\cdot) - f(x)|$ is a convex function over $[a, b]$. Then*

$$|T_N(f)(x) - f(x)| \leq \omega_1\left(f, (C)\int_a^b |t-x|\, d\mu_{N,x}(t)\right), \ \forall N \in \mathbb{N}. \tag{8.141}$$

If $(C)\int_a^b |t-x|\, d\mu_{N,x}(t) \to 0$, *then* $T_N(f)(x) \to f(x)$, *as* $N \to +\infty$.

Proof By Theorem 8.7. ∎

Theorem 8.66 *Let* $f \in C^n([a,b], \mathbb{R}_+)$, $n \in \mathbb{N}$, $x \in (a,b)$, $0 < (C)\int_a^b |t-x|^{n+1} d\mu_{N,x}(t) \leq \min(x-a, b-x)$, $\forall N \in \mathbb{N}$, *and* $\left|f^{(n)}\right|$ *is convex over* $[a,b]$. *Assume that* $f^{(k)}(x) = 0$, $k = 1, \ldots, n$. *Then*

$$|T_N(f)(x) - f(x)| \leq \frac{\omega_1\left(f^{(n)}, (C)\int_a^b |t-x|^{n+1} d\mu_{N,x}(t)\right)}{(n+1)!}, \ \forall N \in \mathbb{N}. \tag{8.142}$$

If $(C)\int_a^b |t-x|^{n+1} d\mu_{N,x}(t) \to 0$, *then* $T_N(f)(x) \to f(x)$, *as* $N \to +\infty$.

Proof By Theorem 8.8. ∎

We continue with

Theorem 8.67 *Let* Q *with* $\|\cdot\|$ *the* l_1*-norm, be a compact and convex subset of* $\mathbb{R}^d$, $d \in \mathbb{N} - \{1\}$, *and* $f \in C_+(Q)$. *Assume that* $(C)\int_Q \|t-x\|\, d\mu_{N,x}(t) > 0$, $\forall N \in \mathbb{N}$. *Then*

$$|T_N(f)(x) - f(x)| \leq 2\omega_1\left(f, (C)\int_Q \|t-x\|\, d\mu_{N,x}(t)\right), \tag{8.143}$$

$\forall N \in \mathbb{N}$, $x \in Q$.

If $(C)\int_Q \|t-x\|\, d\mu_{N,x}(t) \to 0$, *then* $T_N(f)(x) \to f(x)$, *as* $N \to +\infty$.

Proof By Theorem 8.14. ∎

Theorem 8.68 *Let* $(Q, \|\cdot\|)$, *where* $\|\cdot\|$ *is the* l_1*-norm, be a compact and convex subset of* $\mathbb{R}^d$, $d \in \mathbb{N} - \{1\}$, *and let* $x \in Q$ *be fixed. Let* $f \in C^n(Q, \mathbb{R}_+)$, $n \in \mathbb{N}$. *We assume that* $f_\alpha(x) = 0$, *for all* $\alpha : |\alpha| = 1, \ldots, n$. *Assume that* $(C)\int_Q \|t-x\|^{n+1} d\mu_{N,x}(t) > 0$, $\forall N \in \mathbb{N}$. *Then*

$$|T_N(f)(x) - f(x)| \leq \frac{1}{2n!}\left(3 + \frac{n}{4(n+1)}\right) \cdot \tag{8.144}$$

$$\left(\max_{\alpha:|\alpha|=n} \omega_1\left(f_\alpha, \frac{1}{(n+1)}\left((C)\int_Q \|t-x\|^{n+1} d\mu_{N,x}(t)\right)^{\frac{1}{n+1}}\right)\right)$$

$$\left((C)\int_Q \|t-x\|^{n+1} d\mu_{N,x}(t)\right)^{\frac{n}{n+1}}, \ \forall N \in \mathbb{N}.$$

If $(C)\int_Q \|t-x\|^{n+1} d\mu_{N,x}(t) \to 0$, *then* $T_N(f)(x) \to f(x)$, *as* $N \to +\infty$.

Proof By Theorem 8.16. ∎

An application for $n = 1$ follows.

Corollary 8.69 *Let* $(Q, \|\cdot\|)$, *where* $\|\cdot\|$ *is the* l_1*-norm, be a compact and convex subset of* $\mathbb{R}^d$, $d \in \mathbb{N} - \{1\}$, *and let* $x \in Q$ *be fixed. Let* $f \in C^1(Q, \mathbb{R}_+)$. *We assume that* $\frac{\partial f}{\partial x_i}(x) = 0$, $i = 1, ..., d$. *Assume that* $(C) \int_Q \|t - x\|^2 d\mu_{N,x}(t) > 0$, $\forall N \in \mathbb{N}$. *Then*

$$|T_N(f)(x) - f(x)| \leq \frac{25}{16}\cdot$$

$$\left(\max_{i=1,...,d} \omega_1\left(\frac{\partial f}{\partial x_i}, \frac{1}{2}\left((C)\int_Q \|t - x\|^2 d\mu_{N,x}(t)\right)^{\frac{1}{2}}\right)\right)\cdot \tag{8.145}$$

$$\left((C)\int_Q \|t - x\|^2 d\mu_{N,x}(t)\right)^{\frac{1}{2}}, \ \forall N \in \mathbb{N}.$$

If $(C) \int_Q \|t - x\|^2 d\mu_{N,x}(t) \to 0$, *then* $T_N(f)(x) \to f(x)$, *as* $N \to +\infty$.

Proof By Corollary 8.17. ∎

We continue with

Theorem 8.70 *Let* Q *with the* l_1*-norm* $\|\cdot\|$, *be a compact and convex subset of* $\mathbb{R}^d$, $d \in \mathbb{N} - \{1\}$, *and* $f \in C_+(Q) : |f(t) - f(x)|$ *is a convex function in* $t \in Q$ *for a fixed* $x \in Q^o$. *The first modulus of continuity* $\omega_1(f, \cdot)$ *is with respect to* l_1*-norm. We assume that* $(C) \int_Q \|t - x\| d\mu_{N,x}(t) > 0$ *and the ball in* $\mathbb{R}^d$: $B\left(x, (C) \int_Q \|t - x\| d\mu_{N,x}(t)\right) \subset Q$, $\forall N \in \mathbb{N}$. *Then*

$$|T_N(f)(x) - f(x)| \leq \omega_1\left(f, (C)\int_Q \|t - x\| d\mu_{N,x}(t)\right), \ \forall N \in \mathbb{N}. \tag{8.146}$$

If $(C) \int_Q \|t - x\| d\mu_{N,x}(t) \to 0$, *then* $T_N(f)(x) \to f(x)$, *as* $N \to +\infty$.

Proof By Theorem 8.20. ∎

We finish with

Theorem 8.71 *Let* Q *with the* l_1*-norm* $\|\cdot\|$, *be a compact and convex subset of* $\mathbb{R}^d$, $d \geq 1$, *let* $x \in Q^o$ *be fixed. Let* $f \in C^n(Q, \mathbb{R}_+)$, $n \in \mathbb{N}$, *and* f_α *with* $\alpha : |\alpha| = n$, *has a first modulus of continuity* $\omega_1(f_\alpha, \cdot)$ *relative to* Q *with respect to* l_1*-norm. Each* $|f_\alpha|$ *is a convex function over* Q, *all* $\alpha : |\alpha| = n$. *Assume further that* $f_\alpha(x) = 0$, *all* $\alpha : |\alpha| = 1, ..., n$. *We suppose that* $(C) \int_Q \|t - x\|^{n+1} d\mu_{N,x}(t) > 0$ *and the ball in* $\mathbb{R}^d : B\left(x, (C) \int_Q \|t - x\|^{n+1} d\mu_{N,x}(t)\right) \subset Q$, $\forall N \in \mathbb{N}$. *Then*

$$|T_N(f)(x) - f(x)| \leq \frac{\left(\max_{\alpha:|\alpha|=n} \omega_1\left(f_\alpha, (C)\int_Q \|t-x\|^{n+1} d\mu_{N,x}(t)\right)\right)}{(n+1)!}, \ \forall N \in \mathbb{N}. \tag{8.147}$$

If $(C)\int_Q \|t-x\|^{n+1} d\mu_{N,x}(t) \to 0$, *then* $T_N(f)(x) \to f(x)$, *when* $N \to +\infty$.

Proof By Theorem 8.22. ∎

References

1. G. Anastassiou, Approximation by multivariate sublinear and max-product operators, Revista De La real Academia De Ciencias exactas, Fisicas Y Naturales Serie A. Matematicas (RACSAM) (2017). Accepted for publication
2. G. Anastassiou, Approximation by multivariate sublinear and max-product operators under convexity. Demonstr. Math. **51**, 85–105 (2018)
3. G. Anastassiou, Approximation by sublinear and max-product operators using convexity (2018). Submitted for publication
4. G. Anastassiou, Multivariate and convex approximation by Choquet integrals, RACSAM (2018). Accepted for publication
5. G. Choquet, Theory of capacities. Ann. Inst. Fourier (Grenoble) **5**, 131–295 (1954)
6. D. Denneberg, *Non-additive Measure and Integral* (Kluwer, Dordrecht, 1994)
7. D. Dubois, H. Prade, *Possibility Theory* (Plenum Press, New York, 1988)
8. S. Gal, Uniform and pointwise quantitative approximation by Kantorovich-Choquet type integral operators with respect to monotone and submodular set functions. Mediter. J. Math. **14**(5), Art. 205 (2017), 12 pp
9. S. Gal, S. Trifa, Quantitative estimates in uniform and pointwise approximation by Bernstein-Durrmeyer-Choquet operators. Carpathian J. Math. **33**(1), 49–58 (2017)
10. D. Schmeidler, Integral representation without additivity. Proc. Am. Math. Soc. **97**, 255–261 (1986)
11. D. Schmeidler, Subjective probability and expected utility without additivity. Econometrica **57**, 571–587 (1989)
12. L.S. Shapley, A value for n-person games, in *Contributions to the Theory of Games*, ed. by H.W. Kuhn, A.W. Tucker. Annals of Mathematical Studies, vol. 28 (Princeton University Press, Princeton, 1953), pp. 307–317
13. Z. Wang, G.J. Klir, *Generalized Measure Theory* (Springer, New York, 2009)

Chapter 9
Caputo and Canavati Fractional Quantitative Approximation by Choquet Integrals

Here we consider the quantitative Caputo and Canavati fractional approximation of positive sublinear operators to the unit operator. These are given a precise Choquet integral interpretation. Initially we start with the study of the fractional rate of the convergence of the well-known Bernstein–Kantorovich–Choquet and Bernstein–Durrweyer–Choquet polynomial Choquet-integral operators. Then we study the very general comonotonic positive sublinear operators based on the representation theorem of Schmeidler [18]. We finish with the approximation by the very general direct Choquet-integral form positive sublinear operators. All fractional approximations are given via inequalities involving the modulus of continuity of the approximated function fractional order derivative. It follows [7].

9.1 Introduction

G. Choquet [9], introduced the capacities and his integral. Initially these were applied to statistical mechanics and potential theory, and they gave rise to the study of non-additive measure theory. Slowly but steady these ideas of Choquet started to attract economists especially after the very important work of Shapley [20] in the study of cooperative games. Capacities and Choquet integrals became main stream to Decision theorists since 1989 when D. Schmeidler [19] was the first to use them in a model of choice with non-additive beliefs. The expected utility results are strengthened by the use of Choquet capacities instead of probability measures.

In now days Choquet integral has wide applications, among others, to decision making under risk and uncertainty, in finance, in economics, in portofolio problems and in insurance.

Our motivation also comes from the foundations of Bayesian decision theory and subjective probability.

G. A. Anastassiou, *Ordinary and Fractional Approximation by Non-additive Integrals: Choquet, Shilkret and Sugeno Integral Approximators*, Studies in Systems, Decision and Control 190, https://doi.org/10.1007/978-3-030-04287-5_9

Because of the paramount importance of Choquet integral, we decided to research the related positive sublinear operators approximation, part of it is exhibited in this work in Caputo and Canavati fractional sense.

9.2 Background - I

Next we present briefly about the Choquet integral, see also [16].

We make

Definition 9.1 Consider $\Omega \neq \varnothing$ and let $\mathcal{C}$ be a σ-algebra of subsets in Ω.

(i) (see, e.g., [21], p. 63) The set function $\mu : \mathcal{C} \to [0, +\infty]$ is called a monotone set function (or capacity) if $\mu(\varnothing) = 0$ and $\mu(A) \leq \mu(B)$ for all $A, B \in \mathcal{C}$, with $A \subset B$. Also, μ is called submodular if

$$\mu(A \cup B) + \mu(A \cap B) \leq \mu(A) + \mu(B), \text{ for all } A, B \in \mathcal{C}.$$

μ is called bounded if $\mu(\Omega) < +\infty$ and normalized if $\mu(\Omega) = 1$.

(ii) (see, e.g., [21], p. 233, or [9]) If μ is a monotone set function on $\mathcal{C}$ and if $f : \Omega \to \mathbb{R}$ is $\mathcal{C}$-measurable (that is, for any Borel subset $B \subset \mathbb{R}$ it follows $f^{-1}(B) \in \mathcal{C}$), then for any $A \in \mathcal{C}$, the Choquet integral is defined by

$$(C)\int_A f d\mu = \int_0^{+\infty} \mu\left(F_\beta(f) \cap A\right) d\beta + \int_{-\infty}^{0} \left[\mu\left(F_\beta(f) \cap A\right) - \mu(A)\right] d\beta,$$

where we used the notation $F_\beta(f) = \{\omega \in \Omega : f(\omega) \geq \beta\}$. Notice that if $f \geq 0$ on A, then in the above formula we get $\int_{-\infty}^{0} = 0$.

The integrals on the right-hand side are the usual Riemann integral.

The function f will be called Choquet integrable on A if $(C)\int_A f d\mu \in \mathbb{R}$.

Next we list some well known properties of the Choquet integral.

Remark 9.2 If $\mu : \mathcal{C} \to [0, +\infty]$ is a monotone set function, then the following properties hold:

(i) For all $a \geq 0$ we have $(C)\int_A af d\mu = a \cdot (C)\int_A f d\mu$ (if $f \geq 0$ then see, e.g., [21], Theorem 11.2, (5), p. 228 and if f is arbitrary sign, then see, e.g., [10], p. 64, Proposition 5.1, (ii)).

(ii) For all $c \in \mathbb{R}$ and f of arbitrary sign, we have (see, e.g., [21], pp. 232-233, or [10], p. 65) $(C)\int_A (f + c) d\mu = (C)\int_A f d\mu + c \cdot \mu(A)$.

If μ is submodular too, then for all f, g of arbitrary sign and lower bounded, we have (see, e.g., [10], p. 75, Theorem 6.3)

$$(C)\int_A (f + g) d\mu \leq (C)\int_A f d\mu + (C)\int_A g d\mu.$$

(iii) If $f \leq g$ on A then $(C) \int_A f d\mu \leq (C) \int_A g d\mu$ (see, e.g., [21], p. 228, Theorem 11.2, (3) if $f, g \geq 0$ and p. 232 if f, g are of arbitrary sign).

(iv) Let $f \geq 0$. If $A \subset B$ then $(C) \int_A f d\mu \leq (C) \int_B f d\mu$. In addition, if μ is finitely subadditive, then

$$(C) \int_{A \cup B} f d\mu \leq (C) \int_A f d\mu + (C) \int_B f d\mu.$$

(v) It is immediate that $(C) \int_A 1 \cdot d\mu(t) = \mu(A)$.

(vi) The formula $\mu(A) = \gamma(M(A))$, where $\gamma : [0, 1] \to [0, 1]$ is an increasing and concave function, with $\gamma(0) = 0$, $\gamma(1) = 1$ and M is a probability measure (or only finitely additive) on a σ-algebra on Ω (that is, $M(\emptyset) = 0$, $M(\Omega) = 1$ and M is countably additive), gives simple examples of normalized, monotone and submodular set functions (see, e.g., [10], pp. 16-17, Example 2.1). Such of set functions μ are also called distorsions of countably normalized, additive measures (or distorted measures). For a simple example, we can take $\gamma(t) = \frac{2t}{1+t}$, $\gamma(t) = \sqrt{t}$.

If the above γ function is increasing, concave and satisfies only $\gamma(0) = 0$, then for any bounded Borel measure m, $\mu(A) = \gamma(m(A))$ gives a simple example of bounded, monotone and submodular set function.

(vii) If μ is a countably additive bounded measure, then the Choquet integral $(C) \int_A f d\mu$ reduces to the usual Lebesgue type integral (see, e.g., [10], p. 62, or [21], p. 226).

(viii) If $f \geq 0$, then $(C) \int_A f d\mu \geq 0$.

(ix) Let $\mu = \sqrt{M}$, where M is the Lebesgue measure on $[0, +\infty)$, then μ is a monotone and submodular set function, furthermore μ is strictly positive, see [15].

(x) If $\Omega = \mathbb{R}^N$, $N \in \mathbb{N}$, we call μ strictly positive if $\mu(A) > 0$, for any open subset $A \subseteq \mathbb{R}^N$.

We need some possibility theory:

Definition 9.3 ([12]) For the $\Omega \neq \emptyset$, the power set $\mathcal{P}(\Omega)$ denotes the family of all subsets of Ω.

(i) A function $\lambda : \Omega \to [0, 1]$ with the property $\sup\{\lambda(s) : s \in \Omega\} = 1$, is called possibility distribution on Ω.

(ii) $P : \mathcal{P}(\Omega) \to [0, 1]$ is called possibility measure, if it satisfies $P(\emptyset) = 0$, $P(\Omega) = 1$, and $P(\cup_{i \in I} A_i) = \sup\{P(A_i) : i \in I\}$ for all $A_i \subset \Omega$, and any I, an at most countable family of indices. Note that if $A, B \subset \Omega$, $A \subset B$, then the last property implies $P(A) \leq P(B)$ and that $P(A \cup B) \leq P(A) + P(B)$.

Any possibility distribution λ on Ω, induces the possibility measure $P_\lambda : \mathcal{P}(\Omega) \to [0, 1]$, $P_\lambda(A) = \sup\{\lambda(s) : s \in A\}$, $A \subset \Omega$. Also, if $f : \Omega \to \mathbb{R}_+$, then the possibilistic integral of f on $A \subset \Omega$ with respect to P_λ is defined by $(Pos) \int_A f dP_\lambda = \sup\{f(t) \lambda(t) : t \in A\}$ (see [12], chapter 1).

Note that any possibility measure μ is normalized, monotone and submodular. From $\mu(A \cup B) = \max\{\mu(A), \mu(B)\}$ we get monotonicity, and from $\mu(A \cap B) \leq \min\{\mu(A), \mu(B)\}$ we derive the submodularity.

9.3 Background - II

From Caputo fractional calculus:

We need

Definition 9.4 Let $\nu \geq 0$, $n = \lceil \nu \rceil$ ($\lceil \cdot \rceil$ is the ceiling of the number), $f \in AC^n([a,b])$ (space of functions f with $f^{(n-1)} \in AC([a,b])$, absolutely continuous functions). We call left Caputo fractional derivative (see [11], p. 40, [14, 17]) the function

$$D_{*a}^{\nu} f(x) = \frac{1}{\Gamma(n-\nu)} \int_a^x (x-t)^{n-\nu-1} f^{(n)}(t)\, dt, \ \forall\, x \in [a,b], \tag{9.1}$$

where Γ is the gamma function $\Gamma(\nu) = \int_0^{\infty} e^{-t} t^{\nu-1} dt$, $\nu > 0$.

We set $D_{*a}^{0} f(x) = f(x)$, $\forall\, x \in [a,b]$.

Lemma 9.5 ([3]) *Let* $\nu > 0$, $\nu \notin \mathbb{N}$, $n = \lceil \nu \rceil$, $f \in C^{n-1}([a,b])$ *and* $f^{(n)} \in L_{\infty}([a,b])$. *Then* $D_{*a}^{\nu} f(a) = 0$.

We need

Definition 9.6 (*see also* [2, 13, 14]) Let $f \in AC^m([a,b])$, $m = \lceil \alpha \rceil$, $\alpha > 0$. We right Caputo fractional derivative of order $\alpha > 0$ is given by

$$D_{b-}^{\alpha} f(x) = \frac{(-1)^m}{\Gamma(m-\alpha)} \int_x^b (\zeta - x)^{m-\alpha-1} f^{(m)}(\zeta)\, d\zeta, \ \forall\, x \in [a,b]. \tag{9.2}$$

We set $D_{b-}^{0} f(x) = f(x)$.

Lemma 9.7 ([3]) *Let* $f \in C^{m-1}([a,b])$, $f^{(m)} \in L_{\infty}([a,b])$, $m = \lceil \alpha \rceil$, $\alpha > 0$, $\alpha \notin \mathbb{N}$. *Then* $D_{b-}^{\alpha} f(b) = 0$.

Convention 9.8 *We assume that*

$$D_{*x_0}^{a} f(x) = 0, \text{ for } x < x_0, \tag{9.3}$$

and

$$D_{x_0-}^{\alpha} f(x) = 0, \text{ for } x > x_0, \tag{9.4}$$

for all $x, x_0 \in [a,b]$.

We mention

Proposition 9.9 ([3]) *Let* $f \in C^n([a,b])$, $n = \lceil \nu \rceil$, $\nu > 0$. *Then* $D_{*a}^{\nu} f(x)$ *is continuous in* $x \in [a,b]$.

Proposition 9.10 ([3]) *Let* $f \in C^m([a,b])$, $m = \lceil \alpha \rceil$, $\alpha > 0$. *Then* $D_{b-}^{\alpha} f(x)$ *is continuous in* $x \in [a,b]$.

For $f \in C([a,b])$ we define the (first) modulus of continuity:

$$\omega_1(f,\delta) := \sup_{\substack{x,y\in[a,b]:\\ |x-y|\le\delta}} |f(x)-f(y)|, \ \delta>0. \tag{9.5}$$

The modulus of continuity $\omega_1(f,\delta)$ is defined the same way for bounded functions, see (9.5), and it is finite.

We make

Remark 9.11 ([3]) Let $f \in C^{n-1}([a,b])$, $f^{(n)} \in L_\infty([a,b])$, $n=\lceil\nu\rceil$, $\nu>0$, $\nu\notin\mathbb{N}$. Then

$$\omega_1\left(D_{*a}^{\nu}f,\delta\right) \le \frac{2\left\|f^{(n)}\right\|_\infty}{\Gamma(n-\nu+1)}(b-a)^{n-\nu}. \tag{9.6}$$

Similarly, let $f \in C^{m-1}([a,b])$, $f^{(m)} \in L_\infty([a,b])$, $m=\lceil\alpha\rceil$, $\alpha>0$, $\alpha\notin\mathbb{N}$, then

$$\omega_1\left(D_{b-}^{\alpha}f,\delta\right) \le \frac{2\left\|f^{(m)}\right\|_\infty}{\Gamma(m-\alpha+1)}(b-a)^{m-\alpha}. \tag{9.7}$$

That is $\omega_1\left(D_{*a}^{\nu}f,\delta\right)$, $\omega_1\left(D_{b-}^{\alpha}f,\delta\right)$ are finite.

Clearly, above $D_{*a}^{\nu}f$ and $D_{b-}^{\alpha}f$ are bounded, from

$$\left|D_{*a}^{\nu}f(x)\right| \le \frac{\left\|f^{(n)}\right\|_\infty}{\Gamma(n-\nu+1)}(b-a)^{n-\nu}, \ \forall\, x\in[a,b], \tag{9.8}$$

see [3].

We need

Definition 9.12 Let $D_{x_0}^{\alpha}f$ denote any of $D_{x_0-}^{\alpha}f$, $D_{*x_0}^{\alpha}f$, and $\delta>0$. We set

$$\omega_1\left(D_{x_0}^{\alpha}f,\delta\right) := \max\left\{\omega_1\left(D_{x_0-}^{\alpha}f,\delta\right)_{[a,x_0]}, \omega_1\left(D_{*x_0}^{\alpha}f,\delta\right)_{[x_0,b]}\right\}, \tag{9.9}$$

where $x_0\in[a,b]$. Here the moduli of continuity are considered over $[a,x_0]$ and $[x_0,b]$, respectively.

We need

Definition 9.13 Here $C_+([a,b]) := \{f:[a,b]\to\mathbb{R}_+, \text{ continuous functions}\}$. Let $L_N: C_+([a,b])\to C_+([a,b])$, operators, $\forall\, N\in\mathbb{N}$, such that

(i)

$$L_N(\alpha f) = \alpha L_N(f), \ \forall\alpha\ge 0, \forall f\in C_+([a,b]), \tag{9.10}$$

(ii) if $f,g\in C_+([a,b]): f\le g$, then

$$L_N(f)\le L_N(g), \ \forall N\in\mathbb{N}, \tag{9.11}$$

(iii)

$$L_N(f+g) \leq L_N(f) + L_N(g), \ \forall \ f, g \in C_+([a,b]). \tag{9.12}$$

We call $\{L_N\}_{N \in \mathbb{N}}$ positive sublinear operators.

We need:

Theorem 9.14 ([5]) *Let* $\alpha > 0$, $\alpha \notin \mathbb{N}$, $m = \lceil \alpha \rceil$, $x_0 \in [a,b] \subset \mathbb{R}$, $f \in AC^m([a,b], \mathbb{R}_+)$, *and* $f^{(m)} \in L_\infty([a,b])$. *Furthermore we assume that* $f^{(k)}(x_0) = 0$, $k = 1, ..., m-1$. *Let* $L_N : C_+([a,b]) \to C_+([a,b])$, $\forall \ N \in \mathbb{N}$, *be positive sublinear operators, such that* $L_N(1) = 1$, $\forall \ N \in \mathbb{N}$. *Then*

$$|L_N(f)(x_0) - f(x_0)| \leq \frac{\omega_1\left(D_{x_0}^\alpha f, \delta\right)}{\Gamma(\alpha+1)} \cdot$$

$$\left[L_N\left(|\cdot - x_0|^\alpha\right)(x_0) + \frac{L_N\left(|\cdot - x_0|^{\alpha+1}\right)(x_0)}{(\alpha+1)\delta}\right], \tag{9.13}$$

$\delta > 0$, $\forall \ N \in \mathbb{N}$.

In particular (9.13) is true for $\alpha > 1$, $\alpha \notin \mathbb{N}$.

Corollary 9.15 ([5]) *Let* $0 < \alpha < 1$, $x_0 \in [a,b] \subset \mathbb{R}$, $f \in AC([a,b], \mathbb{R}_+)$, *and* $f' \in L_\infty([a,b])$. *Let* $L_N : C_+([a,b]) \to C_+([a,b])$, $\forall \ N \in \mathbb{N}$, *be positive sublinear operators, such that* $L_N(1) = 1$, $\forall \ N \in \mathbb{N}$. *Then (9.13) is valid.*

We also need:

Theorem 9.16 ([5]) *Let* $0 < \alpha < 1$, $x_0 \in [a,b] \subset \mathbb{R}$, $f \in AC([a,b], \mathbb{R}_+)$, *and* $f' \in L_\infty([a,b])$. *Let* L_N *from* $C_+([a,b])$ *into itself be positive sublinear operators, such that* $L_N(1) = 1$, $\forall \ N \in \mathbb{N}$. *Assume that* $L_N\left(|\cdot - x_0|^{\alpha+1}\right)(x_0) > 0$, $\forall \ N \in \mathbb{N}$. *Then*

$$|L_N(f)(x_0) - f(x_0)| \leq$$

$$\frac{(\alpha+2)\,\omega_1\left(D_{x_0}^\alpha f, \left(L_N\left(|\cdot - x_0|^{\alpha+1}\right)(x_0)\right)^{\frac{1}{\alpha+1}}\right)}{\Gamma(\alpha+2)} \left(L_N\left(|\cdot - x_0|^{\alpha+1}\right)(x_0)\right)^{\frac{\alpha}{\alpha+1}}. \tag{9.14}$$

We mention some notions from Canavati fractional calculus:
We need

Remark 9.17 (I) Here see [1], pp. 7-10.

Let $x, x_0 \in [a,b]$ such that $x \geq x_0$, $\nu > 0$, $\nu \notin \mathbb{N}$, such that $p = [\nu]$, $[\cdot]$ the integral part, $\alpha = \nu - p$ $(0 < \alpha < 1)$.

Let $f \in C^p([a,b])$ and define

$$\left(J_\nu^{x_0} f\right)(x) := \frac{1}{\Gamma(\nu)} \int_{x_0}^x (x-t)^{\nu-1} f(t)\,dt, \ x_0 \leq x \leq b. \tag{9.15}$$

the left generalized Riemann–Liouville fractional integral.

Here Γ stands for the gamma function.

Clearly here it holds $\left(J_{\nu}^{x_0} f\right)(x_0) = 0$. We define $\left(J_{\nu}^{x_0} f\right)(x) = 0$ for $x < x_0$. By [1], p. 388, $\left(J_{\nu}^{x_0} f\right)(x)$ is a continuous function in x, for a fixed x_0.

We define the subspace $C_{x_0+}^{\nu}([a,b])$ of $C^p([a,b])$:

$$C_{x_0+}^{\nu}([a,b]) := \left\{ f \in C^p([a,b]) : J_{1-\alpha}^{x_0} f^{(p)} \in C^1([x_0,b]) \right\}. \tag{9.16}$$

So let $f \in C_{x_0+}^{\nu}([a,b])$, we define the left generalized ν-fractional derivative of f over $[x_0, b]$ as

$$D_{x_0+}^{\nu} f = \left(J_{1-\alpha}^{x_0} f^{(p)}\right)', \tag{9.17}$$

that is

$$\left(D_{x_0+}^{\nu} f\right)(x) = \frac{1}{\Gamma(1-\alpha)} \frac{d}{dx} \int_{x_0}^{x} (x-t)^{-\alpha} f^{(p)}(t)\, dt, \tag{9.18}$$

which exists for $f \in C_{x_0+}^{\nu}([a,b])$, for $a \leq x_0 \leq x \leq b$.

Canavati in [8] first introduced this kind of left fractional derivative over $[0,1]$.

(II) Here see [2, 4].

Let $x, x_0 \in [a,b]$ such that $x \leq x_0$, $\nu > 0$, $\nu \notin \mathbb{N}$, such that $p = [\nu]$, $\alpha = \nu - p$ $(0 < \alpha < 1)$.

Let $f \in C^p([a,b])$ and define

$$\left(J_{x_0-}^{\nu} f\right)(x) := \frac{1}{\Gamma(\nu)} \int_{x}^{x_0} (z-x)^{\nu-1} f(z)\, dz, \quad a \leq x \leq x_0. \tag{9.19}$$

the right generalized Riemann–Liouville fractional integral.

Define the subspace of functions

$$C_{x_0-}^{\nu}([a,b]) := \left\{ f \in C^p([a,b]) : J_{x_0-}^{1-\alpha} f^{(p)} \in C^1([a,x_0]) \right\}. \tag{9.20}$$

Define the right generalized ν-fractional derivative of f over $[a, x_0]$ as

$$D_{x_0-}^{\nu} f = (-1)^{p-1} \left(J_{x_0-}^{1-\alpha} f^{(p)}\right)'. \tag{9.21}$$

Notice that

$$J_{x_0-}^{1-\alpha} f^{(p)}(x) = \frac{1}{\Gamma(1-\alpha)} \int_{x}^{x_0} (z-x)^{-\alpha} f^{(p)}(z)\, dz, \tag{9.22}$$

exists for $f \in C_{x_0-}^{\nu}([a,b])$, and

$$\left(D_{x_0-}^{\nu} f\right)(x) = \frac{(-1)^{p-1}}{\Gamma(1-\alpha)} \frac{d}{dx} \int_{x}^{x_0} (z-x)^{-\alpha} f^{(p)}(z)\, dz. \tag{9.23}$$

I.e.

$$\left(D_{x_0-}^{\nu} f\right)(x) = \frac{(-1)^{p-1}}{\Gamma(p-\nu+1)} \frac{d}{dx} \int_x^{x_0} (z-x)^{p-\nu} f^{(p)}(z)\, dz, \tag{9.24}$$

which exists for $f \in C_{x_0-}^{\nu}([a,b])$, for $a \leq x \leq x_0 \leq b$.

We make

Definition 9.18 Let $\overline{D}_{x_0}^{\nu} f$ denote any of $D_{x_0-}^{\nu} f$, $D_{x_{0+}}^{\nu} f$, and $\delta > 0$. We set

$$\omega_1\left(\overline{D}_{x_0}^{\nu} f, \delta\right) := \max\left\{\omega_1\left(D_{x_0-}^{\nu} f, \delta\right)_{[a,x_0]}, \omega_1\left(D_{x_0+}^{\nu} f, \delta\right)_{[x_0,b]}\right\}, \tag{9.25}$$

where $x_0 \in [a,b]$. Here the moduli of continuity are considered over $[a, x_0]$ and $[x_0, b]$, respectively.

We will use

Theorem 9.19 ([6]) *Let $\nu > 1$, $\nu \notin \mathbb{N}$, $p = [\nu]$, $x_0 \in [a,b]$ and $f : [a,b] \to \mathbb{R}_+$, $f \in C_{x_0+}^{\nu}([a,b]) \cap C_{x_0-}^{\nu}([a,b])$. Assume that $f^{(k)}(x_0) = 0$, $k = 1, ..., p-1$, and $\left(D_{x_0+}^{\nu} f\right)(x_0) = \left(D_{x_0-}^{\nu} f\right)(x_0) = 0$. Let $L_N : C_+([a,b]) \to C_+([a,b])$, $\forall\, N \in \mathbb{N}$, be positive sublinear operators, such that $L_N(1) = 1$, $\forall\, N \in \mathbb{N}$. Then*

$$|L_N(f)(x_0) - f(x_0)| \leq \frac{\omega_1\left(\overline{D}_{x_0}^{\nu} f, \delta\right)}{\Gamma(\nu+1)} \cdot$$

$$\left[L_N\left(|\cdot - x_0|^{\nu}\right)(x_0) + \frac{L_N\left(|\cdot - x_0|^{\nu+1}\right)(x_0)}{(\nu+1)\delta}\right], \tag{9.26}$$

$\delta > 0$, $\forall\, N \in \mathbb{N}$.

9.4 Background - III

We mention

Definition 9.20 ([15]) Let $I = [0,1]$, $\mathcal{B}_I$ the σ-algebra of all Borel measurable subsets of I, $\left(\Gamma_{N,x}\right)_{N\in\mathbb{N},\, x\in I}$ will be the collection of the family $\Gamma_{N,x} = \{\mu_{N,k,x}\}_{k=0}^{N}$, of monotone, submodular and strictly positive set functions $\mu_{N,k,x}$ on $\mathcal{B}_I$.

Let $f : [0,1] \to \mathbb{R}_+$ be a B_I-measurable function which is bounded, and call $p_{N,k}(x) = \binom{N}{k} x^k (1-x)^{N-k}$, for any $x \in [0,1]$.

The Bernstein–Kantorovich–Choquet operators are defined by the formula

$$K_{N,\Gamma_{N,x}}(f)(x) = \sum_{k=0}^{N} p_{N,k}(x) \frac{(C)\int_{\frac{k}{(N+1)}}^{\frac{(k+1)}{(N+1)}} f(t)\,d\mu_{N,k,x}(t)}{\mu_{N,k,x}\left(\left[\frac{k}{(N+1)}, \frac{(k+1)}{(N+1)}\right]\right)}, \ \forall\, x \in [0,1]. \quad (9.27)$$

If $\mu_{N,k,x} = \mu$, for all N, x, k, we will denote $K_{N,\Gamma_{N,x}}(f) := K_{N,\mu}(f)$.

Theorem 9.21 ([15]) *Suppose that* $\mu_{N,k,x} = \mu := \sqrt{M}$, *for all* N, k *and* x, *where* M *is the Lebesgue measure on* $[0,1]$. *Then*

$$\left|K_{N,\mu}(f)(x) - f(x)\right| \le 2\omega_1\left(f, \frac{\sqrt{x(1-x)}}{\sqrt{N}} + \frac{1}{N}\right), \quad (9.28)$$

$\forall\, N \in \mathbb{N}$, $x \in [0,1]$, $f \in C_+([0,1])$, *above* ω_1 *is over* $[0,1]$.

Remark 9.22 By [15] we have that

$$K_{N,\mu}(|\cdot - x|)(x) \le \frac{\sqrt{x(1-x)}}{\sqrt{N}} + \frac{1}{N}, \ \forall\, N \in \mathbb{N}. \quad (9.29)$$

Let $m > 1$, notice that $|\cdot - x|^{m-1} \le 1$, therefore

$$|\cdot - x|^m = |\cdot - x|\,|\cdot - x|^{m-1} \le |\cdot - x|,$$

hence

$$K_{N,\mu}\left(|\cdot - x|^m\right)(x) \le K_{N,\mu}(|\cdot - x|)(x),$$

that is

$$K_{N,\mu}\left(|\cdot - x|^m\right)(x) \le \frac{\sqrt{x(1-x)}}{\sqrt{N}} + \frac{1}{N}, \ \forall\, x \in [0,1], \ N \in \mathbb{N}, m \ge 1. \quad (9.30)$$

Notice that $K_{N,\mu}(1) = 1$, $\forall\, N \in \mathbb{N}$.

Clearly $K_{N,\mu}$ operators are positive sublinear operators from $C_+([0,1])$ into itself.

We mention

Definition 9.23 ([16]) Here we consider measures of possibility. Denoting $p_{N,k}(x) = \binom{N}{k} x^k (1-x)^{N-k}$, let us defined

$$\lambda_{N,k}(t) := \frac{p_{N,k}(t)}{k^k N^{-N} (N-k)^{N-k} \binom{N}{k}} = \frac{t^k (1-t)^{N-k}}{k^k N^{-N} (N-k)^{N-k}}, \ k = 0, \ldots, N. \quad (9.31)$$

By convention we assume that $0^0 = 1$, so that the cases $k = 0$, and $k = N$ make sense. By considering the root $\frac{k}{N}$ of $p'_{N,k}(x)$, it is clear that

$$\max\{p_{N,k}(t) : t \in [0,1]\} = k^k N^{-N} (N-k)^{N-k} \binom{N}{k},$$

which implies that each $\lambda_{N,k}$ is a possibility distribution on $[0,1]$.

Denoting by $P_{\lambda_{N,k}}$ the possibility measure induced by $\lambda_{N,k}$ and $\Gamma_{n,x} := \Gamma_N := \{P_{\lambda_{N,k}}\}_{k=0}^{N}$ (that is Γ_N is independent of x), we define the nonlinear Bernstein-Durrmeyer-Choquet polynomial operators with respect to the set functions in Γ_N given by the formula

$$D_{N,\Gamma_N}(f)(x) := \sum_{k=0}^{N} p_{N,k}(x) \frac{(C)\int_0^1 f(t)\, t^k (1-t)^{N-k} dP_{\lambda_{N,k}}(t)}{(C)\int_0^1 t^k (1-t)^{N-k} dP_{\lambda_{N,k}}(t)}, \tag{9.32}$$

$\forall\, x \in [0,1]$, $N \in \mathbb{N}$, $f \in C_+([0,1])$.

Remark 9.24 Above $P_{\lambda_{N,k}}$ is bounded, monotone, submodular and strictly positive, $N \in \mathbb{N}$, $k = 0, 1, \ldots, N$. Notice that $D_{N,\Gamma_N}(1) = 1$, $\forall\, N \in \mathbb{N}$.

Clearly D_{N,Γ_N} operators are positive sublinear operators mapping $C_+([0,1])$ into itself.

We mention

Theorem 9.25 ([16]) *For every $f \in C_+([0,1])$, $x \in [0,1]$ and $N \in \mathbb{N} - \{1\}$, we have*

$$\left|D_{N,\Gamma_N}(f)(x) - f(x)\right| \le 2\omega_1\left(f, \frac{\left(1+\sqrt{2}\right)\sqrt{x(1-x)} + \sqrt{2}\sqrt{x}}{\sqrt{N}} + \frac{1}{N}\right), \tag{9.33}$$

where ω_1 is on $[0,1]$.

Remark 9.26 By [16] we have that

$$D_{N,\Gamma_N}(|\cdot - x|)(x) \le \frac{\left(1+\sqrt{2}\right)\sqrt{x(1-x)} + \sqrt{2}\sqrt{x}}{\sqrt{N}} + \frac{1}{N}, \quad \forall\, N \in \mathbb{N} - \{1\}. \tag{9.34}$$

Let $m > 1$, notice that $|\cdot - x|^{m-1} \le 1$, therefore

$$|\cdot - x|^m = |\cdot - x|\, |\cdot - x|^{m-1} \le |\cdot - x|,$$

hence

$$D_{N,\Gamma_N}\left(|\cdot - x|^m\right)(x) \le D_{N,\Gamma_N}(|\cdot - x|)(x),$$

that is

$$D_{N,\Gamma_N}\left(|\cdot - x|^m\right)(x) \le \frac{\left(1+\sqrt{2}\right)\sqrt{x(1-x)} + \sqrt{2}\sqrt{x}}{\sqrt{N}} + \frac{1}{N}, \tag{9.35}$$

$\forall\, N \in \mathbb{N} - \{1\}, \forall\, x \in [0,1]\,, m \geq 1.$

We make

Remark 9.27 When $x \in [0,1]$, then the max $(x\,(1-x)) = \frac{1}{4}$, at $x = \frac{1}{2}$. Therefore it holds

$$\frac{\sqrt{x\,(1-x)}}{\sqrt{N}} + \frac{1}{N} \leq \frac{1}{2\sqrt{N}} + \frac{1}{N}, \tag{9.36}$$

$\forall\, x \in [0,1]\,, \forall\, N \in \mathbb{N}.$

Similarly, it holds

$$\frac{\left(1+\sqrt{2}\right)\sqrt{x\,(1-x)} + \sqrt{2}\sqrt{x}}{\sqrt{N}} + \frac{1}{N} \leq \frac{1+3\sqrt{2}}{2\sqrt{N}} + \frac{1}{N}, \tag{9.37}$$

$\forall\, x \in [0,1]\,, \forall\, N \in \mathbb{N} - \{1\}.$

Corollary 9.28 (To Theorem 9.21) *It holds*

$$\left\| K_{N,\mu}(f) - f \right\|_{\infty} \leq 2\omega_1\left(f, \frac{1}{2\sqrt{N}} + \frac{1}{N}\right), \tag{9.38}$$

$\forall\, N \in \mathbb{N},\, f \in C_+([0,1])\,.$

Corollary 9.29 (To Theorem 9.25) *It holds*

$$\left\| D_{N,\Gamma_N}(f) - f \right\|_{\infty} \leq 2\omega_1\left(f, \frac{1+3\sqrt{2}}{2\sqrt{N}} + \frac{1}{N}\right), \tag{9.39}$$

$\forall\, N \in \mathbb{N} - \{1\},\, f \in C_+([0,1])\,.$

The Bernstein–Kantorovich–Choquet operators $K_{N,\mu}$, where $\mu := \sqrt{M}$, with M the Lebesgue measure on $[0,1]$ have the following exact form:

$$K_{N,\mu}(f)(x) = \sum_{k=0}^{N} p_{N,k}(x) \frac{(C)\int_{\frac{k}{(N+1)}}^{\frac{(k+1)}{(N+1)}} f(t)\,d\mu(t)}{\mu\left(\left[\frac{k}{(N+1)}, \frac{(k+1)}{(N+1)}\right]\right)}, \tag{9.40}$$

$\forall\, x \in [0,1]\,, \forall\, N \in \mathbb{N},\, f \in C_+([0,1])\,.$

9.5 Background - IV

Here we talk about representations of positive sublinear operators by Choquet integrals:

We need

Definition 9.30 Let Ω be a set, and let $f, g : \Omega \to \mathbb{R}$ be bounded functions. We say that f and g are comonotonic, if for every $\omega, \omega' \in \Omega$,

$$\left(f(\omega) - f(\omega')\right)\left(g(\omega) - g(\omega')\right) \geq 0. \tag{9.41}$$

We also need the famous Schmeidler's Representation Theorem (Schmeidler 1986)

Theorem 9.31 ([18]) *Denote with $\mathcal{L}_\infty(\mathcal{A})$ the vector space of $\mathcal{A}$-measurable bounded real valued functions on Ω, where $\mathcal{A} \subset 2^\Omega$ is a σ-algebra. Given a real functional $\Gamma : \mathcal{L}_\infty(\mathcal{A}) \to \mathbb{R}$, assume that for $f, g \in \mathcal{L}_\infty(\mathcal{A})$:*
(i) $\Gamma(cf) = c\Gamma(f)$, $\forall\, c > 0$,
(ii) $f \leq g$, implies $\Gamma(f) \leq \Gamma(g)$,
and
(iii) $\Gamma(f + g) = \Gamma(f) + \Gamma(g)$, for any comonotonic f, g.
Then $\gamma(A) := \Gamma(1_A)$, $\forall\, A \in \mathcal{A}$, defines a finite monotone set function on $\mathcal{A}$, and Γ is the Choquet integral with respect to γ, i.e.

$$\Gamma(f) = (C)\int_\Omega f(t)\, d\gamma(t), \ \forall\, f \in \mathcal{L}_\infty(\mathcal{A}). \tag{9.42}$$

Above 1_A denotes the characteristic function on A.

We make

Remark 9.32 Consider here $[a, b] \subset \mathbb{R}$, $\mathcal{B} = \mathcal{B}([a, b])$ is the Borel σ-algebra on $[a, b]$, and $\mathcal{L}_\infty(\mathcal{B})$ is the vector space of $\mathcal{B}$-measurable bounded real valued functions on $[a, b]$. Let $(L_N)_{N\in\mathbb{N}}$ be a sequence of positive sublinear operators from $\mathcal{L}_\infty(\mathcal{B})$ into $C_+([a, b])$, and $x \in [a, b]$. That is here L_N fulfills the positive homogeneity, monotonicity and subadditivity properties, see (9.10)–(9.12).

Assume $L_N(1) = 1$, $\forall\, N \in \mathbb{N}$. Clearly here $\mathcal{L}_\infty(\mathcal{B}) \supset C_+([a, b])$. In particular we treat $L_N|_{C_+([a,b])}$, just denoted for simplicity by L_N, $\forall\, N \in \mathbb{N}$.

It is clear that $L_N(\cdot)(x) : \mathcal{L}_\infty(\mathcal{B}) \to \mathbb{R}$ is a functional, $\forall\, N \in \mathbb{N}$. It has the properties:

(i)

$$L_N(cf)(x) = cL_N(f)(x), \ \forall\, c > 0, \ \forall\, f \in \mathcal{L}_\infty(\mathcal{B}), \tag{9.43}$$

(ii)

$$f \leq g, \text{ implies } L_N(f)(x) \leq L_N(g)(x), \quad \text{where } f, g \in \mathcal{L}_\infty(\mathcal{B}), \tag{9.44}$$

and

(iii)

$$L_N(f + g)(x) \leq L_N(f)(x) + L_N(g)(x), \quad \forall\, f, g \in \mathcal{L}_\infty(\mathcal{B}). \tag{9.45}$$

For comonotonic $f, g \in \mathcal{L}_{\infty}(\mathcal{B})$, we further assume that

$$L_N(f+g)(x) = L_N(f)(x) + L_N(g)(x). \tag{9.46}$$

In that case L_N is called comonotonic.

By Theorem 9.31 we get that:

$$\gamma_{N,x}(A) := L_N(1_A)(x), \ \forall\, A \in \mathcal{B}, \forall\, N \in \mathbb{N}, \tag{9.47}$$

defines a finite monotone set function on $\mathcal{B}$, and

$$L_N(f)(x) = (C)\int_a^b f(t)\, d\gamma_{N,x}(t), \tag{9.48}$$

$\forall\, f \in \mathcal{L}_{\infty}(\mathcal{B}), \forall\, N \in \mathbb{N}$.

In particular (9.47) is valid for any $f \in C_+([a,b])$. Furthermore $\gamma_{N,x}$ is normalized, that is $\gamma_{N,x}([a,b]) = 1, \forall\, N \in \mathbb{N}$.

A different kind of general operators follows:

We make

Remark 9.33 Consider again $[a,b] \subset \mathbb{R}$, $\mathcal{B} = \mathcal{B}([a,b])$ is the Borel σ-algebra on $[a,b]$. For each $N \in \mathbb{N}$ and each $x \in [a,b]$ consider the monotone set functions $\mu_{N,x}; \mathcal{B} \to \mathbb{R}_+$. We assume that all $\mu_{N,x}$ are normalized, that is $\mu_{N,x}([a,b]) = 1$, and submodular. Here we consider the operators $T_N : C_+([a,b]) \to C_+([a,b])$ given by the formula

$$T_N(f)(x) = (C)\int_{[a,b]} f(t)\, d\mu_{N,x}(t), \tag{9.49}$$

$\forall\, N \in \mathbb{N}, \forall\, x \in [a,b]$.

Infact here $\mu_{N,x}$ are chosen so that $T_N(C_+([a,b])) \subseteq C_+([a,b])$.

We notice here that hold:

(i)

$$T_N(\alpha f)(x) = \alpha T_N(f)(x), \ \forall\, \alpha \geq 0, \tag{9.50}$$

(ii)

$$f \leq g, \text{ implies } T_N(f)(x) \leq T_N(g)(x), \tag{9.51}$$

and

(iii)

$$T_N(f+g)(x) \leq T_N(f)(x) + T_N(g)(x), \tag{9.52}$$

$\forall\, N \in \mathbb{N}, \forall\, x \in [a,b], \forall\, f, g \in C_+([a,b])$.

Clearly T_N are positive sublinear operators, compare to (9.10)–(9.12). We also have that $T_N(1) = 1, \forall\, N \in \mathbb{N}$.

9.6 Main Results

We give

Theorem 9.34 *Let* $\alpha > 1$, $\alpha \notin \mathbb{N}$, $m = \lceil \alpha \rceil$, $x_0 \in [0,1]$, $f \in AC^m([0,1], \mathbb{R}_+)$, *and* $f^{(m)} \in L_\infty([0,1])$. *Furthermore we assume that* $f^{(k)}(x_0) = 0$, $k = 1, ..., m-1$. *Then*

$$\left|K_{N,\mu}(f)(x_0) - f(x_0)\right| \leq \frac{\omega_1\left(D_{x_0}^\alpha f, \left(\frac{1}{2\sqrt{N}} + \frac{1}{N}\right)^{\frac{1}{\alpha+1}}\right)}{\Gamma(\alpha+1)} \cdot$$

$$\left[\left(\frac{1}{2\sqrt{N}} + \frac{1}{N}\right) + \frac{1}{(\alpha+1)}\left(\frac{1}{2\sqrt{N}} + \frac{1}{N}\right)^{\frac{\alpha}{\alpha+1}}\right], \ \forall\, N \in \mathbb{N}. \tag{9.53}$$

If $N \to \infty$, *we get that* $K_{N,\mu}(f)(x_0) \to f(x_0)$.

Proof By (9.13) we get ($\delta > 0$)

$$\left|K_{N,\mu}(f)(x_0) - f(x_0)\right| \leq \frac{\omega_1\left(D_{x_0}^\alpha f, \delta\right)}{\Gamma(\alpha+1)} \cdot$$

$$\left[K_{N,\mu}\left(|\cdot - x_0|^\alpha\right)(x_0) + \frac{K_{N,\mu}\left(|\cdot - x_0|^{\alpha+1}\right)(x_0)}{(\alpha+1)\delta}\right] \overset{(9.30.)}{\leq}$$

$$\frac{\omega_1\left(D_{x_0}^\alpha f, \delta\right)}{\Gamma(\alpha+1)}\left[\left(\frac{\sqrt{x_0(1-x_0)}}{\sqrt{N}} + \frac{1}{N}\right) + \frac{1}{(\alpha+1)\delta}\left(\frac{\sqrt{x_0(1-x_0)}}{\sqrt{N}} + \frac{1}{N}\right)\right] \overset{(9.36.)}{\leq}$$

$$\frac{\omega_1\left(D_{x_0}^\alpha f, \delta\right)}{\Gamma(\alpha+1)}\left[\left(\frac{1}{2\sqrt{N}} + \frac{1}{N}\right) + \frac{1}{(\alpha+1)\delta}\left(\frac{1}{2\sqrt{N}} + \frac{1}{N}\right)\right] \tag{9.54}$$

(setting $\delta := \left(\frac{1}{2\sqrt{N}} + \frac{1}{N}\right)^{\frac{1}{\alpha+1}}$, then $\delta^{\alpha+1} = \frac{1}{2\sqrt{N}} + \frac{1}{N}$, and $\delta^\alpha = \left(\frac{1}{2\sqrt{N}} + \frac{1}{N}\right)^{\frac{\alpha}{\alpha+1}}$)

$$= \frac{\omega_1\left(D_{x_0}^\alpha f, \left(\frac{1}{2\sqrt{N}} + \frac{1}{N}\right)^{\frac{1}{\alpha+1}}\right)}{\Gamma(\alpha+1)}\left[\left(\frac{1}{2\sqrt{N}} + \frac{1}{N}\right) + \frac{1}{(\alpha+1)}\left(\frac{1}{2\sqrt{N}} + \frac{1}{N}\right)^{\frac{\alpha}{\alpha+1}}\right], \tag{9.55}$$

proving the claim. ■

We continue with

Theorem 9.35 *Let* $\nu > 1$, $\nu \notin \mathbb{N}$, $p = [\nu]$, $x_0 \in [0,1]$ *and* $f : [0,1] \to \mathbb{R}_+$, $f \in C_{x_0+}^\nu([0,1]) \cap C_{x_0-}^\nu([0,1])$. *Assume that* $f^{(k)}(x_0) = 0$, $k = 1, ..., p-1$, *and* $\left(D_{x_0+}^\nu f\right)(x_0) = \left(D_{x_0-}^\nu f\right)(x_0) = 0$. *Then*

$$\left|K_{N,\mu}(f)(x_0) - f(x_0)\right| \le \frac{\omega_1\left(\overline{D}_{x_0}^{\nu} f, \left(\frac{1}{2\sqrt{N}} + \frac{1}{N}\right)^{\frac{1}{\nu+1}}\right)}{\Gamma(\nu+1)} \cdot$$

$$\left[\left(\frac{1}{2\sqrt{N}} + \frac{1}{N}\right) + \frac{1}{(\nu+1)}\left(\frac{1}{2\sqrt{N}} + \frac{1}{N}\right)^{\frac{\nu}{\nu+1}}\right], \ \forall\, N \in \mathbb{N}. \tag{9.56}$$

If $N \to \infty$, we get that $K_{N,\mu}(f)(x_0) \to f(x_0)$.

Proof By (9.26) we get ($\delta > 0$)

$$\left|K_{N,\mu}(f)(x_0) - f(x_0)\right| \le \frac{\omega_1\left(\overline{D}_{x_0}^{\nu} f, \delta\right)}{\Gamma(\nu+1)} \cdot \tag{9.57}$$

$$\left[K_{N,\mu}\left(\left|\cdot - x_0\right|^{\nu}\right)(x_0) + \frac{K_{N,\mu}\left(\left|\cdot - x_0\right|^{\nu+1}\right)(x_0)}{(\nu+1)\,\delta}\right] \overset{\text{(by (9.30.), (9.36.))}}{\le}$$

$$\frac{\omega_1\left(\overline{D}_{x_0}^{\nu} f, \delta\right)}{\Gamma(\nu+1)}\left[\left(\frac{1}{2\sqrt{N}} + \frac{1}{N}\right) + \frac{1}{(\nu+1)\,\delta}\left(\frac{1}{2\sqrt{N}} + \frac{1}{N}\right)\right]$$

(setting $\delta := \left(\frac{1}{2\sqrt{N}} + \frac{1}{N}\right)^{\frac{1}{\nu+1}}$, then $\delta^{\nu+1} = \frac{1}{2\sqrt{N}} + \frac{1}{N}$, and $\delta^{\nu} = \left(\frac{1}{2\sqrt{N}} + \frac{1}{N}\right)^{\frac{\nu}{\nu+1}}$)

$$= \frac{\omega_1\left(\overline{D}_{x_0}^{\nu} f, \left(\frac{1}{2\sqrt{N}} + \frac{1}{N}\right)^{\frac{1}{\nu+1}}\right)}{\Gamma(\nu+1)}\left[\left(\frac{1}{2\sqrt{N}} + \frac{1}{N}\right) + \frac{1}{(\nu+1)}\left(\frac{1}{2\sqrt{N}} + \frac{1}{N}\right)^{\frac{\nu}{\nu+1}}\right], \tag{9.58}$$

proving the claim. ■

We present

Theorem 9.36 *Let $\alpha > 1$, $\alpha \notin \mathbb{N}$, $m = \lceil \alpha \rceil$, $x_0 \in [0,1]$, $f \in AC^m([0,1], \mathbb{R}_+)$, and $f^{(m)} \in L_\infty([0,1]) : f^{(k)}(x_0) = 0$, $k = 1, ..., m-1$. Then*

$$\left|D_{N,\Gamma_N}(f)(x_0) - f(x_0)\right| \le \frac{\omega_1\left(D_{x_0}^{\alpha} f, \left(\frac{1+3\sqrt{2}}{2\sqrt{N}} + \frac{1}{N}\right)^{\frac{1}{\alpha+1}}\right)}{\Gamma(\alpha+1)} \cdot$$

$$\left[\left(\frac{1+3\sqrt{2}}{2\sqrt{N}} + \frac{1}{N}\right) + \frac{1}{(\alpha+1)}\left(\frac{1+3\sqrt{2}}{2\sqrt{N}} + \frac{1}{N}\right)^{\frac{\alpha}{\alpha+1}}\right], \ \forall\, N \in \mathbb{N} - \{1\}. \tag{9.59}$$

If $N \to \infty$, we get that $D_{N,\Gamma_N}(f)(x_0) \to f(x_0)$.

Proof By (9.13) we obtain ($\delta > 0$)

$$\left|D_{N,\Gamma_N}(f)(x_0) - f(x_0)\right| \leq \frac{\omega_1\left(D^{\alpha}_{x_0} f, \delta\right)}{\Gamma(\alpha+1)}\cdot$$

$$\left[D_{N,\Gamma_N}\left(|\cdot - x_0|^{\alpha}\right)(x_0) + \frac{D_{N,\Gamma_N}\left(|\cdot - x_0|^{\alpha+1}\right)(x_0)}{(\alpha+1)\delta}\right] \overset{\text{(by (9.35.), (9.37.))}}{\leq}$$

$$\frac{\omega_1\left(D^{\alpha}_{x_0} f, \delta\right)}{\Gamma(\alpha+1)}\left[\left(\frac{1+3\sqrt{2}}{2\sqrt{N}} + \frac{1}{N}\right) + \frac{1}{(\alpha+1)\delta}\left(\frac{1+3\sqrt{2}}{2\sqrt{N}} + \frac{1}{N}\right)\right]$$

(setting $\delta := \left(\frac{1+3\sqrt{2}}{2\sqrt{N}} + \frac{1}{N}\right)^{\frac{1}{\alpha+1}}$)

$$= \frac{\omega_1\left(D^{\alpha}_{x_0} f, \left(\frac{1+3\sqrt{2}}{2\sqrt{N}} + \frac{1}{N}\right)^{\frac{1}{\alpha+1}}\right)}{\Gamma(\alpha+1)}\cdot$$

$$\left[\left(\frac{1+3\sqrt{2}}{2\sqrt{N}} + \frac{1}{N}\right) + \frac{1}{(\alpha+1)}\left(\frac{1+3\sqrt{2}}{2\sqrt{N}} + \frac{1}{N}\right)^{\frac{\alpha}{\alpha+1}}\right], \tag{9.60}$$

proving the claim. ∎

We also present

Theorem 9.37 *Let $\nu > 1$, $\nu \notin \mathbb{N}$, $p = [\nu]$, $x_0 \in [0,1]$ and $f : [0,1] \to \mathbb{R}_+$, $f \in C^{\nu}_{x_0+}([0,1]) \cap C^{\nu}_{x_0-}([0,1])$. Assume that $f^{(k)}(x_0) = 0$, $k = 1, ..., p-1$, and $\left(D^{\nu}_{x_0+} f\right)(x_0) = \left(D^{\nu}_{x_0-} f\right)(x_0) = 0$. Then*

$$\left|D_{N,\Gamma_N}(f)(x_0) - f(x_0)\right| \leq \frac{\omega_1\left(\overline{D}^{\nu}_{x_0} f, \left(\frac{1+3\sqrt{2}}{2\sqrt{N}} + \frac{1}{N}\right)^{\frac{1}{\nu+1}}\right)}{\Gamma(\nu+1)}\cdot$$

$$\left[\left(\frac{1+3\sqrt{2}}{2\sqrt{N}} + \frac{1}{N}\right) + \frac{1}{(\nu+1)}\left(\frac{1+3\sqrt{2}}{2\sqrt{N}} + \frac{1}{N}\right)^{\frac{\nu}{\nu+1}}\right], \ \forall\, N \in \mathbb{N} - \{1\}. \tag{9.61}$$

If $N \to \infty$, we get that $D_{N,\Gamma_N}(f)(x_0) \to f(x_0)$.

Proof By (9.26) we get ($\delta > 0$)

$$\left|D_{N,\Gamma_N}(f)(x_0) - f(x_0)\right| \leq \frac{\omega_1\left(\overline{D}^{\nu}_{x_0} f, \delta\right)}{\Gamma(\nu+1)}\cdot$$

$$\left[D_{N,\Gamma_N}\left(|\cdot - x_0|^{\nu}\right)(x_0) + \frac{1}{(\nu+1)\,\delta} D_{N,\Gamma_N}\left(|\cdot - x_0|^{\nu+1}\right)(x_0)\right] \overset{\text{(by (9.35.), (9.37.))}}{\leq}$$

$$\frac{\omega_1\left(\overline{D}_{x_0}^{\nu} f, \delta\right)}{\Gamma(\nu+1)} \left[\left(\frac{1+3\sqrt{2}}{2\sqrt{N}} + \frac{1}{N}\right) + \frac{1}{(\nu+1)\,\delta}\left(\frac{1+3\sqrt{2}}{2\sqrt{N}} + \frac{1}{N}\right)\right]$$

(setting $\delta := \left(\frac{1+3\sqrt{2}}{2\sqrt{N}} + \frac{1}{N}\right)^{\frac{1}{\nu+1}}$)

$$= \frac{\omega_1\left(\overline{D}_{x_0}^{\nu} f, \left(\frac{1+3\sqrt{2}}{2\sqrt{N}} + \frac{1}{N}\right)^{\frac{1}{\nu+1}}\right)}{\Gamma(\nu+1)} \cdot$$

$$\left[\left(\frac{1+3\sqrt{2}}{2\sqrt{N}} + \frac{1}{N}\right) + \frac{1}{(\nu+1)}\left(\frac{1+3\sqrt{2}}{2\sqrt{N}} + \frac{1}{N}\right)^{\frac{\nu}{\nu+1}}\right], \tag{9.62}$$

proving the claim. ■

Based on Theorem 9.31 and Remark 9.32 we present

Theorem 9.38 *Let $\alpha > 0$, $\alpha \notin \mathbb{N}$, $m = \lceil \alpha \rceil$, $x_0 \in [a,b] \subset \mathbb{R}$, $f \in AC^m([a,b], \mathbb{R}_+)$, and $f^{(m)} \in L_\infty([a,b])$. Furthermore we assume that $f^{(k)}(x_0) = 0$, $k = 1, ..., m-1$. Let $L_N : \mathcal{L}_\infty(\mathcal{B}([a,b])) \to C_+([a,b])$, $\forall\, N \in \mathbb{N}$, be positive sublinear comonotonic operators, such that $L_N(1) = 1$, $\forall\, N \in \mathbb{N}$. Then*

$$|L_N(f)(x_0) - f(x_0)| \leq \frac{\omega_1\left(D_{x_0}^{\alpha} f, \delta\right)}{\Gamma(\alpha+1)} \cdot$$

$$\left[\left((C)\int_a^b |t - x_0|^{\alpha}\, d\gamma_{N,x_0}(t)\right) + \frac{1}{(\alpha+1)\,\delta}\left((C)\int_a^b |t - x_0|^{\alpha+1}\, d\gamma_{N,x_0}(t)\right)\right], \tag{9.63}$$

$\delta > 0$, $\forall\, N \in \mathbb{N}$.

Proof By Theorem 9.14. ■

Theorem 9.39 *Let $0 < \alpha < 1$, $x_0 \in [a,b] \subset \mathbb{R}$, $f \in AC([a,b], \mathbb{R}_+)$, and $f' \in L_\infty([a,b])$. Let $L_N : \mathcal{L}_\infty(\mathcal{B}([a,b])) \to C_+([a,b])$, $\forall\, N \in \mathbb{N}$, be positive sublinear comonotonic operators, such that $L_N(1) = 1$, $\forall\ N \in \mathbb{N}$. Assume that $(C)\int_a^b |t - x_0|^{\alpha+1}\, d\gamma_{N,x_0}(t) > 0$, $\forall\, N \in \mathbb{N}$. Then*

$$|L_N(f)(x_0) - f(x_0)| \leq \frac{(\alpha+2)\,\omega_1\left(D_{x_0}^{\alpha} f, \left((C)\int_a^b |t - x_0|^{\alpha+1}\, d\gamma_{N,x_0}(t)\right)^{\frac{1}{\alpha+1}}\right)}{\Gamma(\alpha+2)}$$

$$\left((C)\int_a^b |t-x_0|^{\alpha+1} d\gamma_{N,x_0}(t)\right)^{\frac{\alpha}{\alpha+1}}, \tag{9.64}$$

$\forall\, N \in \mathbb{N}$.

If $(C)\int_a^b |t-x_0|^{\alpha+1} d\gamma_{N,x_0}(t) \to 0$, *then* $L_N(f)(x_0) \to f(x_0)$, *as* $N \to \infty$.

Proof By Theorem 9.16. ∎

Theorem 9.40 *Let* $\nu > 1$, $\nu \notin \mathbb{N}$, $p = [\nu]$, $x_0 \in [a,b]$ *and* $f : [a,b] \to \mathbb{R}_+$, $f \in C^{\nu}_{x_0+}([a,b]) \cap C^{\nu}_{x_0-}([a,b])$. *Assume that* $f^{(k)}(x_0) = 0$, $k = 1, ..., p-1$, *and* $\left(D^{\nu}_{x_0+}f\right)(x_0) = \left(D^{\nu}_{x_0-}f\right)(x_0) = 0$. *Let* $L_N : \mathcal{L}_\infty(\mathcal{B}([a,b])) \to C_+([a,b])$, $\forall\, N \in \mathbb{N}$, *be positive sublinear comonotonic operators, such that* $L_N(1) = 1$, $\forall\, N \in \mathbb{N}$. *Then*

$$|L_N(f)(x_0) - f(x_0)| \le \frac{\omega_1\left(\overline{D}^{\nu}_{x_0} f, \delta\right)}{\Gamma(\nu+1)}\cdot$$

$$\left[(C)\int_a^b |t-x_0|^{\nu} d\gamma_{N,x_0}(t) + \frac{1}{(\nu+1)\delta}\left((C)\int_a^b |t-x_0|^{\nu+1} d\gamma_{N,x_0}(t)\right)\right], \tag{9.65}$$

$\forall\, N \in \mathbb{N}$.

Proof By Theorem 9.19. ∎

We finish by giving related results based on Remark 9.33.

We give

Theorem 9.41 *Let* $\alpha > 0$, $\alpha \notin \mathbb{N}$, $m = \lceil \alpha \rceil$, $x_0 \in [a,b] \subset \mathbb{R}$, $f \in AC^m([a,b], \mathbb{R}_+)$, *and* $f^{(m)} \in L_\infty([a,b])$. *Furthermore we assume that* $f^{(k)}(x_0) = 0$, $k = 1, ..., m-1$. *Then*

$$|T_N(f)(x_0) - f(x_0)| \le \frac{\omega_1\left(D^{\alpha}_{x_0} f, \delta\right)}{\Gamma(\alpha+1)}\cdot$$

$$\left[\left((C)\int_a^b |t-x_0|^{\alpha} d\mu_{N,x_0}(t)\right) + \frac{1}{(\alpha+1)\delta}\left((C)\int_a^b |t-x_0|^{\alpha+1} d\mu_{N,x_0}(t)\right)\right], \tag{9.66}$$

$\delta > 0$, $\forall\, N \in \mathbb{N}$.

Proof By Theorem 9.14. ∎

Theorem 9.42 *Let* $0 < \alpha < 1$, $x_0 \in [a,b] \subset \mathbb{R}$, $f \in AC([a,b], \mathbb{R}_+)$, *and* $f' \in L_\infty([a,b])$. *Assume that* $(C)\int_a^b |t-x_0|^{\alpha+1} d\mu_{N,x_0}(t) > 0$, $\forall\, N \in \mathbb{N}$. *Then*

$$|T_N(f)(x_0) - f(x_0)| \le \frac{(\alpha+2)\,\omega_1\left(D^{\alpha}_{x_0} f, \left((C)\int_a^b |t-x_0|^{\alpha+1} d\mu_{N,x_0}(t)\right)^{\frac{1}{\alpha+1}}\right)}{\Gamma(\alpha+2)}$$

$$\left((C) \int_a^b |t - x_0|^{\alpha+1} d\mu_{N,x_0}(t) \right)^{\frac{\alpha}{\alpha+1}}, \tag{9.67}$$

$\forall\, N \in \mathbb{N}$.

If $(C) \int_a^b |t - x_0|^{\alpha+1} d\mu_{N,x_0}(t) \to 0$, *then* $T_N(f)(x_0) \to f(x_0)$, *as* $N \to \infty$.

Proof By Theorem 9.16. ■

Theorem 9.43 *Let* $\nu > 1$, $\nu \notin \mathbb{N}$, $p = [\nu]$, $x_0 \in [a, b]$ *and* $f : [a, b] \to \mathbb{R}_+$, $f \in C^{\nu}_{x_0+}([a, b]) \cap C^{\nu}_{x_0-}([a, b])$. *Assume that* $f^{(k)}(x_0) = 0$, $k = 1, ..., p - 1$, *and* $\left(D^{\nu}_{x_0+} f\right)(x_0) = \left(D^{\nu}_{x_0-} f\right)(x_0) = 0$. *Then*

$$|T_N(f)(x_0) - f(x_0)| \le \frac{\omega_1\left(\overline{D}^{\nu}_{x_0} f, \delta\right)}{\Gamma(\nu + 1)}\cdot$$

$$\left[(C) \int_a^b |t - x_0|^{\nu} d\mu_{N,x_0}(t) + \frac{1}{(\nu + 1)\delta} \left((C) \int_a^b |t - x_0|^{\nu+1} d\mu_{N,x_0}(t) \right) \right], \tag{9.68}$$

$\delta > 0$, $\forall\, N \in \mathbb{N}$.

Proof By Theorem 9.19. ■

References

1. G. Anastassiou, *Fractional Differentiation Inequalities* (Springer, Heidelberg, 2009)
2. G. Anastassiou, On right fractional calculus. Chaos Solitons Fractals **42**, 365–376 (2009)
3. G. Anastassiou, Fractional Korovkin theory. Chaos Solitons Fractals **42**, 2080–2094 (2009)
4. G. Anastassiou, *Intelligent Mathematics: Computational Analysis* (Springer, Heidelberg, 2011)
5. G. Anastassiou, *Caputo Fractional Approximation by Sublinear Operators* (2017). Submitted
6. G. Anastassiou, Canavati fractional approximation by max-product operator. Prog. Fract. Differ. Appl. **4**(3), 1–17 (2018)
7. G. Anastassiou, Caputo and Canavati fractional approximation by Choquet integrals, *Progress in Fractional Differentiation and Applications* (2018). Accepted
8. J.A. Canavati, The Riemann-Liouville integral. Nieuw Archif Voor Wiskunde **5**(1), 53–75 (1987)
9. G. Choquet, Theory of capacities. Ann. Inst. Fourier (Grenoble) **5**, 131–295 (1954)
10. D. Denneberg, *Non-Additive Measure and Integral* (Kluwer, Dordrecht, 1994)
11. K. Diethelm, *The Analysis of Fractional Differential Equations* (Springer, Heidelberg, 2010)
12. D. Dubois, H. Prade, *Possibility Theory* (Plenum Press, New York, 1988)
13. A.M.A. El-Sayed, M. Gaber, On the finite Caputo and finite Riesz derivatives. Electron. J. Theor. Phys. **3**(12), 81–95 (2006)
14. G.S. Frederico, D.F.M. Torres, Fractional optimal control in the sense of Caputo and the fractional Noether's theorem. Int. Math. Forum **3**(10), 479–493 (2008)
15. S. Gal, Uniform and pointwise quantitative approximation by Kantorovich-Choquet type integral operators with respect to monotone and submodular set functions. Mediterr. J. Math. **14**(5), 12 (2017). Art. 205

16. S. Gal, S. Trifa, Quantitative estimates in uniform and pointwise approximation by Bernstein-Durrmeyer-Choquet operators. Carpathian J. Math. **33**(1), 49–58 (2017)
17. S.G. Samko, A.A. Kilbas, O.I. Marichev, *Fractional Integrals and Derivatives, Theory and Applications* (Gordon and Breach, Amsterdam, 1993). [English translation from the Russian, Integrals and Derivatives of Fractional Order and Some of Their Applications (Nauka i Tekhnika, Minsk, 1987)]
18. D. Schmeidler, Integral representation without additivity. Proc. Am. Math. Soc. **97**, 255–261 (1986)
19. D. Schmeidler, Subjective probability and expected utility without additivity. Econometrica **57**, 571–587 (1989)
20. L.S. Shapley, A value for n-person games, in *Contributions to the Theory of Games*, ed. by H.W. Kuhn, A.W. Tucker. Annals of Mathematical Studies, vol. 28 (Princeton University Press, Princeton, 1953), pp. 307–317
21. Z. Wang, G.J. Klir, *Generalized Measure Theory* (Springer, New York, 2009)

Chapter 10
Mixed Conformable and Iterated Fractional Quantitative Approximation by Choquet Integrals

Here we study the quantitative mixed conformable and iterated fractional approximation of positive sublinear operators to the unit operator. These are given a precise Choquet integral interpretation. Initially we start with the research of the mixed conformable and iterated fractional rate of the convergence of the well-known Bernstein–Kantorovich Choquet and Bernstein–Durrweyer–Choquet polynomial Choquet-integral operators. Then we study the very general comonotonic positive sublinear operators based on the representation theorem of Schmeidler [14]. We finish with the approximation by the very general direct Choquet-integral form positive sublinear operators. All mixed conformable and iterated fractional approximations are given via inequalities involving the modulus of continuity of the approximated function mixed conformable and iterated fractional order derivative. It follows [3].

10.1 Introduction

G. Choquet [7], introduced the capacities and his integral. Initially these were applied to statistical mechanics and potential theory, and they gave rise to the study of non-additive measure theory. Slowly but steady these ideas of Choquet started to attract economists especially after the very important work of Shapley [16] in the study of cooperative games. Capacities and Choquet integrals became main stream to Decision theorists since 1989 when D. Schmeidler [15] was the first to use them in a model of choice with non-additive beliefs. The expected utility results are strengthned by the use of Choquet capacities instead of probability measures.

In now days Choquet integral has wide applications, among others, to decision making under risk and uncertainty, in finance, in economics, in portofolio problems and in insurance.

Our motivation also comes from the foundations of Bayesian decision theory and subjective probability.

G. A. Anastassiou, *Ordinary and Fractional Approximation by Non-additive Integrals: Choquet, Shilkret and Sugeno Integral Approximators*, Studies in Systems, Decision and Control 190, https://doi.org/10.1007/978-3-030-04287-5_10

Because of the paramount importance of Choquet integral, we decided to research the related positive sublinear operators approximation, part of it is exhibited in this work in mixed conformable and iterated fractional sense.

10.2 Background - I

Next we present briefly about the Choquet integral, see also [11].

We make

Definition 10.1 Consider $\Omega \neq \varnothing$ and let $\mathcal{C}$ be a σ-algebra of subsets in Ω.

(i) (see, e.g., [17], p. 63) The set function $\mu : \mathcal{C} \rightarrow [0, +\infty]$ is called a monotone set function (or capacity) if $\mu(\varnothing) = 0$ and $\mu(A) \leq \mu(B)$ for all $A, B \in \mathcal{C}$, with $A \subset B$. Also, μ is called submodular if

$$\mu(A \cup B) + \mu(A \cap B) \leq \mu(A) + \mu(B), \text{ for all } A, B \in \mathcal{C}.$$

μ is called bounded if $\mu(\Omega) < +\infty$ and normalized if $\mu(\Omega) = 1$.

(ii) (see, e.g., [17], p. 233, or [7]) If μ is a monotone set function on $\mathcal{C}$ and if $f : \Omega \rightarrow \mathbb{R}$ is $\mathcal{C}$-measurable (that is, for any Borel subset $B \subset \mathbb{R}$ it follows $f^{-1}(B) \in \mathcal{C}$), then for any $A \in \mathcal{C}$, the Choquet integral is defined by

$$(C)\int_A f d\mu = \int_0^{+\infty} \mu\left(F_\beta(f) \cap A\right) d\beta + \int_{-\infty}^0 \left[\mu\left(F_\beta(f) \cap A\right) - \mu(A)\right] d\beta,$$

where we used the notation $F_\beta(f) = \{\omega \in \Omega : f(\omega) \geq \beta\}$. Notice that if $f \geq 0$ on A, then in the above formula we get $\int_{-\infty}^0 = 0$.

The integrals on the right-hand side are the usual Riemann integral.

The function f will be called Choquet integrable on A if $(C)\int_A f d\mu \in \mathbb{R}$.

Next we list some well known properties of the Choquet integral.

Remark 10.2 If $\mu : \mathcal{C} \rightarrow [0, +\infty]$ is a monotone set function, then the following properties hold:

(i) For all $a \geq 0$ we have $(C)\int_A af d\mu = a \cdot (C)\int_A f d\mu$ (if $f \geq 0$ then see, e.g., [17], Theorem 11.2, (5), p. 228 and if f is arbitrary sign, then see, e.g., [8], p. 64, Proposition 5.1, (ii)).

(ii) For all $c \in \mathbb{R}$ and f of arbitrary sign, we have (see, e.g., [17], pp. 232-233, or [8], p. 65) $(C)\int_A (f + c) d\mu = (C)\int_A f d\mu + c \cdot \mu(A)$.

If μ is submodular too, then for all f, g of arbitrary sign and lower bounded, we have (see, e.g., [8], p. 75, Theorem 6.3)

$$(C)\int_A (f + g) d\mu \leq (C)\int_A f d\mu + (C)\int_A g d\mu.$$

(iii) If $f \le g$ on A then $(C)\int_A f d\mu \le (C)\int_A g d\mu$ (see, e.g., [17], p. 228, Theorem 11.2, (3) if $f, g \ge 0$ and p. 232 if f, g are of arbitrary sign).

(iv) Let $f \ge 0$. If $A \subset B$ then $(C)\int_A f d\mu \le (C)\int_B f d\mu$. In addition, if μ is finitely subadditive, then

$$(C)\int_{A\cup B} f d\mu \le (C)\int_A f d\mu + (C)\int_B f d\mu.$$

(v) It is immediate that $(C)\int_A 1 \cdot d\mu(t) = \mu(A)$.

(vi) The formula $\mu(A) = \gamma(M(A))$, where $\gamma : [0,1] \to [0,1]$ is an increasing and concave function, with $\gamma(0) = 0$, $\gamma(1) = 1$ and M is a probability measure (or only finitely additive) on a σ-algebra on Ω (that is, $M(\emptyset) = 0$, $M(\Omega) = 1$ and M is countably additive), gives simple examples of normalized, monotone and submodular set functions (see, e.g., [8], pp. 16-17, Example 2.1). Such of set functions μ are also called distorsions of countably normalized, additive measures (or distorted measures). For a simple example, we can take $\gamma(t) = \frac{2t}{1+t}$, $\gamma(t) = \sqrt{t}$.

If the above γ function is increasing, concave and satisfies only $\gamma(0) = 0$, then for any bounded Borel measure m, $\mu(A) = \gamma(m(A))$ gives a simple example of bounded, monotone and submodular set function.

(vii) If μ is a countably additive bounded measure, then the Choquet integral $(C)\int_A f d\mu$ reduces to the usual Lebesgue type integral (see, e.g., [8], p. 62, or [17], p. 226).

(viii) If $f \ge 0$, then $(C)\int_A f d\mu \ge 0$.

(ix) Let $\mu = \sqrt{M}$, where M is the Lebesgue measure on $[0, +\infty)$, then μ is a monotone and submodular set function, furthermore μ is strictly positive, see [10].

(x) If $\Omega = \mathbb{R}^N$, $N \in \mathbb{N}$, we call μ strictly positive if $\mu(A) > 0$, for any open subset $A \subseteq \mathbb{R}^N$.

We need some possibility theory:

Definition 10.3 ([9]) For the $\Omega \neq \emptyset$, the power set $\mathcal{P}(\Omega)$ denotes the family of all subsets of Ω.

(i) A function $\lambda : \Omega \to [0,1]$ with the property $\sup\{\lambda(s) : s \in \Omega\} = 1$, is called possibility distribution on Ω.

(ii) $P : \mathcal{P}(\Omega) \to [0,1]$ is called possibility measure, if it satisfies $P(\emptyset) = 0$, $P(\Omega) = 1$, and $P(\cup_{i\in I} A_i) = \sup\{P(A_i) : i \in I\}$ for all $A_i \subset \Omega$, and any I, an at most countable family of indices. Note that if $A, B \subset \Omega$, $A \subset B$, then the last property implies $P(A) \le P(B)$ and that $P(A \cup B) \le P(A) + P(B)$.

Any possibility distribution λ on Ω, induces the possibility measure $P_\lambda : \mathcal{P}(\Omega) \to [0,1]$, $P_\lambda(A) = \sup\{\lambda(s) : s \in A\}$, $A \subset \Omega$. Also, if $f : \Omega \to \mathbb{R}_+$, then the possibilistic integral of f on $A \subset \Omega$ with respect to P_λ is defined by $(Pos)\int_A f dP_\lambda = \sup\{f(t)\lambda(t) : t \in A\}$ (see [9], Chapter 1).

Note that any possiblity measure μ is normalized, monotone and submodular. From $\mu(A \cup B) = \max\{\mu(A), \mu(B)\}$ we get monotonicity, and from $\mu(A \cap B) \le \min\{\mu(A), \mu(B)\}$ we derive the submodularity.

10.3 Background - II

Here we follow [1] for the basics of Conformable fractional calculus, see also [12].

We need

Definition 10.4 ([1]) Let $a, b \in \mathbb{R}$. The left conformable fractional derivative starting from a of a function $f : [a, \infty) \to \mathbb{R}$ of order $0 < \alpha \leq 1$ is defined by

$$\left(T_{\alpha}^{a} f\right)(t) = \lim_{\varepsilon \to 0} \frac{f\left(t + \varepsilon (t-a)^{1-\alpha}\right) - f(t)}{\varepsilon}. \tag{10.1}$$

If $\left(T_{\alpha}^{a} f\right)(t)$ exists on (a, b), then

$$\left(T_{\alpha}^{a} f\right)(a) = \lim_{t \to a+} \left(T_{\alpha}^{a} f\right)(t). \tag{10.2}$$

The right conformable fractional derivative of order $0 < \alpha \leq 1$ terminating at b of $f : (-\infty, b] \to \mathbb{R}$ is defined by

$$\left({}_{\alpha}^{b} T f\right)(t) = -\lim_{\varepsilon \to 0} \frac{f\left(t + \varepsilon (b-t)^{1-\alpha}\right) - f(t)}{\varepsilon}. \tag{10.3}$$

If $\left({}_{\alpha}^{b} T f\right)(t)$ exists on (a, b), then

$$\left({}_{\alpha}^{b} T f\right)(b) = \lim_{t \to b-} \left({}_{\alpha}^{b} T f\right)(t). \tag{10.4}$$

Note that if f is differentiable then

$$\left(T_{\alpha}^{a} f\right)(t) = (t-a)^{1-\alpha} f'(t), \tag{10.5}$$

and

$$\left({}_{\alpha}^{b} T f\right)(t) = -(b-t)^{1-\alpha} f'(t). \tag{10.6}$$

Denote by

$$\left(I_{\alpha}^{a} f\right)(t) = \int_{a}^{t} (x-a)^{\alpha-1} f(x)\, dx, \tag{10.7}$$

and

$$\left({}^{b} I_{\alpha} f\right)(t) = \int_{t}^{b} (b-x)^{\alpha-1} f(x)\, dx, \tag{10.8}$$

these are the left and right conformable fractional integrals of order $0 < \alpha \leq 1$.

In the higher order case we can generalize things as follows:

Definition 10.5 ([1]) Let $\alpha \in (n, n+1]$, and set $\beta = \alpha - n$. Then, the left conformable fractional derivative starting from a of a function $f : [a, \infty) \to \mathbb{R}$ of order α, where $f^{(n)}(t)$ exists, is defined by

$$\left(\mathbf{T}_{\alpha}^{a} f\right)(t) = \left(T_{\beta}^{a} f^{(n)}\right)(t), \tag{10.9}$$

The right conformable fractional derivative of order α terminating at b of $f : (-\infty, b] \to \mathbb{R}$, where $f^{(n)}(t)$ exists, is defined by

$$\left({}_{\alpha}^{b}\mathbf{T} f\right)(t) = (-1)^{n+1} \left({}_{\beta}^{b} T f^{(n)}\right)(t). \tag{10.10}$$

If $\alpha = n + 1$ then $\beta = 1$ and $\mathbf{T}_{n+1}^{a} f = f^{(n+1)}$.

If n is odd, then ${}_{n+1}^{b}\mathbf{T} f = -f^{(n+1)}$, and if n is even, then ${}_{n+1}^{b}\mathbf{T} f = f^{(n+1)}$.

When $n = 0$ (or $\alpha \in (0, 1]$), then $\beta = \alpha$, and (10.9), (10.10) collapse to $\{(10.1)$–$(10.4)\}$, respectively.

Lemma 10.6 ([1]) *Let $f : (a, b) \to \mathbb{R}$ be continuously differentiable and* $0 < \alpha \leq 1$. *Then, for all $t > a$ we have*

$$I_{\alpha}^{a} T_{\alpha}^{a}(f)(t) = f(t) - f(a). \tag{10.11}$$

We need

Definition 10.7 (*see also* [1]) If $\alpha \in (n, n+1]$, then the left fractional integral of order α starting at a is defined by

$$\left(\mathbf{I}_{\alpha}^{a} f\right)(t) = \frac{1}{n!} \int_{a}^{t} (t - x)^{n} (x - a)^{\beta - 1} f(x)\, dx. \tag{10.12}$$

Similarly, (author's definition, see [5]) the right fractional integral of order α terminating at b is defined by

$$\left({}^{b}\mathbf{I}_{\alpha} f\right)(t) = \frac{1}{n!} \int_{t}^{b} (x - t)^{n} (b - x)^{\beta - 1} f(x)\, dx. \tag{10.13}$$

We need

Proposition 10.8 ([1]) *Let $\alpha \in (n, n+1]$ and $f : [a, \infty) \to \mathbb{R}$ be $(n+1)$ times continuously differentiable for $t > a$. Then, for all $t > a$ we have*

$$\mathbf{I}_{\alpha}^{a} \mathbf{T}_{\alpha}^{a}(f)(t) = f(t) - \sum_{k=0}^{n} \frac{f^{(k)}(a)(t - a)^{k}}{k!}. \tag{10.14}$$

We also have

Proposition 10.9 ([5]) *Let $\alpha \in (n, n+1]$ and $f : (-\infty, b] \to \mathbb{R}$ be $(n+1)$ times continuously differentiable for $t < b$. Then, for all $t < b$ we have*

$$-{}^{b}\mathbf{I}_{\alpha}\, {}^{b}_{a}\mathbf{T}(f)(t) = f(t) - \sum_{k=0}^{n} \frac{f^{(k)}(b)(t-b)^k}{k!}. \tag{10.15}$$

If $n = 0$ or $0 < \alpha \leq 1$, then (see also [1])

$${}^{b}I_{\alpha}\, {}^{b}_{\alpha}T(f)(t) = f(t) - f(b). \tag{10.16}$$

In conclusion we derive

Theorem 10.10 ([5]) *Let $\alpha \in (n, n+1]$ and $f \in C^{n+1}([a,b])$, $n \in \mathbb{N}$. Then*
(1)

$$f(t) - \sum_{k=0}^{n} \frac{f^{(k)}(a)(t-a)^k}{k!} = \frac{1}{n!} \int_a^t (t-x)^n (x-a)^{\beta-1} \left(\mathbf{T}_{\alpha}^{a}(f)\right)(x)\, dx, \tag{10.17}$$

and
(2)

$$f(t) - \sum_{k=0}^{n} \frac{f^{(k)}(b)(t-b)^k}{k!} = -\frac{1}{n!} \int_t^b (b-x)^{\beta-1} (x-t)^n \left({}^{b}_{\alpha}\mathbf{T}(f)\right)(x)\, dx, \tag{10.18}$$

$\forall\, t \in [a,b]$.

We need

Remark 10.11 ([5]) We notice the following: let $\alpha \in (n, n+1]$ and $f \in C^{n+1}([a,b])$, $n \in \mathbb{N}$. Then ($\beta := \alpha - n$, $0 < \beta \leq 1$)

$$\left(\mathbf{T}_{\alpha}^{a}(f)\right)(x) = \left(T_{\beta}^{\alpha} f^{(n)}\right)(x) = (x-a)^{1-\beta} f^{(n+1)}(x), \tag{10.19}$$

and

$$\left({}^{b}_{\alpha}\mathbf{T}(f)\right)(x) = (-1)^{n+1} \left({}^{b}_{\beta}T f^{(n)}\right)(x) =$$

$$(-1)^{n+1}(-1)(b-x)^{1-\beta} f^{(n+1)}(x) = (-1)^n (b-x)^{1-\beta} f^{(n+1)}(x). \tag{10.20}$$

Consequently we get that

$$\left(\mathbf{T}_{\alpha}^{a}(f)\right)(x),\ \left({}^{b}_{\alpha}\mathbf{T}(f)\right)(x) \in C([a,b]).$$

Furthermore it is obvious that

$$\left(\mathbf{T}_{\alpha}^{a}(f)\right)(a)=\left({}_{\alpha}^{b}\mathbf{T}(f)\right)(b)=0, \tag{10.21}$$

when $0<\beta<1$, i.e. when $\alpha\in(n,n+1)$.

If $f^{(k)}(a)=0$, $k=1,\ldots,n$, then

$$f(t)-f(a)=\frac{1}{n!}\int_{a}^{t}(t-x)^{n}(x-a)^{\beta-1}\left(\mathbf{T}_{\alpha}^{a}(f)\right)(x)\,dx, \tag{10.22}$$

$\forall\, t\in[a,b]$.

If $f^{(k)}(b)=0$, $k=1,\ldots,n$, then

$$f(t)-f(b)=-\frac{1}{n!}\int_{t}^{b}(b-x)^{\beta-1}(x-t)^{n}\left({}_{\alpha}^{b}\mathbf{T}(f)\right)(x)\,dx, \tag{10.23}$$

$\forall\, t\in[a,b]$.

For $f\in C([a,b])$, $\delta>0$, denote by

$$\omega_{1}(f,\delta)=\sup_{\substack{x,y\in[a,b]:\\|x-y|\leq\delta}}|f(x)-f(y)|$$

the (first) modulus of continuity of f.

We mention

Theorem 10.12 ([5]) *Let $\alpha\in(n,n+1)$, $n\in\mathbb{N}$, and $f\in C^{n+1}([a,b])$, $x\in[a,b]$ and $f^{(k)}(x)=0$, $k=1,\ldots,n$. Denote*

$$\omega_{1}\left({}^{x}\mathbf{T}_{\alpha}f,\delta\right):=\max\left\{\omega_{1}\left(\mathbf{T}_{\alpha}^{x}f,\delta\right)_{[x,b]},\omega_{1}\left({}_{\alpha}^{x}\mathbf{T}f,\delta\right)_{[a,x]}\right\}. \tag{10.24}$$

Then, over $[a,b]$, we have

$$|f(\cdot)-f(x)|\leq\frac{\omega_{1}\left({}^{x}\mathbf{T}_{\alpha}f,\delta\right)}{\prod_{j=0}^{n-1}(\alpha-j)}\left[\frac{|\cdot-x|^{\alpha}}{(\alpha-n)}+\frac{|\cdot-x|^{\alpha+1}}{(\alpha+1)\,\delta}\right],\ \delta>0. \tag{10.25}$$

We need

Definition 10.13 Here $C_{+}([a,b]):=\{f:[a,b]\to\mathbb{R}_{+},\text{ continuous functions}\}$. Let $L_{N}:C_{+}([a,b])\to C_{+}([a,b])$, operators, $\forall\, N\in\mathbb{N}$, such that

(i)

$$L_{N}(\alpha f)=\alpha L_{N}(f),\ \forall\,\alpha\geq0,\ \forall\, f\in C_{+}([a,b]), \tag{10.26}$$

(ii) if $f,g\in C_{+}([a,b]):f\leq g$, then

$$L_{N}(f)\leq L_{N}(g),\ \forall\, N\in\mathbb{N}, \tag{10.27}$$

(iii)

$$L_N(f+g) \leq L_N(f) + L_N(g), \ \forall\, f, g \in C_+([a,b]). \tag{10.28}$$

We call $\{L_N\}_{N\in\mathbb{N}}$ positive sublinear operators.

We will use

Theorem 10.14 ([5]) *Let $\alpha \in (n, n+1)$, $n \in \mathbb{N}$, and $f \in C^{n+1}([a,b], \mathbb{R}_+)$, $x \in [a,b]$ and $f^{(k)}(x) = 0$, $k = 1, \ldots, n$. Let $L_N : C_+([a,b]) \to C_+([a,b])$, $\forall\, N \in \mathbb{N}$, be positive sublinear operators, such that $L_N(1) = 1$, $\forall\, N \in \mathbb{N}$. Then*

$$|L_N(f)(x) - f(x)| \leq \frac{\omega_1\left({}^x\mathbf{T}_\alpha f, \delta\right)}{\prod_{j=0}^{n-1}(\alpha - j)}\left[\frac{L_N\left(|\cdot - x|^\alpha\right)(x)}{(\alpha - n)} + \frac{L_N\left(|\cdot - x|^{\alpha+1}\right)(x)}{(\alpha+1)\,\delta}\right], \tag{10.29}$$

$\delta > 0$.

We also mention

Theorem 10.15 ([5]) *Let $f \in C^1([a,b])$, $\alpha \in (0,1)$, $x \in [a,b]$. Denote*

$$\omega_1\left({}^xT_\alpha f, \delta\right) := \max\left\{\omega_1\left(T_\alpha^x f, \delta\right)_{[x,b]}, \omega_1\left({}_\alpha^x T f, \delta\right)_{[a,x]}\right\}, \ \delta > 0. \tag{10.30}$$

Then over $[a,b]$ we have

$$|f(\cdot) - f(x)| \leq \omega_1\left({}^xT_\alpha f, \delta\right)\left[\frac{|\cdot - x|^\alpha}{\alpha} + \frac{|\cdot - x|^{\alpha+1}}{(\alpha+1)\,\delta}\right], \ \delta > 0. \tag{10.31}$$

We mention

Theorem 10.16 ([5]) *Let $f \in C^1([a,b], \mathbb{R}_+)$, $\alpha \in (0,1)$, and let $L_N : C_+([a,b]) \to C_+([a,b])$, $\forall\, N \in \mathbb{N}$, be positive sublinear operators, such that $L_N(1) = 1$, $\forall\, N \in \mathbb{N}$. Then*

$$|L_N(f)(x) - f(x)| \leq \omega_1\left({}^xT_\alpha f, \delta\right)\left[\frac{L_N\left(|\cdot - x|^\alpha\right)(x)}{\alpha} + \frac{L_N\left(|\cdot - x|^{\alpha+1}\right)(x)}{(\alpha+1)\,\delta}\right], \tag{10.32}$$

$\forall\, N \in \mathbb{N}$, $\forall\, x \in [a,b]$, $\delta > 0$.

Also we will use

Theorem 10.17 ([5]) *Let $f \in C^1([a,b], \mathbb{R}_+)$, $\alpha \in (0,1)$, $x \in [a,b]$. Let $L_N : C_+([a,b]) \to C_+([a,b])$, $\forall\, N \in \mathbb{N}$ be positive sublinear operators, such that $L_N(1) = 1$, and $L_N\left(|\cdot - x|^{\alpha+1}\right)(x) > 0$, $\forall\, N \in \mathbb{N}$. Then*

$$|L_N(f)(x) - f(x)| \leq$$

$$\frac{(2\alpha+1)}{\alpha(\alpha+1)}\omega_1\left({}^{x}T_{\alpha}f,\left(L_N\left(|\cdot-x|^{\alpha+1}\right)(x)\right)^{\frac{1}{\alpha+1}}\right)\left(L_N\left(|\cdot-x|^{\alpha+1}\right)(x)\right)^{\frac{\alpha}{\alpha+1}}, \tag{10.33}$$

$\forall\, N\in\mathbb{N}$.

10.4 Background - III

Here we mention some Caputo fractional calculus basics.

We make

Remark 10.18 Let $f:[a,b]\to\mathbb{R}$ such that $f'\in L_{\infty}([a,b])$, $x_0\in[a,b]$, $0<\alpha<1$, the left Caputo fractional derivative of order α is defined as follows

$$\left(D_{*x_0}^{\alpha}f\right)(x_0)=\frac{1}{\Gamma(1-\alpha)}\int_{x_0}^{x}(x-t)^{-\alpha}f'(t)\,dt, \tag{10.34}$$

where Γ is the gamma function for all $x_0\leq x\leq b$.

We observe that

$$\left|\left(D_{*x_0}^{\alpha}f\right)(x)\right|=\frac{1}{\Gamma(1-\alpha)}\int_{x_0}^{x}(x-t)^{-\alpha}\left|f'(t)\right|dt$$

$$\leq\frac{\left\|f'\right\|_{\infty}}{\Gamma(1-\alpha)}\int_{x_0}^{x}(x-t)^{-\alpha}dt=\frac{\left\|f'\right\|_{\infty}}{\Gamma(1-\alpha)}\frac{(x-x_0)^{1-\alpha}}{(1-\alpha)}=\frac{\left\|f'\right\|_{\infty}(x-x_0)^{1-\alpha}}{\Gamma(2-\alpha)}. \tag{10.35}$$

I.e.

$$\left|\left(D_{*x_0}^{\alpha}f\right)(x)\right|\leq\frac{\left\|f'\right\|_{\infty}(x-x_0)^{1-\alpha}}{\Gamma(2-\alpha)}\leq\frac{\left\|f'\right\|_{\infty}(b-x_0)^{1-\alpha}}{\Gamma(2-\alpha)}<+\infty, \tag{10.36}$$

$\forall\, x\in[x_0,b]$.

Clearly, then

$$\left(D_{*x_0}^{\alpha}f\right)(x_0)-0. \tag{10.37}$$

We define $\left(D_{*x_0}^{\alpha}f\right)(x)=0$, for $a\leq x<x_0$.

Let $n\in\mathbb{N}$, we denote the iterated fractional derivative $D_{*x_0}^{n\alpha}=D_{*x_0}^{\alpha}D_{*x_0}^{\alpha}...D_{*x_0}^{\alpha}$ (n-times).

Let us assume that

$$D_{*x_0}^{k\alpha}f\in C([x_0,b]),\ k=0,1,\ldots,n+1;\ n\in\mathbb{N},0<\alpha<1.$$

By [6, 13], pp. 156-158, we have the following generalized fractional Caputo type Taylor's formula:

$$f(x) = \sum_{i=0}^{n} \frac{(x - x_0)^{i\alpha}}{\Gamma(i\alpha + 1)} \left(D_{*x_0}^{i\alpha} f\right)(x_0) + \tag{10.38}$$

$$\frac{1}{\Gamma((n+1)\alpha)} \int_{x_0}^{x} (x - t)^{(n+1)\alpha - 1} \left(D_{*x_0}^{(n+1)\alpha} f\right)(t)\, dt,$$

$\forall\, x \in [x_0, b]$.

Based on the above (10.37) and (10.38), we derive

$$f(x) - f(x_0) = \sum_{i=2}^{n} \frac{(x - x_0)^{i\alpha}}{\Gamma(i\alpha + 1)} \left(D_{*x_0}^{i\alpha} f\right)(x_0) + \tag{10.39}$$

$$\frac{1}{\Gamma((n+1)\alpha)} \int_{x_0}^{x} (x - t)^{(n+1)\alpha - 1} \left(D_{*x_0}^{(n+1)\alpha} f\right)(t)\, dt,$$

$\forall\, x \in [x_0, b]$, $0 < \alpha < 1$.

In case of $\left(D_{*x_0}^{i\alpha} f\right)(x_0) = 0$, $i = 2, 3, \ldots, n+1$, we get

$$f(x) - f(x_0) =$$

$$\frac{1}{\Gamma((n+1)\alpha)} \int_{x_0}^{x} (x - t)^{(n+1)\alpha - 1} \left(\left(D_{*x_0}^{(n+1)\alpha} f\right)(t) - \left(D_{*x_0}^{(n+1)\alpha} f\right)(x_0)\right) dt, \tag{10.40}$$

$\forall\, x \in [x_0, b]$, $0 < \alpha < 1$.

We make

Remark 10.19 Let $f : [a, b] \to \mathbb{R}$ such that $f' \in L_\infty([a, b])$, $x_0 \in [a, b]$, $0 < \alpha < 1$, the right Caputo fractional derivative of order α is defined as follows

$$\left(D_{x_0-}^{\alpha} f\right)(x_0) = \frac{-1}{\Gamma(1 - \alpha)} \int_{x}^{x_0} (z - x)^{-\alpha} f'(z)\, dz, \tag{10.41}$$

$\forall\, x \in [a, x_0]$.

We observe that

$$\left|\left(D_{x_0-}^{\alpha} f\right)(x)\right| = \frac{1}{\Gamma(1 - \alpha)} \int_{x}^{x_0} (z - x)^{-\alpha} \left|f'(z)\right| dz \leq$$

$$\frac{\|f'\|_\infty}{\Gamma(1 - \alpha)} \left(\int_{x}^{x_0} (z - x)^{-\alpha} dz\right) = \frac{\|f'\|_\infty}{\Gamma(1 - \alpha)} \frac{(x_0 - x)^{1-\alpha}}{(1 - \alpha)} = \frac{\|f'\|_\infty}{\Gamma(2 - \alpha)} (x_0 - x)^{1-\alpha}. \tag{10.42}$$

That is

$$\left|\left(D_{x_0-}^{\alpha} f\right)(x)\right| \leq \frac{\|f'\|_{\infty}}{\Gamma(2-\alpha)}(x_0-x)^{1-\alpha} \leq \frac{\|f'\|_{\infty}}{\Gamma(2-\alpha)}(x_0-a)^{1-\alpha} < \infty, \tag{10.43}$$

$\forall\, x \in [a, x_0]$.

In particular we have

$$\left(D_{x_0-}^{\alpha} f\right)(x_0) = 0. \tag{10.44}$$

We define $\left(D_{x_0-}^{\alpha} f\right)(x) = 0$, for $x_0 < x \leq b$.

For $n \in \mathbb{N}$, denote the iterated fractional derivative $D_{x_0-}^{n\alpha} = D_{x_0-}^{\alpha} D_{x_0-}^{\alpha} \ldots D_{x_0-}^{\alpha}$ (n-times).

In [4], we proved the following right generalized fractional Taylor's formula: Suppose that

$$D_{x_0-}^{k\alpha} f \in C([a, x_0]), \quad \text{for } k = 0, 1, \ldots, n+1, 0 < \alpha < 1.$$

Then

$$f(x) = \sum_{i=0}^{n} \frac{(x_0-x)^{i\alpha}}{\Gamma(i\alpha+1)} \left(D_{x_0-}^{i\alpha} f\right)(x_0) + \tag{10.45}$$

$$\frac{1}{\Gamma((n+1)\alpha)} \int_x^{x_0} (z-x)^{(n+1)\alpha-1} \left(D_{x_0-}^{(n+1)\alpha} f\right)(z)\, dz,$$

$\forall\, x \in [a, x_0]$.

Based on (10.44) and (10.45), we derive

$$f(x) - f(x_0) = \sum_{i=2}^{n} \frac{(x_0-x)^{i\alpha}}{\Gamma(i\alpha+1)} \left(D_{x_0-}^{i\alpha} f\right)(x_0) + \tag{10.46}$$

$$\frac{1}{\Gamma((n+1)\alpha)} \int_x^{x_0} (z-x)^{(n+1)\alpha-1} \left(D_{x_0-}^{(n+1)\alpha} f\right)(z)\, dz,$$

$\forall\, x \in [a, x_0]$, $0 < \alpha < 1$.

In case of $\left(D_{x_0-}^{i\alpha} f\right)(x_0) = 0$, for $i = 2, 3, \ldots, n+1$, we get

$$f(x) - f(x_0) =$$

$$\frac{1}{\Gamma((n+1)\alpha)} \int_x^{x_0} (z-x)^{(n+1)\alpha-1} \left(\left(D_{x_0-}^{(n+1)\alpha} f\right)(z) - \left(D_{x_0-}^{(n+1)\alpha} f\right)(x_0)\right) dz, \tag{10.47}$$

$\forall\, x \in [a, x_0]$, $0 < \alpha < 1$.

We need

Definition 10.20 Let $D_{x_0}^{(n+1)\alpha} f$ denote any of $D_{*x_0}^{(n+1)\alpha} f$, $D_{x_0-}^{(n+1)\alpha} f$, and $\delta > 0$. We set

$$\omega_1 \left(D_{x_0}^{(n+1)\alpha} f, \delta\right) = \max \left\{\omega_1 \left(D_{*x_0}^{(n+1)\alpha} f, \delta\right)_{[x_0,b]}, \omega_1 \left(D_{x_0-}^{(n+1)\alpha} f, \delta\right)_{[a,x_0]}\right\}, \tag{10.48}$$

where $x_0 \in [a, b]$. Here the moduli of continuity are considered over $[x_0, b]$ and $[a, x_0]$, respectively.

We mention

Theorem 10.21 ([2]) *Let* $0 < \alpha < 1$, $f : [a, b] \to \mathbb{R}$, $f' \in L_\infty ([a, b])$, $x_0 \in [a, b]$. *Assume that* $D_{*x_0}^{k\alpha} f \in C ([x_0, b])$, $k = 0, 1, \dots, n + 1$; $n \in \mathbb{N}$, *and* $\left(D_{*x_0}^{i\alpha} f\right)(x_0) = 0$, $i = 2, 3, \dots, n + 1$. *Also, suppose that* $D_{x_0-}^{k\alpha} f \in C ([a, x_0])$, *for* $k = 0, 1, \dots, n + 1$, *and* $\left(D_{x_0-}^{i\alpha} f\right)(x_0) = 0$, *for* $i = 2, 3, \dots, n + 1$. *Then*

$$|f (x) - f (x_0)| \le \frac{\omega_1 \left(D_{x_0}^{(n+1)\alpha} f, \delta\right)}{\Gamma ((n + 1) \alpha + 1)} \left[|x - x_0|^{(n+1)\alpha} + \frac{|x - x_0|^{(n+1)\alpha+1}}{\delta ((n + 1) \alpha + 1)}\right], \tag{10.49}$$

$\forall\, x \in [a, b]$, $\delta > 0$.

We will use

Theorem 10.22 ([2]) *Let* $\frac{1}{n+1} < \alpha < 1$, $n \in \mathbb{N}$, $f : [a, b] \to \mathbb{R}_+$, $f' \in L_\infty ([a, b])$, $x_0 \in [a, b]$. *Assume that* $D_{*x_0}^{k\alpha} f \in C ([x_0, b])$, $k = 0, 1, \dots, n + 1$, *and* $\left(D_{*x_0}^{i\alpha} f\right)(x_0) = 0$, $i = 2, 3, \dots, n + 1$. *Also, suppose that* $D_{x_0-}^{k\alpha} f \in C ([a, x_0])$, *for* $k = 0, 1, \dots, n + 1$, *and* $\left(D_{x_0-}^{i\alpha} f\right)(x_0) = 0$, *for* $i = 2, 3, \dots, n + 1$. *Denote* $\lambda = (n + 1) \alpha > 1$. *Let* $L_N : C_+ ([a, b]) \to C_+ ([a, b])$, $\forall\, N \in \mathbb{N}$, *be positive sublinear operators, such that* $L_N (1) = 1$, $\forall\, N \in \mathbb{N}$. *Then*

$$|L_N (f) (x_0) - f (x_0)| \le \frac{\omega_1 \left(D_{x_0}^{(n+1)\alpha} f, \delta\right)}{\Gamma (\lambda + 1)} \cdot$$

$$\left[L_N \left(|\cdot - x_0|^\lambda\right)(x_0) + \frac{L_N \left(|\cdot - x_0|^{\lambda+1}\right)(x_0)}{(\lambda + 1)\, \delta}\right], \tag{10.50}$$

$\delta > 0$, $\forall\, N \in \mathbb{N}$.

Note: ([2]) Theorem 10.22 is also true when $0 < \alpha \le \frac{1}{n+1}$.

We will also use

Theorem 10.23 ([2]) *Let* $0 < \alpha \le \frac{1}{n+1}$, $n \in \mathbb{N}$, $f : [a, b] \to \mathbb{R}_+$, $f' \in L_\infty ([a, b])$, $x_0 \in [a, b]$. *Assume that* $D_{*x_0}^{k\alpha} f \in C ([x_0, b])$, $k = 0, 1, \dots, n + 1$, *and* $\left(D_{*x_0}^{i\alpha} f\right)(x_0) = 0$, $i = 2, 3, \dots, n + 1$. *Also, suppose that* $D_{x_0-}^{k\alpha} f \in C ([a, x_0])$, *for* $k = 0, 1, \dots, n + 1$, *and* $\left(D_{x_0-}^{i\alpha} f\right)(x_0) = 0$, *for* $i = 2, 3, \dots, n + 1$. *Denote* $\lambda :=$

$(n+1)\alpha \leq 1$. Let $L_N : C_+([a,b]) \to C_+([a,b])$, $\forall\, N \in \mathbb{N}$, be positive sublinear operators, such that $L_N\left(|\cdot - x_0|^{\lambda+1}\right)(x_0) > 0$ and $L_N(1) = 1$, $\forall\, N \in \mathbb{N}$. Then

$$|L_N(f)(x_0) - f(x_0)| \leq \frac{\omega_1\left(D_{x_0}^{(n+1)\alpha} f, \delta\right)}{\Gamma(\lambda+1)} \cdot \tag{10.51}$$

$$\left[\left(L_N\left(|\cdot - x_0|^{\lambda+1}\right)(x_0)\right)^{\frac{\lambda}{\lambda+1}} + \frac{L_N\left(|\cdot - x_0|^{\lambda+1}\right)(x_0)}{(\lambda+1)\delta}\right],$$

$\delta > 0$, $\forall\, N \in \mathbb{N}$.

We will also use

Theorem 10.24 ([2]) *Let $0 < \alpha \leq \frac{1}{n+1}$, $n \in \mathbb{N}$, $f : [a,b] \to \mathbb{R}_+$, $f' \in L_\infty([a,b])$, $x_0 \in [a,b]$. Assume that $D_{*x_0}^{k\alpha} f \in C([x_0,b])$, $k = 0, 1, \ldots, n+1$, and $\left(D_{*x_0}^{i\alpha} f\right)(x_0) = 0$, $i = 2, 3, \ldots, n+1$. Also, suppose that $D_{x_0-}^{k\alpha} f \in C([a,x_0])$, for $k = 0, 1, \ldots, n+1$, and $\left(D_{x_0-}^{i\alpha} f\right)(x_0) = 0$, for $i = 2, 3, \ldots, n+1$. Denote $\lambda := (n+1)\alpha \leq 1$. Let $L_N : C_+([a,b]) \to C_+([a,b])$, $\forall\, N \in \mathbb{N}$, be positive sublinear operators, such that $L_N\left(|\cdot - x_0|^{\lambda+1}\right)(x_0) > 0$ and $L_N(1) = 1$, $\forall\, N \in \mathbb{N}$. Then*

$$|L_N(f)(x_0) - f(x_0)| \leq \frac{(\lambda+2)\,\omega_1\left(D_{x_0}^{(n+1)\alpha} f, \left(L_N\left(|\cdot - x_0|^{\lambda+1}\right)(x_0)\right)^{\frac{1}{\lambda+1}}\right)}{\Gamma(\lambda+2)} \cdot \tag{10.52}$$

$$\left(L_N\left(|\cdot - x_0|^{\lambda+1}\right)(x_0)\right)^{\frac{\lambda}{\lambda+1}}, \quad \forall\, N \in \mathbb{N}.$$

Note: ([2]) From (10.52) we get that: if $L_N\left(|\cdot - x_0|^{\lambda+1}\right)(x_0) \to 0$, as $N \to +\infty$, then $L_N(f)(x_0) \to f(x_0)$, as $N \to +\infty$.

10.5 Background - IV

Here, we mention the operators we will deal with:

We need

Definition 10.25 ([10]) Let $I = [0,1]$, $\mathcal{B}_I$ the σ-algebra of all Borel measurable subsets of I, $\left(\Gamma_{N,x}\right)_{N \in \mathbb{N},\, x \in I}$ will be the collection of the family $\Gamma_{N,x} = \{\mu_{N,k,x}\}_{k=0}^{N}$, of monotone, submodular and strictly positive set functions $\mu_{N,k,x}$ on $\mathcal{B}_I$.

Let $f : [0,1] \to \mathbb{R}_+$ be a B_I-measurable function which is bounded, and call $p_{N,k}(x) = \binom{N}{k} x^k (1-x)^{N-k}$, for any $x \in [0,1]$.

The Bernstein–Kantorovich–Choquet operators are defined by the formula

$$K_{N,\Gamma_{N,x}}(f)(x) = \sum_{k=0}^{N} p_{N,k}(x) \frac{(C)\int_{\frac{k}{(N+1)}}^{\frac{(k+1)}{(N+1)}} f(t)\, d\mu_{N,k,x}(t)}{\mu_{N,k,x}\left(\left[\frac{k}{(N+1)}, \frac{(k+1)}{(N+1)}\right]\right)}, \ \forall x \in [0,1]. \quad (10.53)$$

If $\mu_{N,k,x} = \mu$, for all N, x, k, we will denote $K_{N,\Gamma_{N,x}}(f) := K_{N,\mu}(f)$.

It holds

Theorem 10.26 ([10]) *Suppose that* $\mu_{N,k,x} = \mu := \sqrt{M}$, *for all* N, k *and* x, *where* M *is the Lebesgue measure on* $[0,1]$. *Then*

$$\left|K_{N,\mu}(f)(x) - f(x)\right| \leq 2\omega_1\left(f, \frac{\sqrt{x(1-x)}}{\sqrt{N}} + \frac{1}{N}\right), \quad (10.54)$$

$\forall\, N \in \mathbb{N}$, $x \in [0,1]$, $f \in C_+([0,1])$, *above* ω_1 *is over* $[0,1]$.

We make

Remark 10.27 By [10] we have that

$$K_{N,\mu}(|\cdot - x|)(x) \leq \frac{\sqrt{x(1-x)}}{\sqrt{N}} + \frac{1}{N}, \ \forall\, N \in \mathbb{N}. \quad (10.55)$$

Let $m > 1$, notice that $|\cdot - x|^{m-1} \leq 1$, therefore

$$|\cdot - x|^m = |\cdot - x|\,|\cdot - x|^{m-1} \leq |\cdot - x|,$$

hence

$$K_{N,\mu}\left(|\cdot - x|^m\right)(x) \leq K_{N,\mu}(|\cdot - x|)(x),$$

that is

$$K_{N,\mu}\left(|\cdot - x|^m\right)(x) \leq \frac{\sqrt{x(1-x)}}{\sqrt{N}} + \frac{1}{N}, \ \forall x \in [0,1], \ N \in \mathbb{N}, \ m \geq 1. \quad (10.56)$$

Notice that $K_{N,\mu}(1) = 1$, $\forall\, N \in \mathbb{N}$.

Clearly $K_{N,\mu}$ operators are positive sublinear operators from $C_+([0,1])$ into itself.

We mention

Definition 10.28 ([11]) Here we consider measures of possibility. Denoting $p_{N,k}(x) = \binom{N}{k} x^k (1-x)^{N-k}$, let us defined

$$\lambda_{N,k}(t) := \frac{p_{N,k}(t)}{k^k N^{-N}(N-k)^{N-k}\binom{N}{k}} = \frac{t^k(1-t)^{N-k}}{k^k N^{-N}(N-k)^{N-k}},\ k = 0, \ldots, N. \tag{10.57}$$

By convention we assume that $0^0 = 1$, so that the cases $k = 0$, and $k = N$ make sense. By considering the root $\frac{k}{N}$ of $p'_{N,k}(x)$, it is clear that

$$\max\{p_{N,k}(t) : t \in [0, 1]\} = k^k N^{-N}(N-k)^{N-k}\binom{N}{k},$$

which implies that each $\lambda_{N,k}$ is a possibility distribution on $[0, 1]$.

Denoting by $P_{\lambda_{N,k}}$ the possibility measure induced by $\lambda_{N,k}$ and $\Gamma_{n,x} := \Gamma_N := \{P_{\lambda_{N,k}}\}_{k=0}^N$ (that is Γ_N is independent of x), we define the nonlinear Bernstein–Durrmeyer–Choquet polynomial operators with respect to the set functions in Γ_N given by the formula

$$D_{N,\Gamma_N}(f)(x) := \sum_{k=0}^{N} p_{N,k}(x)\frac{(C)\int_0^1 f(t)\,t^k(1-t)^{N-k}\,dP_{\lambda_{N,k}}(t)}{(C)\int_0^1 t^k(1-t)^{N-k}\,dP_{\lambda_{N,k}}(t)}, \tag{10.58}$$

$\forall\, x \in [0, 1]$, $N \in \mathbb{N}$, $f \in C_+([0, 1])$.

Remark 10.29 Above $P_{\lambda_{N,k}}$ is bounded, monotone, submodular and strictly positive, $N \in \mathbb{N}$, $k = 0, 1, \ldots, N$. Notice that $D_{N,\Gamma_N}(1) = 1$, $\forall\, N \in \mathbb{N}$.

Clearly D_{N,Γ_N} operators are positive sublinear operators mapping $C_+([0, 1])$ into itself.

It holds

Theorem 10.30 ([11]) *For every $f \in C_+([0, 1])$, $x \in [0, 1]$ and $N \in \mathbb{N} - \{1\}$, we have*

$$\left|D_{N,\Gamma_N}(f)(x) - f(x)\right| \le 2\omega_1\left(f, \frac{\left(1+\sqrt{2}\right)\sqrt{x(1-x)} + \sqrt{2}\sqrt{x}}{\sqrt{N}} + \frac{1}{N}\right), \tag{10.59}$$

where ω_1 is on $[0, 1]$.

We make

Remark 10.31 By [11] we have that

$$D_{N,\Gamma_N}(|\cdot - x|)(x) \le \frac{\left(1+\sqrt{2}\right)\sqrt{x(1-x)} + \sqrt{2}\sqrt{x}}{\sqrt{N}} + \frac{1}{N},\ \forall\, N \in \mathbb{N} - \{1\}. \tag{10.60}$$

Let $m > 1$, notice that $|\cdot - x|^{m-1} \leq 1$, therefore

$$|\cdot - x|^m = |\cdot - x| \, |\cdot - x|^{m-1} \leq |\cdot - x| \,,$$

hence

$$D_{N,\Gamma_N}\left(|\cdot - x|^m\right)(x) \leq D_{N,\Gamma_N}\left(|\cdot - x|\right)(x) \,,$$

that is

$$D_{N,\Gamma_N}\left(|\cdot - x|^m\right)(x) \leq \frac{\left(1+\sqrt{2}\right)\sqrt{x\,(1-x)} + \sqrt{2}\sqrt{x}}{\sqrt{N}} + \frac{1}{N}, \tag{10.61}$$

$\forall\, N \in \mathbb{N} - \{1\}$, $m \geq 1$, $\forall\, x \in [0,1]$.

We make

Remark 10.32 When $x \in [0,1]$, then the max $(x\,(1-x)) = \frac{1}{4}$, at $x = \frac{1}{2}$. Therefore it holds

$$\frac{\sqrt{x\,(1-x)}}{\sqrt{N}} + \frac{1}{N} \leq \frac{1}{2\sqrt{N}} + \frac{1}{N}, \tag{10.62}$$

$\forall\, x \in [0,1]$, $\forall\, N \in \mathbb{N}$.

Similarly, it holds

$$\frac{\left(1+\sqrt{2}\right)\sqrt{x\,(1-x)} + \sqrt{2}\sqrt{x}}{\sqrt{N}} + \frac{1}{N} \leq \frac{1+3\sqrt{2}}{2\sqrt{N}} + \frac{1}{N}, \tag{10.63}$$

$\forall\, x \in [0,1]$, $\forall\, N \in \mathbb{N} - \{1\}$.

Corollary 10.33 (to Theorem 10.26) *It holds*

$$\left\| K_{N,\mu}(f) - f \right\|_\infty \leq 2\omega_1\left(f, \frac{1}{2\sqrt{N}} + \frac{1}{N}\right), \tag{10.64}$$

$\forall\, N \in \mathbb{N}$, $f \in C_+([0,1])$.

Corollary 10.34 (to Theorem 10.30) *It holds*

$$\left\| D_{N,\Gamma_N}(f) - f \right\|_\infty \leq 2\omega_1\left(f, \frac{1+3\sqrt{2}}{2\sqrt{N}} + \frac{1}{N}\right), \tag{10.65}$$

$\forall\, N \in \mathbb{N} - \{1\}$, $f \in C_+([0,1])$.

The Bernstein–Kantorovich–Choquet operators $K_{N,\mu}$, where $\mu := \sqrt{M}$, with M the Lebesgue measure on [0, 1] have the following exact form:

$$K_{N,\mu}(f)(x) = \sum_{k=0}^{N} p_{N,k}(x) \frac{(C)\int_{\frac{k}{(N+1)}}^{\frac{(k+1)}{(N+1)}} f(t)\, d\mu(t)}{\mu\left(\left[\frac{k}{(N+1)}, \frac{(k+1)}{(N+1)}\right]\right)}, \tag{10.66}$$

$\forall\, x \in [0, 1]$, $\forall\, N \in \mathbb{N}$, $f \in C_+([0, 1])$.

Next we talk about representations of positive sublinear operators by Choquet integrals:

We need

Definition 10.35 Let Ω be a set, and let $f, g : \Omega \to \mathbb{R}$ be bounded functions. We say that f and g are comonotonic, if for every $\omega, \omega' \in \Omega$,

$$\left(f(\omega) - f(\omega')\right)\left(g(\omega) - g(\omega')\right) \geq 0. \tag{10.67}$$

We also need the famous Schmeidler's Representation Theorem (Schmeidler 1986)

Theorem 10.36 ([14]) *Denote with $\mathcal{L}_\infty(\mathcal{A})$ the vector space of $\mathcal{A}$-measurable bounded real valued functions on Ω, where $\mathcal{A} \subset 2^\Omega$ is a σ-algebra. Given a real functional $\Gamma : \mathcal{L}_\infty(\mathcal{A}) \to \mathbb{R}$, assume that for $f, g \in \mathcal{L}_\infty(\mathcal{A})$:*

(i) $\Gamma(cf) = c\Gamma(f)$, $\forall\, c > 0$,

(ii) $f \leq g$, implies $\Gamma(f) \leq \Gamma(g)$,

and

(iii) $\Gamma(f + g) = \Gamma(f) + \Gamma(g)$, for any comonotonic f, g.

Then $\gamma(A) := \Gamma(1_A)$, $\forall\, A \in \mathcal{A}$, defines a finite monotone set function on $\mathcal{A}$, and Γ is the Choquet integral with respect to γ, i.e.

$$\Gamma(f) = (C)\int_\Omega f(t)\, d\gamma(t), \ \forall\, f \in \mathcal{L}_\infty(\mathcal{A}). \tag{10.68}$$

Above 1_A denotes the characteristic function on A.

We make

Remark 10.37 Consider here $[a, b] \subset \mathbb{R}$, $\mathcal{B} = \mathcal{B}([a, b])$ is the Borel σ-algebra on $[a, b]$, and $\mathcal{L}_\infty(\mathcal{B})$ is the vector space of $\mathcal{B}$-measurable bounded real valued functions on $[a, b]$. Let $(L_N)_{N\in\mathbb{N}}$ be a sequence of positive sublinear operators from $\mathcal{L}_\infty(\mathcal{B})$ into $C_+([a, b])$, and $x \in [a, b]$. That is here L_N fulfills the positive homogenuity, monotonicity and subadditivity properties, see (10.26), (10.27), (10.28).

Assume $L_N(1) = 1$, $\forall\, N \in \mathbb{N}$. Clearly here $\mathcal{L}_\infty(\mathcal{B}) \supset C_+([a, b])$. In particular we treat $L_N|_{C_+([a,b])}$, just denoted for simplicity by L_N, $\forall\, N \in \mathbb{N}$.

It is clear that $L_N(\cdot)(x) : \mathcal{L}_\infty(\mathcal{B}) \to \mathbb{R}$ is a functional, $\forall\, N \in \mathbb{N}$. It has the properties:

(i)

$$L_N(cf)(x) = cL_N(f)(x), \ \forall c > 0, \forall f \in \mathcal{L}_\infty(\mathcal{B}), \tag{10.69}$$

(ii)

$$f \le g, \text{ implies } L_N(f)(x) \le L_N(g)(x), \quad \text{where } f, g \in \mathcal{L}_\infty(\mathcal{B}), \tag{10.70}$$

and

(iii)

$$L_N(f+g)(x) \le L_N(f)(x) + L_N(g)(x), \ \forall f, g \in \mathcal{L}_\infty(\mathcal{B}). \tag{10.71}$$

For comonotonic $f, g \in \mathcal{L}_\infty(\mathcal{B})$, we further assume that

$$L_N(f+g)(x) = L_N(f)(x) + L_N(g)(x). \tag{10.72}$$

In that case L_N is called comonotonic.

By Theorem 10.36 we get that:

$$\gamma_{N,x}(A) := L_N(1_A)(x), \forall A \in \mathcal{B}, \forall N \in \mathbb{N}, \tag{10.73}$$

defines a finite monotone set function on $\mathcal{B}$, and

$$L_N(f)(x) = (C)\int_a^b f(t)\, d\gamma_{N,x}(t), \tag{10.74}$$

$\forall f \in \mathcal{L}_\infty(\mathcal{B}), \forall N \in \mathbb{N}$.

In particular (10.74) is valid for any $f \in C_+([a,b])$. Furthermore $\gamma_{N,x}$ is normalized, that is $\gamma_{N,x}([a,b]) = 1, \forall N \in \mathbb{N}$.

A different kind of general operators follows:

We make

Remark 10.38 Consider again $[a,b] \subset \mathbb{R}$, $\mathcal{B} = \mathcal{B}([a,b])$ is the Borel σ-algebra on $[a,b]$. For each $N \in \mathbb{N}$ and each $x \in [a,b]$ consider the monotone set functions $\mu_{N,x}; \mathcal{B} \to \mathbb{R}_+$. We assume that all $\mu_{N,x}$ are normalized, that is $\mu_{N,x}([a,b]) = 1$, and submodular. Here we consider the operators $T_N : C_+([a,b]) \to C_+([a,b])$ given by the formula

$$T_N(f)(x) = (C)\int_{[a,b]} f(t)\, d\mu_{N,x}(t), \tag{10.75}$$

$\forall N \in \mathbb{N}, \forall x \in [a,b]$.

Infact here $\mu_{N,x}$ are chosen so that $T_N(C_+([a,b])) \subseteq C_+([a,b])$.

We notice here that hold:

(i)

$$T_N(\alpha f)(x) = \alpha T_N(f)(x), \ \forall \alpha \geq 0, \tag{10.76}$$

(ii)

$$f \leq g, \text{ implies } T_N(f)(x) \leq T_N(g)(x), \tag{10.77}$$

and

(iii)

$$T_N(f+g)(x) \leq T_N(f)(x) + T_N(g)(x), \tag{10.78}$$

$\forall N \in \mathbb{N}, \forall x \in [a,b], \forall f, g \in C_+([a,b])$.

Clearly T_N are positive sublinear operators, compare to (10.26)–(10.28). We also have that $T_N(1) = 1, \forall N \in \mathbb{N}$.

10.6 Main Results

We present

Theorem 10.39 *Let $\alpha \in (n, n+1)$, $n \in \mathbb{N}$, and $f \in C^{n+1}([0,1], \mathbb{R}_+)$, $x \in [0,1]$ and $f^{(k)}(x) = 0$, $k = 1, \ldots, n$. Then*

$$\left|K_{N,\mu}(f)(x) - f(x)\right| \leq \frac{\omega_1\left({}^{x}T_\alpha f, \left(\frac{1}{2\sqrt{N}} + \frac{1}{N}\right)^{\frac{1}{\alpha+1}}\right)}{\prod_{j=0}^{n-1}(\alpha - j)} \cdot$$

$$\left[\frac{1}{(\alpha - n)}\left(\frac{1}{2\sqrt{N}} + \frac{1}{N}\right) + \frac{1}{(\alpha+1)}\left(\frac{1}{2\sqrt{N}} + \frac{1}{N}\right)^{\frac{\alpha}{\alpha+1}}\right], \ \forall N \in \mathbb{N}. \tag{10.79}$$

We notice that $\lim_{N\to\infty} K_{N,\mu}(f)(x) = f(x)$.

Proof Using (10.29) we would have ($\delta > 0$):

$$\left|K_{N,\mu}(f)(x) - fx\right| \leq \frac{\omega_1\left({}^{x}T_\alpha f, \delta\right)}{\prod_{j=0}^{n-1}(\alpha - j)} \cdot$$

$$\left[\frac{K_{N,\mu}\left(|\cdot - x|^\alpha\right)(x)}{(\alpha - n)} + \frac{K_{N,\mu}\left(|\cdot - x|^{\alpha+1}\right)(x)}{(\alpha+1)\,\delta}\right] \overset{(10.56)}{\leq}$$

$$\frac{\omega_1\left({}^{x}T_{\alpha}f,\delta\right)}{\prod_{j=0}^{n-1}(\alpha-j)}\left[\frac{1}{(\alpha-n)}\left(\frac{\sqrt{x(1-x)}}{\sqrt{N}}+\frac{1}{N}\right)+\frac{1}{(\alpha+1)\delta}\left(\frac{\sqrt{x(1-x)}}{\sqrt{N}}+\frac{1}{N}\right)\right]\overset{(10.62)}{\leq}$$

$$(10.80)$$

$$\frac{\omega_1\left({}^{x}T_{\alpha}f,\delta\right)}{\prod_{j=0}^{n-1}(\alpha-j)}\left[\frac{1}{(\alpha-n)}\left(\frac{1}{2\sqrt{N}}+\frac{1}{N}\right)+\frac{1}{(\alpha+1)\delta}\left(\frac{1}{2\sqrt{N}}+\frac{1}{N}\right)\right]$$

(choosing $\delta := \left(\frac{1}{2\sqrt{N}}+\frac{1}{N}\right)^{\frac{1}{\alpha+1}}$, then $\delta^{\alpha+1}=\frac{1}{2\sqrt{N}}+\frac{1}{N}$, and $\delta^{\alpha}=\left(\frac{1}{2\sqrt{N}}+\frac{1}{N}\right)^{\frac{\alpha}{\alpha+1}}$)

$$=\frac{\omega_1\left({}^{x}T_{\alpha}f,\left(\frac{1}{2\sqrt{N}}+\frac{1}{N}\right)^{\frac{1}{\alpha+1}}\right)}{\prod_{j=0}^{n-1}(\alpha-j)}\cdot$$

$$\left[\frac{1}{(\alpha-n)}\left(\frac{1}{2\sqrt{N}}+\frac{1}{N}\right)+\frac{1}{(\alpha+1)}\left(\frac{1}{2\sqrt{N}}+\frac{1}{N}\right)^{\frac{\alpha}{\alpha+1}}\right], \tag{10.81}$$

proving the claim. ■

We also give

Theorem 10.40 *Let $\alpha \in (n, n+1)$, $n \in \mathbb{N}$, and $f \in C^{n+1}([0,1], \mathbb{R}_{+})$, $x \in [0,1]$ and $f^{(k)}(x) = 0$, $k = 1, \ldots, n$. Then*

$$\left|D_{N,\Gamma_N}(f)(x)-f(x)\right| \leq \frac{\omega_1\left({}^{x}T_{\alpha}f,\left(\frac{1+3\sqrt{2}}{2\sqrt{N}}+\frac{1}{N}\right)^{\frac{1}{\alpha+1}}\right)}{\prod_{j=0}^{n-1}(\alpha-j)}\cdot$$

$$\left[\frac{1}{(\alpha-n)}\left(\frac{1+3\sqrt{2}}{2\sqrt{N}}+\frac{1}{N}\right)+\frac{1}{(\alpha+1)}\left(\frac{1+3\sqrt{2}}{2\sqrt{N}}+\frac{1}{N}\right)^{\frac{\alpha}{\alpha+1}}\right], \tag{10.82}$$

$\forall\, N \in \mathbb{N} - \{1\}$.

We notice that $\lim_{N\to\infty} D_{N,\Gamma_N}(f)(x) = f(x)$.

Proof Using (10.29) we would have $(\delta > 0)$:

$$\left|D_{N,\Gamma_N}(f)(x)-f(x)\right| \leq \frac{\omega_1\left({}^{x}T_{\alpha}f,\delta\right)}{\prod_{j=0}^{n-1}(\alpha-j)}\cdot$$

$$\left[\frac{D_{N,\Gamma_N}\left(|\cdot - x|^{\alpha}\right)(x)}{(\alpha - n)} + \frac{D_{N,\Gamma_N}\left(|\cdot - x|^{\alpha+1}\right)(x)}{(\alpha+1)\,\delta}\right] \overset{(10.61)}{\leq}$$

$$\frac{\omega_1\left({}^{x}T_{\alpha}f, \delta\right)}{\prod_{j=0}^{n-1}(\alpha - j)}\left[\frac{1}{(\alpha - n)}\left[\frac{\left(1+\sqrt{2}\right)\sqrt{x(1-x)}+\sqrt{2x}}{\sqrt{N}} + \frac{1}{N}\right] + \right.$$

$$\left.\frac{1}{(\alpha+1)\,\delta}\left[\frac{\left(1+\sqrt{2}\right)\sqrt{x(1-x)}+\sqrt{2x}}{\sqrt{N}} + \frac{1}{N}\right]\right] \overset{(10.63)}{\leq}$$

$$\frac{\omega_1\left({}^{x}T_{\alpha}f, \delta\right)}{\prod_{j=0}^{n-1}(\alpha - j)}\left[\frac{1}{(\alpha - n)}\left(\frac{1+3\sqrt{2}}{2\sqrt{N}} + \frac{1}{N}\right) + \frac{1}{(\alpha+1)\,\delta}\left(\frac{1+3\sqrt{2}}{2\sqrt{N}} + \frac{1}{N}\right)\right] \tag{10.83}$$

(choosing $\delta := \left(\frac{1+3\sqrt{2}}{2\sqrt{N}} + \frac{1}{N}\right)^{\frac{1}{\alpha+1}}$, then $\delta^{\alpha+1} = \frac{1+3\sqrt{2}}{2\sqrt{N}} + \frac{1}{N}$, and $\delta^{\alpha} = \left(\frac{1+3\sqrt{2}}{2\sqrt{N}} + \frac{1}{N}\right)^{\frac{\alpha}{\alpha+1}}$)

$$= \frac{\omega_1\left({}^{x}T_{\alpha}f, \left(\frac{1+3\sqrt{2}}{2\sqrt{N}} + \frac{1}{N}\right)^{\frac{1}{\alpha+1}}\right)}{\prod_{j=0}^{n-1}(\alpha - j)} \cdot$$

$$\left[\frac{1}{(\alpha - n)}\left(\frac{1+3\sqrt{2}}{2\sqrt{N}} + \frac{1}{N}\right) + \frac{1}{(\alpha+1)}\left(\frac{1+3\sqrt{2}}{2\sqrt{N}} + \frac{1}{N}\right)^{\frac{\alpha}{\alpha+1}}\right], \tag{10.84}$$

$\forall\, N \in \mathbb{N} - \{1\}$,
proving the claim. ∎

We continue with

Theorem 10.41 *Let* $\frac{1}{n+1} < \alpha < 1$, $n \in \mathbb{N}$, $f : [0, 1] \to \mathbb{R}_+$, $f' \in L_\infty([0, 1])$, $x_0 \in [0, 1]$. *Assume that* $D^{k\alpha}_{*x_0} f \in C([x_0, 1])$, $k = 0, 1, \ldots, n+1$, *and* $\left(D^{i\alpha}_{*x_0} f\right)(x_0) = 0$, $i = 2, 3, \ldots, n+1$. *Also, suppose that* $D^{k\alpha}_{x_0-} f \in C([0, x_0])$, *for* $k = 0, 1, \ldots, n+1$, *and* $\left(D^{i\alpha}_{x_0-} f\right)(x_0) = 0$, *for* $i = 2, 3, \ldots, n+1$. *Denote by* $\lambda = (n+1)\,\alpha > 1$. *Then*

$$\left|K_{N,\mu}(f)(x_0) - f(x_0)\right| \leq \frac{\omega_1\left(D^{(n+1)\alpha}_{x_0} f, \left(\frac{1}{2\sqrt{N}} + \frac{1}{N}\right)^{\frac{1}{\lambda+1}}\right)}{\Gamma(\lambda+1)} \cdot$$

$$\left[\left(\frac{1}{2\sqrt{N}}+\frac{1}{N}\right)+\frac{1}{(\lambda+1)}\left(\frac{1}{2\sqrt{N}}+\frac{1}{N}\right)^{\frac{\lambda}{\lambda+1}}\right],\ \forall N\in\mathbb{N}. \tag{10.85}$$

We notice that $\lim_{N\to\infty}K_{N,\mu}(f)(x_0)=f(x_0)$.

Proof By (10.50) we get that

$$\left|K_{N,\mu}(f)(x_0)-f(x_0)\right|\leq\frac{\omega_1\left(D_{x_0}^{(n+1)\alpha}f,\delta\right)}{\Gamma(\lambda+1)}\cdot$$

$$\left[K_{N,\mu}\left(|\cdot-x_0|^{\lambda}\right)(x_0)+\frac{1}{(\lambda+1)\,\delta}K_{N,\mu}\left(|\cdot-x_0|^{\lambda+1}\right)(x_0)\right]\overset{\text{(by (10.56), (10.62))}}{\leq}$$

$$\frac{\omega_1\left(D_{x_0}^{(n+1)\alpha}f,\delta\right)}{\Gamma(\lambda+1)}\left[\left(\frac{1}{2\sqrt{N}}+\frac{1}{N}\right)+\frac{1}{(\lambda+1)\,\delta}\left(\frac{1}{2\sqrt{N}}+\frac{1}{N}\right)\right]$$

choosing $\delta:=\left(\frac{1}{2\sqrt{N}}+\frac{1}{N}\right)^{\frac{1}{\lambda+1}}$, then $\delta^{\lambda+1}=\frac{1}{2\sqrt{N}}+\frac{1}{N}$, and $\delta^{\lambda}=\left(\frac{1}{2\sqrt{N}}+\frac{1}{N}\right)^{\frac{\lambda}{\lambda+1}}$)

$$=\frac{\omega_1\left(D_{x_0}^{(n+1)\alpha}f,\left(\frac{1}{2\sqrt{N}}+\frac{1}{N}\right)^{\frac{1}{\lambda+1}}\right)}{\Gamma(\lambda+1)}\cdot$$

$$\left[\left(\frac{1}{2\sqrt{N}}+\frac{1}{N}\right)+\frac{1}{(\lambda+1)}\left(\frac{1}{2\sqrt{N}}+\frac{1}{N}\right)^{\frac{\lambda}{\lambda+1}}\right], \tag{10.86}$$

proving the claim. ∎

Theorem 10.42 *Same assumptions as in Theorem 10.41. Then*

$$\left|D_{N,\Gamma_N}(f)(x_0)-f(x_0)\right|\leq\frac{\omega_1\left(D_{x_0}^{(n+1)\alpha}f,\left(\frac{1+3\sqrt{2}}{2\sqrt{N}}+\frac{1}{N}\right)^{\frac{1}{\lambda+1}}\right)}{\Gamma(\lambda+1)}\cdot$$

$$\left[\left(\frac{1+3\sqrt{2}}{2\sqrt{N}}+\frac{1}{N}\right)+\frac{1}{(\lambda+1)}\left(\frac{1+3\sqrt{2}}{2\sqrt{N}}+\frac{1}{N}\right)^{\frac{\lambda}{\lambda+1}}\right],\ \forall N\in\mathbb{N}-\{1\}. \tag{10.87}$$

We notice that $\lim_{N\to\infty}D_{N,\Gamma_N}(f)(x_0)=f(x_0)$.

Proof By (10.50) we get that

$$\left|D_{N,\Gamma_N}(f)(x_0)-f(x_0)\right|\leq\frac{\omega_1\left(D_{x_0}^{(n+1)\alpha}f,\delta\right)}{\Gamma(\lambda+1)}\cdot$$

$$\left[D_{N,\Gamma_N}\left(|\cdot - x_0|^{\lambda}\right)(x_0) + \frac{1}{(\lambda+1)\delta} D_{N,\Gamma_N}\left(|\cdot - x_0|^{\lambda+1}\right)(x_0)\right] \overset{\text{(by (10.61), (10.63))}}{\leq}$$

$$\frac{\omega_1\left(D_{x_0}^{(n+1)\alpha} f, \delta\right)}{\Gamma(\lambda+1)}\left[\left(\frac{1+3\sqrt{2}}{2\sqrt{N}} + \frac{1}{N}\right) + \frac{1}{(\lambda+1)\delta}\left(\frac{1+3\sqrt{2}}{2\sqrt{N}} + \frac{1}{N}\right)\right] \tag{10.88}$$

(choosing $\delta := \left(\frac{1+3\sqrt{2}}{2\sqrt{N}} + \frac{1}{N}\right)^{\frac{1}{\lambda+1}}$, then $\delta^{\lambda+1} = \frac{1+3\sqrt{2}}{2\sqrt{N}} + \frac{1}{N}$, and $\delta^{\lambda} = \left(\frac{1+3\sqrt{2}}{2\sqrt{N}} + \frac{1}{N}\right)^{\frac{\lambda}{\lambda+1}}$)

$$= \frac{\omega_1\left(D_{x_0}^{(n+1)\alpha} f, \left(\frac{1+3\sqrt{2}}{2\sqrt{N}} + \frac{1}{N}\right)^{\frac{1}{\lambda+1}}\right)}{\Gamma(\lambda+1)} \cdot$$

$$\left[\left(\frac{1+3\sqrt{2}}{2\sqrt{N}} + \frac{1}{N}\right) + \frac{1}{(\lambda+1)}\left(\frac{1+3\sqrt{2}}{2\sqrt{N}} + \frac{1}{N}\right)^{\frac{\lambda}{\lambda+1}}\right],$$

$\forall\, N \in \mathbb{N} - \{1\}$, proving the claim. ■

Next we are acting according to Theorem 10.36 and Remark 10.37.
We give

Theorem 10.43 *Let $\alpha \in (n, n+1)$, $n \in \mathbb{N}$, and $f \in C^{n+1}([a,b], \mathbb{R}_+)$, $x \in [a,b]$ and $f^{(k)}(x) = 0$, $k = 1, \ldots, n$. Let $L_N : \mathcal{L}_{\infty}(\mathcal{B}([a,b])) \to C_+([a,b])$, $\forall\, N \in \mathbb{N}$, be positive sublinear comonotonic operators, such that $L_N(1) = 1$, $\forall\, N \in \mathbb{N}$. Then*

$$|L_N(f)(x) - f(x)| \leq \frac{\omega_1\left({}^{x}T_{\alpha} f, \delta\right)}{\prod_{j=0}^{n-1}(\alpha - j)} \cdot \tag{10.89}$$

$$\left[\frac{1}{(\alpha-n)}\left((C)\int_a^b |t-x|^{\alpha}\, d\gamma_{N,x}(t)\right) + \frac{1}{(\alpha+1)\delta}\left((C)\int_a^b |t-x|^{\alpha+1}\, d\gamma_{N,x}(t)\right)\right],$$

$\delta > 0$.

Proof By Theorem 10.14. ■

Theorem 10.44 *Let $f \in C^1([a,b], \mathbb{R}_+)$, $\alpha \in (0,1)$, and let $L_N : \mathcal{L}_{\infty}(\mathcal{B}([a,b])) \to C_+([a,b])$, $\forall\, N \in \mathbb{N}$, be positive sublinear comonotonic operators, such that $L_N(1) = 1$, $\forall\, N \in \mathbb{N}$. Then*

$$|L_N(f)(x) - f(x)| \leq \omega_1\left({}^{x}T_{\alpha} f, \delta\right) \cdot$$

$$\left[\frac{1}{\alpha}\left((C)\int_a^b |t-x|^{\alpha} d\gamma_{N,x}(t)\right)+\frac{1}{(\alpha+1)\delta}\left((C)\int_a^b |t-x|^{\alpha+1} d\gamma_{N,x}(t)\right)\right]; \tag{10.90}$$

$\delta > 0$, $\forall\, N \in \mathbb{N}$, $\forall\, x \in [a,b]$.

Proof By Theorem 10.16. ■

Theorem 10.45 *Let* $f \in C^1([a,b], \mathbb{R}_+)$, $\alpha \in (0,1)$, $x \in [a,b]$. *Let* $L_N : \mathcal{L}_\infty(\mathcal{B}([a,b])) \to C_+([a,b])$, $\forall\, N \in \mathbb{N}$, *be positive sublinear comonotonic operators, such that* $L_N(1) = 1$, $\forall\, N \in \mathbb{N}$, *and* $(C)\int_a^b |t-x|^{\alpha+1} d\gamma_{N,x}(t) > 0$, $\forall\, N \in \mathbb{N}$. *Then*

$$|L_N(f)(x) - f(x)| \le \frac{(2\alpha+1)}{\alpha(\alpha+1)}\omega_1\left({}^{x}T_\alpha f, \left((C)\int_a^b |t-x|^{\alpha+1} d\gamma_{N,x}(t)\right)^{\frac{1}{\alpha+1}}\right)\cdot$$

$$\left((C)\int_a^b |t-x|^{\alpha+1} d\gamma_{N,x}(t)\right)^{\frac{\alpha}{\alpha+1}}, \ \forall\, N \in \mathbb{N}. \tag{10.91}$$

If $(C)\int_a^b |t-x|^{\alpha+1} d\gamma_{N,x}(t) \to 0$, *then* $L_N(f)(x) \to f(x)$, *as* $N \to \infty$.

Proof By Theorem 10.17. ■

We continue with

Theorem 10.46 *Let* $\frac{1}{n+1} < \alpha < 1$, $n \in \mathbb{N}$, $f : [a,b] \to \mathbb{R}_+$, $f' \in L_\infty([a,b])$, $x_0 \in [a,b]$. *Assume that* $D_{*x_0}^{k\alpha} f \in C([x_0,b])$, $k = 0, 1, \ldots, n+1$, *and* $\left(D_{*x_0}^{i\alpha} f\right)(x_0) = 0$, $i = 2, 3, \ldots, n+1$. *Also, suppose that* $D_{x_0-}^{k\alpha} f \in C([a,x_0])$, *for* $k = 0, 1, \ldots, n+1$, *and* $\left(D_{x_0-}^{i\alpha} f\right)(x_0) = 0$, *for* $i = 2, 3, \ldots, n+1$. *Denote* $\lambda = (n+1)\alpha > 1$. *Let* $L_N : \mathcal{L}_\infty(\mathcal{B}([a,b])) \to C_+([a,b])$, $\forall\, N \in \mathbb{N}$, *be positive sublinear comonotonic operators, such that* $L_N(1) = 1$, $\forall\, N \in \mathbb{N}$. *Then*

$$|L_N(f)(x_0) - f(x_0)| \le \frac{\omega_1\left(D_{x_0}^{(n+1)\alpha} f, \delta\right)}{\Gamma(\lambda+1)}\cdot \tag{10.92}$$

$$\left[\left((C)\int_a^b |t-x|^{\lambda} d\gamma_{N,x_0}(t)\right)+\frac{1}{(\lambda+1)\delta}\left((C)\int_a^b |t-x|^{\lambda+1} d\gamma_{N,x_0}(t)\right)\right],$$

$\delta > 0$, $\forall\, N \in \mathbb{N}$.

Proof By Theorem 10.22. ■

Note: Theorem 10.46 is true also when $0 < \alpha \le \frac{1}{n+1}$.

Theorem 10.47 *Let* $0 < \alpha \le \frac{1}{n+1}$, $n \in \mathbb{N}$, $f : [a,b] \to \mathbb{R}_+$, $f' \in L_\infty([a,b])$, $x_0 \in [a,b]$. *Assume that* $D_{*x_0}^{k\alpha} f \in C([x_0,b])$, $k = 0, 1, \ldots, n+1$, *and*

$\left(D_{*x_0}^{i\alpha} f\right)(x_0) = 0$, $i = 2, 3, \ldots, n+1$. *Also, suppose that* $D_{x_0-}^{k\alpha} f \in C([a, x_0])$, *for* $k = 0, 1, \ldots, n+1$, *and* $\left(D_{x_0-}^{i\alpha} f\right)(x_0) = 0$, *for* $i = 2, 3, \ldots, n+1$. *Denote* $\lambda = (n+1)\alpha \leq 1$. *Let* $L_N : \mathcal{L}_\infty(\mathcal{B}([a, b])) \to C_+([a, b])$, $\forall\, N \in \mathbb{N}$, *be positive sublinear comonotonic operators, such that* $L_N(1) = 1$, $\forall\, N \in \mathbb{N}$, *and* $(C)\int_a^b |t - x_0|^{\lambda+1} d\gamma_{N,x_0}(t) > 0$, $\forall\, N \in \mathbb{N}$. *Then*

$$|L_N(f)(x_0) - f(x_0)| \leq \frac{\omega_1\left(D_{x_0}^{(n+1)\alpha} f, \delta\right)}{\Gamma(\lambda+1)} \cdot \tag{10.93}$$

$$\left[\left((C)\int_a^b |t - x_0|^{\lambda+1} d\gamma_{N,x_0}(t)\right)^{\frac{\lambda}{\lambda+1}} + \frac{1}{(\lambda+1)\delta}\left((C)\int_a^b |t - x_0|^{\lambda+1} d\gamma_{N,x_0}(t)\right)\right],$$

$\delta > 0$, $\forall\, N \in \mathbb{N}$.

Proof By Theorem 10.23. ∎

Theorem 10.48 *All as in Theorem 10.47. Then*

$$|L_N(f)(x_0) - f(x_0)| \leq \frac{(\lambda+2)\,\omega_1\left(D_{x_0}^{(n+1)\alpha} f, \left((C)\int_a^b |t - x_0|^{\lambda+1} d\gamma_{N,x_0}(t)\right)^{\frac{1}{\lambda+1}}\right)}{\Gamma(\lambda+2)}$$

$$\left((C)\int_a^b |t - x_0|^{\lambda+1} d\gamma_{N,x_0}(t)\right)^{\frac{\lambda}{\lambda+1}}, \quad \forall\, N \in \mathbb{N}. \tag{10.94}$$

If $(C)\int_a^b |t - x_0|^{\lambda+1} d\gamma_{N,x_0}(t) \to 0$, *then* $L_N(f)(x_0) \to f(x_0)$, *as* $N \to \infty$.

Proof By Theorem 10.24. ∎

Next we present results based on Remark 10.38.

Theorem 10.49 *All as in Theorem 10.14, with* $L_N = T_N$, $\forall\, N \in \mathbb{N}$. *Then*

$$|T_N(f)(x) - f(x)| \leq \frac{\omega_1\left({}^xT_\alpha f, \delta\right)}{\prod_{j=0}^{n-1}(\alpha - j)} \cdot$$

$$\left[\frac{1}{(\alpha-n)}\left((C)\int_a^b |t - x|^{\alpha} d\mu_{N,x}(t)\right) + \frac{1}{(\alpha+1)\delta}\left((C)\int_a^b |t - x|^{\alpha+1} d\mu_{N,x}(t)\right)\right], \tag{10.95}$$

$\delta > 0$, $\forall\, N \in \mathbb{N}$.

Proof By Theorem 10.14. ∎

Theorem 10.50 *All as in Theorem 10.16, with $L_N = T_N$, $\forall\, N \in \mathbb{N}$. Then*

$$|T_N(f)(x) - f(x)| \leq \omega_1\left({}^{x}T_{\alpha} f, \delta\right) \cdot$$

$$\left[\frac{1}{\alpha}\left((C)\int_a^b |t-x|^{\alpha}\, d\mu_{N,x}(t)\right) + \frac{1}{(\alpha+1)\,\delta}\left((C)\int_a^b |t-x|^{\alpha+1}\, d\mu_{N,x}(t)\right)\right], \tag{10.96}$$

$\forall\, N \in \mathbb{N}$, $\forall\, x \in [a,b]$, $\delta > 0$.

Proof By Theorem 10.16. ■

Theorem 10.51 *All as in Theorem 10.17, with $L_N = T_N$, with $(C)\int_a^b |t-x|^{\alpha+1} d\mu_{N,x}(t) > 0$, $\forall\, N \in \mathbb{N}$. Then*

$$|T_N(f)(x) - f(x)| \leq \frac{(2\alpha+1)}{\alpha(\alpha+1)}\omega_1\left({}^{x}T_{\alpha} f, \left((C)\int_a^b |t-x|^{\alpha+1}\, d\mu_{N,x}(t)\right)^{\frac{1}{\alpha+1}}\right)$$

$$\cdot\left((C)\int_a^b |t-x|^{\alpha+1}\, d\mu_{N,x}(t)\right)^{\frac{\alpha}{\alpha+1}}, \tag{10.97}$$

$\forall\, N \in \mathbb{N}$.

If $(C)\int_a^b |t-x|^{\alpha+1} d\mu_{N,x}(t) \to 0$, then $T_N(f)(x) \to f(x)$, as $N \to \infty$.

Proof By Theorem 10.17. ■

We continue with

Theorem 10.52 *All as in Theorem 10.22, with $L_N = T_N$, $\forall\, N \in \mathbb{N}$. Then*

$$|T_N(f)(x_0) - f(x_0)| \leq \frac{\omega_1\left(D_{x_0}^{(n+1)\alpha} f, \delta\right)}{\Gamma(\lambda+1)} \cdot$$

$$\left[(C)\int_a^b |t-x_0|^{\lambda}\, d\mu_{N,x_0}(t) + \frac{1}{(\lambda+1)\,\delta}(C)\int_a^b |t-x_0|^{\lambda+1}\, d\mu_{N,x_0}(t)\right], \tag{10.98}$$

$\delta > 0$, $\forall\, N \in \mathbb{N}$.

Proof By Theorem 10.22. ■

Note: Theorem 10.52 is also valid when $0 < \alpha \leq \frac{1}{n+1}$.

Theorem 10.53 *All as in Theorem 10.23, with $L_N = T_N$, with $(C)\int_a^b |t-x_0|^{\lambda+1} d\mu_{N,x_0}(t) > 0$, $\forall\, N \in \mathbb{N}$. Then*

$$|T_N(f)(x_0) - f(x_0)| \leq \frac{\omega_1\left(D_{x_0}^{(n+1)\alpha} f, \delta\right)}{\Gamma(\lambda+1)} \cdot$$

$$\left[\left((C)\int_a^b |t-x_0|^{\lambda+1}\,d\mu_{N,x_0}(t)\right)^{\frac{\lambda}{\lambda+1}}+\frac{1}{(\lambda+1)\,\delta}\left((C)\int_a^b |t-x_0|^{\lambda+1}\,d\mu_{N,x_0}(t)\right)\right], \tag{10.99}$$

$\delta>0$, $\forall\, N\in\mathbb{N}$.

Proof By Theorem 10.23. ■

Theorem 10.54 *All as in Theorem 10.53. Then*

$$|T_N(f)(x_0)-f(x_0)|\leq\frac{(\lambda+2)\,\omega_1\left(D_{x_0}^{(n+1)\alpha}f,\left((C)\int_a^b |t-x_0|^{\lambda+1}\,d\mu_{N,x_0}(t)\right)^{\frac{1}{\lambda+1}}\right)}{\Gamma(\lambda+2)}$$

$$\cdot\left((C)\int_a^b |t-x_0|^{\lambda+1}\,d\mu_{N,x_0}(t)\right)^{\frac{\lambda}{\lambda+1}},\ \forall\, N\in\mathbb{N}. \tag{10.100}$$

If $(C)\int_a^b |t-x_0|^{\lambda+1}\,d\mu_{N,x_0}(t)\to 0$, *then* $T_N(f)(x_0)\to f(x_0)$, *as* $N\to\infty$.

Proof By Theorem 10.24. ■

References

1. T. Abdeljawad, On conformable fractional calculus. J. Comput. Appl. Math. **279**, 57–66 (2015)
2. G. Anastassiou, Iterated fractional approximation by Max-Product operators (2017). submitted
3. G. Anastassiou, *Mixed Conformable and Iterated fractional Approximation by Choquet integrals*, Progress in Fractional Differentiation and Applications (2018). Accepted
4. G. Anastassiou, Advanced fractional Taylor's formulae. J. Comput. Anal. Appl. **21**(7), 1185–1204 (2016)
5. G. Anastassiou, Mixed conformable fractional approximation by sublinear operators. Indian J. Math. **60**(1), 107–140 (2018)
6. G. Anastassiou, I. Argyros, *Intelligent Numerical Methods: Applications to Fractional Calculus* (Springer, New York, 2016)
7. G. Choquet, Theory of capacities. Ann. Inst. Fourier (Grenoble) **5**, 131–295 (1954)
8. D. Denneberg, *Non-additive Measure and Integral* (Kluwer, Dordrecht, 1994)
9. D. Dubois, H. Prade, *Possibility Theory* (Plenum Press, New York, 1988)
10. S. Gal, Uniform and Pointwise quantitative approximation by Kantorovich-Choquet type integral Operators with respect to monotone and submodular set functions. Mediterr. J. Math. **14**(5), 12 (2017). Art. 205
11. S. Gal, S. Trifa, Quantitative estimates in uniform and pointwise approximation by Bernstein-Durrmeyer-Choquet operators. Carpathian J. Math. **33**(1), 49–58 (2017)
12. R. Khalil, M. Al Horani, A. Yousef, M. Sababheh, A new definition of fractional derivative. J. Comput. Appl. Math. **264**, 65–70 (2014)
13. Z.M. Odibat, N.J. Shawagleh, Generalized Taylor's formula. Appl. Math. Comput. **186**, 286–293 (2007)
14. D. Schmeidler, Integral representation without additivity. Proc. Am. Math. Soc. **97**, 255–261 (1986)

15. D. Schmeidler, Subjective probability and expected utility without additivity. Econometrica **57**, 571–587 (1989)
16. L.S. Shapley, A value for n-person games, in *Contributions to the Theory of Games*, Annals of mathematical studies 28, ed. by H.W. Kuhn, A.W. Tucker (Princeton University Press, 1953), pp. 307–317
17. Z. Wang, G.J. Klir, *Generalized Measure Theory* (Springer, New York, 2009)

Chapter 11
Approximation by a Kantorovich–Shilkret Quasi-interpolation Neural Network Operator

In this chapter we present multivariate basic approximation by a Kantorovich–Shilkret type quasi-interpolation neural network operator with respect to supremum norm. This is done with rates using the multivariate modulus of continuity. We approximate continuous and bounded functions on $\mathbb{R}^N$, $N \in \mathbb{N}$. When they are additionally uniformly continuous we derive pointwise and uniform convergences. It follows [3].

11.1 Introduction

The author here performs multivariate error function based neural network approximation to continuous functions over $\mathbb{R}^N$, $N \in \mathbb{N}$, and then he extends his results to complex valued functions. The convergences here are with rates expressed via the multivariate modulus of continuity of the involved function and give by very tight Jackson type inequalities.

The author comes up with the "right" precisely defined flexible quasi-interpolation Baskakov-Shilkret type integral coefficient neural network operator associated to the error function.

Feed-forward neural network (FNNs) with one hidden layer with deal with, are expressed mathematical as

$$N_n(x) = \sum_{j=0}^{n} c_j \sigma\left(\langle a_j \cdot x \rangle + b_j\right), \quad x \in \mathbb{R}^s, \ s \in \mathbb{N},$$

where for $0 \leq j \leq n$, $b_j \in \mathbb{R}$ are the thresholds, $a_j \in \mathbb{R}^s$ are the connection weights, $c_j \in \mathbb{R}$ are the coefficients, $\langle a_j \cdot x \rangle$ is the inner product of a_j and x, and σ is the activation function of the network. In many fundamental neural network models the activation function is error function generated.

G. A. Anastassiou, *Ordinary and Fractional Approximation by Non-additive Integrals: Choquet, Shilkret and Sugeno Integral Approximators*, Studies in Systems, Decision and Control 190, https://doi.org/10.1007/978-3-030-04287-5_11

About neural networks in general you may read [5–7]. In recent years non-additive integrals, like the N. Shilkret one [8], have become fashionable and more useful in Economic theory, etc.

11.2 Background

Here we follow [8].

Let $\mathcal{F}$ be a σ-field of subsets of an arbitrary set Ω. An extended non-negative real valued function μ on $\mathcal{F}$ is called maxitive if $\mu(\varnothing) = 0$ and

$$\mu\left(\cup_{i\in I} E_i\right) = \sup_{i\in I} \mu\left(E_i\right), \tag{11.1}$$

where the set I is of cardinality at most countable. We also call μ a maxitive measure. Here f stands for a non-negative measurable function on Ω. In [8], Niel Shilkret developed his non-additive integral defined as follows:

$$\left(N^*\right) \int_D f d\mu := \sup_{y\in Y} \left\{ y \cdot \mu\left(D \cap \{f \geq y\}\right)\right\}, \tag{11.2}$$

where $Y = [0, m]$ or $Y = [0, m)$ with $0 < m \leq \infty$, and $D \in \mathcal{F}$. Here we take $Y = [0, \infty)$.

It is easily proved that

$$\left(N^*\right) \int_D f d\mu = \sup_{y>0} \left\{ y \cdot \mu\left(D \cap \{f > y\}\right)\right\}. \tag{11.3}$$

The Shilkret integral takes values in $[0, \infty]$.

The Shilkret integral [8] has the following properties:

$$\left(N^*\right) \int_\Omega \chi_E d\mu = \mu(E), \tag{11.4}$$

where χ_E is the indicator function on $E \in \mathcal{F}$,

$$(N*) \int_D c f d\mu = c \left(N^*\right) \int_D f d\mu, \ c \geq 0, \tag{11.5}$$

$$\left(N^*\right) \int_D \sup_{n\in\mathbb{N}} f_n d\mu = \sup_{n\in\mathbb{N}} \left(N^*\right) \int_D f_n d\mu, \tag{11.6}$$

where f_n, $n \in \mathbb{N}$, is an increasing sequence of elementary (countably valued) functions converging uniformly to f. Furthermore we have

$$\left(N^{*}\right)\int_{D} f d\mu \geq 0, \tag{11.7}$$

$$f \geq g \text{ implies } \left(N^{*}\right)\int_{D} f d\mu \geq \left(N^{*}\right)\int_{D} g d\mu, \tag{11.8}$$

where $f, g : \Omega \to [0, \infty]$ are measurable.

Let $a \leq f(\omega) \leq b$ for almost every $\omega \in E$, then

$$a\mu(E) \leq \left(N^{*}\right)\int_{E} f d\mu \leq b\mu(E);$$

$$\left(N^{*}\right)\int_{E} 1 d\mu = \mu(E);$$

$f > 0$ almost everywhere and $(N^{*})\int_{E} f d\mu = 0$ imply $\mu(E) = 0$;

$(N^{*})\int_{\Omega} f d\mu = 0$ if and only $f = 0$ almost everywhere;

$(N^{*})\int_{\Omega} f d\mu < \infty$ implies that

$$\overline{N}(f) := \{\omega \in \Omega | f(\omega) \neq 0\} \text{ has } \sigma\text{-finite measure}; \tag{11.9}$$

$$\left(N^{*}\right)\int_{D} (f+g) d\mu \leq \left(N^{*}\right)\int_{D} f d\mu + \left(N^{*}\right)\int_{D} g d\mu;$$

and

$$\left|\left(N^{*}\right)\int_{D} f d\mu - \left(N^{*}\right)\int_{D} g d\mu\right| \leq \left(N^{*}\right)\int_{D} |f-g| d\mu. \tag{11.10}$$

From now on in this chapter we assume that $\mu : \mathcal{F} \to [0, +\infty)$.

11.3 Main Results

We consider here the (Gauss) error special function ([1, 4])

$$erf(x) = \frac{2}{\sqrt{\pi}}\int_{0}^{x} e^{-t^{2}} dt, \quad x \in \mathbb{R}, \tag{11.11}$$

which is a sigmoidal type function and a strictly increasing function.

It has the properties

$$erf(0) = 0, \; erf(-x) = erf(x), erf(+\infty) = 1, \; erf(-\infty) = -1,$$

and

$$(erf\ (x))' = \frac{2}{\sqrt{\pi}} e^{-x^2}, \ x \in \mathbb{R},$$

$$\int erf\ (x)\, dx = x erf\ (x) + \frac{e^{-x^2}}{\sqrt{\pi}} + C,$$

where C is a constant.

The error function is related to the cumulative probability distribution function of the standard normal distribution

$$\Phi (x) = \frac{1}{2} + \frac{1}{2} erf \left(\frac{x}{\sqrt{2}} \right).$$

We consider the activation function

$$\chi (x) = \frac{1}{4} \left(erf\ (x+1) - erf\ (x-1) \right), \ x \in \mathbb{R}, \tag{11.12}$$

and we notice that

$$\chi (-x) = \chi (x), \tag{11.13}$$

and even function.

Clearly $\chi (x) > 0$, all $x \in \mathbb{R}$.

We see that

$$\chi (0) = \frac{erf\ (1)}{2} \simeq 0.4215. \tag{11.14}$$

Let $x > 0$, we have that

$$\chi' (x) < 0, \ \text{for } x > 0. \tag{11.15}$$

That is χ is strictly decreasing on $[0, \infty)$ and is strictly increasing on $(-\infty, 0]$, and $\chi' (0) = 0$.

Clearly the x-axis is the horizontal asymptote on χ.

Conclusion, χ is a bell symmetric function with maximum $\chi (0) \simeq 0.4215$.

We further need

Theorem 11.1 ([2]) *We have that*

$$\sum_{i=-\infty}^{\infty} \chi (x - i) = 1, \ all\ x \in \mathbb{R}. \tag{11.16}$$

Theorem 11.2 ([2]) *It holds*

$$\int_{-\infty}^{\infty} \chi(x)\, dx = 1. \tag{11.17}$$

So $\chi(x)$ is a density function on $\mathbb{R}$.

Theorem 11.3 ([2]) *Let* $0 < \alpha < 1$, *and* $n \in \mathbb{N}$ *with* $n^{1-\alpha} \geq 3$.

It holds

$$\sum_{\begin{cases} k=-\infty \\ : |nx-k| \geq n^{1-\alpha} \end{cases}}^{\infty} \chi(nx-k) < \frac{1}{2\sqrt{\pi}\left(n^{1-\alpha}-2\right) e^{\left(n^{1-\alpha}-2\right)^2}}. \tag{11.18}$$

Remark 11.4 We introduce

$$Z(x_1, ..., x_N) := Z(x) := \prod_{i=1}^{N} \chi(x_i), \tag{11.19}$$

$x = (x_1, ..., x_N) \in \mathbb{R}^N$, $N \in \mathbb{N}$.

It has the properties:

(i)

$$Z(x) > 0, \ \forall\, x \in \mathbb{R}^N, \tag{11.20}$$

(ii)

$$\sum_{k=-\infty}^{\infty} Z(x-k) := \sum_{k_1=-\infty}^{\infty} \sum_{k_2=-\infty}^{\infty} \cdots \sum_{k_N=-\infty}^{\infty} Z(x_1-k_1, ..., x_N-k_N) = 1, \tag{11.21}$$

where $k := (k_1, ..., k_N) \in \mathbb{Z}^N$, $\forall\, x \in \mathbb{R}^N$,

hence

(iii)

$$\sum_{k=-\infty}^{\infty} Z(nx-k) = 1, \forall\, x \in \mathbb{R}^N, n \in \mathbb{N}, \tag{11.22}$$

and

(iv)

$$\int_{\mathbb{R}^N} Z(x)\, dx = 1, \tag{11.23}$$

that is Z is a multivariate density function.

Here $\|x\|_\infty := \max\{|x_1|, ..., |x_N|\}$, $x \in \mathbb{R}^N$, also set $\infty := (\infty, ..., \infty)$, $-\infty = (-\infty, ..., -\infty)$ upon the multivariate context.

It is also clear that (see (11.18))
(v)

$$\sum_{\begin{cases} k=-\infty \\ \left\| \frac{k}{n} - x \right\|_{\infty} > \frac{1}{n^{\beta}} \end{cases}}^{\infty} Z(nx-k) \leq \frac{1}{2\sqrt{\pi}\left(n^{1-\beta}-2\right)e^{\left(n^{1-\beta}-2\right)^{2}}}, \tag{11.24}$$

$0 < \beta < 1, n \in \mathbb{N} : n^{1-\beta} \geq 3, x \in \mathbb{R}^{N}$.

For $f \in C_{B}^{+}\left(\mathbb{R}^{N}\right)$ (continuous and bounded functions from $\mathbb{R}^{N}$ into $\mathbb{R}_{+}$), we define the first modulus of continuity

$$\omega_{1}(f,h) := \sup_{\substack{x,y\in\mathbb{R}^{N} \\ \|x-y\|_{\infty}\leq h}} |f(x)-f(y)|, \quad h>0. \tag{11.25}$$

Given that $f \in C_{U}^{+}\left(\mathbb{R}^{N}\right)$ (uniformly continuous from $\mathbb{R}^{N}$ into $\mathbb{R}_{+}$), we have that

$$\lim_{h\to 0}\omega_{1}(f,h) = 0. \tag{11.26}$$

We make

Definition 11.5 Let $\mathcal{L}$ be the Lebesgue σ-algebra on $\mathbb{R}^{N}$, $N \in \mathbb{N}$, and the maxitive measure $\mu : \mathcal{L} \to [0, +\infty)$, such that for any $A \in \mathcal{L}$ with $A \neq \varnothing$, we get $\mu(A) > 0$.

For $f \in C_{B}^{+}\left(\mathbb{R}^{N}\right)$, we define the multivariate Kantorovich–Shilkret type neural network operator for any $x \in \mathbb{R}^{N}$:

$$T_{n}^{\mu}(f,x) = T_{n}^{\mu}(f,x_{1},...,x_{N}) :=$$

$$\sum_{k=-\infty}^{\infty}\left(\frac{(N^{*})\int_{\left[0,\frac{1}{N}\right]^{N}} f\left(t+\frac{k}{n}\right)d\mu(t)}{\mu\left(\left[0,\frac{1}{n}\right]^{N}\right)}\right)Z(nx-k) =$$

$$\sum_{k_{1}=-\infty}^{\infty}\sum_{k_{2}=-\infty}^{\infty}\cdots\sum_{k_{N}=-\infty}^{\infty}\left(\frac{(N^{*})\int_{0}^{\frac{1}{n}}\cdots\int_{0}^{\frac{1}{n}} f\left(t_{1}+\frac{k_{1}}{n},t_{2}+\frac{k_{2}}{n},...,t_{N}+\frac{k_{N}}{n}\right)d\mu(t_{1},...,t_{N})}{\mu\left(\left[0,\frac{1}{n}\right]^{N}\right)}\right) \tag{11.27}$$

$$\cdot\left(\prod_{i=1}^{N} Z(nx_{i}-k_{i})\right),$$

where $x = (x_{1},...,x_{N}) \in \mathbb{R}^{N}$, $k = (k_{1},...,k_{N})$, $t = (t_{1},...,t_{N})$, $n \in \mathbb{N}$.

Clearly here $\mu\left(\left[0,\frac{1}{n}\right]^{N}\right) > 0$, $\forall\, n \in \mathbb{N}$.

Above we notice that

$$\left\| T_n^{\mu}(f) \right\|_{\infty} \leq \|f\|_{\infty}, \tag{11.28}$$

so that $T_n^{\mu}(f, x)$ is well-defined.

Remark 11.6 Let $t \in \left[0, \frac{1}{n}\right]^N$ and $x \in \mathbb{R}^N$, then

$$f\left(t + \frac{k}{n}\right) = f\left(t + \frac{k}{n}\right) - f(x) + f(x) \leq \left| f\left(t + \frac{k}{n}\right) - f(x) \right| + f(x), \tag{11.29}$$

hence

$$(N^*) \int_{[0,\frac{1}{n}]^N} f\left(t + \frac{k}{n}\right) d\mu(t) \leq$$

$$(N^*) \int_{[0,\frac{1}{n}]^N} \left| f\left(t + \frac{k}{n}\right) - f(x) \right| d\mu(t) + f(x)\, \mu\left(\left[0, \frac{1}{n}\right]^N\right). \tag{11.30}$$

That is

$$(N^*) \int_{[0,\frac{1}{n}]^N} f\left(t + \frac{k}{n}\right) d\mu(t) - f(x)\, \mu\left(\left[0, \frac{1}{n}\right]^N\right) \leq \tag{11.31}$$

$$(N^*) \int_{[0,\frac{1}{n}]^N} \left| f\left(t + \frac{k}{n}\right) - f(x) \right| d\mu(t).$$

Similarly we have

$$f(x) = f(x) - f\left(t + \frac{k}{n}\right) + f\left(t + \frac{k}{n}\right) \leq \left| f\left(t + \frac{k}{n}\right) - f(x) \right| + f\left(t + \frac{k}{n}\right),$$

hence

$$(N^*) \int_{[0,\frac{1}{n}]^N} f(x)\, d\mu(t) \leq (N^*) \int_{[0,\frac{1}{n}]^N} \left| f\left(t + \frac{k}{n}\right) - f(x) \right| d\mu(t)$$

$$+ (N^*) \int_{[0,\frac{1}{n}]^N} f\left(t + \frac{k}{n}\right) d\mu(t).$$

That is

$$f(x)\, \mu\left(\left[0, \frac{1}{n}\right]^N\right) - (N^*) \int_{[0,\frac{1}{n}]^N} f\left(t + \frac{k}{n}\right) d\mu(t) \leq \tag{11.32}$$

$$\left(N^*\right)\int_{\left[0,\frac{1}{n}\right]^N}\left|f\left(t+\frac{k}{n}\right)-f(x)\right|d\mu(t).$$

By (11.31) and (11.32) we derive

$$\left|\left(N^*\right)\int_{\left[0,\frac{1}{n}\right]^N}f\left(t+\frac{k}{n}\right)d\mu(t)-f(x)\,\mu\left(\left[0,\frac{1}{n}\right]^N\right)\right|\leq$$

$$\left(N^*\right)\int_{\left[0,\frac{1}{n}\right]^N}\left|f\left(t+\frac{k}{n}\right)-f(x)\right|d\mu(t). \tag{11.33}$$

In particular it holds

$$\left|\frac{\left(N^*\right)\int_{\left[0,\frac{1}{n}\right]^N}f\left(t+\frac{k}{n}\right)d\mu(t)}{\mu\left(\left[0,\frac{1}{n}\right]^N\right)}-f(x)\right|\leq$$

$$\frac{\left(N^*\right)\int_{\left[0,\frac{1}{n}\right]^N}\left|f\left(t+\frac{k}{n}\right)-f(x)\right|d\mu(t)}{\mu\left(\left[0,\frac{1}{n}\right]^N\right)}. \tag{11.34}$$

We present

Theorem 11.7 *Let $f\in C_B^+\left(\mathbb{R}^N\right)$, $0<\beta<1$, $x\in\mathbb{R}^N$; $N,n\in\mathbb{N}$ with $n^{1-\beta}\geq 3$. Then*

(i)

$$\sup_{\mu}\left|T_n^{\mu}(f,x)-f(x)\right|\leq\omega_1\left(f,\frac{1}{n}+\frac{1}{n^{\beta}}\right)+\frac{\|f\|_{\infty}}{\sqrt{\pi}\left(n^{1-\beta}-2\right)e^{\left(n^{1-\beta}-2\right)^2}}=:\lambda_n, \tag{11.35}$$

(ii)

$$\sup_{\mu}\left\|T_n^{\mu}(f)-f\right\|_{\infty}\leq\lambda_n. \tag{11.36}$$

Given that $f\in\left(C_U^+\left(\mathbb{R}^N\right)\cap C_B^+\left(\mathbb{R}^N\right)\right)$, we obtain $\lim_{n\to\infty}T_n^{\mu}(f)=f$, uniformly.

Proof We observe that

$$\left|T_n^{\mu}(f,x)-f(x)\right|=$$

$$\left|\sum_{k=-\infty}^{\infty}\left(\frac{\left(N^*\right)\int_{\left[0,\frac{1}{n}\right]^N}f\left(t+\frac{k}{n}\right)d\mu(t)}{\mu\left(\left[0,\frac{1}{n}\right]^N\right)}\right)Z(nx-k)-\sum_{k=-\infty}^{\infty}f(x)\,Z(nx-k)\right|=$$

$$\left|\sum_{k=-\infty}^{\infty}\left(\left(\frac{(N^*)\int_{[0,\frac{1}{n}]^N} f\left(t+\frac{k}{n}\right)d\mu(t)}{\mu\left(\left[0,\frac{1}{n}\right]^N\right)}\right)-f(x)\right)Z(nx-k)\right| \leq \tag{11.37}$$

$$\sum_{k=-\infty}^{\infty}\left|\left(\frac{(N^*)\int_{[0,\frac{1}{n}]^N} f\left(t+\frac{k}{n}\right)d\mu(t)}{\mu\left(\left[0,\frac{1}{n}\right]^N\right)}\right)-f(x)\right|Z(nx-k) \overset{(11.34)}{\leq}$$

$$\sum_{k=-\infty}^{\infty}\left(\frac{(N^*)\int_{[0,\frac{1}{n}]^N}\left|f\left(t+\frac{k}{n}\right)-f(x)\right|d\mu(t)}{\mu\left(\left[0,\frac{1}{n}\right]^N\right)}\right)Z(nx-k) = \tag{11.38}$$

$$\sum_{\begin{cases} k=-\infty \\ :\left\|\frac{k}{n}-x\right\|_\infty \leq \frac{1}{n^\beta}\end{cases}}^{\infty}\left(\frac{(N^*)\int_{[0,\frac{1}{n}]^N}\left|f\left(t+\frac{k}{n}\right)-f(x)\right|d\mu(t)}{\mu\left(\left[0,\frac{1}{n}\right]^N\right)}\right)Z(nx-k)+$$

$$\sum_{\begin{cases} k=-\infty \\ :\left\|\frac{k}{n}-x\right\|_\infty > \frac{1}{n^\beta}\end{cases}}^{\infty}\left(\frac{(N^*)\int_{[0,\frac{1}{n}]^N}\left|f\left(t+\frac{k}{n}\right)-f(x)\right|d\mu(t)}{\mu\left(\left[0,\frac{1}{n}\right]^N\right)}\right)Z(nx-k) \leq$$

$$\sum_{\begin{cases} k=-\infty \\ :\left\|\frac{k}{n}-x\right\|_\infty \leq \frac{1}{n^\beta}\end{cases}}^{\infty}\left(\frac{(N^*)\int_{[0,\frac{1}{n}]^N}\omega_1\left(f,\|t\|_\infty+\left\|\frac{k}{n}-x\right\|_\infty\right)d\mu(t)}{\mu\left(\left[0,\frac{1}{n}\right]^N\right)}\right)Z(nx-k) \tag{11.39}$$

$$+2\|f\|_\infty\left(\sum_{\begin{cases} k=-\infty \\ :\left\|\frac{k}{n}-x\right\|_\infty > \frac{1}{n^\beta}\end{cases}}^{\infty} Z(nx-k)\right)\overset{(11.24)}{\leq}$$

$$\omega_1\left(f,\frac{1}{n}+\frac{1}{n^\beta}\right)+\frac{\|f\|_\infty}{\sqrt{\pi}\left(n^{1-\beta}-2\right)e^{\left(n^{1-\beta}-2\right)^2}}, \tag{11.40}$$

proving the claim. ■

Additionally we give

Definition 11.8 Denote by $C_B^+\left(\mathbb{R}^N,\mathbb{C}\right)=\{f:\mathbb{R}^N\to\mathbb{C}\,|\,f=f_1+if_2,$ where $f_1,f_2\in C_B^+\left(\mathbb{R}^N\right), N\in\mathbb{N}\}$. We set for $f\in C_B^+\left(\mathbb{R}^N,\mathbb{C}\right)$ that

$$T_n^{\mu}(f,x) := T_n^{\mu}(f_1,x) + iT_n^{\mu}(f_2,x), \tag{11.41}$$

$\forall\, n \in \mathbb{N}, x \in \mathbb{R}^N, i = \sqrt{-1}$.

Theorem 11.9 *Let* $f \in C_B^{+}\left(\mathbb{R}^N, \mathbb{C}\right)$, $f = f_1 + if_2$, $N \in \mathbb{N}$, $0 < \beta < 1$, $x \in \mathbb{R}^N$; $n \in \mathbb{N}$ *with* $n^{1-\beta} \geq 3$. *Then*

(i)

$$\sup_{\mu}\left|T_n^{\mu}(f,x) - f(x)\right| \leq \left[\omega_1\left(f_1, \frac{1}{n} + \frac{1}{n^{\beta}}\right) + \omega_1\left(f_2, \frac{1}{n} + \frac{1}{n^{\beta}}\right)\right]$$

$$+ \frac{\left(\|f_1\|_{\infty} + \|f_2\|_{\infty}\right)}{\sqrt{\pi}\left(n^{1-\beta} - 2\right)e^{\left(n^{1-\beta}-2\right)^2}} =: \psi_n, \tag{11.42}$$

and

(ii)

$$\sup_{\mu}\left\|T_n^{\mu}(f) - f\right\| \leq \psi_n. \tag{11.43}$$

Proof

$$\left|T_n^{\mu}(f,x) - f(x)\right| = \left|T_n^{\mu}(f_1,x) + iT_n^{\mu}(f_2,x) - f_1(x) - if_2(x)\right| =$$

$$\left|\left(T_n^{\mu}(f_1,x) - f_1(x)\right) + i\left(T_n^{\mu}(f_2,x) - f_2(x)\right)\right| \leq$$

$$\left|T_n^{\mu}(f_1,x) - f_1(x)\right| + \left|T_n^{\mu}(f_2,x) - f_2(x)\right| \overset{(11.35)}{\leq} \tag{11.44}$$

$$\left(\omega_1\left(f_1, \frac{1}{n} + \frac{1}{n^{\beta}}\right) + \frac{\|f_1\|_{\infty}}{\sqrt{\pi}\left(n^{1-\beta} - 2\right)e^{\left(n^{1-\beta}-2\right)^2}}\right) +$$

$$\left(\omega_1\left(f_2, \frac{1}{n} + \frac{1}{n^{\beta}}\right) + \frac{\|f_2\|_{\infty}}{\sqrt{\pi}\left(n^{1-\beta} - 2\right)e^{\left(n^{1-\beta}-2\right)^2}}\right) =$$

$$\left[\omega_1\left(f_1, \frac{1}{n} + \frac{1}{n^{\beta}}\right) + \omega_1\left(f_2, \frac{1}{n} + \frac{1}{n^{\beta}}\right)\right] +$$

$$\frac{\left(\|f_1\|_{\infty} + \|f_2\|_{\infty}\right)}{\sqrt{\pi}\left(n^{1-\beta} - 2\right)e^{\left(n^{1-\beta}-2\right)^2}}, \tag{11.45}$$

proving the claim. ∎

References

1. M. Abramowitz, I.A. Stegun (eds.), *Handbook of Mathematical Functions with Formulas, Graphs, and Mathematical Tables* (Dover Publication, New York, 1972)
2. G.A. Anastassiou, Univariate error function based neural network approximation. Indian J. Math. **57**(2), 243–291 (2015)
3. G. Anastassiou, Quantitative approximation by a Kantorovich-Shilkret quasi-interpolation neural network operator (2018). submitted
4. L.C. Andrews, *Special Functions of Mathematics for Engineers*, 2nd edn. (Mc Graw-Hill, New York, 1992)
5. I.S. Haykin, *Neural Networks: A Comprehensive Foundation*, 2nd edn. (Prentice Hall, New York, 1998)
6. W. McCulloch, W. Pitts, A logical calculus of the ideas immanent in nervous activity. Bull. Math. Biophys. **7**, 115–133 (1943)
7. T.M. Mitchell, *Machine Learning* (WCB-McGraw-Hill, New York, 1997)
8. Niel Shilkret, Maxitive measure and integration. Indag. Math. **33**, 109–116 (1971)

Chapter 12
Multivariate Quantitative Approximation by Perturbed Kantorovich–Shilkret Neural Network Operators

This chapter deals with the determination of the rate of convergence to the unit of Perturbed Kantorovich–Shilkret multivariate normalized neural network operators of one hidden layer. These are given through the multivariate modulus of continuity of the engaged multivariate function or its high order partial derivatives and that appears in the associated multivariate Jackson type inequalities. The activation function is very general and it can derive from any multivariate sigmoid or bell-shaped functions. The right hand sides of our Jackson type inequalities do not depend on the activation function. The sample functionals are Kantorovich–Shilkret type. We provide an application for the first partial derivatives of the involved function. It follows [4].

12.1 Introduction

Our work is motivated by [2, 3]. The basic ingredient of our perturbed multivariate neural network operators here is the Shilkret [6] integral.

Our operators fall into the category of feed-forward neural networks (FNNs) with one hidden layer, that are mathematically expressed as

$$N_n(x) = \sum_{j=0}^{n} c_j \sigma\left(\left\langle a_j \cdot x\right\rangle + b_j\right), x \in \mathbb{R}^s, s \in \mathbb{N},$$

where for $0 \leq j \leq n$, $b_j \in \mathbb{R}$ are the thresholds, $a_j \in \mathbb{R}^s$ are the connection weights, $c_j \in \mathbb{R}$ are the coefficients, $\left\langle a_j \cdot x\right\rangle$ is the inner product of a_j and x, and σ is the activation function of the network.

G. A. Anastassiou, *Ordinary and Fractional Approximation by Non-additive Integrals: Choquet, Shilkret and Sugeno Integral Approximators*, Studies in Systems, Decision and Control 190, https://doi.org/10.1007/978-3-030-04287-5_12

In many fundamental network models, the activation function is the sigmoid function of logistic type or other sigmoid function or bell-shaped function. It is well known that FNN's are universal approximators. Theoretically any continuous function defined on a compact set can be approximated to any desired degree of accuracy by increasing the number of hidden neurons. Similar approximations are possible for measurable functions in the L_p norm. P. Cardaliaguet and G. Euvrad were the first [5] to describe precisely and study neural network approximation operators to the unit operator. The author in [1–3] was the first to put neural network approximation in a quantitative form and produce Jackson type inequalities with moduli of continuity.

The author here continues these studies by employing the non-additive Shilkret integral [6] and going very generally by using general activation functions of compact support under perturbation. Because Shilkret integral is so general and simple and non-additive, it has great applications to economic theory.

12.2 Background

Here we follow [6].

Let $\mathcal{F}$ be a σ-field of subsets of an arbitrary set Ω. An extended non-negative real valued function μ on $\mathcal{F}$ is called maxitive if $\mu(\varnothing) = 0$ and

$$\mu\left(\cup_{i\in I} E_i\right) = \sup_{i\in I} \mu\left(E_i\right), \tag{12.1}$$

where the set I is of cardinality at most countable. We also call μ a maxitive measure. Here f stands for a non-negative measurable function on Ω. In [6], Niel Shilkret developed his non-additive integral defined as follows:

$$\left(N^*\right) \int_D f d\mu := \sup_{y\in Y} \left\{y \cdot \mu\left(D \cap \{f \geq y\}\right)\right\}, \tag{12.2}$$

where $Y = [0, m]$ or $Y = [0, m)$ with $0 < m \leq \infty$, and $D \in \mathcal{F}$. Here we take $Y = [0, \infty)$.

It is easily proved that

$$\left(N^*\right) \int_D f d\mu = \sup_{y>0} \left\{y \cdot \mu\left(D \cap \{f > y\}\right)\right\}. \tag{12.3}$$

The Shilkret integral takes values in $[0, \infty]$.

The Shilkret integral [6] has the following properties:

$$\left(N^*\right) \int_\Omega \chi_E d\mu = \mu\left(E\right), \tag{12.4}$$

where χ_E is the indicator function on $E \in \mathcal{F}$,

$$\left(N^*\right) \int_D cf d\mu = c \left(N^*\right) \int_D f d\mu, c \geq 0, \tag{12.5}$$

$$\left(N^*\right) \int_D \sup_{n \in \mathbb{N}} f_n d\mu = \sup_{n \in \mathbb{N}} \left(N^*\right) \int_D f_n d\mu, \tag{12.6}$$

where f_n, $n \in \mathbb{N}$, is an increasing sequence of elementary (countably valued) functions converging uniformly to f. Furthermore we have

$$\left(N^*\right) \int_D f d\mu \geq 0, \tag{12.7}$$

$$f \geq g \text{ implies } \left(N^*\right) \int_D f d\mu \geq \left(N^*\right) \int_D g d\mu, \tag{12.8}$$

where $f, g : \Omega \to [0, \infty]$ are measurable.

Let $a \leq f(\omega) \leq b$ for almost every $\omega \in E$, then

$$a\mu(E) \leq \left(N^*\right) \int_E f d\mu \leq b\mu(E);$$

$$\left(N^*\right) \int_E 1 d\mu = \mu(E);$$

$f > 0$ almost everywhere and $(N^*) \int_E f d\mu = 0$ imply $\mu(E) = 0$;

$(N^*) \int_\Omega f d\mu = 0$ if and only $f = 0$ almost everywhere;

$(N^*) \int_\Omega f d\mu < \infty$ implies that

$$\overline{N}(f) := \{\omega \in \Omega | f(\omega) \neq 0\} \text{ has } \sigma\text{-finite measure;} \tag{12.9}$$

$$\left(N^*\right) \int_D (f + g) d\mu \leq \left(N^*\right) \int_D f d\mu + \left(N^*\right) \int_D g d\mu;$$

and

$$\left|\left(N^*\right) \int_D f d\mu - \left(N^*\right) \int_D g d\mu\right| \leq \left(N^*\right) \int_D |f - g| d\mu. \tag{12.10}$$

From now on in this chapter we assume that $\mu : \mathcal{F} \to [0, +\infty)$.

We need

Proposition 12.1 ([1]) *Let $a \leq b$, $a, b \in \mathbb{R}$. Let $card(K)$ (≥ 0) be the maximum number of integers contained in $[a, b]$. Then*

$$\max(0, (b - a) - 1) \leq card(K) \leq (b - a) + 1. \tag{12.11}$$

12.3 Main Results

Here the activation function $b : \mathbb{R}^d \to \mathbb{R}_+$, $d \in \mathbb{N}$, is of compact support $B := \prod_{i=1}^{d} [-T_i, T_i]$, $T_i > 0$, $i = 1, ..., d$. That is $b(x) > 0$ for any $x \in B$, and clearly b may have jump discontinuities. Also the shape of the graph of b is immaterial.

Typically in neural networks approximation we take b to be a d -dimensional bell-shaped function (i.e. per coordinate is a centered bell-shaped function), or a product of univariate centered bell-shaped functions, or a product of sigmoid functions, in our case all of them are of compact support B.

Example Take $b(x) = \beta(x_1)\beta(x_2)...\beta(x_d)$, where β is any of the following functions, $i = 1, ..., d$:

(i) $\beta(x_j)$ is the characteristic function on $[-1, 1]$,

(ii) $\beta(x_j)$ is the hat function over $[-1+\varepsilon, 1-\varepsilon]$, where $\varepsilon > 0$ is small, that is,

$$\beta(x_j) = \begin{cases} 1 + x_j, -1 \le x_j \le 0, \\ 1 - x_j, 0 < x_j \le 1, \\ 0, \text{ elsewhere,} \end{cases}$$

(iii) the truncated sigmoids

$$\beta(x_j) = \begin{cases} \frac{1}{1+e^{-x_j}} \text{ or } \tanh x_j \text{ or } erf(x_j), \text{ for } x_j \in [-T_j, T_j], \text{ with large } T_j > 0, \\ 0, x_j \in \mathbb{R} - [-T_j, T_j], \end{cases}$$

(iv) the truncated Gompertz function

$$\beta(x_j) = \begin{cases} e^{-\alpha e^{-\beta x j}}, x_j \in [-T_j, T_j]; \ \alpha, \beta > 0; \text{ large } T_j > 0, \\ 0, x_j \in \mathbb{R} - [-T_j, T_j], \end{cases}$$

Thus the general activation function b we will be using here includes all kinds of activation functions in neural network approximations.

Here we consider functions $f : \mathbb{R}^d \to \mathbb{R}_+$ that either continuous and bounded, or uniformly continuous.

Let here the parameters: $0 < \alpha < 1$, $x = (x_1, ..., x_d) \in \mathbb{R}^d$, $n \in \mathbb{N}$; $r = (r_1, ..., r_d) \in \mathbb{N}^d$, $i = (i_1, ..., i_d) \in \mathbb{N}^d$, with $i_j = 1, 2, ..., r_j$, $j = 1, ..., d$; also let $w_i = w_{i_1,...,i_d} \ge 0$, such that $\sum_{i_1=1}^{r_1} \sum_{i_2=1}^{r_2} ... \sum_{i_d=1}^{r_d} w_{i_1,...,i_d} = 1$, in brief written as $\sum_{i=1}^{r} w_i = 1$. We further consider the parameters $k = (k_1, ..., k_d) \in \mathbb{Z}^d$; $\mu_i = (\mu_{i_1}, ..., \mu_{i_d}) \in \mathbb{R}^d_+$, $\nu_i = (\nu_{i_1}, ..., \nu_{i_d}) \in \mathbb{R}^d_+$; and $\lambda_i = \lambda_{i_1,...,i_d}$, $\rho_i = \rho_{i_1,...,i_d} \ge 0$.

We use here the first multivariate modulus of continuity, with $\delta > 0$,

$$\omega_1(f, \delta) := \sup_{\substack{x, y \in \mathbb{R}^d \\ \|x - y\|_\infty \le \delta}} |f(x) - f(y)|,$$

where $\|x\|_{\infty} = \max(|x_1|, ..., |x_d|)$.

Given that f is uniformly continuous we get $\lim_{\delta \to 0} \omega_1(f, \delta) = 0$.

So in this section mainly we study the pointwise convergence with rates over $\mathbb{R}^d$, to the unit operator, of the following one hidden layer multivariate normalized neural network perturbed operator:

Definition 12.2 Let $\mathcal{L}$ be the Lebesgue σ-algebra on $\mathbb{R}^N$, $N \in \mathbb{N}$, and the maxitive measure $\mu : \mathcal{L} \to [0, +\infty)$, such that for any $A \in \mathcal{L}$ with $A \neq \varnothing$, we get $\mu(A) > 0$. We define the multivariate Kantorovich–Shilkret type neural network operators for any $x \in \mathbb{R}^N$, $n \in \mathbb{N}$:

$$\left(M_n^{\mu}(f)\right)(x) =$$

$$\frac{\left[\sum_{k=-n^2}^{n^2} \left(\sum_{i=1}^{r} w_i \frac{1}{\mu\left(\left[0, \frac{1}{n+\rho_i}\right]^d\right)} (N^*) \int_0^{\frac{1}{n+\rho_i}} f\left(t + \frac{k+\lambda_i}{n+\rho_i}\right) d\mu(t) \right) b\left(n^{1-\alpha}\left(x - \frac{k}{n}\right)\right)\right]}{\sum_{k=-n^2}^{n^2} b\left(n^{1-\alpha}\left(x - \frac{k}{n}\right)\right)} \tag{12.12}$$

$$= \sum_{k_1=-n^2}^{n^2} \cdots \sum_{k_d=-n^2}^{n^2} \left(\sum_{i_1=1}^{r_1} \cdots \sum_{i_d=1}^{r_d} w_{i_1,\ldots,i_d} \frac{1}{\mu\left(\left[0, \frac{1}{n+\rho_{i_1,\ldots,i_d}}\right]^d\right)} \right.$$

$$\left. (N^*) \underbrace{\int \cdots \int \cdots \int}_{\left[0, \frac{1}{n+\rho_{i_1,\ldots,i_d}}\right]^d} f\left(t_1 + \frac{k_1 + \lambda_{i_1,\ldots,i_d}}{n + \rho_{i_1,\ldots,i_d}}, \ldots, t_d + \frac{k_d + \lambda_{i_1,\ldots,i_d}}{n + \rho_{i_1,\ldots,i_d}}\right) d\mu(t_1, \ldots, t_d) \right)$$

$$\frac{b\left(n^{1-\alpha}\left(x_1 - \frac{k_1}{n}\right), \ldots, n^{1-\alpha}\left(x_d - \frac{k_d}{n}\right)\right)}{\sum_{k_1=-n^2}^{n^2} \cdots \sum_{k_d=-n^2}^{n^2} b\left(n^{1-\alpha}\left(x_1 - \frac{k_1}{n}\right), \ldots, n^{1-\alpha}\left(x_d - \frac{k_d}{n}\right)\right)}.$$

For f continuous and bounded we get:

$$\left\|M_n^{\mu}(f)\right\|_{\infty} \leq \|f\|_{\infty}, \text{ and } M_n^{\mu}(1) = 1.$$

Remark 12.3 The terms in the ratio of sums in (12.12) can be nonzero, iff simultaneously

$$\left| n^{1-\alpha}\left(x_j - \frac{k_j}{n}\right) \right| \leq T_j, \text{ all } j = 1, \ldots, d, \tag{12.13}$$

i.e. $\left|x_j - \frac{k_j}{n}\right| \leq \frac{T_j}{n^{1-\alpha}}$, all $j = 1, \ldots, d$, iff

$$nx_j - T_j n^{\alpha} \leq k_j \leq nx_j + T_j n^{\alpha}, \text{ all } j = 1, ..., d. \tag{12.14}$$

To have the order

$$-n^2 \leq nx_j - T_j n^{\alpha} \leq k_j \leq nx_j + T_j n^{\alpha} \leq n^2, \tag{12.15}$$

we need $n \geq T_j + |x_j|$, all $j = 1, ..., d$. So (12.15) is true when we take

$$n \geq \max_{j \in \{1,...,d\}} \left(T_j + |x_j| \right). \tag{12.16}$$

When $x \in B$ in order to have (12.15) it is enough to assume that $n \geq 2T^*$, where $T^* := \max\{T_1, ..., T_d\} > 0$. Consider

$$\widetilde{I}_j := \left[nx_j - T_j n^{\alpha}, nx_j + T_j n^{\alpha} \right], \; j = 1, ..., d, \; n \in \mathbb{N}.$$

The length of $\widetilde{I}_j$ is $2T_j n^{\alpha}$. By Proposition 12.1, we get that the cardinality of $k_j \in \mathbb{Z}$ that belong to $\widetilde{I}_j := card\left(k_j\right) \geq \max\left(2T_j n^{\alpha} - 1, 0\right)$, any $j \in \{1, ..., d\}$. In order to have $card\left(k_j\right) \geq 1$, we need $2T_j n^{\alpha} - 1 \geq 1$ if and only if $n \geq T_j^{-\frac{1}{\alpha}}$, any $j \in \{1, ..., d\}$.

Therefore, a sufficient condition in order to obtain the order (12.15) along with the interval $\widetilde{I}_j$ to contain at least one integer for all $j = 1, ..., d$ is that

$$n \geq \max_{j \in \{1,...,d\}} \left\{ T_j + |x_j|, T_j^{-\frac{1}{\alpha}} \right\}. \tag{12.17}$$

Clearly as $n \to +\infty$ we get that $card\left(k_j\right) \to +\infty$, all $j = 1, ..., d$. Also notice that $card\left(k_j\right)$ equals to the cardinality of integers in $\left[\lceil nx_j - T_j n^{\alpha} \rceil, \left[nx_j + T_j n^{\alpha}\right]\right]$ for all $j = 1, ..., d$. Here $[\cdot]$ denotes the integral part of the number while $\lceil \cdot \rceil$ denotes its ceiling.

From now on, in this section we assume (12.17).

We shall denote by $T = (T_1, ..., T_d)$, $[nx + Tn^{\alpha}] = ([nx_1 + T_1 n^{\alpha}], ..., [nx_d + T_d n^{\alpha}])$, and $\lceil nx - Tn^{\alpha} \rceil = (\lceil nx_1 - T_1 n^{\alpha} \rceil, ..., \lceil nx_d - T_d n^{\alpha} \rceil)$. Furthermore it holds

$$\left(M_n^{\mu}(f)\right)(x) =$$

$$\frac{\sum_{k=\lceil nx - Tn^{\alpha} \rceil}^{[nx+Tn^{\alpha}]} \left(\sum_{i=1}^{r} w_i \frac{1}{\mu\left(\left[0, \frac{1}{n+\rho_i}\right]^d\right)} (N^*) \int_0^{\frac{1}{n+\rho_i}} f\left(t + \frac{k+\lambda_i}{n+\rho_i}\right) d\mu(t) \right) b\left(n^{1-\alpha}\left(x - \frac{k}{n}\right)\right)}{\sum_{k=\lceil nx - Tn^{\alpha} \rceil}^{[nx+Tn^{\alpha}]} b\left(n^{1-\alpha}\left(x - \frac{k}{n}\right)\right)} \tag{12.18}$$

$$= \sum_{k_1=\lceil nx_1-T_1n^{\alpha}\rceil}^{[nx_1+T_1n^{\alpha}]} \cdots \sum_{k_d=\lceil nx_d-T_dn^{\alpha}\rceil}^{[nx_d+T_dn^{\alpha}]} \left(\sum_{i_1=1}^{r_1} \cdots \sum_{i_d=1}^{r_d} w_{i_1,\ldots,i_d} \frac{1}{\mu\left(\left[0, \frac{1}{n+\rho_{i_1,\ldots,i_d}}\right]^d\right)} \right.$$

$$\left. (N^*) \int \ldots \int_{\left[0, \frac{1}{n+\rho_{i_1,\ldots,i_d}}\right]^d} \ldots \int f\left(t_1 + \frac{k_1+\lambda_{i_1,\ldots,i_d}}{n+\rho_{i_1,\ldots,i_d}}, \ldots, t_d + \frac{k_d+\lambda_{i_1,\ldots,i_d}}{n+\rho_{i_1,\ldots,i_d}}\right) d\mu(t_1,\ldots,t_d) \right)$$

$$\frac{b\left(n^{1-\alpha}\left(x_1-\frac{k_1}{n}\right), \ldots, n^{1-\alpha}\left(x_d-\frac{k_d}{n}\right)\right)}{\sum_{k_1=\lceil nx_1-T_1n^{\alpha}\rceil}^{[nx_1+T_1n^{\alpha}]} \cdots \sum_{k_d=\lceil nx_d-T_dn^{\alpha}\rceil}^{[nx_d+T_dn^{\alpha}]} b\left(n^{1-\alpha}\left(x_1-\frac{k_1}{n}\right), \ldots, n^{1-\alpha}\left(x_d-\frac{k_d}{n}\right)\right)}.$$

So if $\left|n^{1-\alpha}\left(x_j-\frac{k_j}{n}\right)\right| \leq T_j$, all $j = 1, \ldots, d$, we get that

$$\left\|x - \frac{k}{n}\right\|_{\infty} \leq \frac{T^*}{n^{1-\alpha}}. \tag{12.19}$$

For convenience we call

$$V(x) := \sum_{k=\lceil nx-Tn^{\alpha}\rceil}^{[nx+Tn^{\alpha}]} b\left(n^{1-\alpha}\left(x-\frac{k}{n}\right)\right) =$$

$$\sum_{k_1=\lceil nx_1-T_1n^{\alpha}\rceil}^{[nx_1+T_1n^{\alpha}]} \cdots \sum_{k_d=\lceil nx_d-T_dn^{\alpha}\rceil}^{[nx_d+T_dn^{\alpha}]} b\left(n^{1-\alpha}\left(x_1-\frac{k_1}{n}\right), \ldots, n^{1-\alpha}\left(x_d-\frac{k_d}{n}\right)\right). \tag{12.20}$$

We make

Remark 12.4 Here always k is as in (12.15).

We have for

$$0 \leq t_j \leq \frac{1}{n+\rho_{i_1,\ldots,i_d}}, \; j = 1, \ldots, d, \tag{12.21}$$

that

$$\left\|t + \frac{k+\lambda_{i_1,\ldots,i_d}}{n+\rho_{i_1,\ldots,i_d}} - x\right\|_{\infty} \leq \frac{1}{n+\rho_{i_1,\ldots,i_d}} + \left\|\frac{k+\lambda_{i_1,\ldots,i_d}}{n+\rho_{i_1,\ldots,i_d}} - x\right\|_{\infty} \leq \tag{12.22}$$

$$\frac{1+\lambda_{i_1,\ldots,i_d}}{n+\rho_{i_1,\ldots,i_d}} + \left\|\frac{k}{n+\rho_{i_1,\ldots,i_d}} - x\right\|_{\infty} \leq$$

$$\frac{1+\lambda_{i_1,\dots,i_d}}{n+\rho_{i_1,\dots,i_d}}+\left\|\frac{k}{n+\rho_{i_1,\dots,i_d}}-\frac{k}{n}\right\|_\infty+\left\|\frac{k}{n}-x\right\|_\infty\leq$$

$$\frac{1+\lambda_{i_1,\dots,i_d}}{n+\rho_{i_1,\dots,i_d}}+\frac{T^*}{n^{1-\alpha}}+\frac{\rho_{i_1,\dots,i_d}\|k\|_\infty}{\left(n+\rho_{i_1,\dots,i_d}\right)n}\leq$$

$$\frac{1+\lambda_{i_1,\dots,i_d}}{n+\rho_{i_1,\dots,i_d}}+\frac{T^*}{n^{1-\alpha}}+\frac{\rho_{i_1,\dots,i_d}}{n\left(n+\rho_{i_1,\dots,i_d}\right)}\left(n\|x\|_\infty+T^*n^\alpha\right)= \tag{12.23}$$

$$\frac{1+\lambda_{i_1,\dots,i_d}}{n+\rho_{i_1,\dots,i_d}}+\frac{T^*}{n^{1-\alpha}}+\left(\frac{\rho_{i_1,\dots,i_d}}{n+\rho_{i_1,\dots,i_d}}\right)\left(\|x\|_\infty+\frac{T^*}{n^{1-\alpha}}\right).$$

We have found that

$$\left\|t+\frac{k+\lambda_{i_1,\dots,i_d}}{n+\rho_{i_1,\dots,i_d}}-x\right\|_\infty\leq$$

$$\left(\frac{\rho_{i_1,\dots,i_d}\|x\|_\infty+\lambda_{i_1,\dots,i_d}+1}{n+\rho_{i_1,\dots,i_d}}\right)+\left(1+\frac{\rho_{i_1,\dots,i_d}}{n+\rho_{i_1,\dots,i_d}}\right)\frac{T^*}{n^{1-\alpha}}. \tag{12.24}$$

So when $0\leq t_j\leq\frac{1}{n+\rho_{i_1,\dots,i_d}}$, $j=1,\dots,d$, we get that

$$\omega_1\left(f,\left\|t+\frac{k+\lambda_{i_1,\dots,i_d}}{n+\rho_{i_1,\dots,i_d}}-x\right\|_\infty\right)\leq$$

$$\omega_1\left(f,\left(\frac{\rho_{i_1,\dots,i_d}\|x\|_\infty+\lambda_{i_1,\dots,i_d}+1}{n+\rho_{i_1,\dots,i_d}}\right)+\left(1+\frac{\rho_{i_1,\dots,i_d}}{n+\rho_{i_1,\dots,i_d}}\right)\frac{T^*}{n^{1-\alpha}}\right), \tag{12.25}$$

with dominant speed $\frac{1}{n^{1-\alpha}}$.

We present

Theorem 12.5 *Let $f:\mathbb{R}^d\to\mathbb{R}_+$ be continuous and bounded or uniformly continuous. Let $x\in\mathbb{R}^d$ and $n\in\mathbb{N}$ such that $n\geq\max_{j\in\{1,\dots,d\}}\left(T_j+|x_j|,T_j^{-\frac{1}{\alpha}}\right)$, $T_j>0$, $0<\alpha<1$. Then*

$$\sup_\mu\left|\left(M_n^\mu(f)\right)(x)-f(x)\right|\leq$$

$$\sum_{i=1}^r w_i\omega_1\left(f,\left(\frac{\rho_i\|x\|_\infty+\lambda_i+1}{n+\rho_i}\right)+\left(1+\frac{\rho_i}{n+\rho_i}\right)\frac{T^*}{n^{1-\alpha}}\right)= \tag{12.26}$$

$$\sum_{i_1=1}^{r_1}\cdots\sum_{i_d=1}^{r_d}w_{i_1,\dots,i_d}\omega_1\left(f,\left(\frac{\rho_{i_1,\dots,i_d}\|x\|_\infty+\lambda_{i_1,\dots,i_d}+1}{n+\rho_{i_1,\dots,i_d}}\right)+\right.$$

$$\left(1+\frac{\rho_{i_1,\ldots,i_d}}{n+\rho_{i_1,\ldots,i_d}}\right)\frac{T^*}{n^{1-\alpha}}\right).$$

Proof We notice the following:

$$f\left(t+\frac{k+\lambda_i}{n+\rho_i}\right)=f\left(t+\frac{k+\lambda_i}{n+\rho_i}\right)-f(x)+f(x)\le$$

$$\left|f\left(t+\frac{k+\lambda_i}{n+\rho_i}\right)-f(x)\right|+f(x). \tag{12.27}$$

Hence

$$\frac{1}{\mu\left(\left[0,\frac{1}{n+\rho_i}\right]^d\right)}(N^*)\int_0^{\frac{1}{n+\rho_i}}f\left(t+\frac{k+\lambda_i}{n+\rho_i}\right)d\mu(t)\le$$

$$\frac{1}{\mu\left(\left[0,\frac{1}{n+\rho_i}\right]^d\right)}(N^*)\int_0^{\frac{1}{n+\rho_i}}\left|f\left(t+\frac{k+\lambda_i}{n+\rho_i}\right)-f(x)\right|d\mu(t)+f(x). \tag{12.28}$$

That is

$$\frac{1}{\mu\left(\left[0,\frac{1}{n+\rho_i}\right]^d\right)}(N^*)\int_0^{\frac{1}{n+\rho_i}}f\left(t+\frac{k+\lambda_i}{n+\rho_i}\right)d\mu(t)-f(x)\le$$

$$\frac{1}{\mu\left(\left[0,\frac{1}{n+\rho_i}\right]^d\right)}(N^*)\int_0^{\frac{1}{n+\rho_i}}\left|f\left(t+\frac{k+\lambda_i}{n+\rho_i}\right)-f(x)\right|d\mu(t). \tag{12.29}$$

Similarly, we have

$$f(x)=f(x)-f\left(t+\frac{k+\lambda_i}{n+\rho_i}\right)+f\left(t+\frac{k+\lambda_i}{n+\rho_i}\right)\le$$

$$\left|f\left(t+\frac{k+\lambda_i}{n+\rho_i}\right)-f(x)\right|+f\left(t+\frac{k+\lambda_i}{n+\rho_i}\right). \tag{12.30}$$

Hence

$$f(x)\le\frac{1}{\mu\left(\left[0,\frac{1}{n+\rho_i}\right]^d\right)}(N^*)\int_0^{\frac{1}{n+\rho_i}}\left|f\left(t+\frac{k+\lambda_i}{n+\rho_i}\right)-f(x)\right|d\mu(t)+$$

$$\frac{1}{\mu\left(\left[0,\frac{1}{n+\rho_i}\right]^d\right)}\left(N^*\right)\int_0^{\frac{1}{n+\rho_i}} f\left(t+\frac{k+\lambda_i}{n+\rho_i}\right)d\mu\left(t\right). \tag{12.31}$$

That is,

$$f\left(x\right)-\frac{1}{\mu\left(\left[0,\frac{1}{n+\rho_i}\right]^d\right)}\left(N^*\right)\int_0^{\frac{1}{n+\rho_i}} f\left(t+\frac{k+\lambda_i}{n+\rho_i}\right)d\mu\left(t\right)\leq \tag{12.32}$$

$$\frac{1}{\mu\left(\left[0,\frac{1}{n+\rho_i}\right]^d\right)}\left(N^*\right)\int_0^{\frac{1}{n+\rho_i}} \left|f\left(t+\frac{k+\lambda_i}{n+\rho_i}\right)-f\left(x\right)\right|d\mu\left(t\right).$$

We have proved that

$$\left|\frac{1}{\mu\left(\left[0,\frac{1}{n+\rho_i}\right]^d\right)}\left(N^*\right)\int_0^{\frac{1}{n+\rho_i}} f\left(t+\frac{k+\lambda_i}{n+\rho_i}\right)d\mu\left(t\right)-f\left(x\right)\right|\leq \tag{12.33}$$

$$\frac{1}{\mu\left(\left[0,\frac{1}{n+\rho_i}\right]^d\right)}\left(N^*\right)\int_0^{\frac{1}{n+\rho_i}} \left|f\left(t+\frac{k+\lambda_i}{n+\rho_i}\right)-f\left(x\right)\right|d\mu\left(t\right).$$

We observe the following

$$\left|\left(M_n^{\mu}\left(f\right)\right)\left(x\right)-f\left(x\right)\right|\overset{(12.18)}{=}$$

$$\left|\sum_{k=\lceil nx-Tn^{\alpha}\rceil}^{[nx+Tn^{\alpha}]}\left(\sum_{i=1}^{r}w_i\frac{1}{\mu\left(\left[0,\frac{1}{n+\rho_i}\right]^d\right)}\left(N^*\right)\int_0^{\frac{1}{n+\rho_i}} f\left(t+\frac{k+\lambda_i}{n+\rho_i}\right)d\mu\left(t\right)\right)\cdot\right. \tag{12.34}$$

$$\left.\frac{b\left(n^{1-\alpha}\left(x-\frac{k}{n}\right)\right)}{V\left(x\right)}-f\left(x\right)\right|=$$

$$\left|\sum_{k=\lceil nx-Tn^{\alpha}\rceil}^{[nx+Tn^{\alpha}]}\left(\sum_{i=1}^{r}w_i\frac{1}{\mu\left(\left[0,\frac{1}{n+\rho_i}\right]^d\right)}\left(N^*\right)\int_0^{\frac{1}{n+\rho_i}} f\left(t+\frac{k+\lambda_i}{n+\rho_i}\right)d\mu\left(t\right)\right)\cdot\right.$$

$$\left. \frac{b\left(n^{1-\alpha}\left(x-\frac{k}{n}\right)\right)}{V(x)} - \frac{f(x)\,V(x)}{V(x)} \right| =$$

$$\left| \frac{\sum_{k=\lceil nx-Tn^{\alpha}\rceil}^{[nx+Tn^{\alpha}]} \left[\left(\sum_{i=1}^{r} w_i \frac{1}{\mu\left(\left[0,\frac{1}{n+\rho_i}\right]^d\right)} (N^*) \int_0^{\frac{1}{n+\rho_i}} f\left(t+\frac{k+\lambda_i}{n+\rho_i}\right) d\mu(t) \right) - f(x) \right]}{V(x)} \right.$$

$$\left. \cdot b\left(n^{1-\alpha}\left(x-\frac{k}{n}\right)\right) \right| \leq$$

$$\frac{\sum_{k=\lceil nx-Tn^{\alpha}\rceil}^{[nx+Tn^{\alpha}]} \left[\sum_{i=1}^{r} w_i \left| \frac{1}{\mu\left(\left[0,\frac{1}{n+\rho_i}\right]^d\right)} (N^*) \int_0^{\frac{1}{n+\rho_i}} f\left(t+\frac{k+\lambda_i}{n+\rho_i}\right) d\mu(t) - f(x) \right| \right]}{V(x)}$$

$$\cdot\, b\left(n^{1-\alpha}\left(x-\frac{k}{n}\right)\right) \leq \tag{12.35}$$

$$\frac{\sum_{k=\lceil nx-Tn^{\alpha}\rceil}^{[nx+Tn^{\alpha}]} \left[\sum_{i=1}^{r} w_i \left[\frac{1}{\mu\left(\left[0,\frac{1}{n+\rho_i}\right]^d\right)} (N^*) \int_0^{\frac{1}{n+\rho_i}} \left| f\left(t+\frac{k+\lambda_i}{n+\rho_i}\right) - f(x) \right| d\mu(t) \right] \right]}{V(x)}$$

$$\cdot b\left(n^{1-\alpha}\left(x-\frac{k}{n}\right)\right) \leq$$

$$\frac{\sum_{k=\lceil nx-Tn^{\alpha}\rceil}^{[nx+Tn^{\alpha}]} \left[\sum_{i=1}^{r} w_i \left[\frac{1}{\mu\left(\left[0,\frac{1}{n+\rho_i}\right]^d\right)} (N^*) \int_0^{\frac{1}{n+\rho_i}} \omega_1\left(f, \left\| t+\frac{k+\lambda_i}{n+\rho_i} - x \right\|_{\infty}\right) d\mu(t) \right] \right]}{V(x)}$$

$$\cdot b\left(n^{1-\alpha}\left(x-\frac{k}{n}\right)\right) \overset{\text{(by (12.25))}}{\leq}$$

$$\frac{\sum_{k=\lceil nx-Tn^{\alpha}\rceil}^{[nx+Tn^{\alpha}]} \left[\sum_{i=1}^{r} w_i \left[\frac{1}{\mu\left(\left[0,\frac{1}{n+\rho_i}\right]^d\right)} (N^*) \int_0^{\frac{1}{n+\rho_i}} \omega_1\left(f, \left(\frac{\rho_i \|x\|_{\infty}+\lambda_i+1}{n+\rho_i}\right) + \right.\right.\right.}{V(x)}$$

$$\left(1+\frac{\rho_i}{n+\rho_i}\right)\frac{T^*}{n^{1-\alpha}}\right)d\mu(t)\Bigg]\Bigg]\cdot b\left(n^{1-\alpha}\left(x-\frac{k}{n}\right)\right)=$$

$$\frac{\sum_{k=\lceil nx-Tn^{\alpha}\rceil}^{[nx+Tn^{\alpha}]}\left[\sum_{i=1}^{r}w_i\omega_1\left(f,\left(\frac{\rho_i\|x\|_{\infty}+\lambda_i+1}{n+\rho_i}\right)+\left(1+\frac{\rho_i}{n+\rho_i}\right)\frac{T^*}{n^{1-\alpha}}\right)\right]}{V(x)}$$

$$\cdot b\left(n^{1-\alpha}\left(x-\frac{k}{n}\right)\right)=$$

$$\sum_{i=1}^{r}w_i\omega_1\left(f,\left(\frac{\rho_i\|x\|_{\infty}+\lambda_i+1}{n+\rho_i}\right)+\left(1+\frac{\rho_i}{n+\rho_i}\right)\frac{T^*}{n^{1-\alpha}}\right), \tag{12.36}$$

proving the claim. ∎

Corollary 12.6 *Let $f:\mathbb{R}^d\to\mathbb{R}_+$ continuous and bounded or uniformly continuous function. Let $x\in\prod_{j=1}^{d}\left[-\gamma_j,\gamma_j\right]\subset\mathbb{R}^d$, $\gamma_j>0$, $\gamma^*=\max\{\gamma_1,\ldots,\gamma_d\}$ and $n\in\mathbb{N}$: $n\geq\max_{j\in\{1,\ldots,d\}}\left\{T_j+\gamma_j,T_j^{-\frac{1}{\alpha}}\right\}$. Then*

$$\sup_{\mu}\left\|M_n^{\mu}(f)-f\right\|_{\infty,\prod_{j=1}^{d}[-\gamma_j,\gamma_j]}\leq$$

$$\sum_{i=1}^{r}w_i\omega_1\left(f,\left(\frac{\rho_i\gamma^*+\lambda_i+1}{n+\rho_i}\right)+\left(1+\frac{\rho_i}{n+\rho_i}\right)\frac{T^*}{n^{1-\alpha}}\right)= \tag{12.37}$$

$$\sum_{i_1=1}^{r_1}\cdots\sum_{i_d=1}^{r_d}w_{i_1,\ldots,i_d}\omega_1\left(f,\left(\frac{\rho_{i_1,\ldots,i_d}\gamma^*+\lambda_{i_1,\ldots,i_d}+1}{n+\rho_{i_1,\ldots,i_d}}\right)+\left(1+\frac{\rho_{i_1,\ldots,i_d}}{n+\rho_{i_1,\ldots,i_d}}\right)\frac{T^*}{n^{1-\alpha}}\right).$$

Proof By (12.26). ∎

Next here that the right hand sides of (12.26) and (12.37) do not depend on μ and b.

We continue with the higher order of approximation results based on the high order differentiability of the approximated function.

Theorem 12.7 *Let $x\in\mathbb{R}^d$ and $n\in\mathbb{N}$ such that $n\geq\max_{j\in\{1,\ldots,d\}}\left(T_j+\left|x_j\right|,T_j^{-\frac{1}{\alpha}}\right)$, $T_j>0$, $0<\alpha<1$. Let also $f\in C^N\left(\mathbb{R}^d,\mathbb{R}_+\right)$, $N\in\mathbb{N}$, such that all of its par-*

tial derivatives $f_{\widetilde{\alpha}}$ *of order* N, $\widetilde{\alpha} : |\widetilde{\alpha}| = \sum_{j=1}^{d} \alpha_j = N$, *are uniformly continuous or continuous and bounded. Then*

$$\sup_{\mu} \left| \left(M_n^{\mu} (f) \right) (x) - f (x) \right| \leq$$

$$\sum_{l=1}^{N} \frac{1}{l!} \left(\sum_{i=1}^{r} w_i \left[\left(\frac{\rho_i \|x\|_{\infty} + \lambda_i + 1}{n + \rho_i} \right) + \left(1 + \frac{\rho_i}{n + \rho_i} \right) \frac{T^*}{n^{1-\alpha}} \right]^l \right) \cdot$$

$$\left(\left(\sum_{j=1}^{d} \left| \frac{\partial}{\partial x_j} \right| \right)^l f (x) \right) + \tag{12.38}$$

$$\frac{d^N}{N!} \sum_{i=1}^{r} w_i \left[\left(\frac{\rho_i \|x\|_{\infty} + \lambda_i + 1}{n + \rho_i} \right) + \left(1 + \frac{\rho_i}{n + \rho_i} \right) \frac{T^*}{n^{1-\alpha}} \right]^N \cdot$$

$$\max_{\widetilde{\alpha}:|\widetilde{\alpha}|=N} \omega_1 \left(f_{\widetilde{\alpha}}, \left[\left(\frac{\rho_i \|x\|_{\infty} + \lambda_i + 1}{n + \rho_i} \right) + \left(1 + \frac{\rho_i}{n + \rho_i} \right) \frac{T^*}{n^{1-\alpha}} \right] \right).$$

Inequality (12.38) implies the pointwise convergence with rates of $\left(M_n^{\mu} (f) \right) (x)$ *to* $f (x)$, *as* $n \to \infty$, *at speed* $\frac{1}{n^{1-\alpha}}$.

Proof As in the proof of Theorem 12.5 we get:

$$\left| \left(M_n^{\mu} (f) \right) (x) - f (x) \right| \leq$$

$$\frac{\sum_{k=\lceil nx - Tn^{\alpha} \rceil}^{[nx+Tn^{\alpha}]} \left[\sum_{i=1}^{r} w_i \left[\frac{1}{\mu\left(\left[0, \frac{1}{n+\rho_i} \right]^d \right)} (N^*) \int_0^{\frac{1}{n+\rho_i}} \left| f \left(t + \frac{k+\lambda_i}{n+\rho_i} \right) - f (x) \right| d\mu (t) \right] \right]}{V (x)} \tag{12.39}$$

$$\cdot b \left(n^{1-\alpha} \left(x - \frac{k}{n} \right) \right).$$

Here we consider $0 \leq t_j \leq \frac{1}{n+\rho_{i_1,\dots,i_d}}$, $j = 1, \dots, d$.

Set

$$g_{t+\frac{k+\lambda_i}{n+\rho_i}} \left(\lambda^* \right) = f \left(x + \lambda^* \left(\frac{k + \lambda_i}{n + \rho_i} - x \right) \right), 0 \leq \lambda^* \leq 1. \tag{12.40}$$

Then we have

$$g^{(l)}_{t+\frac{k+\lambda_i}{n+\rho_i}} \left(\lambda^* \right) =$$

$$\left[\left(\sum_{j=1}^{d}\left(t_j\frac{k_j+\lambda_i}{n+\rho_i}-x_j\right)\frac{\partial}{\partial x_j}\right)^l f\right]\left(x+\lambda^*\left(t+\frac{k+\lambda_i}{n+\rho_i}-x\right)\right), \quad (12.41)$$

and

$$g_{t+\frac{k+\lambda_i}{n+\rho_i}}(0)=f(x). \quad (12.42)$$

By multivariate Taylor's formula, we get

$$f\left(t+\frac{k+\lambda_i}{n+\rho_i}\right)-f(x)=g_{t+\frac{k+\lambda_i}{n+\rho_i}}(1)-f(x)=$$

$$\sum_{l=1}^{N}\frac{g^{(l)}_{t+\frac{k+\lambda_i}{n+\rho_i}}(0)}{l!}+R_N\left(t+\frac{k+\lambda_i}{n+\rho_i},0\right), \quad (12.43)$$

where

$$R_N\left(t+\frac{k+\lambda_i}{n+\rho_i},0\right)=$$

$$\int_0^1\left(\int_0^{\lambda_1^*}\ldots\left(\int_0^{\lambda_{N-1}^*}\left(g^{(N)}_{t+\frac{k+\lambda_i}{n+\rho_i}}(\lambda_N^*)-g^{(N)}_{t+\frac{k+\lambda_i}{n+\rho_i}}(0)\right)d\lambda_N^*\right)\ldots\right)d\lambda_1^*. \quad (12.44)$$

Here we denote by

$$f_{\widetilde{\alpha}}:=\frac{\partial^{\widetilde{\alpha}}f}{\partial x^{\widetilde{\alpha}}},\ \widetilde{\alpha}:=(\alpha_1,\ldots,\alpha_d),\alpha_j\in\mathbb{Z}^+,\ j=1,\ldots,d, \quad (12.45)$$

such that $|\widetilde{\alpha}|=\sum_{j=1}^{d}\alpha_j=N$.

Notice that

$$\left|g^{(l)}_{t+\frac{k+\lambda_i}{n+\rho_i}}(0)\right|\overset{(12.24)}{\leq}\left(\left(\sum_{j=1}^{d}\left|\frac{\partial}{\partial x_j}\right|\right)^l f(x)\right)\cdot$$

$$\left[\left(\frac{\rho_i\|x\|_\infty+\lambda_i+1}{n+\rho_i}\right)+\left(1+\frac{\rho_i}{n+\rho_i}\right)\frac{T^*}{n^{1-\alpha}}\right]^l, \quad (12.46)$$

We observe that $(0\leq\lambda_N^*\leq1)$

$$\left|g^{(N)}_{t+\frac{k+\lambda_i}{n+\rho_i}}(\lambda_N^*)-g^{(N)}_{t+\frac{k+\lambda_i}{n+\rho_i}}(0)\right|=$$

$$|\left[\left(\sum_{j=1}^{d}\left(t_j+\frac{k_j+\lambda_i}{n+\rho_i}-x_j\right)\frac{\partial}{\partial x_j}\right)^N f\right]\left(x+\lambda_N^*\left(t+\frac{k+\lambda_i}{n+\rho_i}-x\right)\right)-$$

$$\left[\left(\sum_{j=1}^{d}\left(t_j+\frac{k_j+\lambda_i}{n+\rho_i}-x_j\right)\frac{\partial}{\partial x_j}\right)^N f\right](x)|\overset{\text{(by (12.24), (12.25))}}{\leq} \tag{12.47}$$

$$d^N\left[\left(\frac{\rho_i\|x\|_\infty+\lambda_i+1}{n+\rho_i}\right)+\left(1+\frac{\rho_i}{n+\rho_i}\right)\frac{T^*}{n^{1-\alpha}}\right]^N\cdot$$

$$\max_{\widetilde{\alpha}:|\widetilde{\alpha}|=N}\omega_1\left(f_{\widetilde{\alpha}},\left[\left(\frac{\rho_i\|x\|_\infty+\lambda_i+1}{n+\rho_i}\right)+\left(1+\frac{\rho_i}{n+\rho_i}\right)\frac{T^*}{n^{1-\alpha}}\right]\right).$$

Thus we find

$$\left|R_N\left(t+\frac{k+\lambda_i}{n+\rho_i},0\right)\right|\leq$$

$$\int_0^1\left(\int_0^{\lambda_1^*}\ldots\left(\int_0^{\lambda_{N-1}^*}\left|g_{t+\frac{k+\lambda_i}{n+\rho_i}}^{(N)}\left(\lambda_N^*\right)-g_{t+\frac{k+\lambda_i}{n+\rho_i}}^{(N)}(0)\right|d\lambda_N^*\right)\ldots\right)d\lambda_1^*\leq \tag{12.48}$$

$$\frac{d^N}{N!}\left[\left(\frac{\rho_i\|x\|_\infty+\lambda_i+1}{n+\rho_i}\right)+\left(1+\frac{\rho_i}{n+\rho_i}\right)\frac{T^*}{n^{1-\alpha}}\right]^N\cdot$$

$$\max_{\widetilde{\alpha}:|\widetilde{\alpha}|=N}\omega_1\left(f_{\widetilde{\alpha}},\left[\left(\frac{\rho_i\|x\|_\infty+\lambda_i+1}{n+\rho_i}\right)+\left(1+\frac{\rho_i}{n+\rho_i}\right)\frac{T^*}{n^{1-\alpha}}\right]\right).$$

Therefore it holds

$$\left|f\left(t+\frac{k+\lambda_i}{n+\rho_i}\right)-f(x)\right|\overset{(12.43)}{\leq}$$

$$\sum_{l=1}^{N}\frac{\left|g_{t+\frac{k+\lambda_i}{n+\rho_i}}^{(l)}(0)\right|}{l!}+\left|R_N\left(t+\frac{k+\lambda_i}{n+\rho_i},0\right)\right|\overset{((12.46),\,(12.48))}{\leq}$$

$$\sum_{l=1}^{N}\frac{1}{l!}\left(\left(\sum_{j=1}^{d}\left|\frac{\partial}{\partial x_j}\right|\right)^l f(x)\right)\left[\left(\frac{\rho_i\|x\|_\infty+\lambda_i+1}{n+\rho_i}\right)+\left(1+\frac{\rho_i}{n+\rho_i}\right)\frac{T^*}{n^{1-\alpha}}\right]^l$$

$$+\frac{d^N}{N!}\left[\left(\frac{\rho_i\|x\|_\infty+\lambda_i+1}{n+\rho_i}\right)+\left(1+\frac{\rho_i}{n+\rho_i}\right)\frac{T^*}{n^{1-\alpha}}\right]^N\cdot \tag{12.49}$$

$$\max_{\widetilde{\alpha}:|\widetilde{\alpha}|=N}\omega_1\left(f_{\widetilde{\alpha}},\left[\left(\frac{\rho_i\|x\|_\infty+\lambda_i+1}{n+\rho_i}\right)+\left(1+\frac{\rho_i}{n+\rho_i}\right)\frac{T^*}{n^{1-\alpha}}\right]\right).$$

Hence we have that

$$\sum_{i=1}^{r}w_i\left[\frac{1}{\mu\left(\left[0,\frac{1}{n+\rho_i}\right]^d\right)}(N^*)\int_0^{\frac{1}{n+\rho_i}}\left|f\left(t+\frac{k+\lambda_i}{n+\rho_i}\right)-f(x)\right|d\mu(t)\right]\overset{(12.49)}{\leq}$$

$$\sum_{i=1}^{r}w_i\left\{\sum_{l=1}^{N}\frac{1}{l!}\left(\left(\sum_{j=1}^{d}\left|\frac{\partial}{\partial x_j}\right|\right)^l f(x)\right)\cdot\right. \tag{12.50}$$

$$\left[\left(\frac{\rho_i\|x\|_\infty+\lambda_i+1}{n+\rho_i}\right)+\left(1+\frac{\rho_i}{n+\rho_i}\right)\frac{T^*}{n^{1-\alpha}}\right]^l$$

$$+\frac{d^N}{N!}\left[\left(\frac{\rho_i\|x\|_\infty+\lambda_i+1}{n+\rho_i}\right)+\left(1+\frac{\rho_i}{n+\rho_i}\right)\frac{T^*}{n^{1-\alpha}}\right]^N\cdot$$

$$\left.\max_{\widetilde{\alpha}:|\widetilde{\alpha}|=N}\omega_1\left(f_{\widetilde{\alpha}},\left[\left(\frac{\rho_i\|x\|_\infty+\lambda_i+1}{n+\rho_i}\right)+\left(1+\frac{\rho_i}{n+\rho_i}\right)\frac{T^*}{n^{1-\alpha}}\right]\right)\right\}.$$

Consequently we derive

$$\left|\left(M_n^\mu(f)\right)(x)-f(x)\right|\overset{((12.39),(12.50))}{\leq}$$

$$\sum_{k=\lceil nx-Tn^\alpha\rceil}^{[nx+Tn^\alpha]}\left[\sum_{l=1}^{N}\frac{1}{l!}\sum_{i=1}^{r}w_i\left[\left(\frac{\rho_i\|x\|_\infty+\lambda_i+1}{n+\rho_i}\right)+\left(1+\frac{\rho_i}{n+\rho_i}\right)\frac{T^*}{n^{1-\alpha}}\right]^l\cdot\right.$$

$$\left(\left(\sum_{j=1}^{d}\left|\frac{\partial}{\partial x_j}\right|\right)^l f(x)\right)+\frac{d^N}{N!}\sum_{i=1}^{r}w_i\left[\left(\frac{\rho_i\|x\|_\infty+\lambda_i+1}{n+\rho_i}\right)+\left(1+\frac{\rho_i}{n+\rho_i}\right)\frac{T^*}{n^{1-\alpha}}\right]^N\cdot \tag{12.51}$$

$$\frac{\left.\max_{\widetilde{\alpha}:|\widetilde{\alpha}|=N}\omega_1\left(f_{\widetilde{\alpha}},\left[\left(\frac{\rho_i\|x\|_\infty+\lambda_i+1}{n+\rho_i}\right)+\left(1+\frac{\rho_i}{n+\rho_i}\right)\frac{T^*}{n^{1-\alpha}}\right]\right)\right]b\left(n^{1-\alpha}\left(x-\frac{k}{n}\right)\right)}{V(x)}=$$

$$\sum_{l=1}^{N}\frac{1}{l!}\sum_{i=1}^{r}w_i\left[\left(\frac{\rho_i\|x\|_\infty+\lambda_i+1}{n+\rho_i}\right)+\left(1+\frac{\rho_i}{n+\rho_i}\right)\frac{T^*}{n^{1-\alpha}}\right]^l\cdot$$

$$\left(\left(\sum_{j=1}^{d}\left|\frac{\partial}{\partial x_j}\right|\right)^{l} f(x)\right)+\frac{d^N}{N!}\sum_{i=1}^{r}w_i\left[\left(\frac{\rho_i\|x\|_\infty+\lambda_i+1}{n+\rho_i}\right)+\left(1+\frac{\rho_i}{n+\rho_i}\right)\frac{T^*}{n^{1-\alpha}}\right]^N\cdot \tag{12.52}$$

$$\max_{\widetilde{\alpha}:|\widetilde{\alpha}|=N}\omega_1\left(f_{\widetilde{\alpha}},\left[\left(\frac{\rho_i\|x\|_\infty+\lambda_i+1}{n+\rho_i}\right)+\left(1+\frac{\rho_i}{n+\rho_i}\right)\frac{T^*}{n^{1-\alpha}}\right]\right),$$

proving the claim. ■

We have

Corollary 12.8 *Let all as in Theorem 12.7. Additionally, assume that* $f_{\widetilde{\alpha}}(x)=0$, $\widetilde{\alpha}:|\widetilde{\alpha}|=\rho$, $1\le\rho\le N$. *Then*

$$\sup_{\mu}\left|\left(M_n^{\mu}(f)\right)(x)-f(x)\right|\le$$

$$\frac{d^N}{N!}\sum_{i=1}^{r}w_i\left[\left(\frac{\rho_i\|x\|_\infty+\lambda_i+1}{n+\rho_i}\right)+\left(1+\frac{\rho_i}{n+\rho_i}\right)\frac{T^*}{n^{1-\alpha}}\right]^N\cdot \tag{12.53}$$

$$\max_{\widetilde{\alpha}:|\widetilde{\alpha}|=N}\omega_1\left(f_{\widetilde{\alpha}},\left[\left(\frac{\rho_i\|x\|_\infty+\lambda_i+1}{n+\rho_i}\right)+\left(1+\frac{\rho_i}{n+\rho_i}\right)\frac{T^*}{n^{1-\alpha}}\right]\right).$$

Inequality (12.53) implies the pointwise convergence with rates of $\left(M_n^{\mu}(f)\right)(x)$ *to* $f(x)$, *as* $n\to\infty$, *at speed* $\frac{1}{n^{(1-\alpha)(N+1)}}$.

Proof By (12.38) we get the desired result. ■

The uniform convergence with rates follows from the following

Corollary 12.9 *All as in Theorem 12.7. Let* $x\in G:=\prod_{j=1}^{d}\left[-\gamma_j,\gamma_j\right]\subset\mathbb{R}^d$, $\gamma_j>0$, $\gamma^*:=\max\{\gamma_1,\ldots,\gamma_d\}$ *and* $n\in\mathbb{N}:n\ge\max_{j\in\{1,\ldots,d\}}\left(T_j+\gamma_j,T_j^{-\frac{1}{\alpha}}\right)$. *Then*

$$\sup_{\mu}\left\|M_n^{\mu}(f)-f\right\|_{\infty,\prod_{j=1}^{d}[-\gamma_j,\gamma_j]}\le\sum_{l=1}^{N}\frac{1}{l!}\left(\left(\sum_{j=1}^{d}\left|\frac{\partial}{\partial x_j}\right|\right)^{l}\|f\|_{\infty,\prod_{j=1}^{d}[-\gamma_j,\gamma_j]}\right)$$

$$\left[\sum_{i=1}^{r}w_i\left[\left(\frac{\rho_i\gamma^*+\lambda_i+1}{n+\rho_i}\right)+\left(1+\frac{\rho_i}{n+\rho_i}\right)\frac{T^*}{n^{1-\alpha}}\right]^{l}\right]+ \tag{12.54}$$

$$\frac{d^N}{N!}\sum_{i=1}^{r}w_i\left[\left(\frac{\rho_i\gamma^*+\lambda_i+1}{n+\rho_i}\right)+\left(1+\frac{\rho_i}{n+\rho_i}\right)\frac{T^*}{n^{1-\alpha}}\right]^N\cdot$$

$$\max_{\widetilde{\alpha}:|\widetilde{\alpha}|=N} \omega_1 \left(f_{\widetilde{\alpha}}, \left[\left(\frac{\rho_i \gamma^* + \lambda_i + 1}{n + \rho_i} \right) + \left(1 + \frac{\rho_i}{n + \rho_i} \right) \frac{T^*}{n^{1-\alpha}} \right] \right).$$

Inequality (12.54) implies the pointwise convergence with rates of $M_n^{\mu}(f)$ to f on G, as $n \to \infty$, at speed $\frac{1}{n^{1-\alpha}}$.

Proof By (12.38) we get the desired result. ■

Corollary 12.10 *All as in Theorem 12.7 with $N = 1$ hold. Then*

$$\sup_{\mu} \left| \left(M_n^{\mu}(f) \right)(x) - f(x) \right| \leq \left(\sum_{j=1}^{d} \left| \frac{\partial f(x)}{\partial x_j} \right| \right) \cdot$$

$$\left[\sum_{i=1}^{r} w_i \left[\left(\frac{\rho_i \|x\|_{\infty} + \lambda_i + 1}{n + \rho_i} \right) + \left(1 + \frac{\rho_i}{n + \rho_i} \right) \frac{T^*}{n^{1-\alpha}} \right] \right] + \tag{12.55}$$

$$d \sum_{i=1}^{r} w_i \left[\left(\frac{\rho_i \|x\|_{\infty} + \lambda_i + 1}{n + \rho_i} \right) + \left(1 + \frac{\rho_i}{n + \rho_i} \right) \frac{T^*}{n^{1-\alpha}} \right] \cdot$$

$$\max_{\widetilde{\alpha}:|\widetilde{\alpha}|=N} \omega_1 \left(f_{\widetilde{\alpha}}, \left[\left(\frac{\rho_i \|x\|_{\infty} + \lambda_i + 1}{n + \rho_i} \right) + \left(1 + \frac{\rho_i}{n + \rho_i} \right) \frac{T^*}{n^{1-\alpha}} \right] \right).$$

Inequality (12.55) implies the pointwise convergence with rates of $\left(M_n^{\mu}(f) \right)(x)$ to $f(x)$, as $n \to \infty$, at speed $\frac{1}{n^{1-\alpha}}$.

Proof By (12.38) we get the desired result. ■

We finish with the following

Note Notice here that the right hand sides of (12.38), (12.53), (12.54) and (12.55), do not depend on μ and b.

References

1. G.A. Anastassiou, *Quantitative Approximations* (Chapman and Hall/CRC, Boca Raton, 2001)
2. G.A. Anastassiou, *Intelligent Systems: Approximation by Artificial Neural Networks* (Springer, Heidelberg, 2011)
3. G.A. Anastassiou, *Intelligent Systems II: Complete Approximation by Neural Network operators* (Springer, Heidelberg, 2016)
4. G. Anastassiou, Multivariate approximation with rates by perturbed Kantorovich-Shilkret neural network operators (2018). Submitted
5. P. Cardaliaguet, G. Euvrard, Approximation of a function and its derivative with a neural network. Neural Netw. **5**, 207–220 (1992)
6. N. Shilkret, Maxitive measure and integration. Indag. Math. **33**, 109–116 (1971)

Chapter 13
Quantitative Approximation by Shift Invariant Univariate Sublinear-Shilkret Operators

A very general positive sublinear Shilkret integral type operator is given through a convolution-like iteration of another general positive sublinear operator with a scaling type function. For it sufficient conditions are given for shift invariance, preservation of global smoothness, convergence to the unit with rates. Additionally, two examples of very general specialized operators are presented fulfilling all the above properties, the higher order of approximation of these operators is also considered. It follows [3].

13.1 Introduction

Let X, Y be function spaces of functions from $\mathbb{R}$ into $\mathbb{R}_{+}$. Let $L_N : X \to Y$, $N \in \mathbb{N}$, be a sequence of operators with the following properties:

(i) (positive homogeneous)

$$L_N(\alpha f) = \alpha L_N(f), \forall\, \alpha \geq 0, \forall\, f \in X.$$

(ii) (Monotonicity) if $f, g \in X$ satisfy $f \leq g$, then $L_N(f) \leq L_N(g)$, $\forall\, N \in \mathbb{N}$, and

(iii) (Subadditivity)

$$L_N(f+g) \leq L_N(f) + L_N(g)\ , \forall\, f, g \in X.$$

We call L_N positive sublinear operators.

In this chapter we deal with sequences of Shilkret positive sublinear operators that are constructed, with the help of Shilkret integral [6]. Our functions spaces are continuous functions from $\mathbb{R}$ into $\mathbb{R}_{+}$. The sequence of operators is generated by a basic operator via dilated translations of convolution type using the Shilkret integral. We prove that our operators possess the following properties: of shift invariance of

G. A. Anastassiou, *Ordinary and Fractional Approximation by Non-additive Integrals: Choquet, Shilkret and Sugeno Integral Approximators*, Studies in Systems, Decision and Control 190, https://doi.org/10.1007/978-3-030-04287-5_13

global smoothness preservation, of convergence to the unit operator with rates. Then we apply our results to two specific families of such Shilkret type operators.

We continue with the higher order of approximation study of these specific operators, and all results are quantitative.

Earlier similar studies have been done by the author, see [5], Chapters 10–17, and [2], Chapters 16, 17. These serve as motivation and inspiration to this work.

13.2 Background

Here we follow [6].

Let $\mathcal{F}$ be a σ-field of subsets of an arbitrary set Ω. An extended non-negative real valued function μ on $\mathcal{F}$ is called maxitive if $\mu(\varnothing) = 0$ and

$$\mu\left(\cup_{i \in I} E_i\right) = \sup_{i \in I} \mu\left(E_i\right), \tag{13.1}$$

where the set I is of cardinality at most countable, where $\{E_i\}_{i \in I}$ is a disjoint collection of sets from $\mathcal{F}$. We notice that μ is monotone and (13.1) is true even $\{E_i\}_{i \in I}$ are not disjoint. For more properties of μ see [6]. We also call μ a maxitive measure. Here f stands for a non-negative measurable function on Ω. In [6], Niel Shilkret developed his non-additive integral defined as follows:

$$\left(N^*\right) \int_D f d\mu := \sup_{y \in Y} \left\{y \cdot \mu\left(D \cap \{f \geq y\}\right)\right\}, \tag{13.2}$$

where $Y = [0, m]$ or $Y = [0, m)$ with $0 < m \leq \infty$, and $D \in \mathcal{F}$. Here we take $Y = [0, \infty)$.

It is easily proved that

$$\left(N^*\right) \int_D f d\mu = \sup_{y > 0} \left\{y \cdot \mu\left(D \cap \{f > y\}\right)\right\}. \tag{13.3}$$

The Shilkret integral takes values in $[0, \infty]$.

The Shilkret integral [6] has the following properties:

$$\left(N^*\right) \int_{\Omega} \chi_E d\mu = \mu(E), \tag{13.4}$$

where χ_E is the indicator function on $E \in \mathcal{F}$,

$$\left(N^*\right) \int_D c f d\mu = c \left(N^*\right) \int_D f d\mu, \ c \geq 0, \tag{13.5}$$

$$\left(N^*\right)\int_D \sup_{n\in\mathbb{N}} f_n d\mu = \sup_{n\in\mathbb{N}}\left(N^*\right)\int_D f_n d\mu, \tag{13.6}$$

where f_n, $n \in \mathbb{N}$, is an increasing sequence of elementary (countably valued) functions converging uniformly to f. Furthermore we have

$$\left(N^*\right)\int_D f d\mu \geq 0, \tag{13.7}$$

$$f \geq g \text{ implies } \left(N^*\right)\int_D f d\mu \geq \left(N^*\right)\int_D g d\mu, \tag{13.8}$$

where $f, g : \Omega \to [0, \infty]$ are measurable.

Let $a \leq f(\omega) \leq b$ for almost every $\omega \in E$, then

$$a\mu(E) \leq \left(N^*\right)\int_E f d\mu \leq b\mu(E); \tag{13.9}$$

$$\left(N^*\right)\int_E 1 d\mu = \mu(E); \tag{13.10}$$

$f > 0$ almost everywhere and $(N^*)\int_E f d\mu = 0$ imply $\mu(E) = 0$;

$(N^*)\int_\Omega f d\mu = 0$ if and only $f = 0$ almost everywhere;

$(N^*)\int_\Omega f d\mu < \infty$ implies that

$$\overline{N}(f) := \{\omega \in \Omega | f(\omega) \neq 0\} \text{ has } \sigma\text{-finite measure;}$$

$$\left(N^*\right)\int_D (f+g) d\mu \leq \left(N^*\right)\int_D f d\mu + \left(N^*\right)\int_D g d\mu; \tag{13.11}$$

and

$$\left|\left(N^*\right)\int_D f d\mu - \left(N^*\right)\int_D g d\mu\right| \leq \left(N^*\right)\int_D |f-g| d\mu. \tag{13.12}$$

From now on in this chapter we assume that $\mu : \mathcal{F} \to [0, +\infty)$.

13.3 Univariate Theory

This section is motivated and inspired by [4, 5].

Let $\mathcal{L}$ be the Lebesgue $\sigma-$ algebra on $\mathbb{R}$, and the set function $\mu : \mathcal{L} \to [0, +\infty]$, which is assumed to be maxitive. Let $C_U(\mathbb{R}, \mathbb{R}_+)$ be the space of uniformly continuous functions from $\mathbb{R}$ into $\mathbb{R}_+$, and $C(\mathbb{R}, \mathbb{R}_+)$ the space of continuous functions from $\mathbb{R}$ into $\mathbb{R}_+$. For any $f \in C_U(\mathbb{R}, \mathbb{R}_+)$ we have $\omega_1(f, \delta) < +\infty$, $\delta > 0$, where

$$\omega_1(f,\delta) := \sup_{\substack{x,y\in\mathbb{R}:\\ |x-y|\leq\delta}} |f(x)-f(y)|, \delta>0,$$

is the first modulus of continuity.

Let $\{t_k\}_{k\in\mathbb{Z}}$ be a sequence of positive sublinear operators that map $C_U(\mathbb{R},\mathbb{R}_+)$ into $C(\mathbb{R},\mathbb{R}_+)$ with the property

$$(t_k(f))(x) := l_0\left(f\left(2^{-k}\cdot\right)\right)(x), \forall\, x\in\mathbb{R}, \forall\, f\in C_U(\mathbb{R},\mathbb{R}_+). \tag{13.13}$$

For a fixed $a>0$ we assume that

$$\sup_{\substack{u,y\in\mathbb{R}:\\ |u-y|\leq a}} |t_0(f,u)-f(y)| \leq \omega_1\left(f, \frac{ma+n}{2^r}\right), \ \forall\, f\in C_U(\mathbb{R},\mathbb{R}_+), \tag{13.14}$$

where $m\in\mathbb{N}$, $n\in\mathbb{Z}_+$, $r\in\mathbb{Z}$.

Let $\psi:\mathbb{R}\to\mathbb{R}_+$ which is Lebesgue measurable, such that

$$\left(N^*\right)\int_{-a}^{a}\psi(u)\,d\mu(u)=1. \tag{13.15}$$

We define the positive sublinear-Shilkret operators

$$(T_0(f))(x) := \left(N^*\right)\int_{-a}^{a}(t_0 f)(x-u)\,\psi(u)\,d\mu(u), \tag{13.16}$$

and

$$(T_k(f))(x) := \left(T_0\left(f\left(2^{-k}\cdot\right)\right)\right)\left(2^k x\right), \forall\, k\in\mathbb{Z}, \forall\, x\in\mathbb{R}. \tag{13.17}$$

Therefore it holds

$$(T_k(f))(x) = \left(N^*\right)\int_{-a}^{a}\left(t_0\left(f\left(2^{-k}\cdot\right)\right)\right)\left(2^k x-u\right)\psi(u)\,d\mu(u) = \tag{13.18}$$

$$\left(N^*\right)\int_{-a}^{a}(t_k(f))\left(2^k x-u\right)\psi(u)\,d\mu(u),$$

$\forall\, x\in\mathbb{R}$, $\forall\, k\in\mathbb{Z}$.

Indeed here we have

$$(T_k(f))(x) \overset{(13.8)}{\leq} \left(N^*\right)\int_{-a}^{a}\left\|t_k(f)\left(2^k x-\cdot\right)\right\|_{\infty,[-a,a]}\psi(u)\,d\mu(u) \overset{(13.5)}{=}$$

$$\left\|t_k(f)\left(2^k x-\cdot\right)\right\|_{\infty,[-a,a]}\left(\left(N^*\right)\int_{-a}^{a}\psi(u)\,d\mu(u)\right) = \tag{13.19}$$

$$\left\| t_k(f)\left(2^k x - \cdot\right)\right\|_{\infty,[-a,a]} < +\infty.$$

Hence $(T_k(f))(x) \in \mathbb{R}_+$ is well-defined.

Let $f, g \in \mathcal{M}(\mathbb{R}, \mathbb{R}_+)$ (Lebesgue measurable functions) where $X \in A$, $A \subset \mathbb{R}$ is a Lebesgue measurable set.

We derive that

$$\left|(N^*)\int_A f(x)\,d\mu(x) - N^*\int_A g(x)\,d\mu(x)\right| \overset{(13.12)}{\leq} (N^*)\int_A |f(x) - g(x)|\,d\mu(x). \tag{13.20}$$

We need

Definition 13.1 Let $f_\alpha(\cdot) := f(\cdot + \alpha)$, $\alpha \in \mathbb{R}$, and Φ be an operator. If $\Phi(f_\alpha) = (\Phi f)_\alpha$, then Φ is called a shift invariant operator.

We give

Theorem 13.2 *Assume that*

$$\left(t_0\left(f\left(2^{-k}\cdot + \alpha\right)\right)\right)\left(2^k u\right) = \left(t_0\left(f\left(2^{-k}\cdot\right)\right)\right)\left(2^k(u + \alpha)\right), \tag{13.21}$$

for all $k \in \mathbb{Z}$, $\alpha \in \mathbb{R}$ fixed, all $u \in \mathbb{R}$ and any $f \in C_U(\mathbb{R}, \mathbb{R}_+)$. Then T_k is a shift invariant operator for all $k \in \mathbb{Z}$.

Proof We have that

$$(T_k(f(\cdot + \alpha)))(x) = (T_k(f_\alpha))(x) \overset{(13.18)}{=}$$

$$(N^*)\int_{-a}^{a}\left(t_0\left(f_\alpha\left(2^{-k}\cdot\right)\right)\right)\left(2^k x - u\right)\psi(u)\,d\mu(u) =$$

$$(N^*)\int_{-a}^{a}\left(t_0\left(f\left(2^{-k}\cdot + \alpha\right)\right)\right)\left(2^k x - u\right)\psi(u)\,d\mu(u) =$$

$$(N^*)\int_{-a}^{a}\left(t_0\left(f\left(2^{-k}\cdot + \alpha\right)\right)\right)\left(2^k\left(x - 2^{-k}u\right)\right)\psi(u)\,d\mu(u) \overset{(13.21)}{=} \tag{13.22}$$

$$(N^*)\int_{-a}^{a}\left(t_0\left(f\left(2^{-k}\cdot\right)\right)\right)\left(2^k\left(x - 2^{-k}u + \alpha\right)\right)\psi(u)\,d\mu(u) =$$

$$(N^*)\int_{-a}^{a}\left(t_0\left(f\left(2^{-k}\cdot\right)\right)\right)\left(2^k(x + \alpha) - u\right)\psi(u)\,d\mu(u) \overset{(13.18)}{=} (T_k(f))(x + \alpha),$$

that is

$$T_k(f_\alpha) = (T_k(f))_\alpha, \tag{13.23}$$

proving the claim. ∎

It follows the global smoothness of the operators T_k.

Theorem 13.3 *For any $f \in C_U(\mathbb{R}, \mathbb{R}_+)$ assume that, for all $u \in \mathbb{R}$,*

$$|(t_0(f))(x-u) - (t_0(f))(y-u)| \leq \omega_1(f, |x-y|), \quad (13.24)$$

for any $x, y \in \mathbb{R}$. Then

$$\omega_1(T_k f, \delta) \leq \omega_1(f, \delta), \forall \delta > 0. \quad (13.25)$$

Proof We observe that

$$|(T_0(f))(x) - (T_0(f))(y)| =$$

$$\left|(N^*)\int_{-a}^{a}(t_0 f)(x-u)\psi(u)\,d\mu(u) - (N^*)\int_{-a}^{a}(t_0 f)(y-u)\psi(u)\,d\mu(u)\right| \overset{(13.20)}{\leq} \quad (13.26)$$

$$(N^*)\int_{-a}^{a}|(t_0 f)(x-u) - (t_0 f)(y-u)|\,\psi(u)\,d\mu(u) \overset{(\text{by } (13.24), (13.5))}{\leq}$$

$$\omega_1(f, |x-y|)\left((N^*)\int_{-a}^{a}\psi(u)\,d\mu(u)\right) \overset{(13.15)}{=} \omega_1(f, |x-y|).$$

So that

$$|(T_0(f))(x) - (T_0(f))(y)| \leq \omega_1(f, |x-y|). \quad (13.27)$$

From (13.17), (13.27) we get

$$|(T_k(f))(x) - (T_k(f))(y)| \overset{(13.17)}{=}$$

$$\left|(T_0(f(2^{-k}\cdot)))(2^k x) - (T_0(f(2^{-k}\cdot)))(2^k y)\right| \leq \quad (13.28)$$

$$\omega_1\left(f(2^{-k}\cdot), 2^k|x-y|\right) = \omega_1(f, |x-y|),$$

i.e. global smoothness for T_k has been proved. ■

The convergence of T_k to the unit operator, as $k \to +\infty$, k with rates follows:

Theorem 13.4 *For $f \in C_U(\mathbb{R}, \mathbb{R}_+)$, under the assumption (13.14), we have*

$$|(T_k(f))(x) - f(x)| \leq \omega_1\left(f, \frac{ma+n}{2^{k+r}}\right), \quad (13.29)$$

where $m \in \mathbb{N}$, $n \in \mathbb{Z}_+$, $k, r \in \mathbb{Z}$.

Proof We notice that

$$\left|\left(T_k(f)\right)(x)-f(x)\right| \overset{(13.17)}{=} \left|\left(T_0\left(f\left(2^{-k}\cdot\right)\right)\right)\left(2^k x\right)-f(x)\right| \overset{(13.18)}{=}$$

$$\left|\left(N^*\right)\int_{-a}^{a}\left(t_0\left(f\left(2^{-k}\cdot\right)\right)\right)\left(2^k x-u\right)\psi(u)\,d\mu(u)-f(x)\right| \overset{(13.15)}{=}$$

$$\left|\left(N^*\right)\int_{-a}^{a}\left(t_0\left(f\left(2^{-k}\cdot\right)\right)\right)\left(2^k x-u\right)\psi(u)\,d\mu(u)-\left(N^*\right)\int_{-a}^{a}f(x)\psi(u)\,d\mu(u)\right| \overset{(13.20)}{\leq}$$

$$\left(N^*\right)\int_{-a}^{a}\left|\left(t_0\left(f\left(2^{-k}\cdot\right)\right)\right)\left(2^k x-u\right)-f(x)\right|\psi(u)\,d\mu(u)= \tag{13.30}$$

$$\left(N^*\right)\int_{-a}^{a}\left|\left(t_0\left(f\left(2^{-k}\cdot\right)\right)\right)\left(2^k x-u\right)-f\left(2^{-k}\cdot\right)\left(2^k x\right)\right|\psi(u)\,d\mu(u) \overset{(13.14)}{\leq}$$

(here $\left|\left(2^k x-u\right)-2^k x\right|=|u|\leq a$)

$$\omega_1\left(f\left(2^{-k}\cdot\right),\frac{ma+n}{2^r}\right)\left(\left(N^*\right)\int_{-a}^{a}\psi(u)\,d\mu(u)\right) \overset{(13.15)}{=}$$

$$\omega_1\left(f\left(2^{-k}\cdot\right),\frac{ma+n}{2^r}\right)\cdot 1=\omega_1\left(f,\frac{ma+n}{2^{k+r}}\right), \tag{13.31}$$

proving the claim. ∎

We give some applications.

For each $k\in\mathbb{Z}$, we define

(i)

$$\left(B_k f\right)(x):=\left(N^*\right)\int_{-a}^{a}f\left(x-\frac{u}{2^k}\right)\psi(u)\,d\mu(u), \tag{13.32}$$

i.e., here

$$\begin{gathered}\left(t_k(f)\right)(u)-f\left(\tfrac{u}{2^k}\right),\\ \text{and}\\ \left(t_0(f)\right)(u)=f(u),\end{gathered} \tag{13.33}$$

are continuous in $u\in\mathbb{R}$.

Also for $k\in\mathbb{Z}$, we define

(ii)

$$\left(\Gamma_k(f)\right)(x):=\left(N^*\right)\int_{-a}^{a}\gamma_k^f\left(2^k x-u\right)\psi(u)\,d\mu(u), \tag{13.34}$$

where

$$(t_k(f))(u) = \gamma_k^f(u) := \sum_{j=0}^{n} w_j f\left(\frac{u}{2^k} + \frac{j}{2^k n}\right), \tag{13.35}$$

$n \in \mathbb{N}$, $w_j \geq 0$, $\sum_{j=0}^{n} w_j = 1$,
is continuous in $u \in \mathbb{R}$.

Notice here that

$$(t_0(f))(u) = \gamma_0^f(u) = \sum_{j=0}^{n} w_j f\left(u + \frac{j}{n}\right) \tag{13.36}$$

is also continuous in $u \in \mathbb{R}$.

Indeed we have

$$(\Gamma_k(f))(x) = \left(N^*\right) \int_{-a}^{a} \left[\sum_{j=0}^{n} w_j f\left(\left(x - \frac{u}{2^k}\right) + \frac{j}{2^k n}\right)\right] \psi(u)\, d\mu(u). \tag{13.37}$$

Clearly here we have

$$\begin{gathered}(B_k(f))(x) = \left(B_0\left(f\left(2^{-k}\cdot\right)\right)\right)\left(2^k x\right), \\ \text{and} \\ (\Gamma_k(f))(x) = \left(\Gamma_0\left(f\left(2^{-k}\cdot\right)\right)\right)\left(2^k x\right),\end{gathered} \tag{13.38}$$

$\forall\, k \in \mathbb{Z}$, $\forall\, x \in \mathbb{R}$.

We give

Proposition 13.5 *B_k, Γ_k are shift invariant operators.*

Proof (i) For B_k operators: Here $t_0 f = f$. Hence

$$\left(t_0\left(f\left(2^{-k}\cdot + \alpha\right)\right)\right)\left(2^k u\right) = f\left(2^{-k} 2^k u + \alpha\right) = f(u + \alpha) = \tag{13.39}$$

$$\left(t_0\left(f\left(2^{-k}\cdot\right)\right)\right)\left(2^k(u + \alpha)\right).$$

(ii) For Γ_k operators:

$$(t_0(f))(u) = \sum_{j=0}^{n} w_j f\left(u + \frac{j}{n}\right).$$

Hence

$$\left(t_0\left(f\left(2^{-k}\cdot+\alpha\right)\right)\right)\left(2^k u\right)=\sum_{j=0}^{n} w_j f\left(2^{-k}\left(2^k u+\frac{j}{n}\right)+\alpha\right)=$$

$$\sum_{j=0}^{n} w_j f\left(2^{-k}\left(2^k\left(u+\alpha\right)+\frac{j}{n}\right)\right)=\left(t_0\left(f\left(2^{-k}\cdot\right)\right)\right)\left(2^k\left(u+\alpha\right)\right), \quad (13.40)$$

proving the claim. ∎

Next we show that the operators B_k, Γ_k possess the property of global smoothness preservation.

Theorem 13.6 *For all $f \in C_U\left(\mathbb{R}, \mathbb{R}_+\right)$ and all $\delta > 0$ we have*

$$\begin{gathered}\omega_1\left(B_k f, \delta\right) \leq \omega_1\left(f, \delta\right), \\ \textit{and} \\ \omega_1\left(\Gamma_k f, \delta\right) \leq \omega_1\left(f, \delta\right).\end{gathered} \quad (13.41)$$

Proof (i) For B_k operators: Here $t_0 f = f$, therefore

$$\left|\left(t_0\left(f\right)\right)\left(x-u\right)-\left(t_0\left(f\right)\right)\left(y-u\right)\right|=\left|f\left(x-u\right)-f\left(y-u\right)\right|\leq\omega_1\left(f,\left|x-y\right|\right). \quad (13.42)$$

(ii) For Γ_k operators: We observe that

$$\left|\left(t_0\left(f\right)\right)\left(x-u\right)-\left(t_0\left(f\right)\right)\left(y-u\right)\right|=\left|\gamma_0^f\left(x-u\right)-\gamma_0^f\left(y-u\right)\right|=$$

$$\left|\sum_{j=0}^{n} w_j\left(f\left(x-u+\frac{j}{n}\right)-f\left(y-u+\frac{j}{n}\right)\right)\right|\leq$$

$$\sum_{j=0}^{n} w_j\left|f\left(x-u+\frac{j}{n}\right)-f\left(y-u+\frac{j}{n}\right)\right|\leq$$

$$\omega_1\left(f,\left|x-y\right|\right)\left(\sum_{j=0}^{n} w_j\right)=\omega_1\left(f,\left|x-y\right|\right), \quad (13.43)$$

proving the claim. ∎

The operators $B_k, \Gamma_k, k \in \mathbb{Z}$, converge to the unit operator with rates presented next.

Theorem 13.7 *For $k \in \mathbb{Z}$,*

$$\left|\left(B_k(f)\right)(x) - f(x)\right| \leq \omega_1\left(f, \tfrac{a}{2^k}\right), \quad \textit{and} \quad \left|\left(\Gamma_k(f)\right)(x) - f(x)\right| \leq \omega_1\left(f, \tfrac{a+1}{2^k}\right). \tag{13.44}$$

Proof (i) For B_k operators: Here $(t_0(f))(u) = f(u)$ and

$$\sup_{\substack{u,y\in\mathbb{R}\\|u-y|\leq a}} \left|\left(t_0(f)\right)(u) - f(y)\right| = \sup_{\substack{u,y\in\mathbb{R}\\|u-y|\leq a}} \left|f(u) - f(y)\right| = \omega_1(f, a), \tag{13.45}$$

and we use Theorem 13.4.

(ii) For Γ_k operators: Here we see that

$$\sup_{\substack{u,y\in\mathbb{R}\\|u-y|\leq a}} \left|\left(t_0(f)\right)(u) - f(y)\right| = \sup_{\substack{u,y\in\mathbb{R}\\|u-y|\leq a}} \left|\sum_{j=0}^{n} w_j f\left(u + \frac{j}{n}\right) - f(y)\right| \leq$$

$$\sup_{\substack{u,y\in\mathbb{R}\\|u-y|\leq a}} \sum_{j=0}^{n} w_j \left|f\left(u + \frac{j}{n}\right) - f(y)\right| \leq \sup_{\substack{u,y\in\mathbb{R}\\|u-y|\leq a}} \sum_{j=0}^{n} w_j \omega_1\left(f, \left|u + \frac{j}{n} - y\right|\right) \leq \tag{13.46}$$

$$\sup_{\substack{u,y\in\mathbb{R}\\|u-y|\leq a}} \sum_{j=0}^{n} w_j \omega_1\left(f, \frac{j}{n} + |u-y|\right) \leq \left(\sum_{j=0}^{n} w_j\right) \omega_1(f, 1+\alpha) = \omega_1(f, \alpha+1).$$

By (13.29) we are done. ∎

13.4 Higher Order of Approximation

Here all are as in Sect. 13.3. See also earlier our work [1, 2], Chapter 16.

We give

Theorem 13.8 *Let $f \in C^N(\mathbb{R}, \mathbb{R}_+)$, $N \geq 1$. Consider the Shilkret-sublinear operators*

$$\left(B_k f\right)(x) = \left(N^*\right) \int_{-a}^{a} f\left(x - \frac{u}{2^k}\right) \psi(u)\, d\mu(u),$$

$\forall\, k \in \mathbb{Z}$, $\forall\, x \in \mathbb{R}$. Then

$$\left|\left(B_k f\right)(x) - f(x)\right| \leq \sum_{i=1}^{N} \frac{\left|f^{(i)}(x)\right|}{i!} \frac{a^i}{2^{ki}} + \frac{a^N}{2^{kN} N!} \omega_1\left(f^{(N)}, \frac{a}{2^k}\right). \tag{13.47}$$

If $f^{(N)}$ is uniformly continuous or bounded and continuous, then as $k \to +\infty$ we obtain that $(B_k f)(x) \to f(x)$ pointwise with rates.

Proof Since $f \in C^N(\mathbb{R}, \mathbb{R}_+)$, $N \geq 1$, by Taylor's formula we have

$$f\left(x - \frac{u}{2^k}\right) - f(x) = \sum_{i=1}^{N} \frac{f^{(i)}(x)}{i!}\left(-\frac{u}{2^k}\right)^i + \tag{13.48}$$

$$\int_x^{x-\frac{u}{2^k}} \left(f^{(N)}(t) - f^{(N)}(x)\right) \frac{\left(x - \frac{u}{2^k} - t\right)^{N-1}}{(N-1)!} dt.$$

Call

$$\Gamma_u(x) := \left| \int_x^{x-\frac{u}{2^k}} \left(f^{(N)}(t) - f^{(N)}(x)\right) \frac{\left(x - \frac{u}{2^k} - t\right)^{N-1}}{(N-1)!} dt \right|. \tag{13.49}$$

Next we estimate $\Gamma_u(x)$, where $u \in [-a, a]$.

(i) Case of $-a \leq u \leq 0$, then $x \leq x - \frac{u}{2^k}$. Then

$$\Gamma_u(x) \leq \int_x^{x-\frac{u}{2^k}} \left|f^{(N)}(t) - f^{(N)}(x)\right| \frac{\left(x - \frac{u}{2^k} - t\right)^{N-1}}{(N-1)!} dt \leq$$

$$\int_x^{x-\frac{u}{2^k}} \omega_1\left(f^{(N)}, |t - x|\right) \frac{\left(x - \frac{u}{2^k} - t\right)^{N-1}}{(N-1)!} dt \leq$$

$$\omega_1\left(f^{(N)}, \frac{|u|}{2^k}\right) \int_x^{x-\frac{u}{2^k}} \frac{\left(x - \frac{u}{2^k} - t\right)^{N-1}}{(N-1)!} dt \leq \tag{13.50}$$

$$\omega_1\left(f^{(N)}, \frac{a}{2^k}\right) \frac{\left(-\frac{u}{2^k}\right)^N}{N!} \leq \omega_1\left(f^{(N)}, \frac{a}{2^k}\right) \frac{a^N}{2^{kN} N!}.$$

That is, when $-a \leq u \leq 0$, then

$$\Gamma_u(x) \leq \omega_1\left(f^{(N)}, \frac{a}{2^k}\right) \frac{a^N}{2^{kN} N!}. \tag{13.51}$$

(ii) Case of $0 \leq u \leq a$, then $x \geq x - \frac{u}{2^k}$. Then

$$\Gamma_u(x) = \left| \int_{x-\frac{u}{2^k}}^{x} \left(f^{(N)}(t) - f^{(N)}(x)\right) \frac{\left(t - x + \frac{u}{2^k}\right)^{N-1}}{(N-1)!} dt \right| \leq$$

$$\int_{x-\frac{u}{2^k}}^{x} \left| f^{(N)}(t) - f^{(N)}(x) \right| \frac{\left(t - x + \frac{u}{2^k}\right)^{N-1}}{(N-1)!} dt \leq \tag{13.52}$$

$$\int_{x-\frac{u}{2^k}}^{x} \omega_1 \left(f^{(N)}, |t - x| \right) \frac{\left(t - x + \frac{u}{2^k}\right)^{N-1}}{(N-1)!} dt \leq$$

$$\omega_1 \left(f^{(N)}, \frac{|u|}{2^k} \right) \int_{x-\frac{u}{2^k}}^{x} \frac{\left(t - x + \frac{u}{2^k}\right)^{N-1}}{(N-1)!} dt \leq$$

$$\omega_1 \left(f^{(N)}, \frac{a}{2^k} \right) \frac{\left(\frac{u}{2^k}\right)^N}{N!} \leq \omega_1 \left(f^{(N)}, \frac{a}{2^k} \right) \frac{a^N}{2^{kN} N!}. \tag{13.53}$$

That is, when $0 \leq u \leq a$, then

$$\Gamma_u(x) \leq \omega_1 \left(f^{(N)}, \frac{a}{2^k} \right) \frac{a^N}{2^{kN} N!}. \tag{13.54}$$

We proved that

$$\Gamma_u(x) \leq \omega_1 \left(f^{(N)}, \frac{a}{2^k} \right) \frac{a^N}{2^{kN} N!} := \rho \geq 0, \tag{13.55}$$

$\forall\, k \in \mathbb{Z}, \forall\, x \in \mathbb{R}, |u| \leq a.$

By (13.48) we get that ($|u| \leq a$)

$$\left| f\left(x - \frac{u}{2^k}\right) - f(x) \right| \leq \sum_{i=1}^{N} \frac{\left| f^{(i)}(x) \right|}{i!} \frac{a^i}{2^{ki}} + \rho. \tag{13.56}$$

We observe that

$$|(B_k f)(x) - f(x)| =$$

$$\left| \left(N^*\right) \int_{-a}^{a} f\left(x - \frac{u}{2^k}\right) \psi(u)\, d\mu(u) - \left(N^*\right) \int_{-a}^{a} f(x)\, \psi(u)\, d\mu(u) \right| \overset{(13.20)}{\leq} \tag{13.57}$$

$$\left(N^*\right) \int_{-a}^{a} \left| f\left(x - \frac{u}{2^k}\right) - f(x) \right| \psi(u)\, d\mu(u) \leq$$

$$\left(\sum_{i=1}^{N} \frac{\left| f^{(i)}(x) \right|}{i!} \frac{a^i}{2^{ki}} + \rho \right) \left(\left(N^*\right) \int_{-a}^{a} \psi(u)\, d\mu(u) \right) \overset{(13.15)}{=}$$

$$\left(\sum_{i=1}^{N}\frac{\left|f^{(i)}(x)\right|}{i!}\frac{a^{i}}{2^{ki}}+\rho\right)\cdot 1= \tag{13.58}$$

$$\sum_{i=1}^{N}\frac{\left|f^{(i)}(x)\right|}{i!}\frac{a^{i}}{2^{ki}}+\frac{a^{N}}{2^{kN}N!}\omega_{1}\left(f^{(N)},\frac{a}{2^{k}}\right),$$

proving the claim. ■

Corollary 13.9 *Let $f\in C^{1}(\mathbb{R},\mathbb{R}_{+})$. Then*

$$\left|\left(B_{k}f\right)(x)-f(x)\right|\leq\frac{a}{2^{k}}\left(\left|f'(x)\right|+\omega_{1}\left(f',\frac{a}{2^{k}}\right)\right), \tag{13.59}$$

$\forall\, k\in\mathbb{Z}$, $\forall\, x\in\mathbb{R}$.

Proof By (13.47) for $N=1$. ■

We also present

Theorem 13.10 *Let $f\in C^{N}(\mathbb{R},\mathbb{R}_{+})$, $N\geq 1$. Consider the Shilkret-sublinear operators*

$$\left(\Gamma_{k}(f)\right)(x)=\left(N^{*}\right)\int_{-a}^{a}\left[\sum_{j=0}^{n}w_{j}f\left(\left(x-\frac{u}{2^{k}}\right)+\frac{j}{2^{k}n}\right)\right]\psi(u)\,d\mu(u), \tag{13.60}$$

$\forall\, k\in\mathbb{Z}$, $\forall\, x\in\mathbb{R}$. Then

$$\left|\left(\Gamma_{k}f\right)(x)-f(x)\right|\leq\sum_{i=1}^{N}\frac{\left|f^{(i)}(x)\right|}{i!}\frac{(a+1)^{i}}{2^{ki}}+\frac{(a+1)^{N}}{N!2^{kN}}\omega_{1}\left(f^{(N)},\frac{a+1}{2^{k}}\right). \tag{13.61}$$

If $f^{(N)}$ is uniformly continuous or bounded and continuous, then as $k\to+\infty$ we obtain that $\left(\Gamma_{k}f\right)(x)\to f(x)$, pointwise with rates.

Corollary 13.11 *Let $f\in C^{1}(\mathbb{R},\mathbb{R}_{+})$. Then*

$$\left|\left(\Gamma_{k}f\right)(x)-f(x)\right|\leq\frac{(a+1)}{2^{k}}\left[\left|f'(x)\right|+\omega_{1}\left(f',\frac{a+1}{2^{k}}\right)\right], \tag{13.62}$$

$\forall\, k\in\mathbb{Z}$, $\forall\, x\in\mathbb{R}$.

Proof By (13.61) for $N=1$. ■

Proof of Theorem 13.10.

Since $f \in C^N(\mathbb{R})$, $N \geq 1$, by Taylor's formula we get

$$\sum_{j=0}^{n} w_j f\left(\left(x - \frac{u}{2^k}\right) + \frac{j}{2^k n}\right) - f(x) =$$

$$\sum_{i=1}^{N} \frac{f^{(i)}(x)}{i!} \sum_{j=0}^{n} w_j \left(-\frac{u}{2^k} + \frac{j}{2^k n}\right)^i + \tag{13.63}$$

$$\sum_{j=0}^{n} w_j \int_x^{\left(x-\frac{u}{2^k}\right)+\frac{j}{2^k n}} \left(f^{(N)}(t) - f^{(N)}(x)\right) \frac{\left(\left(x - \frac{u}{2^k}\right) + \frac{j}{2^k n} - t\right)^{N-1}}{(N-1)!} dt.$$

Call

$$\varepsilon(x, u, j) := \int_x^{\left(x-\frac{u}{2^k}\right)+\frac{j}{2^k n}} \left(f^{(N)}(t) - f^{(N)}(x)\right) \frac{\left(\left(x - \frac{u}{2^k}\right) + \frac{j}{2^k n} - t\right)^{N-1}}{(N-1)!} dt. \tag{13.64}$$

We estimate $\varepsilon(x, u, j)$. Here $|u| \leq a$.

(i) case of $u \leq \frac{j}{n}$, iff $\frac{u}{2^k} \leq \frac{j}{2^k n}$, iff $x \leq x - \frac{u}{2^k} + \frac{j}{2^k n}$.

Hence

$$|\varepsilon(x, u, j)| \leq \int_x^{\left(x-\frac{u}{2^k}\right)+\frac{j}{2^k n}} \left|f^{(N)}(t) - f^{(N)}(x)\right| \frac{\left(\left(x - \frac{u}{2^k}\right) + \frac{j}{2^k n} - t\right)^{N-1}}{(N-1)!} dt \leq \tag{13.65}$$

$$\int_x^{\left(x-\frac{u}{2^k}\right)+\frac{j}{2^k n}} \omega_1\left(f^{(N)}, |t-x|\right) \frac{\left(\left(x - \frac{u}{2^k}\right) + \frac{j}{2^k n} - t\right)^{N-1}}{(N-1)!} dt \leq$$

$$\omega_1\left(f^{(N)}, \left[\frac{j}{2^k n} - \frac{u}{2^k}\right]\right) \int_x^{\left(x-\frac{u}{2^k}\right)+\frac{j}{2^k n}} \frac{\left(\left(x - \frac{u}{2^k}\right) + \frac{j}{2^k n} - t\right)^{N-1}}{(N-1)!} dt \leq$$

$$\omega_1\left(f^{(N)}, \frac{a+1}{2^k}\right) \frac{\left(\frac{j}{2^k n} - \frac{u}{2^k}\right)^N}{N!} \leq \omega_1\left(f^{(N)}, \frac{a+1}{2^k}\right) \frac{(a+1)^N}{2^{kN} N!}. \tag{13.66}$$

For $u \leq \frac{j}{n}$, we have proved that

$$|\varepsilon(x, u, j)| \leq \omega_1\left(f^{(N)}, \frac{a+1}{2^k}\right) \frac{(a+1)^N}{2^{kN} N!}. \tag{13.67}$$

(ii) case of $u \geq \frac{j}{n}$, iff $\frac{u}{2^k} \geq \frac{j}{2^k n}$, iff $x \geq x - \frac{u}{2^k} + \frac{j}{2^k n}$.
We observe that

$$|\varepsilon(x, u, j)| =$$

$$\left| \int_{\left(x-\frac{u}{2^k}\right)+\frac{j}{2^k n}}^{x} \left(f^{(N)}(t) - f^{(N)}(x)\right) \frac{\left(t - \left[\left(x - \frac{u}{2^k}\right) + \frac{j}{2^k n}\right]\right)^{N-1}}{(N-1)!} dt \right| \leq \tag{13.68}$$

$$\int_{\left(x-\frac{u}{2^k}\right)+\frac{j}{2^k n}}^{x} \left|f^{(N)}(t) - f^{(N)}(x)\right| \frac{\left(t - \left[\left(x - \frac{u}{2^k}\right) + \frac{j}{2^k n}\right]\right)^{N-1}}{(N-1)!} dt \leq$$

$$\int_{\left(x-\frac{u}{2^k}\right)+\frac{j}{2^k n}}^{x} \omega_1\left(f^{(N)}, |t-x|\right) \frac{\left(t - \left[\left(x - \frac{u}{2^k}\right) + \frac{j}{2^k n}\right]\right)^{N-1}}{(N-1)!} dt \leq$$

$$\omega_1\left(f^{(N)}, \frac{u}{2^k} - \frac{j}{2^k n}\right) \int_{\left(x-\frac{u}{2^k}\right)+\frac{j}{2^k n}}^{x} \frac{\left(t - \left[\left(x - \frac{u}{2^k}\right) + \frac{j}{2^k n}\right]\right)^{N-1}}{(N-1)!} dt \leq$$

$$\omega_1\left(f^{(N)}, \frac{a+1}{2^k}\right) \frac{\left(\frac{u}{2^k} - \frac{j}{2^k n}\right)^N}{N!} \leq \omega_1\left(f^{(N)}, \frac{a+1}{2^k}\right) \frac{(a+1)^N}{2^{kN} N!}. \tag{13.69}$$

So when $u \geq \frac{j}{n}$, we proved that

$$|\varepsilon(x, u, j)| \leq \omega_1\left(f^{(N)}, \frac{a+1}{2^k}\right) \frac{(a+1)^N}{2^{kN} N!}. \tag{13.70}$$

Therefore it always holds

$$|\varepsilon(x, u, j)| \leq \omega_1\left(f^{(N)}, \frac{a+1}{2^k}\right) \frac{(a+1)^N}{2^{kN} N!}. \tag{13.71}$$

Consequently we derive

$$\sum_{j=0}^{n} w_j |\varepsilon(x, u, j)| \leq \omega_1\left(f^{(N)}, \frac{a+1}{2^k}\right) \frac{(a+1)^N}{2^{kN} N!} := \overline{\psi}. \tag{13.72}$$

By (13.63) we find

$$\left|\sum_{j=0}^{n} w_j f\left(\left(x - \frac{u}{2^k}\right) + \frac{j}{2^k n}\right) - f(x)\right| \leq \sum_{i=1}^{N} \frac{\left|f^{(i)}(x)\right|}{i!} \frac{(a+1)^i}{2^{ki}} + \overline{\psi}. \quad (13.73)$$

Therefore we get

$$|(\Gamma_k(f))(x) - f(x)| =$$

$$\left|\left(N^*\right)\int_{-a}^{a}\left[\sum_{j=0}^{n} w_j f\left(\left(x - \frac{u}{2^k}\right) + \frac{j}{2^k n}\right)\right]\psi(u)\,d\mu(u) - \left(N^*\right)\int_{-a}^{a} f(x)\,\psi(u)\,d\mu(u)\right| \overset{(13.20)}{\leq} \quad (13.74)$$

$$\left(N^*\right)\int_{-a}^{a}\left|\sum_{j=0}^{n} w_j f\left(\left(x - \frac{u}{2^k}\right) + \frac{j}{2^k n}\right) - f(x)\right|\psi(u)\,d\mu(u) \overset{(13.73)}{\leq}$$

$$\left[\sum_{i=1}^{N} \frac{\left|f^{(i)}(x)\right|}{i!} \frac{(a+1)^i}{2^{ki}} + \overline{\psi}\right]\left(N^*\right)\int_{-a}^{a}\psi(u)\,d\mu(u) \overset{(13.15)}{=}$$

$$\left[\sum_{i=1}^{N} \frac{\left|f^{(i)}(x)\right|}{i!} \frac{(a+1)^i}{2^{ki}} + \overline{\psi}\right]\cdot 1 =$$

$$\sum_{i=1}^{N} \frac{\left|f^{(i)}(x)\right|}{i!} \frac{(a+1)^i}{2^{ki}} + \frac{(a+1)^N}{2^{kN} N!}\omega_1\left(f^{(N)}, \frac{a+1}{2^k}\right), \quad (13.75)$$

proving the claim. ■

We finish with

Corollary 13.12 *Let $f \in C^N(\mathbb{R}, \mathbb{R}_+)$, $N \geq 1$, $f^{(i)}(x) = 0$, $i = 1, ..., N$. Then*
(i)

$$|(B_k(f))(x) - f(x)| \leq \frac{a^N}{2^{kN} N!}\omega_1\left(f^{(N)}, \frac{a}{2^k}\right), \quad (13.76)$$

and
(ii)

$$|(\Gamma_k(f))(x) - f(x)| \leq \frac{(a+1)^N}{N! 2^{kN}}\omega_1\left(f^{(N)}, \frac{a+1}{2^k}\right), \quad (13.77)$$

$\forall\, k \in \mathbb{Z}$, $\forall\, x \in \mathbb{R}$.

Proof By (13.47) and (13.61). ■

Corollary 13.13 *Let $f \in C^1(\mathbb{R}, \mathbb{R}_+)$, $f'(x) = 0$. Then*

(i)

$$|(B_k(f))(x) - f(x)| \leq \frac{a}{2^k}\omega_1\left(f', \frac{a}{2^k}\right), \tag{13.78}$$

and

(ii)

$$|(\Gamma_k(f))(x) - f(x)| \leq \left(\frac{a+1}{2^k}\right)\omega_1\left(f', \frac{a+1}{2^k}\right), \tag{13.79}$$

$\forall\, k \in \mathbb{Z}$, $\forall\, x \in \mathbb{R}$.

Proof By (13.59) and (13.62). ∎

In inequalities (13.76)–(13.79) observe the high speed of convergence and approximation.

13.5 Appendix

Let $f \in C_U(\mathbb{R}, \mathbb{R}_+)$, and the positive sublinear Shilkret operator

$$(M(f))(x) := \left(N^*\right)\int_{-a}^{a} f(x+u)\,\psi(u)\,d\mu(u), \forall\, x \in \mathbb{R}. \tag{13.80}$$

We observe the following (for any $x, y \in \mathbb{R}$):

$$|(M(f))(x) - (M(f))(y)| =$$

$$\left|\left(N^*\right)\int_{-a}^{a} f(x+u)\,\psi(u)\,d\mu(u) - \left(N^*\right)\int_{-a}^{a} f(y+u)\,\psi(u)\,d\mu(u)\right| \overset{(13.20)}{\leq}$$

$$\left(N^*\right)\int_{-a}^{a} |f(x+u) - f(y+u)|\,\psi(u)\,d\mu(u) \leq$$

$$\omega_1(f, |x-y|)\left(\left(N^*\right)\int_{-a}^{a} \psi(u)\,d\mu(u)\right) \overset{(13.15)}{=} \omega_1(f, |x-y|) \cdot 1 = \omega_1(f, |x-y|). \tag{13.81}$$

Therefore it holds the global smoothness preservation property:

$$\omega_1(M(f), \delta) \leq \omega_1(f, \delta), \forall\, \delta > 0. \tag{13.82}$$

References

1. G.A. Anastassiou, High order approximation by univariate shift-invariant integral operators, in *Nonlinear Analysis and Applications, 2 Volumes*, vol. I, ed. by R. Agarwal, D. O'Regan (Kluwer, Dordrecht, 2003), pp. 141–164
2. G.A. Anastassiou, *Intelligent Mathematics: Computational Analysis* (Springer, Heidelberg, 2011)
3. G.A. Anastassiou, Approximation by shift invariant univariate sublinear-Shilkret operators (2018). Submitted
4. G.A. Anastassiou, H.H. Gonska, On some shift invariant integral operators, univariate case. Ann. Polon. Math. **LXI**(3), 225–243 (1995)
5. G.A. Anastassiou, S. Gal, *Approximation Theory* (Birkhauser, Boston, 2000)
6. N. Shilkret, Maxitive measure and integration. Indag. Math. **33**, 109–116 (1971)

Chapter 14
Quantitative Approximation by Shift Invariant Multivariate Sublinear-Shilkret Operators

A very general multivariate positive sublinear Shilkret integral type operator is given through a convolution-like iteration of another multivariate general positive sublinear operator with a multivariate scaling type function. For it sufficient conditions are given for shift invariance, preservation of global smoothness, convergence to the unit with rates. Furthermore, two examples of very general multivariate specialized Shilkret operators are presented fulfilling all the above properties, the higher order of multivariate approximation of these operators is also considered. It follows [3].

14.1 Introduction

Let X, Y be function spaces of functions from $\mathbb{R}^d$ into $\mathbb{R}_+$, $d \in \mathbb{N}$. Let $L_N : X \to Y$, $N \subset \mathbb{N}$, be a sequence of operators with the following properties:

(i) (positive homogeneous)

$$L_N(\alpha f) = \alpha L_N(f), \forall \alpha \geq 0, \forall f \in X.$$

(ii) (Monotonicity) if $f, g \in X$ satisfy $f \leq g$, then $L_N(f) \leq L_N(g)$, $\forall N \in \mathbb{N}$, and

(iii) (Subadditivity)

$$L_N(f+g) \leq L_N(f) + L_N(g), \forall f, g \in X.$$

We call L_N multivariate positive sublinear operators.

In this chapter we deal with sequences of Shilkret multivariate positive sublinear operators that are constructed, with the help of Shilkret integral [6]. Our functions spaces are continuous functions from $\mathbb{R}^d$ into $\mathbb{R}_+$, $d \in \mathbb{N}$. The sequence of operators is generated by a basic multivariate operator via dilated translations of convolution type using the Shilkret integral. We prove that our multivariate Shilkret operators possess

G. A. Anastassiou, *Ordinary and Fractional Approximation by Non-additive Integrals: Choquet, Shilkret and Sugeno Integral Approximators*, Studies in Systems, Decision and Control 190, https://doi.org/10.1007/978-3-030-04287-5_14

the following properties: of shift invariance of global smoothness preservation, of convergence to the unit operator with rates. Then we apply our results to two specific families of such multivariate Shilkret type operators.

We continue with the higher order quantitative approximation study of these multivariate Shilkret specific operators.

Earlier similar studies have been done by the author, see [5], Chaps. 10–17, and [2], Chaps. 16, 17. These serve as motivation and inspiration to our current work.

14.2 Background

Here we follow [6].

Let $\mathcal{F}$ be a σ-field of subsets of an arbitrary set Ω. An extended non-negative real valued function μ on $\mathcal{F}$ is called maxitive if $\mu(\varnothing) = 0$ and

$$\mu\left(\cup_{i\in I} E_i\right) = \sup_{i\in I} \mu\left(E_i\right), \tag{14.1}$$

where the set I is of cardinality at most countable, where $\{E_i\}_{i\in I}$ is a disjoint collection of sets from $\mathcal{F}$. We notice that μ is monotone and (14.1) is true even $\{E_i\}_{i\in I}$ are not disjoint. For more properties of μ see [6]. We also call μ a maxitive measure. Here f stands for a non-negative measurable function on Ω. In [6], Niel Shilkret developed his non-additive integral defined as follows:

$$\left(N^*\right)\int_D f d\mu := \sup_{y\in Y}\left\{y\cdot\mu\left(D\cap\{f\geq y\}\right)\right\}, \tag{14.2}$$

where $Y = [0, m]$ or $Y = [0, m)$ with $0 < m \leq \infty$, and $D \in \mathcal{F}$. Here we take $Y = [0, \infty)$.

It is easily proved that

$$\left(N^*\right)\int_D f d\mu = \sup_{y>0}\left\{y\cdot\mu\left(D\cap\{f > y\}\right)\right\}. \tag{14.3}$$

The Shilkret integral takes values in $[0, \infty]$.

The Shilkret integral [6] has the following properties:

$$\left(N^*\right)\int_\Omega \chi_E d\mu = \mu(E), \tag{14.4}$$

where χ_E is the indicator function on $E \in \mathcal{F}$,

$$\left(N^*\right)\int_D cf d\mu = c\left(N^*\right)\int_D f d\mu,\ c \geq 0, \tag{14.5}$$

$$\left(N^*\right)\int_D \sup_{n\in\mathbb{N}} f_n d\mu = \sup_{n\in\mathbb{N}} \left(N^*\right)\int_D f_n d\mu, \tag{14.6}$$

where f_n, $n \in \mathbb{N}$, is an increasing sequence of elementary (countably valued) functions converging uniformly to f. Furthermore we have

$$\left(N^*\right)\int_D f d\mu \geq 0, \tag{14.7}$$

$$f \geq g \text{ implies } \left(N^*\right)\int_D f d\mu \geq \left(N^*\right)\int_D g d\mu, \tag{14.8}$$

where $f, g : \Omega \to [0, \infty]$ are measurable.

Let $a \leq f(\omega) \leq b$ for almost every $\omega \in E$, then

$$a\mu(E) \leq \left(N^*\right)\int_E f d\mu \leq b\mu(E); \tag{14.9}$$

$$\left(N^*\right)\int_E 1 d\mu = \mu(E); \tag{14.10}$$

$f > 0$ almost everywhere and $(N^*)\int_E f d\mu = 0$ imply $\mu(E) = 0$;

$(N^*)\int_\Omega f d\mu = 0$ if and only $f = 0$ almost everywhere;

$(N^*)\int_\Omega f d\mu < \infty$ implies that

$$\overline{N}(f) := \{\omega \in \Omega | f(\omega) \neq 0\} \text{ has } \sigma\text{-finite measure;}$$

$$\left(N^*\right)\int_D (f + g) d\mu \leq \left(N^*\right)\int_D f d\mu + \left(N^*\right)\int_D g d\mu; \tag{14.11}$$

and

$$\left|\left(N^*\right)\int_D f d\mu - \left(N^*\right)\int_D g d\mu\right| \leq \left(N^*\right)\int_D |f - g| d\mu. \tag{14.12}$$

From now on in this chapter we assume that $\mu : \mathcal{F} \to [0, +\infty)$.

14.3 Multivariate Theory

This section is motivated and inspired by [4, 5].

Let $C_U\left(\mathbb{R}^d, \mathbb{R}\right)$, $d \in \mathbb{N}$, the space of uniformly continuous functions from $\mathbb{R}^d$ into $\mathbb{R}$. Let the multivariate first modulus of continuity

$$\omega_1(f,\delta) := \sup_{\substack{x,y\in\mathbb{R}^d \\ \|x-y\|\le\delta}} |f(x) - f(y)|, \delta > 0,$$

where $f \in C_U(\mathbb{R}^d, \mathbb{R})$, and $\|\cdot\|$ an arbitrary norm in $\mathbb{R}^d$. In [4], and [5], pp. 297–298, we proved that $\omega_1(f,\delta)$ is finite for any $\delta > 0$.

We denote by $\omega_{1,\infty}$ the multivariate modulus of continuity with respect to max-norm $\|\cdot\|_\infty$.

Here we deal with the functions spaces $C_U(\mathbb{R}^d, \mathbb{R}_+)$, $C(\mathbb{R}^d, \mathbb{R}_+)$ (continuous functions from $\mathbb{R}^d$ into $\mathbb{R}_+$).

Let $\{t_k\}_{k\in\mathbb{Z}}$ be a sequence of multivariate positive sublinear operators that map $C_U(\mathbb{R}^d, \mathbb{R}_+)$ into $C(\mathbb{R}^d, \mathbb{R}_+)$ with the property:

$$(t_k(f))(x) := \left(t_0\left(f\left(2^{-k}\right)\right)\right)(x), \forall\, x \in \mathbb{R}^d, \forall\, f \in C_U\left(\mathbb{R}^d, \mathbb{R}_+\right). \tag{14.13}$$

For fixed $a > 0$ we assume that

$$\sup_{\substack{u,y\in\mathbb{R}^d \\ \|u-y\|_\infty\le a}} |(t_0(f))(u) - f(y)| \le \omega_{1,\infty}\left(f, \frac{ma+n}{2^r}\right), \tag{14.14}$$

is true for all $f \in C_U(\mathbb{R}^d, \mathbb{R}_+)$; where $m \in \mathbb{N}$, $n \in \mathbb{Z}_+$, $r \in \mathbb{Z}$.

Let $\mathcal{L}^d$ be the Lebesgue σ-algebra on $\mathbb{R}^d$, and the set function $\mu : \mathcal{L}^d \to [0, +\infty]$, which is assumed to be maxitive.

Let $\psi : \mathbb{R}^d \to \mathbb{R}_+$ which is Lebesgue measurable, such that

$$(N^*)\int_{[-a,a]^d} \psi(u)\, d\mu(u) = 1. \tag{14.15}$$

We define the multivariate positive sublinear-Shilkret operators

$$(T_0(f))(x) := (N^*)\int_{[-a,a]^d} (t_0 f)(x-u)\,\psi(u)\, d\mu(u), \tag{14.16}$$

and

$$(T_k(f))(x) := \left(T_0\left(f\left(2^{-k}\cdot\right)\right)\right)\left(2^k x\right), \forall\, k \in \mathbb{Z}, \forall\, x \in \mathbb{R}^d. \tag{14.17}$$

Therefore it holds

$$(T_k(f))(x) = (N^*)\int_{[-a,a]^d} \left(t_0\left(f\left(2^{-k}\cdot\right)\right)\right)\left(2^k x - u\right)\psi(u)\, d\mu(u) = \tag{14.18}$$

$$(N^*)\int_{[-a,a]^d} (t_k(f))\left(2^k x - u\right)\psi(u)\, d\mu(u),$$

$\forall\, x \in \mathbb{R}^d, \forall\, k \in \mathbb{Z}$.

Indeed here we have

$$(T_k(f))(x) \leq (N^*)\int_{[-a,a]^d} \left\| t_k(f)\left(2^k x - \cdot\right)\right\|_{\infty,[-a,a]^d} \psi(u)\, d\mu(u) =$$

$$\left\| t_k(f)\left(2^k x - \cdot\right)\right\|_{\infty,[-a,a]^d} \left((N^*)\int_{[-a,a]^d} \psi(u)\, d\mu(u)\right) \overset{(14.15)}{=} \tag{14.19}$$

$$\left\| t_k(f)\left(2^k x - \cdot\right)\right\|_{\infty,[-a,a]^d} < +\infty.$$

Hence $(T_k(f))(x) \in \mathbb{R}_+$ is well-defined, $\forall\, x \in \mathbb{R}^d$, $\forall\, k \in \mathbb{Z}$.

Let $f, g \in M\left(\mathbb{R}^d, \mathbb{R}_+\right)$ (Lebesgue measurable functions) and $A \subset \mathbb{R}^d$ is a Lebesgue measurable set.

We derive by (14.12):

$$\left|(N^*)\int_A f(x)\, d\mu(x) - (N^*)\int_A g(x)\, d\mu(x)\right| \leq (N^*)\int_A |f(x) - g(x)|\, d\mu(x). \tag{14.20}$$

We need

Definition 14.1 Let $f_\alpha(\cdot) := f(\cdot + \alpha)$, $\alpha \in \mathbb{R}^d$, and Φ be an operator. If $\Phi(f_\alpha) = (\Phi f)_\alpha$, then Φ is called a shift invariant operator.

We give

Theorem 14.2 *Assume that*

$$\left(t_0\left(f\left(2^{-k}\cdot + \alpha\right)\right)\right)\left(2^k u\right) = \left(t_0\left(f\left(2^{-k}\cdot\right)\right)\right)\left(2^k(u+\alpha)\right), \tag{14.21}$$

for all $k \in \mathbb{Z}$, $\alpha \in \mathbb{R}^d$ fixed, all $u \in \mathbb{R}^d$ and any $f \in C_U\left(\mathbb{R}^d, \mathbb{R}_+\right)$. Then T_k is a shift invariant operator for all $k \in \mathbb{Z}$.

Proof We have that

$$(T_k(f(\cdot + \alpha)))(x) = (T_k(f_\alpha))(x) \overset{(14.18)}{=}$$

$$(N^*)\int_{[-a,a]^d} \left(t_0\left(f_\alpha\left(2^{-k}\cdot\right)\right)\right)\left(2^k x - u\right)\psi(u)\, d\mu(u) =$$

$$(N^*)\int_{[-a,a]^d} \left(t_0\left(f\left(2^{-k}\cdot + \alpha\right)\right)\right)\left(2^k x - u\right)\psi(u)\, d\mu(u) =$$

$$(N^*)\int_{[-a,a]^d} \left(t_0\left(f\left(2^{-k}\cdot + \alpha\right)\right)\right)\left(2^k\left(x - 2^{-k}u\right)\right)\psi(u)\, d\mu(u) \overset{(14.21)}{=} \tag{14.22}$$

$$\left(N^{*}\right)\int_{[-a,a]^{d}}\left(t_{0}\left(f\left(2^{-k}\cdot\right)\right)\right)\left(2^{k}\left(x-2^{-k}u+\alpha\right)\right)\psi\left(u\right)d\mu\left(u\right)=$$

$$\left(N^{*}\right)\int_{[-a,a]^{d}}\left(t_{0}\left(f\left(2^{-k}\cdot\right)\right)\right)\left(2^{k}\left(x+\alpha\right)-u\right)\psi\left(u\right)d\mu\left(u\right)\overset{(14.18)}{=}\left(T_{k}\left(f\right)\right)\left(x+\alpha\right),$$

that is

$$T_{k}\left(f_{\alpha}\right)=\left(T_{k}\left(f\right)\right)_{\alpha}, \tag{14.23}$$

proving the claim. ■

It follows the global smoothness of the operators T_k.

Theorem 14.3 *For any $f \in C_U\left(\mathbb{R}^d, \mathbb{R}_+\right)$ assume that, for all $u \in \mathbb{R}^d$,*

$$\left|\left(t_{0}\left(f\right)\right)\left(x-u\right)-\left(t_{0}\left(f\right)\right)\left(y-u\right)\right|\leq\omega_{1}\left(f,\left\|x-y\right\|\right), \tag{14.24}$$

for any $x, y \in \mathbb{R}^d$, where $\|\cdot\|$ is an arbitrary norm in $\mathbb{R}^d$. Then

$$\omega_{1}\left(T_{k}f,\delta\right)\leq\omega_{1}\left(f,\delta\right),\ \forall\,\delta>0. \tag{14.25}$$

Proof We observe that

$$\left|\left(T_{0}\left(f\right)\right)\left(x\right)-\left(T_{0}\left(f\right)\right)\left(y\right)\right|=$$

$$\left|\left(N^{*}\right)\int_{[-a,a]^{d}}\left(t_{0}f\right)\left(x-u\right)\psi\left(u\right)d\mu\left(u\right)-\left(N^{*}\right)\int_{[-a,a]^{d}}\left(t_{0}f\right)\left(y-u\right)\psi\left(u\right)d\mu\left(u\right)\right|\overset{(14.20)}{\leq} \tag{14.26}$$

$$\left(N^{*}\right)\int_{[-a,a]^{d}}\left|\left(t_{0}f\right)\left(x-u\right)-\left(t_{0}f\right)\left(y-u\right)\right|\psi\left(u\right)d\mu\left(u\right)\overset{(14.24)}{\leq}$$

$$\omega_{1}\left(f,\left\|x-y\right\|\right)\left(\left(N^{*}\right)\int_{[-a,a]^{d}}\psi\left(u\right)d\mu\left(u\right)\right)\overset{(14.15)}{=}\omega_{1}\left(f,\left\|x-y\right\|\right).$$

So that

$$\left|\left(T_{0}\left(f\right)\right)\left(x\right)-\left(T_{0}\left(f\right)\right)\left(y\right)\right|\leq\omega_{1}\left(f,\left\|x-y\right\|\right). \tag{14.27}$$

From (14.17), (14.27) we get

$$\left|\left(T_{k}\left(f\right)\right)\left(x\right)-\left(T_{k}\left(f\right)\right)\left(y\right)\right|\overset{(14.17)}{=}$$

$$\left|\left(T_{0}\left(f\left(2^{-k}\cdot\right)\right)\right)\left(2^{k}x\right)-\left(T_{0}\left(f\left(2^{-k}\cdot\right)\right)\right)\left(2^{k}y\right)\right|\leq \tag{14.28}$$

$$\omega_{1}\left(f\left(2^{-k}\cdot\right),2^{k}\left\|x-y\right\|\right)=\omega_{1}\left(f,\left\|x-y\right\|\right),$$

i.e. global smoothness for T_k has been proved. ■

The convergence of T_k to the unit operator, as $k \to +\infty$, k with rates follows:

Theorem 14.4 *For $f \in C_U\left(\mathbb{R}^d, \mathbb{R}_+\right)$, under the assumption* (14.14), *we have*

$$|(T_k(f))(x) - f(x)| \leq \omega_{1,\infty}\left(f, \frac{ma+n}{2^{k+r}}\right), \tag{14.29}$$

where $m \in \mathbb{N}$, $n \in \mathbb{Z}_+$, $k, r \in \mathbb{Z}$.

Proof We notice that

$$|(T_k(f))(x) - f(x)| \overset{(14.17)}{=} \left|\left(T_0\left(f\left(2^{-k}\cdot\right)\right)\right)\left(2^k x\right) - f(x)\right| \overset{(14.18)}{=}$$

$$\left|\left(N^*\right)\int_{[-a,a]^d}\left(t_0\left(f\left(2^{-k}\cdot\right)\right)\right)\left(2^k x - u\right)\psi(u)\,d\mu(u) - f(x)\right| \overset{(14.15)}{=}$$

$$\left|\left(N^*\right)\int_{[-a,a]^d}\left(t_0\left(f\left(2^{-k}\cdot\right)\right)\right)\left(2^k x - u\right)\psi(u)\,d\mu(u) - \left(N^*\right)\int_{[-a,a]^d} f(x)\,\psi(u)\,d\mu(u)\right|$$

$$\overset{(14.20)}{\leq} \left(N^*\right)\int_{[-a,a]^d}\left|\left(t_0\left(f\left(2^{-k}\cdot\right)\right)\right)\left(2^k x - u\right) - f(x)\right|\psi(u)\,d\mu(u) = \tag{14.30}$$

$$\left(N^*\right)\int_{[-a,a]^d}\left|\left(t_0\left(f\left(2^{-k}\cdot\right)\right)\right)\left(2^k x - u\right) - f\left(2^{-k}\cdot\right)\left(2^k x\right)\right|\psi(u)\,d\mu(u) \overset{(14.14)}{\leq}$$

(here $\left\|\left(2^k x - u\right) - 2^k x\right\|_\infty = \|u\|_\infty \leq a$)

$$\omega_{1,\infty}\left(f\left(2^{-k}\cdot\right), \frac{ma+n}{2^r}\right)\left(\left(N^*\right)\int_{[-a,a]^d}\psi(u)\,d\mu(u)\right) \overset{(14.15)}{=}$$

$$\omega_{1,\infty}\left(f\left(2^{-k}\cdot\right), \frac{ma+n}{2^r}\right)\cdot 1 = \omega_{1,\infty}\left(f, \frac{ma+n}{2^{k+r}}\right),$$

proving the claim. ■

We give some applications.

For each $k \in \mathbb{Z}$, we define

(i)

$$\left(A_k f\right)(x) := \left(N^*\right)\int_{[-a,a]^d} f\left(x - \frac{u}{2^k}\right)\psi(u)\,d\mu(u), \forall\, x \in \mathbb{R}^d, \tag{14.31}$$

i.e., here

$$\begin{gathered}\left(t_k(f)\right)(u) = f\left(\tfrac{u}{2^k}\right),\\ \text{and}\\ \left(t_0(f)\right)(u) = f(u),\end{gathered} \tag{14.32}$$

are continuous in $u \in \mathbb{R}^d$.

Also for $k \in \mathbb{Z}$, we define

(ii)

$$(D_k(f))(x) := (N^*) \int_{[-a,a]^d} \gamma_k^f \left(2^k x - u\right) \psi(u)\, d\mu(u), \forall\, x \in \mathbb{R}^d, \tag{14.33}$$

where

$$\gamma_k^f(u) := \sum_{j_1=0}^{n_1} \dots \sum_{j_d=0}^{n_d} w_{j_1,\dots,j_d} f\left(\frac{u_1}{2^k} + \frac{j_1}{2^k n_1}, \dots, \frac{u_d}{2^k} + \frac{j_d}{2^k n_d}\right), \tag{14.34}$$

$(n_1, \dots, n_d) \in \mathbb{N}^d$, $w_{j_1,\dots,j_d} \geq 0$, $\sum_{j_1=0}^{n_1} \dots \sum_{j_d=0}^{n_d} w_{j_1,\dots,j_d} = 1.$

That is, here

$$(t_k(f))(u) = \gamma_k^f(u) \tag{14.35}$$

is continuous in $u \in \mathbb{R}^d$, where $u = (u_1, \dots, u_d)$.

Notice that

$$(t_0(f))(u) = \gamma_0^f(u) = \sum_{j_1=0}^{n_1} \dots \sum_{j_d=0}^{n_d} w_{j_1,\dots,j_d} \cdot f\left(u_1 + \frac{j_1}{n_1}, \dots, u_d + \frac{j_d}{n_d}\right) \tag{14.36}$$

is continuous in $u \in \mathbb{R}^d$.

Setting $x = (x_1, \dots, x_d) \in \mathbb{R}^d$, $n = (n_1, \dots, n_d) \in \mathbb{N}^d$, $j = (j_1, \dots, j_d) \in \mathbb{Z}_+^d$, $w_j = w_{j_1,\dots,j_d} \geq 0$, and $\sum_{j=0}^{n} = \sum_{j_1=0}^{n_1} \dots \sum_{j_d=0}^{n_d}$. That is $\sum_{j=0}^{n} w_j = 1$.

So we can write ($k \in \mathbb{Z}$)

$$(t_k(f))(u) = \gamma_k^f(u) = \sum_{j=0}^{n} w_j f\left(\frac{u}{2^k} + \frac{j}{2^k n}\right), \tag{14.37}$$

and

$$(t_0(f))(u) = \gamma_0^f(u) = \sum_{j=0}^{n} w_j f\left(u + \frac{j}{n}\right). \tag{14.38}$$

Indeed we have

$$(D_k(f))(x) = (N^*) \int_{[-a,a]^d} \left[\sum_{j=0}^{n} w_j f\left(\left(x - \frac{u}{2^k}\right) + \frac{j}{2^k n}\right)\right] \psi(u)\, d\mu(u). \tag{14.39}$$

Clearly here we get

$$\begin{gathered}(A_k(f))(x) = \left(A_0\left(f\left(2^{-k}\cdot\right)\right)\right)\left(2^k x\right),\\ \text{and}\\ (D_k(f))(x) = \left(D_0\left(f\left(2^{-k}\cdot\right)\right)\right)\left(2^k x\right),\end{gathered} \tag{14.40}$$

$\forall\, k \in \mathbb{Z}, \forall\, x \in \mathbb{R}^d$.

We give

Proposition 14.5 *A_k, D_k are shift invariant operators.*

Proof (i) For A_k operators: Here $t_0 f = f$. Hence

$$\left(t_0\left(f\left(2^{-k}\cdot+\alpha\right)\right)\right)\left(2^k u\right) = f\left(2^{-k}2^k u + \alpha\right) = f(u+\alpha) = \tag{14.41}$$

$$\left(t_0\left(f\left(2^{-k}\cdot\right)\right)\right)\left(2^k(u+\alpha)\right).$$

(ii) For D_k operators:

$$(t_0(f))(u) = \sum_{j=0}^{n} w_j f\left(u + \frac{j}{n}\right).$$

Hence

$$\left(t_0\left(f\left(2^{-k}\cdot+\alpha\right)\right)\right)\left(2^k u\right) = \sum_{j=0}^{n} w_j f\left(2^{-k}\left(2^k u + \frac{j}{n}\right) + \alpha\right) =$$

$$\sum_{j=0}^{n} w_j f\left(2^{-k}\left(2^k(u+\alpha) + \frac{j}{n}\right)\right) = \left(t_0\left(f\left(2^{-k}\cdot\right)\right)\right)\left(2^k(u+\alpha)\right), \tag{14.42}$$

proving the claim. ∎

Next we show that the operators A_k, D_k possess the property of global smoothness preservation.

Theorem 14.6 *For all $f \in C_U\left(\mathbb{R}^d, \mathbb{R}_+\right)$ and all $\delta > 0$ we have*

$$\begin{gathered}\omega_1(A_k f, \delta) \le \omega_1(f, \delta),\\ \textit{and}\\ \omega_1(D_k f, \delta) \le \omega_1(f, \delta).\end{gathered} \tag{14.43}$$

Proof (i) For A_k operators: Here $t_0 f = f$, therefore

$$|(t_0(f))(x-u) - (t_0(f))(y-u)| = |f(x-u) - f(y-u)| \le \omega_1(f, \|x-y\|). \tag{14.44}$$

(ii) For D_k operators: We observe that

$$\left|\left(t_0\left(f\right)\right)\left(x-u\right)-\left(t_0\left(f\right)\right)\left(y-u\right)\right| = \left|\gamma_0^f\left(x-u\right)-\gamma_0^f\left(y-u\right)\right| =$$

$$\left|\sum_{j=0}^{n} w_j\left(f\left(x-u+\frac{j}{n}\right)-f\left(y-u+\frac{j}{n}\right)\right)\right| \leq$$

$$\sum_{j=0}^{n} w_j\left|f\left(x-u+\frac{j}{n}\right)-f\left(y-u+\frac{j}{n}\right)\right| \leq$$

$$\omega_1\left(f, \|x-y\|\right)\left(\sum_{j=0}^{n} w_j\right) = \omega_1\left(f, \|x-y\|\right), \tag{14.45}$$

proving the claim. ∎

The operators A_k, D_k, $k \in \mathbb{Z}$, converge to the unit operator, as $k \to +\infty$, with rates presented next.

Theorem 14.7 *For $k \in \mathbb{Z}$, $f \in C_U\left(\mathbb{R}^d, \mathbb{R}_+\right)$, it holds*

$$\begin{gathered}\left|\left(A_k\left(f\right)\right)\left(x\right)-f\left(x\right)\right| \leq \omega_{1,\infty}\left(f, \tfrac{a}{2^k}\right),\\ \textit{and}\\ \left|\left(D_k\left(f\right)\right)\left(x\right)-f\left(x\right)\right| \leq \omega_{1,\infty}\left(f, \tfrac{a+1}{2^k}\right).\end{gathered} \tag{14.46}$$

Proof (i) For A_k operators: Here $\left(t_0\left(f\right)\right)\left(u\right) = f\left(u\right)$ and

$$\sup_{\substack{u,y\in\mathbb{R}^d\\ \|u-y\|_\infty\leq a}} \left|\left(t_0\left(f\right)\right)\left(u\right)-f\left(y\right)\right| = \sup_{\substack{u,y\in\mathbb{R}^d\\ \|u-y\|_\infty\leq a}} \left|f\left(u\right)-f\left(y\right)\right| = \omega_{1,\infty}\left(f, a\right), \tag{14.47}$$

and we use Theorem 14.4.

(ii) For D_k operators: Here we see that

$$\sup_{\substack{u,y\in\mathbb{R}^d\\ \|u-y\|_\infty\leq a}} \left|\left(t_0\left(f\right)\right)\left(u\right)-f\left(y\right)\right| = \sup_{\substack{u,y\in\mathbb{R}^d\\ \|u-y\|_\infty\leq a}} \left|\sum_{j=0}^{n} w_j f\left(u+\frac{j}{n}\right)-f\left(y\right)\right| \leq$$

$$\sup_{\substack{u,y\in\mathbb{R}^d\\ \|u-y\|_\infty\leq a}} \sum_{j=0}^{n} w_j\left|f\left(u+\frac{j}{n}\right)-f\left(y\right)\right| \leq$$

$$\sup_{\substack{u,y\in\mathbb{R}^d \\ \|u-y\|_\infty\le a}} \sum_{j=0}^{n} w_j \omega_{1,\infty}\left(f, \left\|u+\frac{j}{n}-y\right\|_\infty\right) \le \tag{14.48}$$

$$\sup_{\substack{u,y\in\mathbb{R}^d \\ \|u-y\|_\infty\le a}} \sum_{j=0}^{n} w_j \omega_{1,\infty}\left(f, \left\|\frac{j}{n}\right\|_\infty + \|u-y\|_\infty\right) \le$$

$$\left(\sum_{j=0}^{n} w_j\right) \omega_{1,\infty}(f, 1+\alpha) = \omega_{1,\infty}(f, \alpha+1),$$

where $\frac{j}{n} = \left(\frac{j_1}{n_1}, \ldots, \frac{j_d}{n_d}\right)$.

By (14.29) we are done. ■

14.4 Higher Order Multivariate Approximation

Here all are as in Sect. 14.3. See also our earlier work [1], and [2], Chap. 17.

We present

Theorem 14.8 *Let $f \in C^N\left(\mathbb{R}^d, \mathbb{R}_+\right)$, $N, d \in \mathbb{N}$. Consider the multivariate Shilkret-sublinear operators*

$$(A_k f)(x) = \left(N^*\right)\int_{[-a,a]^d} f\left(x - \frac{u}{2^k}\right)\psi(u)\,d\mu(u),$$

$\forall\, x \in \mathbb{R}^d$, $\forall\, k \in \mathbb{Z}$. Here we further assume that all partials $f_{\overline{\alpha}}$, $|\overline{\alpha}| = N$, are uniformly continuous or bounded and continuous on $\mathbb{R}^d$. Then

$$|(A_k f)(x) - f(x)| \le \sum_{j=1}^{N} \frac{a^j}{j!2^{kj}} \left(\left(\sum_{i=1}^{d}\left|\frac{\partial}{\partial x_i}\right|\right)^j f(x)\right)$$

$$+ \frac{a^N d^N}{N!2^{kN}} \max_{\overline{\alpha}:|\overline{\alpha}|=N} \omega_{1,\infty}\left(f_{\overline{\alpha}}, \frac{a}{2^k}\right), \tag{14.49}$$

$\forall\, k \in \mathbb{Z}$, $\forall\, x \in \mathbb{R}^d$.

Remark 14.9 (i) Since the Nth order partials $f_{\overline{\alpha}}$ are uniformly continuous or bounded and continuous, as $k \to +\infty$, we get that $(A_k f)(x) \to f(x)$, pointwise with rates.

(ii) When $N = 1$, inequality (14.49) becomes

$$\left|(A_k f)(x) - f(x)\right| \leq \frac{a}{2^k}\left\{\left(\sum_{i=1}^{d}\left|\frac{\partial f(x)}{\partial x_i}\right|\right) + d \max_{i\in\{1,\ldots,d\}} \omega_{1,\infty}\left(\frac{\partial f}{\partial x_i}, \frac{a}{2^k}\right)\right\}, \tag{14.50}$$

$\forall\, k \in \mathbb{Z}, \forall\, x \in \mathbb{R}^d$.

(iii) When $f_{\overline{\alpha}}(x) = 0, \overline{\alpha} : |\overline{\alpha}| = \rho, 1 \leq \rho \leq N$. Then (14.49) becomes

$$\left|(A_k f)(x) - f(x)\right| \leq \frac{a^N d^N}{N! 2^{kN}} \max_{\overline{\alpha}:|\overline{\alpha}|=N} \omega_{1,\infty}\left(f_{\overline{\alpha}}, \frac{a}{2^k}\right), \forall\, k \in \mathbb{N}. \tag{14.51}$$

(iv) When $\frac{\partial f(x)}{\partial x_i} = 0$, $i = 1, \ldots, d$, then (14.50) becomes

$$\left|(A_k f)(x) - f(x)\right| \leq \frac{a}{2^k} d \max_{i\in\{1,\ldots,d\}} \omega_{1,\infty}\left(\frac{\partial f}{\partial x_i}, \frac{a}{2^k}\right), \tag{14.52}$$

$\forall\, k \in \mathbb{Z}$.

In (14.51), (14.52), we observe the high speed of convergence.

Proof of Theorem 14.8
Set

$$g_{\frac{u}{2^k}}(t) := f\left(x - t\frac{u}{2^k}\right), \text{ all } 0 \leq t \leq 1. \tag{14.53}$$

We notice that

$$g_{\frac{u}{2^k}}(0) = f(x), \text{ and } g_{\frac{u}{2^k}}(1) = f\left(x - \frac{u}{2^k}\right). \tag{14.54}$$

Then, for $j = 1, 2, \ldots, N$, we get that

$$g^{(j)}_{\frac{u}{2^k}}(t) = \left(\left(\sum_{i=1}^{d}\left(-\frac{u_i}{2^k}\right)\frac{\partial}{\partial x_i}\right)^j f\right)\left(x - t\frac{u}{2^k}\right). \tag{14.55}$$

By Taylor's formula we find

$$f\left(x - \frac{u}{2^k}\right) - f(x) = g_{\frac{u}{2^k}}(1) - g_{\frac{u}{2^k}}(0) = \sum_{i=1}^{N}\frac{g^{(j)}_{\frac{u}{2^k}}(0)}{j!} + R_N\left(x - \frac{u}{2^k}, 0\right), \tag{14.56}$$

where

$$R_N\left(x - \frac{u}{2^k}, 0\right) := \int_0^1\left(\int_0^{t_1} \ldots \left(\int_0^{t_{N-1}}\left(g^{(N)}_{\frac{u}{2^k}}(t_N) - g^{(N)}_{\frac{u}{2^k}}(0)\right)dt_N\right)\ldots\right)dt_1. \tag{14.57}$$

Here we consider $u = (u_1, \ldots, u_d) \in \mathbb{R}^d$, with $|u_i| \leq a$, $i = 1, \ldots, d$.

Clearly here it holds

$$\left|g_{\frac{u}{2^k}}^{(j)}(0)\right| \leq \left(\frac{a}{2^k}\right)^j \left(\left(\sum_{i=1}^{d}\left|\frac{\partial}{\partial x_i}\right|\right)^j f(x)\right), \ j = 1, ..., N. \tag{14.58}$$

Next we estimate $\left|R_N\left(x - \frac{u}{2^k}, 0\right)\right|$, $0 \leq t_N \leq 1$. We observe that

$$\left|g_{\frac{u}{2^k}}^{(N)}(t_N) - g_{\frac{u}{2^k}}^{(N)}(0)\right| =$$

$$\left|\left\{\left(\sum_{i=1}^{d}\left(-\frac{u_i}{2^k}\right)\frac{\partial}{\partial x_i}\right)^N f\right\}\left(x - t\frac{u}{2^k}\right) - \left\{\left(\sum_{i=1}^{d}\left(-\frac{u_i}{2^k}\right)\frac{\partial}{\partial x_i}\right)^N f\right\}(x)\right| \leq$$

$$\frac{a^N d^N}{2^{kN}} \max_{\overline{\alpha}:|\overline{\alpha}|=N} \omega_{1,\infty}\left(f_{\overline{\alpha}}, \frac{a}{2^k}\right). \tag{14.59}$$

Therefore we get

$$\left|R_N\left(x - \frac{u}{2^k}, 0\right)\right| \leq \int_0^1 \left(\int_0^{t_1} ... \left(\int_0^{t_{N-1}} \left|g_{\frac{u}{2^k}}^{(N)}(t_N) - g_{\frac{u}{2^k}}^{(N)}(0)\right| dt_N\right) ...\right) dt_1 \overset{(14.59)}{\leq}$$

$$\int_0^1 \left(\int_0^{t_1} ... \left(\int_0^{t_{N-1}} \frac{a^N d^N}{2^{kN}} \max_{\overline{\alpha}:|\overline{\alpha}|=N} \omega_{1,\infty}\left(f_{\overline{\alpha}}, \frac{a}{2^k}\right) dt_N\right) ...\right) dt_1 = \tag{14.60}$$

$$\frac{a^N d^N}{N! 2^{kN}} \max_{\overline{\alpha}:|\overline{\alpha}|=N} \omega_{1,\infty}\left(f_{\overline{\alpha}}, \frac{a}{2^k}\right).$$

That is

$$\left|R_N\left(x - \frac{u}{2^k}, 0\right)\right| \leq \frac{a^N d^N}{N! 2^{kN}} \max_{\overline{\alpha}:|\overline{\alpha}|=N} \omega_{1,\infty}\left(f_{\overline{\alpha}}, \frac{a}{2^k}\right) := \overline{\Psi}. \tag{14.61}$$

By (14.56), (14.58) and (14.61), we get

$$\left|f\left(x - \frac{u}{2^k}\right) - f(x)\right| \leq \sum_{j=1}^{N} \frac{1}{j!}\left(\frac{a}{2^k}\right)^j \left(\left(\sum_{i=1}^{d}\left|\frac{\partial}{\partial x_i}\right|\right)^j f(x)\right) + \overline{\Psi}. \tag{14.62}$$

We further observe

$$|(A_k f)(x) - f(x)| \overset{(14.15)}{=}$$

$$\left|(N^*)\int_{[-a,a]^d} f\left(x - \frac{u}{2^k}\right)\psi(u)\, d\mu(u) - (N^*)\int_{[-a,a]^d} f(x)\psi(u)\, d\mu(u)\right| \overset{(14.20)}{\leq}$$

$$\left(N^*\right)\int_{[-a,a]^d}\left|f\left(x-\frac{u}{2^k}\right)-f(x)\right|\psi(u)\,d\mu(u)\overset{(14.62)}{\leq}$$

$$\left\{\sum_{j=1}^{N}\frac{1}{j!}\left(\frac{a}{2^k}\right)^j\left(\left(\sum_{i=1}^{d}\left|\frac{\partial}{\partial x_i}\right|\right)^j f(x)\right)+\overline{\Psi}\right\}\left(\left(N^*\right)\int_{[-a,a]^d}\psi(u)\,d\mu(u)\right)\overset{(14.15)}{=}$$

$$\sum_{j=1}^{N}\frac{a^j}{j!2^{kj}}\left(\left(\sum_{i=1}^{d}\left|\frac{\partial}{\partial x_i}\right|\right)^j f(x)\right)+\frac{a^N d^N}{N!2^{kN}}\max_{\overline{\alpha}:|\overline{\alpha}|=N}\omega_{1,\infty}\left(f_{\overline{\alpha}},\frac{a}{2^k}\right), \quad (14.63)$$

proving the claim. ■

We also give

Theorem 14.10 *Let $f\in C^N\left(\mathbb{R}^d,\mathbb{R}_+\right)$, $N,d\in\mathbb{N}$. Consider the multivariate Shilkret-sublinear operators*

$$\left(D_k(f)\right)(x)=\left(N^*\right)\int_{[-a,a]^d}\left[\sum_{j=0}^{n}w_j f\left(\left(x-\frac{u}{2^k}\right)+\frac{j}{2^k n}\right)\right]\psi(u)\,d\mu(u),$$

$\forall\, x\in\mathbb{R}^d$, $\forall\, k\in\mathbb{Z}$. Here we further assume that all partials $f_{\overline{\alpha}}$, $|\overline{\alpha}|=N$, are uniformly continuous or bounded and continuous on $\mathbb{R}^d$. Then

$$\left|\left(D_k f\right)(x)-f(x)\right|\leq\sum_{j=1}^{N}\frac{(a+1)^j}{j!2^{kj}}\left(\left(\sum_{i=1}^{d}\left|\frac{\partial}{\partial x_i}\right|\right)^j f(x)\right)+ \quad (14.64)$$

$$\frac{(a+1)^N d^N}{N!2^{kN}}\max_{\overline{\alpha}:|\overline{\alpha}|=N}\omega_{1,\infty}\left(f_{\overline{\alpha}},\frac{a+1}{2^k}\right),$$

$\forall\, x\in\mathbb{R}^d$, $\forall\, k\in\mathbb{Z}$.

Remark 14.11 (i) Since the Nth order partials $f_{\overline{\alpha}}$ are uniformly continuous or bounded and continuous and $k\to+\infty$, we get that $\left(D_k(f)\right)(x)\to f(x)$ pointwise with rates.

(ii) When $N=1$, inequality (14.64) becomes

$$\left|\left(D_k(f)\right)(x)-f(x)\right|\leq$$

$$\left(\frac{a+1}{2^k}\right)\left\{\left(\sum_{i=1}^{d}\left|\frac{\partial f(x)}{\partial x_i}\right|\right)+d\max_{i\in\{1,\dots,d\}}\omega_{1,\infty}\left(\frac{\partial f}{\partial x_i},\frac{a+1}{2^k}\right)\right\}, \quad (14.65)$$

$\forall\, k\in\mathbb{Z}$, $\forall\, x\in\mathbb{R}^d$.

(iii) When $f_{\overline{\alpha}}(x) = 0$, $\overline{\alpha} : |\overline{\alpha}| = \rho$, $1 \leq \rho \leq N$. Then (14.64) becomes

$$|(D_k(f))(x) - f(x)| \leq \frac{(a+1)^N d^N}{N! 2^{kN}} \max_{\overline{\alpha}:|\overline{\alpha}|=N} \omega_{1,\infty}\left(f_{\overline{\alpha}}, \frac{a+1}{2^k}\right), \forall\, k \in \mathbb{Z}. \tag{14.66}$$

(iv) When $\frac{\partial f(x)}{\partial x_i} = 0$, $i = 1, ..., d$, then (14.65) becomes

$$|(D_k(f))(x) - f(x)| \leq \left(\frac{a+1}{2^k}\right) d \max_{i\in\{1,...,d\}} \omega_{1,\infty}\left(\frac{\partial f}{\partial x_i}, \frac{a+1}{2^k}\right), \tag{14.67}$$

$\forall\, k \in \mathbb{Z}$.

In (14.66), (14.67), we observe the high speed of convergence.

Proof of Theorem 14.10

Set

$$g_{\frac{u}{2^k}+\frac{j}{2^k n}}(t) := f\left(x + t\left(-\frac{u}{2^k} + \frac{j}{2^k n}\right)\right), \text{ all } 0 \leq t \leq 1. \tag{14.68}$$

We notice that

$$g_{\frac{u}{2^k}+\frac{j}{2^k n}}(0) = f(x), \text{ and } g_{\frac{u}{2^k}+\frac{j}{2^k n}}(1) = f\left(\left(x - \frac{u}{2^k}\right) + \frac{j}{2^k n}\right). \tag{14.69}$$

Then, for $j^* = 1, 2, ..., N$, we get that

$$g^{(j^*)}_{\frac{u}{2^k}+\frac{j}{2^k n}}(t) = \left\{\left(\sum_{i=1}^{d}\left(-\frac{u_i}{2^k} + \frac{j_i}{2^k n_i}\right)\frac{\partial}{\partial x_i}\right)^{j^*} f\right\}\left(x + t\left(-\frac{u}{2^k} + \frac{j}{2^k n}\right)\right). \tag{14.70}$$

By Taylor's formula we find

$$f\left(\left(x - \frac{u}{2^k}\right) + \frac{j}{2^k n}\right) - f(x) = g_{\frac{u}{2^k}+\frac{j}{2^k n}}(1) - g_{\frac{u}{2^k}+\frac{j}{2^k n}}(0) = \tag{14.71}$$

$$\sum_{j^*=1}^{N} \frac{g^{(j^*)}_{\frac{u}{2^k}+\frac{j}{2^k n}}(0)}{j^*!} + R_N\left(\left(x - \frac{u}{2^k}\right) + \frac{j}{2^k n}, 0\right),$$

where

$$R_N\left(\left(x - \frac{u}{2^k}\right) + \frac{j}{2^k n}, 0\right) :=$$

$$\int_0^1\left(\int_0^{t_1} ... \left(\int_0^{t_{N-1}}\left(g^{(N)}_{\frac{u}{2^k}+\frac{j}{2^k n}}(t_N) - g^{(N)}_{\frac{u}{2^k}+\frac{j}{2^k n}}(0)\right) dt_N\right) ...\right) dt_1. \tag{14.72}$$

Here we consider $u = (u_1, ..., u_d) \in \mathbb{R}^d$, with $|u_i| \leq a$, $i = 1, ..., d$.

Clearly here it holds

$$\left| g^{(j^*)}_{\frac{u}{2^k}+\frac{j}{2^k n}}(0) \right| \leq \left(\frac{a+1}{2^k} \right)^{j^*} \left(\left(\sum_{i=1}^{d} \left| \frac{\partial}{\partial x_i} \right| \right)^{j^*} f(x) \right), \; j^* = 1, ..., N. \quad (14.73)$$

Next we estimate $\left| R_N \left(\left(x - \frac{u}{2^k} \right) + \frac{j}{2^k n}, 0 \right) \right|$, $0 \leq t_N \leq 1$. We observe that

$$\left| g^{(N)}_{\frac{u}{2^k}+\frac{j}{2^k n}}(t_N) - g^{(N)}_{\frac{u}{2^k}+\frac{j}{2^k n}}(0) \right| =$$

$$\left| \left\{ \left(\sum_{i=1}^{d} \left(-\frac{u_i}{2^k} + \frac{j_i}{2^k n_i} \right) \frac{\partial}{\partial x_i} \right)^N f \right\} \left(x + t_N \left(-\frac{u}{2^k} + \frac{j}{2^k n} \right) \right) \right.$$

$$\left. - \left\{ \left(\sum_{i=1}^{d} \left(-\frac{u_i}{2^k} + \frac{j_i}{2^k n_i} \right) \frac{\partial}{\partial x_i} \right)^N f \right\} (x) \right| \leq$$

$$\frac{(a+1)^N d^N}{2^{kN}} \max_{\overline{\alpha}:|\overline{\alpha}|=N} \omega_{1,\infty} \left(f_{\overline{\alpha}}, \frac{a+1}{2^k} \right). \quad (14.74)$$

Therefore we get

$$\left| R_N \left(\left(x - \frac{u}{2^k} \right) + \frac{j}{2^k n}, 0 \right) \right| \leq$$

$$\int_0^1 \left(\int_0^{t_1} ... \left(\int_0^{t_{N-1}} \left| g^{(N)}_{\frac{u}{2^k}+\frac{j}{2^k n}}(t_N) - g^{(N)}_{\frac{u}{2^k}+\frac{j}{2^k n}}(0) \right| dt_N \right) ... \right) dt_1 \overset{(14.74)}{\leq}$$

$$\int_0^1 \left(\int_0^{t_1} ... \left(\int_0^{t_{N-1}} \frac{(a+1)^N d^N}{2^{kN}} \max_{\overline{\alpha}:|\overline{\alpha}|=N} \omega_{1,\infty} \left(f_{\overline{\alpha}}, \frac{a+1}{2^k} \right) dt_N \right) ... \right) dt_1 =$$

$$\frac{(a+1)^N d^N}{N! 2^{kN}} \max_{\overline{\alpha}:|\overline{\alpha}|=N} \omega_{1,\infty} \left(f_{\overline{\alpha}}, \frac{a+1}{2^k} \right). \quad (14.75)$$

That is

$$\left| R_N \left(\left(x - \frac{u}{2^k} \right) + \frac{j}{2^k n}, 0 \right) \right| \leq \frac{(a+1)^N d^N}{N! 2^{kN}} \max_{\overline{\alpha}:|\overline{\alpha}|=N} \omega_{1,\infty} \left(f_{\overline{\alpha}}, \frac{a+1}{2^k} \right) := \theta. \quad (14.76)$$

By (14.71)

$$\sum_{j=0}^{n} w_j f\left(\left(x-\frac{u}{2^k}\right)+\frac{j}{2^k n}\right)-f(x)=$$

$$\sum_{j^*=1}^{N}\frac{1}{j^*!}\sum_{j=0}^{n} w_j\left(g^{(j^*)}_{\frac{u}{2^k}+\frac{j}{2^k n}}(0)\right)+\sum_{j=0}^{n} w_j R_N\left(\left(x-\frac{u}{2^k}\right)+\frac{j}{2^k n},0\right). \tag{14.77}$$

By (14.73) and (14.76), (14.77) we derive

$$\left|\sum_{j=0}^{n} w_j f\left(\left(x-\frac{u}{2^k}\right)+\frac{j}{2^k n}\right)-f(x)\right|\leq$$

$$\sum_{j^*=1}^{N}\frac{1}{j^*!}\left(\frac{a+1}{2^k}\right)^{j^*}\left(\left(\sum_{i=1}^{d}\left|\frac{\partial}{\partial x_i}\right|\right)^{j^*} f(x)\right)+\theta, \tag{14.78}$$

above we applied that $\sum_{j=0}^{n} w_j = 1$.

We further observe that

$$\left|(D_k(f))(x)-f(x)\right| \overset{(14.15)}{=}$$

$$\left|(N^*)\int_{[-a,a]^d}\left[\sum_{j=0}^{n} w_j f\left(\left(x-\frac{u}{2^k}\right)+\frac{j}{2^k n}\right)\right]\psi(u)\,d\mu(u)\right.$$

$$\left.-(N^*)\int_{[-a,a]^d} f(x)\,\psi(u)\,d\mu(u)\right| \overset{(14.20)}{\leq}$$

$$(N^*)\int_{[-a,a]^d}\left|\sum_{j=0}^{n} w_j f\left(\left(x-\frac{u}{2^k}\right)+\frac{j}{2^k n}\right)-f(x)\right|\psi(u)\,d\mu(u) \overset{(14.78)}{\leq} \tag{14.79}$$

$$\left\{\sum_{j^*=1}^{N}\frac{1}{j^*!}\left(\frac{a+1}{2^k}\right)^{j^*}\left(\left(\sum_{i=1}^{d}\left|\frac{\partial}{\partial x_i}\right|\right)^{j^*} f(x)\right)+\theta\right\}\left((N^*)\int_{[-a,a]^d}\psi(u)\,d\mu(u)\right) \overset{(14.15)}{=}$$

$$\sum_{j^*=1}^{N}\frac{(a+1)^{j^*}}{j^*!2^{kj^*}}\left(\left(\sum_{i=1}^{d}\left|\frac{\partial}{\partial x_i}\right|\right)^{j^*} f(x)\right)+\frac{(a+1)^N d^N}{N!2^{kN}}\max_{\overline{\alpha}:|\overline{\alpha}|=N}\omega_{1,\infty}\left(f_{\overline{\alpha}},\frac{a+1}{2^k}\right),$$

proving the claim. ■

14.5 Appendix

Let $f \in C_U\left(\mathbb{R}^d, \mathbb{R}_+\right)$, and the multivariate positive sublinear Shilkret operator

$$(M(f))(x) := \left(N^*\right)\int_{[-a,a]^d} f(x+u)\,\psi(u)\,d\mu(u), \forall\, x \in \mathbb{R}. \quad (14.80)$$

We observe the following (for any $x, y \in \mathbb{R}^d$):

$$|(M(f))(x) - (M(f))(y)| =$$

$$\left|\left(N^*\right)\int_{[-a,a]^d} f(x+u)\,\psi(u)\,d\mu(u) - \left(N^*\right)\int_{[-a,a]^d} f(y+u)\,\psi(u)\,d\mu(u)\right| \overset{(14.20)}{\leq}$$

$$\left(N^*\right)\int_{[-a,a]^d} |f(x+u) - f(y+u)|\,\psi(u)\,d\mu(u) \leq$$

$$\omega_1(f, \|x-y\|)\left(\left(N^*\right)\int_{[-a,a]^d} \psi(u)\,d\mu(u)\right) \overset{(14.15)}{=} \omega_1(f, \|x-y\|)\cdot 1 = \omega_1(f, \|x-y\|). \quad (14.81)$$

Therefore it holds the multivariate global smoothness preservation property:

$$\omega_1(M(f), \delta) \leq \omega_1(f, \delta), \forall\, \delta > 0. \quad (14.82)$$

References

1. G.A. Anastassiou, High order approximation by multivariate shift-invariant convolution type operators, in *Computers and Mathematics with Applications*, ed. by G.A. Anastassiou. Special issue on Computational Methods in Analysis, vol. 48 (2004), pp. 1245–1261
2. G.A. Anastassiou, *Intelligent Mathematics: Computational Analysis* (Springer, Heidelberg, 2011)
3. G.A. Anastassiou, Approximation by shift invariant multivariate sublinear-Shilkret operators (2018). Submitted
4. G.A. Anastassiou, H.H. Gonska, On some shift invariant integral operators, multivariate case, in *Proceedings of International Conference on Approximation, Probability, and Related Fields, U.C.S.B., Santa Barbara, CA, 20–22 May 1993*, ed. G.A. Anastassiou, S.T. Rachev (Plenum Press, New York, 1993), pp. 41–64
5. G.A. Anastassiou, S. Gal, *Approximation Theory* (Birkhauser, Boston, 2000)
6. N. Shilkret, Maxitive measure and integration. Indag. Math. **33**, 109–116 (1971)

Chapter 15
Ostrowski–Sugeno Type Fuzzy Integral Inequalities

Here we present Ostrowski–Sugeno Fuzzy type inequalities. These are Ostrowski-like inequalities in the context of Sugeno fuzzy integral and its special properties. They give tight upper bounds to the deviation of a function from its Sugeno-fuzzy averages. This work is greatly inspired by [1, 4]. It follows [2].

15.1 Introduction

The famous Ostrowski [4] inequality motivates this work and has as follows:

$$\left| \frac{1}{b-a} \int_a^b f(y)\, dy - f(x) \right| \leq \left(\frac{1}{4} + \frac{\left(x - \frac{a+b}{2}\right)^2}{(b-a)^2} \right) (b-a) \left\| f' \right\|_\infty ,$$

where $f \in C'([a,b])$, $x \in [a,b]$, and it is a sharp inequality. One can easily notice that

$$\left(\frac{1}{4} + \frac{\left(x - \frac{a+b}{2}\right)^2}{(b-a)^2} \right) (b-a) = \frac{(x-a)^2 + (b-x)^2}{2(b-a)}.$$

Another motivation is author's article [1].

Here first we give a survey about Sugeno fuzzy integral and its basic properties. Then we derive a series of Ostrowski-like inequalities to all directions in the context of Sugeno integral and its basic important particular properties. We also give applications to special cases of our problem we deal with.

G. A. Anastassiou, *Ordinary and Fractional Approximation by Non-additive Integrals: Choquet, Shilkret and Sugeno Integral Approximators*, Studies in Systems, Decision and Control 190, https://doi.org/10.1007/978-3-030-04287-5_15

15.2 Background

In this section, some definitions and basic important properties of the Sugeno integral which will be used in the next section are presented.

Definition 15.1 (*Fuzzy measure* [6, 8]) Let Σ be a σ-algebra of subsets of X, and let $\mu : \Sigma \to [0, +\infty]$ be a non-negative extended real-valued set function. We say that μ is a fuzzy measure iff:

(1) $\mu(\varnothing) = 0$,

(2) $E, F \in \Sigma : E \subseteq F$ imply $\mu(E) \le \mu(F)$ (monotonicity),

(3) $E_n \in \Sigma$ $(n \in \mathbb{N})$, $E_1 \subset E_2 \subset ...$, imply $\lim_{n\to\infty} \mu(E_n) = \mu\left(\cup_{n=1}^{\infty} E_n\right)$ (continuity from below);

(4) $E_n \in \Sigma$ $(n \in \mathbb{N})$, $E_1 \supset E_2 \supset ...$, $\mu(E_1) < \infty$, imply $\lim_{n\to\infty} \mu(E_n) = \mu\left(\cap_{n=1}^{\infty} E_n\right)$ (continuity from above).

Let (X, Σ, μ) be a fuzzy measure space and f be a non-negative real-valued function on X. We denote by $\mathcal{F}_+$ the set of all non-negative real valued measurable functions, and by $L_\alpha f$ the set: $L_\alpha f := \{x \in X : f(x) \ge \alpha\}$, the α-level of f for $\alpha \ge 0$.

Definition 15.2 Let (X, Σ, μ) be a fuzzy measure space. If $f \in \mathcal{F}_+$ and $A \in \Sigma$, then the Sugeno integral (fuzzy integral) [7] of f on A with respect to the fuzzy measure μ is defined by

$$(S)\int_A f d\mu := \vee_{\alpha\ge 0}\left(\alpha \wedge \mu\left(A \cap L_\alpha f\right)\right), \tag{15.1}$$

where $\vee$ and $\wedge$ denote the sup and inf on $[0, \infty]$, respectively.

The basic properties of Sugeno integral follow:

Theorem 15.3 ([5, 8]) *Let (X, Σ, μ) be a fuzzy measure space with $A, B \in \Sigma$ and $f, g \in \mathcal{F}_+$. Then*

(1) $(S)\int_A f d\mu \le \mu(A)$;

(2) $(S)\int_A k d\mu = k \wedge \mu(A)$ *for a non-negative constant k;*

(3) if $f \le g$ on A, then $(S)\int_A f d\mu \le (S)\int_A g d\mu$;

(4) if $A \subset B$, then $(S)\int_A f d\mu \le (S)\int_B f d\mu$;

(5) $\mu(A \cap L_\alpha f) \le \alpha \Rightarrow (S)\int_A f d\mu \le \alpha$;

(6) if $\mu(A) < \infty$, then $\mu(A \cap L_\alpha f) \ge \alpha \Leftrightarrow (S)\int_A f d\mu \ge \alpha$;

(7) when $A = X$, then $(S)\int_A f d\mu = \vee_{\alpha\ge 0}(\alpha \wedge \mu(L_\alpha f))$;

(8) if $\alpha \le \beta$, then $L_\beta f \subseteq L_\alpha f$;

(9) $(S)\int_A f d\mu \ge 0$.

Theorem 15.4 ([8], p. 135) *Here $f \in \mathcal{F}_+$, the class of all finite nonnegative measurable functions on (X, Σ, μ). Then*

(1) if $\mu(A) = 0$*, then* $(S)\int_A f d\mu = 0$*, for any* $f \in \mathcal{F}_+$*;*
(2) if $(S)\int_A f d\mu = 0$*, then* $\mu(A \cap \{x | f(x) > 0\}) = 0$*;*
(3) $(S)\int_A f d\mu = (S)\int_A f \cdot \chi_A d\mu$*, where* χ_A *is the characteristic function of* A*;*
(4) $(S)\int_A (f + a) d\mu \leq (S)\int_A f d\mu + (S)\int_A a d\mu$*, for any constant* $a \in [0, \infty)$*.*

Corollary 15.5 ([8], p. 136) *Here* $f, f_1, f_2 \in \mathcal{F}_+$*. Then*
(1) $(S)\int_A (f_1 \vee f_2) d\mu \geq (S)\int_A f_1 d\mu \vee (S)\int_A f_2 d\mu$*;*
(2) $(S)\int_A (f_1 \wedge f_2) d\mu \leq (S)\int_A f_1 d\mu \wedge (S)\int_A f_2 d\mu$*;*
(3) $(S)\int_{A \cup B} f d\mu \geq (S)\int_A f d\mu \vee (S)\int_B f d\mu$*;*
(4) $(S)\int_{A \cap B} f d\mu \leq (S)\int_A f d\mu \wedge (S)\int_B f d\mu$*.*

In general we have

$$(S)\int_A (f_1 + f_2) d\mu \neq (S)\int_A f_1 d\mu + (S)\int_A f_2 d\mu,$$

and

$$(S)\int_A a f d\mu \neq a (S)\int_A f d\mu, \text{ where } a \in \mathbb{R},$$

see [8], p. 137.

Lemma 15.6 ([8], p. 138) $(S)\int_A f d\mu = \infty$ *iff* $\mu(A \cap L_\alpha f) = \infty$ *for any* $\alpha \in [0, \infty)$.

We need

Definition 15.7 ([3]) A fuzzy measure μ is subadditive iff $\mu(A \cup B) \leq \mu(A) + \mu(B)$, for all $A, B \in \Sigma$.

We mention

Theorem 15.8 ([3]) *If* μ *is subadditive, then*

$$(S)\int_X (f + g) d\mu \leq (S)\int_X f d\mu + (S)\int_X g d\mu, \tag{15.2}$$

for all measurable functions $f, g : X \to [0, \infty)$*.*

Moreover, if (15.2) holds for all measurable functions $f, g : X \to [0, \infty)$ *and* $\mu(X) < \infty$*, then* μ *is subadditive.*

Notice here in (15.1) we have that $\alpha \in [0, \infty)$*.*

We have

Corollary 15.9 *If* μ *is subadditive,* $n \in \mathbb{N}$*, and* $f : X \to [0, \infty)$ *is a measurable function, then*

$$(S)\int_X n f d\mu \leq n (S)\int_X f d\mu, \tag{15.3}$$

in particular it holds

$$(S)\int_A nf d\mu \le n\,(S)\int_A f d\mu, \tag{15.4}$$

for any $A \in \Sigma$.

Proof By (15.2). ■

A very important property of Sugeno integral follows.

Theorem 15.10 *If* μ *is subadditive measure, and* $f : X \to [0, \infty)$ *is a measurable function, and* $c > 0$*, then*

$$(S)\int_A cf d\mu \le (c+1)\,(S)\int_A f d\mu, \tag{15.5}$$

for any $A \in \Sigma$.

Proof Let the ceiling $\lceil c \rceil = m \in \mathbb{N}$, then by Theorem 15.3 (3) and (4) we get

$$(S)\int_A cf d\mu \le (S)\int_A mf d\mu \le m\,(S)\int_A f d\mu \le (c+1)\,(S)\int_A f d\mu,$$

proving (15.5). ■

15.3 Main Results

From now on in this chapter we work on the fuzzy measure space $([a,b], \mathcal{B}, \mu)$, where $[a,b] \subset \mathbb{R}$, $\mathcal{B}$ is the Borel σ-algebra on $[a,b]$, and μ is a finite fuzzy measure on $\mathcal{B}$. Typically we take it to be subadditive.

The functions f we deal with here are continuous from $[a,b]$ into $\mathbb{R}_+$.

We make

Remark 15.11 Let $f \in C^1([a,b], \mathbb{R}_+)$, and μ is a subadditive fuzzy measure such that $\mu([a,b]) > 0$, $x \in [a,b]$. We will estimate

$$E := \left| (S)\int_{[a,b]} f(x)\, d\mu(t) - \mu([a,b]) \wedge f(x) \right| \tag{15.6}$$

(by Theorem 15.3 (2))

$$= \left| (S)\int_{[a,b]} f(t)\, d\mu(t) - (S)\int_{[a,b]} f(x)\, d\mu(t) \right|.$$

We notice that

$$f(t) = f(t) - f(x) + f(x) \leq |f(t) - f(x)| + f(x),$$

then (by Theorem 15.3 (3) and Theorem 15.4 (4))

$$(S)\int_{[a,b]} f(t)\, d\mu(t) \leq (S)\int_{[a,b]} |f(t) - f(x)|\, d\mu(t) + (S)\int_{[a,b]} f(x)\, d\mu(t), \tag{15.7}$$

that is

$$(S)\int_{[a,b]} f(t)\, d\mu(t) - (S)\int_{[a,b]} f(x)\, d\mu(t) \leq (S)\int_{[a,b]} |f(t) - f(x)|\, d\mu(t). \tag{15.8}$$

Similarly, we have

$$f(x) = f(x) - f(t) + f(t) \leq |f(t) - f(x)| + f(t),$$

then (by Theorem 15.3 (3) and Theorem 15.8)

$$(S)\int_{[a,b]} f(x)\, d\mu(t) \leq (S)\int_{[a,b]} |f(t) - f(x)|\, d\mu(t) + (S)\int_{[a,b]} f(t)\, d\mu(t),$$

that is

$$(S)\int_{[a,b]} f(x)\, d\mu(t) - (S)\int_{[a,b]} f(t)\, d\mu(t) \leq (S)\int_{[a,b]} |f(t) - f(x)|\, d\mu(t). \tag{15.9}$$

By (15.8) and (15.9) we derive that

$$\left|(S)\int_{[a,b]} f(t)\, d\mu(t) - (S)\int_{[a,b]} f(x)\, d\mu(t)\right| \leq (S)\int_{[a,b]} |f(t) - f(x)|\, d\mu(t). \tag{15.10}$$

Consequently it holds

$$E \overset{\text{(by (15.6), (15.10))}}{\leq} (S)\int_{[a,b]} |f(t) - f(x)|\, d\mu(t)$$

(and by $|f(t) - f(x)| \leq \|f'\|_\infty |t - x|$)

$$\leq (S)\int_{[a,b]} \|f'\|_\infty |t - x|\, d\mu(t) \overset{\text{(by (15.5))}}{\leq} \left(\|f'\|_\infty + 1\right)(S)\int_{[a,b]} |t - x|\, d\mu(t). \tag{15.11}$$

We have proved the following Ostrowski-like inequality

$$\left|\frac{1}{\mu([a,b])}(S)\int_{[a,b]} f(t)\, d\mu(t) - \frac{\mu([a,b] \wedge f(x))}{\mu([a,b])}\right| \leq \tag{15.12}$$

$$\frac{\left(\|f'\|_{\infty}+1\right)}{\mu([a,b])}(S)\int_{[a,b]}|t-x|\,d\mu(t).$$

The last can be better written as follows:

$$\left|\frac{1}{\mu([a,b])}(S)\int_{[a,b]}f(t)\,d\mu(t)-\left(1\wedge\frac{f(x)}{\mu([a,b])}\right)\right|\leq$$

$$\frac{\left(\|f'\|_{\infty}+1\right)}{\mu([a,b])}(S)\int_{[a,b]}|t-x|\,d\mu(t). \qquad (15.13)$$

Notice here that $\left(1\wedge\frac{f(x)}{\mu([a,b])}\right)\leq 1$, and $\frac{1}{\mu([a,b])}(S)\int_{[a,b]}f(t)\,d\mu(t)\leq\frac{\mu([a,b])}{\mu([a,b])}=1$, where $(S)\int_{[a,b]}f(t)\,d\mu(t)\geq 0$.

If $f:[a,b]\to\mathbb{R}_{+}$ is a Lipschitz function of order $0<\alpha\leq 1$, i.e. $|f(x)-f(y)|\leq K|x-y|^{\alpha}$, $\forall\, x,y\in[a,b]$, where $K>0$, denoted by $f\in Lip_{\alpha,K}([a,b],\mathbb{R}_{+})$, then we get similarly the following Ostrowski-like inequality:

$$\left|\frac{1}{\mu([a,b])}(S)\int_{[a,b]}f(t)\,d\mu(t)-\left(1\wedge\frac{f(x)}{\mu([a,b])}\right)\right|\leq$$

$$\frac{(K+1)}{\mu([a,b])}(S)\int_{[a,b]}|t-x|^{\alpha}\,d\mu(t). \qquad (15.14)$$

We have proved the following Ostrowski–Sugeno inequalities:

Theorem 15.12 *Here μ is a fuzzy subadditive measure with $\mu([a,b])>0$, $x\in[a,b]$.*

(1) Let $f\in C^{1}([a,b],\mathbb{R}_{+})$, then

$$\left|\frac{1}{\mu([a,b])}(S)\int_{[a,b]}f(t)\,d\mu(t)-\left(1\wedge\frac{f(x)}{\mu([a,b])}\right)\right|\leq$$

$$\frac{\left(\|f'\|_{\infty}+1\right)}{\mu([a,b])}(S)\int_{[a,b]}|t-x|\,d\mu(t). \qquad (15.15)$$

(2) Let $f\in Lip_{\alpha,K}([a,b],\mathbb{R}_{+})$, $0<\alpha\leq 1$, then

$$\left|\frac{1}{\mu([a,b])}(S)\int_{[a,b]}f(t)\,d\mu(t)-\left(1\wedge\frac{f(x)}{\mu([a,b])}\right)\right|\leq$$

$$\frac{(K+1)}{\mu([a,b])}(S)\int_{[a,b]}|t-x|^{\alpha}\,d\mu(t). \qquad (15.16)$$

We make

Remark 15.13 Let $f \in C^1([a,b], \mathbb{R}_+)$ and $g \in C^1([a,b])$, by Cauchy's mean value theorem we get that

$$(f(t) - f(x)) g'(c) = (g(t) - g(x)) f'(c),$$

for some c between t and x; for any $t, x \in [a,b]$.

If $g'(c) \neq 0$, we have

$$(f(t) - f(x)) = \left(\frac{f'(c)}{g'(c)}\right)(g(t) - g(x)).$$

Here we assume that $g'(t) \neq 0$, $\forall\, t \in [a,b]$. Hence it holds

$$|f(t) - f(x)| \leq \left\| \frac{f'}{g'} \right\|_\infty |g(t) - g(x)|, \quad (15.17)$$

$\forall\, t, x \in [a,b]$.

We have again as before (see (15.11))

$$E \leq (S) \int_{[a,b]} |f(t) - f(x)|\, d\mu(t) \overset{\text{(by (15.17))}}{\leq}$$

$$(S) \int_{[a,b]} \left\| \frac{f'}{g'} \right\|_\infty |g(t) - g(x)|\, d\mu(t) \overset{\text{(by (15.5))}}{\leq}$$

$$\left(\left\| \frac{f'}{g'} \right\|_\infty + 1 \right) (S) \int_{[a,b]} |g(t) - g(x)|\, d\mu(t). \quad (15.18)$$

We have established the following general Ostrowski–Sugeno inequality:

Theorem 15.14 *Here μ is a fuzzy subadditive measure with $\mu([a,b]) > 0$, $x \in [a,b]$. Let $f \in C^1([a,b], \mathbb{R}_+)$ and $g \in C^1([a,b])$ with $g'(t) \neq 0$, $\forall\, t \in [a,b]$. Then*

$$\left| \frac{1}{\mu([a,b])} (S) \int_{[a,b]} f(t)\, d\mu(t) - \left(1 \wedge \frac{f(x)}{\mu([a,b])}\right) \right| \leq$$

$$\frac{\left(\left\| \frac{f'}{g'} \right\|_\infty + 1 \right)}{\mu([a,b])} (S) \int_{[a,b]} |g(t) - g(x)|\, d\mu(t). \quad (15.19)$$

We give for $g(t) = e^t$ the next result

Corollary 15.15 *Here μ is a fuzzy subadditive measure with $\mu([a,b]) > 0$, $x \in [a,b]$. Let $f \in C^1([a,b], \mathbb{R}_+)$, then*

$$\left|\frac{1}{\mu([a,b])}(S)\int_{[a,b]} f(t)\,d\mu(t) - \left(1 \wedge \frac{f(x)}{\mu([a,b])}\right)\right| \le$$

$$\frac{\left(\left\|\frac{f'}{e^t}\right\|_\infty + 1\right)}{\mu([a,b])}(S)\int_{[a,b]} \left|e^t - e^x\right| d\mu(t). \tag{15.20}$$

When $g(t) = \ln t$ we get

Corollary 15.16 *Here μ is a fuzzy subadditive measure with $\mu([a,b]) > 0$, $x \in [a,b]$ and $a > 0$. Let $f \in C^1([a,b], \mathbb{R}_+)$. Then*

$$\left|\frac{1}{\mu([a,b])}(S)\int_{[a,b]} f(t)\,d\mu(t) - \left(1 \wedge \frac{f(x)}{\mu([a,b])}\right)\right| \le$$

$$\frac{\left(\left\|tf'(t)\right\|_\infty + 1\right)}{\mu([a,b])}(S)\int_{[a,b]} \left|\ln\frac{t}{x}\right| d\mu(t). \tag{15.21}$$

Many other applications of Theorem 15.14 could follow but we stop it here.
We make

Remark 15.17 Let $f \in \left[C([a,b], \mathbb{R}_+) \cap C^{n+1}([a,b])\right]$, $n \in \mathbb{N}$, $x \in [a,b]$. Then by Taylor's theorem we get

$$f(y) - f(x) = \sum_{k=1}^{n} \frac{f^{(k)}(x)}{k!}(y-x)^k + R_n(x,y), \tag{15.22}$$

where the remainder

$$R_n(x,y) := \int_x^y \left(f^{(n)}(t) - f^{(n)}(x)\right)\frac{(y-t)^{n-1}}{(n-1)!}dt; \tag{15.23}$$

here y can be $\ge x$ or $\le x$.
By [1] we get that

$$|R_n(x,y)| \le \frac{\left\|f^{(n+1)}\right\|_\infty}{(n+1)!}|y-x|^{n+1}, \quad \text{for all } x, y \in [a,b]. \tag{15.24}$$

Here we assume $f^{(k)}(x) = 0$, for all $k = 1, ..., n$.
Therefore it holds

$$|f(t) - f(x)| \le \frac{\left\|f^{(n+1)}\right\|_\infty}{(n+1)!}|t-x|^{n+1}, \quad \text{for all } t, x \in [a,b]. \tag{15.25}$$

Here we have again

$$E \le (S) \int_{[a,b]} |f(t) - f(x)| \, d\mu(t) \overset{\text{(by Theorem 15.3 (3) and (15.25))}}{\le}$$

$$(S) \int_{[a,b]} \frac{\left\| f^{(n+1)} \right\|_{\infty}}{(n+1)!} |t - x|^{n+1} \, d\mu(t) \overset{\text{(by (15.5))}}{\le}$$

$$\left(\frac{\left\| f^{(n+1)} \right\|_{\infty}}{(n+1)!} + 1 \right) (S) \int_{[a,b]} |t - x|^{n+1} \, d\mu(t) . \quad (15.26)$$

We have derived the following high order Ostrowski–Sugeno inequality:

Theorem 15.18 *Let $f \in \left[C([a,b], \mathbb{R}_+) \cap C^{n+1}([a,b]) \right]$, $n \in \mathbb{N}$, $x \in [a,b]$. We assume that $f^{(k)}(x) = 0$, all $k = 1, \ldots, n$. Here μ is subadditive with $\mu([a,b]) > 0$. Then*

$$\left| \frac{1}{\mu([a,b])} (S) \int_{[a,b]} f(t) \, d\mu(t) - \left(1 \wedge \frac{f(x)}{\mu([a,b])} \right) \right| \le$$

$$\frac{\left(\frac{\| f^{(n+1)} \|_{\infty}}{(n+1)!} + 1 \right)}{\mu([a,b])} (S) \int_{[a,b]} |t - x|^{n+1} \, d\mu(t) , \quad (15.27)$$

which generalizes (15.15).

When $x = \frac{a+b}{2}$ we get

Corollary 15.19 *Let $f \in \left[C([a,b], \mathbb{R}_+) \cap C^{n+1}([a,b]) \right]$, $n \in \mathbb{N}$. Assume that $f^{(k)}\left(\frac{a+b}{2}\right) = 0$, $k = 1, \ldots, n$. Here μ is subadditive with $\mu([a,b]) > 0$. Then*

$$\left| \frac{1}{\mu([a,b])} (S) \int_{[a,b]} f(t) \, d\mu(t) - \left(1 \wedge \frac{f\left(\frac{a+b}{2}\right)}{\mu([a,b])} \right) \right| \le$$

$$\frac{\left(\frac{\| f^{(n+1)} \|_{\infty}}{(n+1)!} + 1 \right)}{\mu([a,b])} (S) \int_{[a,b]} \left| t - \frac{a+b}{2} \right|^{n+1} d\mu(t) . \quad (15.28)$$

References

1. G.A. Anastassiou, Ostrowski type inequalities **123**(12), 3775–3781 (1995)
2. G.A. Anastassiou, Ostrowski-Sugeno fuzzy inequalities (2018). Submitted
3. M. Boczek, M. Kaluszka, On the Minkowaki-Hölder type inequalities for generalized Sugeno integrals with an application. Kybernetica **52**(3), 329–347 (2016)

4. A. Ostrowski, Über die Absolutabweichung einer differentiebaren Funktion von ihrem Integralmittelwert. Comment. Math. Helv. **10**, 226–227 (1938)
5. E. Pap, *Null-Additive Set Functions* (Kluwer Academic, Dordrecht, 1995)
6. D. Ralescu, G. Adams, The fuzzy integral. J. Math. Anal. Appl. **75**, 562–570 (1980)
7. M. Sugeno, Theory of fuzzy integrals and its applications. Ph.D. thesis, Tokyo Institute of Technology (1974)
8. Z. Wang, G.J. Klir, *Fuzzy Measure Theory* (Plenum, New York, 1992)

Chapter 16
Fractional Ostrowski–Sugeno Type Fuzzy Integral Univariate Inequalities

Here we present fractional univariate Ostrowski-Sugeno Fuzzy type inequalities. These are of Ostrowski-like inequalities in the setting of Sugeno fuzzy integral and its special-particular properties. In a fractional environment, they give tight upper bounds to the deviation of a function from its Sugeno-fuzzy averages. The fractional derivatives we use are of Canavati and Caputo types. This work is greatly inspired by [2, 4, 9]. It follows [1].

16.1 Introduction

The famous Ostrowski ([9]) inequality motivates this work and has as follows:

$$\left|\frac{1}{b-a}\int_a^b f(y)\,dy - f(x)\right| \leq \left(\frac{1}{4} + \frac{\left(x - \frac{a+b}{2}\right)^2}{(b-a)^2}\right)(b-a)\left\|f'\right\|_\infty,$$

where $f \in C'([a,b])$, $x \in [a,b]$, and it is a sharp inequality.

Another motivation is author's next fractional result, see [4], p. 44:

Let $[a,b] \subset \mathbb{R}$, $\alpha > 0$, $m = \lceil \alpha \rceil$ ($\lceil \cdot \rceil$ ceiling of the number), $f \in AC^m([a,b])$ (i.e. $f^{(m-1)}$ is absolutely continuous), and $\left\|D_{x_0-}^\alpha f\right\|_{\infty,[a,x_0]}, \left\|D_{*x_0}^\alpha f\right\|_{\infty,[x_0,b]} < \infty$ (where $D_{x_0-}^\alpha f$, $D_{*x_0}^\alpha f$ are the right and left Caputo fractional derivatives of f of order α, *respectively*), $x_0 \in [a,b]$. Assume $f^{(k)}(x_0) = 0$, $k = 1, ..., m-1$. Then

$$\left|\frac{1}{b-a}\int_a^b f(x)\,dx - f(x_0)\right| \leq \frac{1}{(b-a)\,\Gamma(\alpha+2)}\cdot$$

$$\left\{\left\|D_{x_0-}^\alpha f\right\|_{\infty,[a,x_0]}(x_0-a)^{\alpha+1} + \left\|D_{*x_0}^\alpha f\right\|_{\infty,[x_0,b]}(b-x_0)^{\alpha+1}\right\} \leq$$

$$\frac{1}{\Gamma(\alpha+2)}\max\left\{\left\|D_{x_0-}^\alpha f\right\|_{\infty,[a,x_0]}, \left\|D_{*x_0}^\alpha f\right\|_{\infty,[x_0,b]}\right\}(b-a)^\alpha.$$

G. A. Anastassiou, *Ordinary and Fractional Approximation by Non-additive Integrals: Choquet, Shilkret and Sugeno Integral Approximators*, Studies in Systems, Decision and Control 190, https://doi.org/10.1007/978-3-030-04287-5_16

Another great source for inspiration is [5].

Here first we give a complete survey about Sugeno fuzzy integral and its basic properties. Then we derive a series of Ostrowski-Sugeno type fractional inequalities in the univariate case to all directions in the context of Sugeno integral and its basic important particular properties. We employ right and left fractional derivatives of Canavati and Caputo types. At the end we present a reverse fractional Polya-Sugeno inequality.

16.2 Background - I

In this section, some definitions and basic important properties of the Sugeno integral which will be used in the next section are presented. Also a preparation for the main results in Sect. 16.5 is given.

Definition 16.1 (*Fuzzy measure* [11, 13]) Let Σ be a σ-algebra of subsets of X, and let $\mu : \Sigma \to [0, +\infty]$ be a non-negative extended real-valued set function. We say that μ is a fuzzy measure iff:

(1) $\mu(\varnothing) = 0$,

(2) $E, F \in \Sigma : E \subseteq F$ imply $\mu(E) \leq \mu(F)$ (monotonicity),

(3) $E_n \in \Sigma$ $(n \in \mathbb{N})$, $E_1 \subset E_2 \subset ...$, imply $\lim_{n\to\infty} \mu(E_n) = \mu\left(\cup_{n=1}^{\infty} E_n\right)$ (continuity from below);

(4) $E_n \in \Sigma$ $(n \in \mathbb{N})$, $E_1 \supset E_2 \supset ...$, $\mu(E_1) < \infty$, imply $\lim_{n\to\infty} \mu(E_n) = \mu\left(\cap_{n=1}^{\infty} E_n\right)$ (continuity from above).

Let (X, Σ, μ) be a fuzzy measure space and f be a non-negative real-valued function on X. We denote by $\mathcal{F}_+$ the set of all non-negative real valued measurable functions, and by $L_\alpha f$ the set: $L_\alpha f := \{x \in X : f(x) \geq \alpha\}$, the α-level of f for $\alpha \geq 0$.

Definition 16.2 Let (X, Σ, μ) be a fuzzy measure space. If $f \in \mathcal{F}_+$ and $A \in \Sigma$, then the Sugeno integral (fuzzy integral) [12] of f on A with respect to the fuzzy measure μ is defined by

$$(S)\int_A f d\mu := \vee_{\alpha\geq 0}\left(\alpha \wedge \mu\left(A \cap L_\alpha f\right)\right), \tag{16.1}$$

where $\vee$ and $\wedge$ denote the sup and inf on $[0, \infty]$, respectively.

The basic properties of Sugeno integral follow:

Theorem 16.3 ([10, 13]) *Let* (X, Σ, μ) *be a fuzzy measure space with* $A, B \in \Sigma$ *and* $f, g \in \mathcal{F}_+$. *Then*

(1) $(S)\int_A f d\mu \leq \mu(A)$;

(2) $(S)\int_A kd\mu = k \wedge \mu(A)$ *for a non-negative constant* k;
(3) *if* $f \leq g$ *on* A, *then* $(S)\int_A fd\mu \leq (S)\int_A gd\mu$;
(4) *if* $A \subset B$, *then* $(S)\int_A fd\mu \leq (S)\int_B fd\mu$;
(5) $\mu(A \cap L_\alpha f) \leq \alpha \Rightarrow (S)\int_A fd\mu \leq \alpha$;
(6) *if* $\mu(A) < \infty$, *then* $\mu(A \cap L_\alpha f) \geq \alpha \Leftrightarrow (S)\int_A fd\mu \geq \alpha$;
(7) *when* $A = X$, *then* $(S)\int_X fd\mu = \vee_{\alpha \geq 0}(\alpha \wedge \mu(L_\alpha f))$;
(8) *if* $\alpha \leq \beta$, *then* $L_\beta f \subseteq L_\alpha f$;
(9) $(S)\int_A fd\mu \geq 0$.

Theorem 16.4 ([13], p. 135) *Here* $f \in \mathcal{F}_+$, *the class of all finite nonnegative measurable functions on* (X, Σ, μ). *Then*
(1) *if* $\mu(A) = 0$, *then* $(S)\int_A fd\mu = 0$, *for any* $f \in \mathcal{F}_+$;
(2) *if* $(S)\int_A fd\mu = 0$, *then* $\mu(A \cap \{x | f(x) > 0\}) = 0$;
(3) $(S)\int_A fd\mu = (S)\int_X f \cdot \chi_A d\mu$, *where* χ_A *is the characteristic function of* A;
(4) $(S)\int_A (f + a) d\mu \leq (S)\int_A fd\mu + (S)\int_A ad\mu$, *for any constant* $a \in [0, \infty)$.

Corollary 16.5 ([13], p. 136) *Here* $f, f_1, f_2 \in \mathcal{F}_+$. *Then*
(1) $(S)\int_A (f_1 \vee f_2) d\mu \geq (S)\int_A f_1 d\mu \vee (S)\int_A f_2 d\mu$;
(2) $(S)\int_A (f_1 \wedge f_2) d\mu \leq (S)\int_A f_1 d\mu \wedge (S)\int_A f_2 d\mu$;
(3) $(S)\int_{A \cup B} fd\mu \geq (S)\int_A fd\mu \vee (S)\int_B fd\mu$;
(4) $(S)\int_{A \cap B} fd\mu \leq (S)\int_A fd\mu \wedge (S)\int_B fd\mu$.

In general we have

$$(S)\int_A (f_1 + f_2) d\mu \neq (S)\int_A f_1 d\mu + (S)\int_A f_2 d\mu,$$

and

$$(S)\int_A afd\mu \neq a(S)\int_A fd\mu, \text{ where } a \in \mathbb{R},$$

see [13], p. 137.

Lemma 16.6 ([13], p. 138) $(S)\int_A fd\mu = \infty$ *iff* $\mu(A \cap L_\alpha f) = \infty$ *for any* $\alpha \in [0, \infty)$.

We need

Definition 16.7 ([6]) A fuzzy measure μ is subadditive iff $\mu(A \cup B) \leq \mu(A) + \mu(B)$, for all $A, B \in \Sigma$.

We mention

Theorem 16.8 ([6]) *If* μ *is subadditive, then*

$$(S)\int_X (f + g) d\mu \leq (S)\int_X fd\mu + (S)\int_X gd\mu, \tag{16.2}$$

for all measurable functions $f, g : X \to [0, \infty)$.

Moreover, if (16.2) holds for all measurable functions $f, g : X \to [0, \infty)$ *and* $\mu(X) < \infty$, *then* μ *is subadditive.*

Notice here in (16.1) we have that $\alpha \in [0, \infty)$.

We have

Corollary 16.9 *If* μ *is subadditive,* $n \in \mathbb{N}$, *and* $f : X \to [0, \infty)$ *is a measurable function, then*

$$(S) \int_X nf d\mu \leq n (S) \int_X f d\mu, \tag{16.3}$$

in particular it holds

$$(S) \int_A nf d\mu \leq n (S) \int_A f d\mu, \tag{16.4}$$

for any $A \in \Sigma$.

Proof By (16.2). ■

A very important property of Sugeno integral follows.

Theorem 16.10 *If* μ *is subadditive measure, and* $f : X \to [0, \infty)$ *is a measurable function, and* $c > 0$, *then*

$$(S) \int_A cf d\mu \leq (c+1) (S) \int_A f d\mu, \tag{16.5}$$

for any $A \in \Sigma$.

Proof Let the ceiling $\lceil c \rceil = m \in \mathbb{N}$, then by Theorem 16.3 (3) and (4) we get

$$(S) \int_A cf d\mu \leq (S) \int_A mf d\mu \leq m (S) \int_A f d\mu \leq (c+1) (S) \int_A f d\mu,$$

proving (16.5). ■

From now on in this chapter we work on the fuzzy measure space $([a, b], \mathcal{B}, \mu)$, where $[a, b] \subset \mathbb{R}$, $\mathcal{B}$ is the Borel σ-algebra on $[a, b]$, and μ is a finite fuzzy measure on $\mathcal{B}$. Typically we take it to be subadditive.

The functions f we deal with here are continuous from $[a, b]$ into $\mathbb{R}_+$.

We make

Remark 16.11 Let $f \in C([a, b], \mathbb{R}_+)$, and μ is a subadditive fuzzy measure such that $\mu([a, b]) > 0$, $x \in [a, b]$. We will estimate

$$E(x) := \left| (S) \int_{[a,b]} f(x) d\mu(t) - \mu([a, b]) \wedge f(x) \right| \tag{16.6}$$

(by Theorem 16.3 (2))

$$= \left| (S) \int_{[a,b]} f(t) \, d\mu(t) - (S) \int_{[a,b]} f(x) \, d\mu(t) \right|.$$

We notice that

$$f(t) = f(t) - f(x) + f(x) \leq |f(t) - f(x)| + f(x),$$

then (by Theorem 16.3 (3) and Theorem 16.4 (4))

$$(S) \int_{[a,b]} f(t) \, d\mu(t) \leq (S) \int_{[a,b]} |f(t) - f(x)| \, d\mu(t) + (S) \int_{[a,b]} f(x) \, d\mu(t), \tag{16.7}$$

that is

$$(S) \int_{[a,b]} f(t) \, d\mu(t) - (S) \int_{[a,b]} f(x) \, d\mu(t) \leq (S) \int_{[a,b]} |f(t) - f(x)| \, d\mu(t). \tag{16.8}$$

Similarly, we have

$$f(x) = f(x) - f(t) + f(t) \leq |f(t) - f(x)| + f(t),$$

then (by Theorem 16.3 (3) and Theorem 16.8)

$$(S) \int_{[a,b]} f(x) \, d\mu(t) \leq (S) \int_{[a,b]} |f(t) - f(x)| \, d\mu(t) + (S) \int_{[a,b]} f(t) \, d\mu(t),$$

that is

$$(S) \int_{[a,b]} f(x) \, d\mu(t) - (S) \int_{[a,b]} f(t) \, d\mu(t) \leq (S) \int_{[a,b]} |f(t) - f(x)| \, d\mu(t). \tag{16.9}$$

By (16.8) and (16.9) we derive that

$$\left| (S) \int_{[a,b]} f(t) \, d\mu(t) - (S) \int_{[a,b]} f(x) \, d\mu(t) \right| \leq (S) \int_{[a,b]} |f(t) - f(x)| \, d\mu(t). \tag{16.10}$$

Consequently it holds

$$E(x) \overset{\text{(by (16.6), (16.10))}}{\leq} (S) \int_{[a,b]} |f(t) - f(x)| \, d\mu(t). \tag{16.11}$$

We will later use (16.11).

16.3 Background - II

We need

Remark 16.12 Here $[\cdot]$ denotes the integral part of the number. Let $\alpha > 0, m = [\alpha]$, $\beta = \alpha - m, 0 < \beta < 1, f \in C([a,b]), [a,b] \subset \mathbb{R}, x \in [a,b]$. The gamma function Γ is given by $\Gamma(\alpha) = \int_0^\infty e^{-t} t^{\alpha-1} dt$. We define the left Riemann-Liouville integral

$$\left(J_\alpha^{a+} f\right)(x) = \frac{1}{\Gamma(\alpha)} \int_a^x (x-t)^{\alpha-1} f(t)\, dt, \tag{16.12}$$

$a \leq x \leq b$. We define the subspace $C_{\alpha+}^\alpha([a,b])$ of $C^m([a,b])$:

$$C_{a+}^\alpha([a,b]) = \{f \in C^m([a,b]) : J_{1-\beta}^{a+} f^{(m)} \in C^1([a,b])\}. \tag{16.13}$$

For $f \in C_{a+}^\alpha([a,b])$, we define the left generalized α-fractional derivative of f over $[a,b]$ as

$$D_{a+}^\alpha f := \left(J_{1-\beta}^{a+} f^{(m)}\right)', \tag{16.14}$$

see [2], p. 24. Canavati first in [7] introduced the above over [0, 1]. Notice that $D_{a+}^\alpha f \in C([a,b])$.

We need the following left fractional Taylor's formula, see [2], pp. 8-10, and in [7] the same over [0, 1] that appeared first.

Let $f \in C_{a+}^\alpha([a,b])$:

(i) If $\alpha \geq 1$, then

$$f(x) = f(a) + f'(a)(x-a) + f''(a)\frac{(x-a)^2}{2} + \ldots + f^{(m-1)}(a)\frac{(x-a)^{m-1}}{(m-1)!} +$$

$$\frac{1}{\Gamma(\alpha)} \int_a^x (x-t)^{\alpha-1} \left(D_{a+}^\alpha f\right)(t)\, dt, \tag{16.15}$$

all $x \in [a,b]$.

(ii) If $0 < \alpha < 1$, we have

$$f(x) = \frac{1}{\Gamma(\alpha)} \int_a^x (x-t)^{\alpha-1} \left(D_{a+}^\alpha f\right)(t)\, dt, \tag{16.16}$$

all $x \in [a,b]$.

Notice that

$$\int_a^x (x-t)^{\alpha-1} \left(D_{a+}^\alpha f\right)(t)\, dt = \int_a^x \left(D_{a+}^\alpha f\right)(t)\, d\left(\frac{(x-t)^\alpha}{-\alpha}\right) \tag{16.17}$$

$$= \left(D_{a+}^{\alpha} f\right)(\xi_x) \frac{(x-a)^{\alpha}}{\alpha}, \text{ where } \xi_x \in [a, x],$$

by first integral mean value theorem.

Hence, when $\alpha \geq 1$ and $f^{(i)}(a) = 0$, $i = 0, 1, ..., m-1$ or when $0 < \alpha < 1$, we get

$$f(x) = \left(D_{a+}^{\alpha} f\right)(\xi_x) \frac{(x-a)^{\alpha}}{\Gamma(\alpha+1)}, \text{ all } x \in [a, b]. \tag{16.18}$$

Furthermore we need:

Let again $\alpha > 0$, $m = [\alpha]$, $\beta = \alpha - m$, $f \in C([a, b])$, call the right Riemann-Liouville fractional integral operator by

$$\left(J_{b-}^{\alpha} f\right)(x) := \frac{1}{\Gamma(\alpha)} \int_x^b (t-x)^{\alpha-1} f(t)\, dt, \tag{16.19}$$

$x \in [a, b]$, see also [3, pp. 333, 345].

Define the subspace of functions

$$C_{b-}^{\alpha}([a, b]) = \{f \in C^m([a, b]) : J_{b-}^{1-\beta} f^{(m)} \in C^1([a, b])\}. \tag{16.20}$$

Define the right generalized α-fractional derivative of f over $[a, b]$ as

$$D_{b-}^{\alpha} f = (-1)^{m-1} \left(J_{b-}^{1-\beta} f^{(m)}\right)', \tag{16.21}$$

see [3, p. 345]. We set $D_{b-}^{0} f = f$.

Notice that $D_{b-}^{\alpha} f \in C([a, b])$.

From [3, p. 348], we need the following right Taylor fractional formula:

Let $f \in C_{b-}^{\alpha}([a, b])$, $\alpha > 0$, $m = [\alpha]$. Then

(i) If $\alpha \geq 1$, we get

$$f(x) = \sum_{k=0}^{m-1} \frac{f^{(k)}(b)}{k!} (x-b)^k + \left(J_{b-}^{\alpha} D_{b-}^{\alpha} f\right)(x), \tag{16.22}$$

all $x \in [a, b]$.

(ii) If $0 < \alpha < 1$, we get

$$f(x) = J_{b-}^{\alpha} D_{b-}^{\alpha} f(x) = \frac{1}{\Gamma(\alpha)} \int_x^b (t-x)^{\alpha-1} \left(D_{b-}^{\alpha} f\right)(t)\, dt, \tag{16.23}$$

all $x \in [a, b]$.

Notice that

$$\int_x^b (t-x)^{\alpha-1} \left(D_{b-}^{\alpha} f\right)(t)\, dt = \int_x^b \left(D_{b-}^{\alpha} f\right)(t)\, d\left(\frac{(t-x)^{\alpha}}{\alpha}\right)$$

$$= \left(D_{b-}^{\alpha} f\right)(\eta_x) \frac{(b-x)^{\alpha}}{\alpha}, \text{ where } \eta_x \in [x, b], \tag{16.24}$$

by first integral mean value theorem.

Hence, when $\alpha \geq 1$ and $f^{(k)}(b) = 0, k = 0, 1, ..., m-1$ or $0 < \alpha < 1$, we obtain

$$f(x) = \left(D_{b-}^{\alpha} f\right)(\eta_x) \frac{(b-x)^{\alpha}}{\Gamma(\alpha+1)}, \text{ all } x \in [a, b]. \tag{16.25}$$

Let $f \in C_{a+}^{\alpha}([a, b])$, $\alpha \geq 1$, and $f^{(i)}(a) = 0$, $i = 1, ..., m-1$, then

$$|f(x) - f(a)| \leq \left\|D_{a+}^{\alpha} f\right\|_{\infty} \frac{(x-a)^{\alpha}}{\Gamma(\alpha+1)}, \tag{16.26}$$

all $x \in [a, b]$, by (16.15).

Again let $f \in C_{a+}^{\alpha}([a, b])$, $\alpha \geq 1$, and $f^{(i)}(a) = 0$, $i = 1, ..., m-1$, then by (16.15) we have

$$f(x) - f(a) = \frac{1}{\Gamma(\alpha)} \int_a^x (x-t)^{\alpha-1} \left(D_{a+}^{\alpha} f\right)(t)\, dt, \tag{16.27}$$

hence

$$|f(x) - f(a)| \leq \frac{(x-a)^{\alpha-1}}{\Gamma(\alpha)} \left\|D_{a+}^{\alpha} f\right\|_{L_1([a,b])}, \tag{16.28}$$

all $x \in [a, b]$.

Let $p, q > 1 : \frac{1}{p} + \frac{1}{q} = 1$, continuing from (16.27), $\alpha \geq 1$, we get

$$|f(x) - f(a)| \leq \frac{1}{\Gamma(\alpha)} \left(\int_a^x (x-t)^{p(\alpha-1)}\, dt\right)^{\frac{1}{p}} \left\|D_{a+}^{\alpha} f\right\|_{L_q([a,b])} =$$

$$\frac{(x-a)^{\frac{(p(\alpha-1)+1)}{p}}}{\Gamma(\alpha)(p(\alpha-1)+1)^{\frac{1}{p}}} \left\|D_{a+}^{\alpha} f\right\|_{L_q([a,b])}, \tag{16.29}$$

$\forall\, x \in [a, b]$.

Let $f \in C_{b-}^{\alpha}([a, b])$, $\alpha \geq 1$, $m = [\alpha]$, $f^{(k)}(b) = 0$, $k = 1, ..., m-1$, then by (16.22) we get:

$$f(x) - f(b) = \frac{1}{\Gamma(\alpha)} \int_x^b (t-x)^{\alpha-1} \left(D_{b-}^{\alpha} f\right)(t)\, dt. \quad (16.30)$$

We derive the following estimates:

(1)

$$|f(x) - f(b)| \leq \frac{(b-x)^{\alpha-1}}{\Gamma(\alpha+1)} \left\| D_{b-}^{\alpha} f \right\|_{\infty}, \quad (16.31)$$

(2)

$$|f(x) - f(b)| \leq \frac{(b-x)^{\alpha-1}}{\Gamma(\alpha)} \left\| D_{b-}^{\alpha} f \right\|_{L_1([a,b])}, \quad (16.32)$$

(3) let $p, q > 1 : \frac{1}{p} + \frac{1}{q} = 1$, then

$$|f(x) - f(b)| \leq \frac{(b-x)^{\frac{(p(\alpha-1)+1)}{p}}}{\Gamma(\alpha)\,(p(\alpha-1)+1)^{\frac{1}{p}}} \left\| D_{b-}^{\alpha} f \right\|_{L_q([a,b])}, \quad (16.33)$$

$\forall\, x \in [a, b]$.

16.4 Background - III

We need

Remark 16.13 Let $\nu > 0, n := \lceil \nu \rceil$, $\lceil \cdot \rceil$ is the ceiling of the number, $f \in AC^n([a,b])$ (i.e. $f^{(n-1)}$ is absolutely continuous on $[a,b]$). We call the left Caputo fractional derivative [8]

$$D_{*a}^{\nu} f(x) := \frac{1}{\Gamma(n-\nu)} \int_a^x (x-t)^{n-\nu-1} f^{(n)}(t)\, dt, \quad (16.34)$$

$\forall\, x \in [a, b]$.

The above function $D_{*a}^{\nu} f(x)$ exists almost everywhere for $x \in [a, b]$.

If $\nu \in \mathbb{N}$, then $D_{*a}^{\nu} f = f^{(\nu)}$ the ordinary derivative, it is also $D_{*a}^{0} f = f$.

We have the left fractional Taylor formula for left Caputo fractional derivatives [8, p. 40].

Assume $\nu > 0$, $n = \lceil \nu \rceil$, and $f \in AC^n([a,b])$. Then

$$f(x) = \sum_{k=0}^{n-1} \frac{f^{(k)}(a)}{k!} (x-a)^k + \frac{1}{\Gamma(\nu)} \int_a^x (x-t)^{\nu-1} D_{*a}^{\nu} f(t)\, dt, \quad (16.35)$$

$\forall\, x \in [a, b]$.

Additionally assume that

$$f^{(k)}(a) = 0, \; k = 1, ..., n-1;$$

then

$$f(x) - f(a) = \frac{1}{\Gamma(\nu)} \int_a^x (x-t)^{\nu-1} D_{*a}^{\nu} f(t)\, dt, \tag{16.36}$$

$\forall\, x \in [a, b]$.

We get the following estimates:

(1) if $D_{*a}^{\nu} f \in L_{\infty}([a, b])$, then

$$|f(x) - f(a)| \leq \frac{\left\| D_{*a}^{\nu} f \right\|_{\infty}}{\Gamma(\nu+1)} (x-a)^{\nu}, \tag{16.37}$$

$\forall\, x \in [a, b]$, see [2], p. 619;

(2) if $\nu \geq 1$, and $D_{*a}^{\nu} f \in L_1([a, b])$, then

$$|f(x) - f(a)| \leq \frac{\left\| D_{*a}^{\nu} f \right\|_{L_1([a,b])}}{\Gamma(\nu)} (x-a)^{\nu-1}, \tag{16.38}$$

$\forall\, x \in [a, b]$, see [2], p. 620;

(3) let $p, q > 1 : \frac{1}{p} + \frac{1}{q} = 1$, and $\nu > \frac{1}{q}$, and $D_{*a}^{\nu} f \in L_q([a, b])$, then

$$|f(x) - f(a)| \leq \frac{\left\| D_{*a}^{\nu} f \right\|_{L_q([a,b])}}{\Gamma(\nu)\,(p(\nu-1)+1)^{\frac{1}{p}}} (x-a)^{\nu-\frac{1}{q}}, \tag{16.39}$$

$\forall\, x \in [a, b]$, see [2], p. 621.

Furthermore we need:

Let $f \in AC^m([a, b])$ ($f^{(m-1)}$ is absolutely continuous on $[a, b]$), $m \in \mathbb{N}$, $m = \lceil \alpha \rceil$, $\alpha > 0$. We define the right Caputo fractional derivative of order $\alpha > 0$ by

$$D_{cb-}^{\alpha} f(x) = \frac{(-1)^m}{\Gamma(m-\alpha)} \int_x^b (J-x)^{m-\alpha-1} f^{(m)}(J)\, dJ, \tag{16.40}$$

$\forall\, x \in [a, b]$, see [3, p. 336].

If $\alpha = m \in \mathbb{N}$, then

$$D_{cb-}^{\alpha} f(x) = (-1)^m f^{(m)}(x), \; \forall\, x \in [a, b]. \tag{16.41}$$

If $x > b$ we define $D_{cb-}^{\alpha} f(x) = 0$.

We also need:

Let $f \in AC^m([a, b])$, $\alpha > 0$, $m = \lceil \alpha \rceil$. Then

$$f(x) = \sum_{k=0}^{m-1} \frac{f^{(k)}(b)}{k!}(x-b)^k + \frac{1}{\Gamma(\alpha)} \int_x^b (J-x)^{\alpha-1} D_{cb-}^{\alpha} f(J)\, dJ, \quad (16.42)$$

$\forall\, x \in [a, b]$, the right Caputo fractional Taylor formula with integral remainder, see [3, p. 338].

Additionally assume that

$$f^{(k)}(b) = 0,\ k = 1, \ldots, m-1,$$

then

$$f(x) - f(b) = \frac{1}{\Gamma(\alpha)} \int_x^b (J-x)^{\alpha-1} D_{cb-}^{\alpha} f(J)\, dJ, \quad (16.43)$$

$\forall\, x \in [a, b]$.

Following (16.43) we get the following estimates:

(1) if $D_{cb-}^{\alpha} f \in L_{\infty}([a, b])$, then

$$|f(x) - f(b)| \leq \frac{(b-x)^{\alpha}}{\Gamma(\alpha+1)} \left\| D_{cb-}^{\alpha} f \right\|_{\infty}, \quad (16.44)$$

$\forall\, x \in [a, b]$, see [4], p. 23;

(2) if $D_{cb-}^{\alpha} f \in L_1([a, b])$, $\alpha \geq 1$, then

$$|f(x) - f(b)| \leq \frac{\left\| D_{cb-}^{\alpha} f \right\|_{L_1([a,b])}}{\Gamma(\alpha)} (b-x)^{\alpha-1}, \quad (16.45)$$

$\forall\, x \in [a, b]$, see [4], p. 24;

(3) let $p, q > 1 : \frac{1}{p} + \frac{1}{q} = 1$, and $\alpha > \frac{1}{q}$, $m = \lceil \alpha \rceil$, $D_{cb-}^{\alpha} f \in L_q([a, b])$, then

$$|f(x) - f(b)| \leq \frac{\left\| D_{cb-}^{\alpha} f \right\|_{L_q([a,b])}}{\Gamma(\alpha)(p(\alpha-1)+1)^{\frac{1}{p}}} (b-x)^{\alpha-\frac{1}{q}}, \quad (16.46)$$

$\forall\, x \in [a, b]$, see [4], p. 25.

16.5 Main Results

We make

Remark 16.14 Let $\alpha \geq 1$, $f \in C_{a+}^{\alpha}([a, b])$ and $f^{(i)}(a) = 0, i = 1, \ldots, m-1$; $m = \lceil \alpha \rceil$. By (16.11), for $x = a$, we get that

$$E(a) = \left| (S) \int_{[a,b]} f(t)\, d\mu(t) - \mu([a,b]) \wedge f(a) \right| \leq \tag{16.47}$$

$$(S) \int_{[a,b]} |f(t) - f(a)|\, d\mu(t) =: \Delta(a).$$

By (16.26) we get

$$\Delta(a) \leq (S) \int_{[a,b]} \frac{\left\| D_{a+}^{\alpha} f \right\|_{\infty}}{\Gamma(\alpha+1)} (t-a)^{\alpha} d\mu(t) \overset{(16.5)}{\leq}$$

$$\left(\frac{\left\| D_{a+}^{\alpha} f \right\|_{\infty}}{\Gamma(\alpha+1)} + 1 \right) (S) \int_{[a,b]} (t-a)^{\alpha} d\mu(t). \tag{16.48}$$

By (16.28) we obtain

$$\Delta(a) \leq (S) \int_{[a,b]} \frac{\left\| D_{a+}^{\alpha} f \right\|_{L_1([a,b])}}{\Gamma(\alpha)} (t-a)^{\alpha-1} d\mu(t) \overset{(16.5)}{\leq}$$

$$\left(\frac{\left\| D_{a+}^{\alpha} f \right\|_{L_1([a,b])}}{\Gamma(\alpha)} + 1 \right) (S) \int_{[a,b]} (t-a)^{\alpha-1} d\mu(t). \tag{16.49}$$

And by (16.29) ($p, q > 1 : \frac{1}{p} + \frac{1}{q} = 1$) we derive

$$\Delta(a) \leq (S) \int_{[a,b]} \frac{\left\| D_{a+}^{\alpha} f \right\|_{L_q([a,b])}}{\Gamma(\alpha)(p(\alpha-1)+1)^{\frac{1}{p}}} (t-a)^{\frac{(p(\alpha-1)+1)}{p}} d\mu(t) \overset{(16.5)}{\leq}$$

$$\left(\frac{\left\| D_{a+}^{\alpha} f \right\|_{L_q([a,b])}}{\Gamma(\alpha)(p(\alpha-1)+1)^{\frac{1}{p}}} + 1 \right) (S) \int_{[a,b]} (t-a)^{\left(\alpha-1+\frac{1}{p}\right)} d\mu(t). \tag{16.50}$$

We have proved that

$$\left| (S) \int_{[a,b]} f(t)\, d\mu(t) - \mu([a,b]) \wedge f(a) \right| \leq$$

$$\min \left\{ \left(\frac{\left\| D_{a+}^{\alpha} f \right\|_{\infty}}{\Gamma(\alpha+1)} + 1 \right) (S) \int_{[a,b]} (t-a)^{\alpha} d\mu(t), \right.$$

$$\left(\frac{\left\| D_{a+}^{\alpha} f \right\|_{L_1([a,b])}}{\Gamma(\alpha)} + 1 \right) (S) \int_{[a,b]} (t-a)^{\alpha-1} d\mu(t),$$

$$\left(\frac{\left\|D_{a+}^{\alpha}f\right\|_{L_q([a,b])}}{\Gamma(\alpha)\left(p(\alpha-1)+1\right)^{\frac{1}{p}}}+1\right)(S)\int_{[a,b]}(t-a)^{\alpha-\frac{1}{q}}\,d\mu(t)\Bigg\}. \tag{16.51}$$

We have established the following left generalized fractional Ostrowski-Sugeno inequality.

Theorem 16.15 *Here μ is a fuzzy subadditive measure with $\mu([a,b])>0$.*

Let $p,q>1:\frac{1}{p}+\frac{1}{q}=1$; $\alpha\geq1$. Let $f\in C_{a+}^{\alpha}([a,b])$ with $f^{(i)}(a)=0$, $i=1,...,m-1$; $m=[\alpha]$; and f is $\mathbb{R}_+$-valued. Then

$$\left|\frac{1}{\mu([a,b])}(S)\int_{[a,b]}f(t)\,d\mu(t)-\left(1\wedge\frac{f(a)}{\mu([a,b])}\right)\right|\leq$$

$$\frac{1}{\mu([a,b])}\min\Bigg\{\left(\frac{\left\|D_{a+}^{\alpha}f\right\|_{\infty}}{\Gamma(\alpha+1)}+1\right)(S)\int_{[a,b]}(t-a)^{\alpha}\,d\mu(t),$$

$$\left(\frac{\left\|D_{a+}^{\alpha}f\right\|_{L_1([a,b])}}{\Gamma(\alpha)}+1\right)(S)\int_{[a,b]}(t-a)^{\alpha-1}\,d\mu(t),$$

$$\left(\frac{\left\|D_{a+}^{\alpha}f\right\|_{L_q([a,b])}}{\Gamma(\alpha)\left(p(\alpha-1)+1\right)^{\frac{1}{p}}}+1\right)(S)\int_{[a,b]}(t-a)^{\alpha-\frac{1}{q}}\,d\mu(t)\Bigg\}. \tag{16.52}$$

Similarly (as in Remark 16.14), we get the right generalized fractional Ostrowski-Sugeno inequality (use of (16.11) for $x=b$, and (16.31), (16.32), (16.33)).

Theorem 16.16 *Here μ is a fuzzy subadditive measure with $\mu([a,b])>0$.*

Let $p,q>1:\frac{1}{p}+\frac{1}{q}=1$; $\alpha\geq1$. Let $f\in C_{b-}^{\alpha}([a,b])$ with $f^{(k)}(b)=0$, $k=1,...,m-1$; $[\alpha]=m$; and f is $\mathbb{R}_+$-valued. Then

$$\left|\frac{1}{\mu([a,b])}(S)\int_{[a,b]}f(t)\,d\mu(t)-\left(1\wedge\frac{f(b)}{\mu([a,b])}\right)\right|\leq$$

$$\frac{1}{\mu([a,b])}\min\Bigg\{\left(\frac{\left\|D_{b-}^{\alpha}f\right\|_{\infty}}{\Gamma(\alpha+1)}+1\right)(S)\int_{[a,b]}(b-t)^{\alpha}\,d\mu(t),$$

$$\left(\frac{\left\|D_{b-}^{\alpha}f\right\|_{L_1([a,b])}}{\Gamma(\alpha)}+1\right)(S)\int_{[a,b]}(b-t)^{\alpha-1}\,d\mu(t),$$

$$\left(\frac{\left\|D_{b-}^{\alpha}f\right\|_{L_q([a,b])}}{\Gamma(\alpha)\left(p(\alpha-1)+1\right)^{\frac{1}{p}}}+1\right)(S)\int_{[a,b]}(b-t)^{\alpha-\frac{1}{q}}\,d\mu(t)\Bigg\}. \tag{16.53}$$

We present the following left Caputo fractional Ostrowski-Sugano inequalities:

Theorem 16.17 *Here μ is a fuzzy subadditive measure with $\mu([a,b]) > 0$. Let $f : [a,b] \to \mathbb{R}_+$ such that $f \in AC^n([a,b])$, where $n = \lceil \nu \rceil$, $\nu > 0$. Assume $f^{(k)}(a) = 0$, $k = 1, ..., n-1$. We have*

(1) if $D_{*a}^{\nu} f \in L_{\infty}([a,b])$, then

$$\left| \frac{1}{\mu([a,b])} (S) \int_{[a,b]} f(t) \, d\mu(t) - \left(1 \wedge \frac{f(a)}{\mu([a,b])} \right) \right| \leq$$

$$\frac{1}{\mu([a,b])} \left(\frac{\left\| D_{*a}^{\nu} f \right\|_{\infty}}{\Gamma(\nu+1)} + 1 \right) (S) \int_{[a,b]} (t-a)^{\nu} \, d\mu(t), \tag{16.54}$$

(2) if $\nu \geq 1$, and $D_{*a}^{\nu} f \in L_1([a,b])$, then

$$\left| \frac{1}{\mu([a,b])} (S) \int_{[a,b]} f(t) \, d\mu(t) - \left(1 \wedge \frac{f(a)}{\mu([a,b])} \right) \right| \leq$$

$$\frac{1}{\mu([a,b])} \left(\frac{\left\| D_{*a}^{\nu} f \right\|_{L_1([a,b])}}{\Gamma(\nu)} + 1 \right) (S) \int_{[a,b]} (t-a)^{\nu-1} \, d\mu(t), \tag{16.55}$$

and

(3) let $p, q > 1 : \frac{1}{p} + \frac{1}{q} = 1$, and $\nu > \frac{1}{q}$, and $D_{*a}^{\nu} f \in L_q([a,b])$, then

$$\left| \frac{1}{\mu([a,b])} (S) \int_{[a,b]} f(t) \, d\mu(t) - \left(1 \wedge \frac{f(a)}{\mu([a,b])} \right) \right| \leq$$

$$\frac{1}{\mu([a,b])} \left(\frac{\left\| D_{*a}^{\nu} f \right\|_{L_q([a,b])}}{\Gamma(\nu)(p(\nu-1)+1)^{\frac{1}{p}}} + 1 \right) (S) \int_{[a,b]} (t-a)^{\nu-\frac{1}{q}} \, d\mu(t). \tag{16.56}$$

Proof By the use of (16.37), (16.38) and (16.39), acting as in the proof of Theorem 16.15. ■

Next, we give the following right Caputo fractional Ostrowski-Sugeno inequalities:

Theorem 16.18 *Here μ is a fuzzy subadditive measure with $\mu([a,b]) > 0$. Let $f : [a,b] \to \mathbb{R}_+$ such that $f \in AC^m([a,b])$, $m \in \mathbb{N}$, $m = \lceil \alpha \rceil$, $\alpha > 0$. Assume $f^{(k)}(b) = 0$, $k = 1, ..., m-1$. We have*

(1) if $D_{cb-}^{\alpha} f \in L_{\infty}([a,b])$, then

$$\left| \frac{1}{\mu([a,b])} (S) \int_{[a,b]} f(t) \, d\mu(t) - \left(1 \wedge \frac{f(b)}{\mu([a,b])} \right) \right| \leq$$

$$\frac{1}{\mu([a,b])}\left(\frac{\left\|D_{cb-}^{\alpha}f\right\|_{\infty}}{\Gamma(\alpha+1)}+1\right)(S)\int_{[a,b]}(b-t)^{\alpha}\,d\mu(t), \tag{16.57}$$

(2) if $\alpha \geq 1$, *and* $D_{cb-}^{\alpha}f \in L_1([a,b])$, *then*

$$\left|\frac{1}{\mu([a,b])}(S)\int_{[a,b]}f(t)\,d\mu(t)-\left(1\wedge\frac{f(b)}{\mu([a,b])}\right)\right|\leq$$

$$\frac{1}{\mu([a,b])}\left(\frac{\left\|D_{cb-}^{\alpha}f\right\|_{L_1([a,b])}}{\Gamma(\alpha)}+1\right)(S)\int_{[a,b]}(b-t)^{\alpha-1}\,d\mu(t), \tag{16.58}$$

(3) if $p,q>1:\frac{1}{p}+\frac{1}{q}=1$, $\alpha>\frac{1}{q}$, $D_{cb-}^{\alpha}f\in L_q([a,b])$, *then*

$$\left|\frac{1}{\mu([a,b])}(S)\int_{[a,b]}f(t)\,d\mu(t)-\left(1\wedge\frac{f(b)}{\mu([a,b])}\right)\right|\leq$$

$$\frac{1}{\mu([a,b])}\left(\frac{\left\|D_{cb-}^{\alpha}f\right\|_{L_q([a,b])}}{\Gamma(\alpha)(p(\alpha-1)+1)^{\frac{1}{p}}}+1\right)(S)\int_{[a,b]}(b-t)^{\alpha-\frac{1}{q}}\,d\mu(t). \tag{16.59}$$

Proof Use of (16.44), (16.45) and (16.46), acting again as in the proof of Theorem 16.15. ∎

Remark 16.19 Let $x_0 \in [a,b]$. Of interest will be to estimate the quantity

$$\left|(S)\int_{[a,x_0]}f(t)\,d\mu(t)+(S)\int_{[x_0,b]}f(t)\,d\mu(t)-\right.$$

$$\left.[\mu([a,x_0])\wedge f(x_0)+\mu([x_0,b])\wedge f(x_0)]\right|\leq$$

$$\left|(S)\int_{[a,x_0]}f(t)\,d\mu(t)-\mu([a,x_0])\wedge f(x_0)\right|+$$

$$\left|(S)\int_{[x_0,b]}f(t)\,d\mu(t)-\mu([x_0,b])\wedge f(x_0)\right|. \tag{16.60}$$

An important special case is when $x_0=\frac{a+b}{2}$.

The above can be done with the use of our earlier results.

Remark 16.20 Here μ is just a finite fuzzy measure on $[a,b]$ which is positive on non-empty closed subsets of $[a,b]$.

Let $f:[a,b]\to\mathbb{R}_+$ be continuous, such that $f\in C_{a+}^{\alpha}\left(\left[a,\frac{a+b}{2}\right]\right)$, $0<\alpha<1$. By (16.18) we get that

$$f(x) \geq \left(\inf_{x \in \left[a, \frac{a+b}{2}\right]} \left| D_{a+}^{\alpha} f \right| \right) \frac{(x-a)^{\alpha}}{\Gamma(\alpha+1)}, \text{ all } x \in \left[a, \frac{a+b}{2}\right]. \tag{16.61}$$

Assume also $f \in C_{b-}^{\alpha}\left(\left[\frac{a+b}{2}, b\right]\right)$, $0 < \alpha < 1$.

By (16.25) we find that

$$f(x) \geq \left(\inf_{x \in \left[\frac{a+b}{2}, b\right]} \left| D_{b-}^{\alpha} f \right| \right) \frac{(b-x)^{\alpha}}{\Gamma(\alpha+1)}, \text{ all } x \in \left[\frac{a+b}{2}, b\right]. \tag{16.62}$$

We notice that

$$(S)\int_{[a,b]} f(t)\, d\mu(t) = (S)\int_{\left[a, \frac{a+b}{2}\right] \cup \left[\frac{a+b}{2}, b\right]} f(t)\, d\mu(t) \overset{\text{(by Cor. 6.5 (3))}}{\geq}$$

$$(S)\int_{\left[a, \frac{a+b}{2}\right]} f(t)\, d\mu(t) \vee (S)\int_{\left[\frac{a+b}{2}, b\right]} f(t)\, d\mu(t) \geq \tag{16.63}$$

$$\frac{1}{\Gamma(\alpha+1)} \left[(S)\int_{\left[a, \frac{a+b}{2}\right]} \left(\inf_{t \in \left[a, \frac{a+b}{2}\right]} \left| D_{a+}^{\alpha} f \right| \right) (t-a)^{\alpha}\, d\mu(t) \vee \right.$$

$$\left. (S)\int_{\left[\frac{a+b}{2}, b\right]} \left(\inf_{t \in \left[\frac{a+b}{2}, b\right]} \left| D_{b-}^{\alpha} f \right| \right) (b-t)^{\alpha}\, d\mu(t) \right].$$

We have proved a reverse fractional Polya-Sugeno type inequality:

$$(S)\int_{[a,b]} f(t)\, d\mu(t) \geq$$

$$\frac{1}{\Gamma(\alpha+1)} \left[(S)\int_{\left[a, \frac{a+b}{2}\right]} \left(\inf_{t \in \left[a, \frac{a+b}{2}\right]} \left| D_{a+}^{\alpha} f \right| \right) (t-a)^{\alpha}\, d\mu(t) \vee \right. \tag{16.64}$$

$$\left. (S)\int_{\left[\frac{a+b}{2}, b\right]} \left(\inf_{t \in \left[\frac{a+b}{2}, b\right]} \left| D_{b-}^{\alpha} f \right| \right) (b-t)^{\alpha}\, d\mu(t) \right].$$

References

1. G.A. Anastassiou, Fractional Ostrowski-Sugeno Fuzzy univariate inequalities (2018). Submitted
2. G.A. Anastassiou, *Fractional Differentiation Inequalities* (Springer, New York, 2009)

3. G.A. Anastassiou, *Intelligent Mathematics: Computational Analysis* (Springer, Heidelberg, New York, 2011)
4. G.A. Anastassiou, *Advances on Fractional Inequalities* (Springer, New York, 2011)
5. G.A. Anastassiou, *Intelligent Comparisons: Analytic Inequalities* (Springer, New York, 2016)
6. M. Boczek, M. Kaluszka, On the Minkowaki-Hölder type inequalities for generalized Sugeno integrals with an application. Kybernetica **52**(3), 329–347 (2016)
7. J.A. Canavati, The Riemann-Liouville integral. Nieuw Archief Voor Wiskunde **5**(1), 53–75 (1987)
8. K. Diethelm, *The Analysis of Fractional Differential Equations*, 1st edn., Lecture notes in mathematics, vol 2004 (Springer, New York, 2010)
9. A. Ostrowski, Über die Absolutabweichung einer differentiebaren Funktion von ihrem Integralmittelwert. Comment. Math. Helv. **10**, 226–227 (1938)
10. E. Pap, *Null-Additive Set functions* (Kluwer Academic, Dordrecht, 1995)
11. D. Ralescu, G. Adams, The fuzzy integral. J. Math. Anal. Appl. **75**, 562–570 (1980)
12. M. Sugeno, *Theory of fuzzy integrals and its applications*, Ph.D. thesis, Tokyo Institute of Technology (1974)
13. Z. Wang, G.J. Klir, *Fuzzy Measure Theory* (Plenum, New York, 1992)

Chapter 17
Multidimensional Ostrowski–Sugeno Type Fuzzy Integral Inequalities

Here we present multivariate Ostrowski–Sugeno Fuzzy type inequalities. These are multivariate Ostrowski-like inequalities in the context of Sugeno fuzzy integral and its special properties. They give tight upper bounds to the deviation of a multivariate function from its Sugeno-fuzzy multivariate averages. It follows [3].

17.1 Introduction

The famous Ostrowski [5] inequality motivates this work and has as follows:

$$\left|\frac{1}{b-a}\int_a^b f(y)\,dy - f(x)\right| \leq \left(\frac{1}{4}+\frac{\left(x-\frac{a+b}{2}\right)^2}{(b-a)^2}\right)(b-a)\left\|f'\right\|_\infty,$$

where $f \in C^1([a,b])$, $x \in [a,b]$, and it is a sharp inequality.

Another motivation comes from author's [2], pp. 507–508, see also [1]:

Let $f \in C^1\left(\prod_{i=1}^k [a_i, b_i]\right)$, where $a_i < b_i$; $a_i, b_i \in \mathbb{R}$, $i = 1, ..., k$, and let $x_0 := (x_{01}, ..., x_{0k}) \in \prod_{i=1}^k [a_i, b_i]$ be fixed. Then

$$\left|\frac{1}{\prod_{i=1}^k (b_i - a_i)}\int_{a_1}^{b_1}\cdots\int_{a_i}^{b_i}\cdots\int_{a_k}^{b_k} f(z_1, ..., z_k)\,dz_1...dz_k - f(x_0)\right| \leq$$

$$\sum_{i=1}^k \left(\frac{(x_{0i}-a_i)^2+(b_i-x_{0i})^2}{2(b_i-a_i)}\right)\left\|\frac{\partial f}{\partial z_i}\right\|_\infty.$$

G. A. Anastassiou, *Ordinary and Fractional Approximation by Non-additive Integrals: Choquet, Shilkret and Sugeno Integral Approximators*, Studies in Systems, Decision and Control 190, https://doi.org/10.1007/978-3-030-04287-5_17

The last inequality is sharp, the optimal function is

$$f^*(z_1, ..., z_k) := \sum_{i=1}^{k} |z_i - x_{0i}|^{\alpha_i}, \alpha_i > 1.$$

Here first we give a survey about Sugeno fuzzy integral and its basic special properties. Then we derive a set of multivariate Ostrowski-like inequalities to all directions in the context of Sugeno integral within its basic important properties. We finish with an application to a special multivariate case.

17.2 Background

In this section, some definitions and basic important properties of the Sugeno integral which will be used in the next section are presented. Also a preparation for the main results Sect. 17.3 is given.

Definition 17.1 (*Fuzzy measure* [7, 9]) Let Σ be a σ-algebra of subsets of X, and let $\mu : \Sigma \to [0, +\infty]$ be a non-negative extended real-valued set function. We say that μ is a fuzzy measure iff:

(1) $\mu(\varnothing) = 0$,

(2) $E, F \in \Sigma : E \subseteq F$ imply $\mu(E) \leq \mu(F)$ (monotonicity),

(3) $E_n \in \Sigma$ $(n \in \mathbb{N})$, $E_1 \subset E_2 \subset ...$, imply $\lim_{n\to\infty} \mu(E_n) = \mu\left(\cup_{n=1}^{\infty} E_n\right)$ (continuity from below);

(4) $E_n \in \Sigma$ $(n \in \mathbb{N})$, $E_1 \supset E_2 \supset ...$, $\mu(E_1) < \infty$, imply $\lim_{n\to\infty} \mu(E_n) = \mu\left(\cap_{n=1}^{\infty} E_n\right)$ (continuity from above).

Let (X, Σ, μ) be a fuzzy measure space and f be a non-negative real-valued function on X. We denote by $\mathcal{F}_+$ the set of all non-negative real valued measurable functions, and by $L_\alpha f$ the set: $L_\alpha f := \{x \in X : f(x) \geq \alpha\}$, the α-level of f for $\alpha \geq 0$.

Definition 17.2 Let (X, Σ, μ) be a fuzzy measure space. If $f \in \mathcal{F}_+$ and $A \in \Sigma$, then the Sugeno integral (fuzzy integral) [8] of f on A with respect to the fuzzy measure μ is defined by

$$(S)\int_A f d\mu := \vee_{\alpha\geq 0} (\alpha \wedge \mu(A \cap L_\alpha f)), \quad (17.1)$$

where $\vee$ and $\wedge$ denote the sup and inf on $[0, \infty]$, respectively.

The basic properties of Sugeno integral follow:

Theorem 17.3 ([6, 9]) *Let (X, Σ, μ) be a fuzzy measure space with $A, B \in \Sigma$ and $f, g \in \mathcal{F}_+$. Then*

(1) $(S)\int_A f d\mu \leq \mu(A)$;
(2) $(S)\int_A k d\mu = k \wedge \mu(A)$ *for a non-negative constant* k;
(3) if $f \leq g$ *on* A*, then* $(S)\int_A f d\mu \leq (S)\int_A g d\mu$;
(4) if $A \subset B$*, then* $(S)\int_A f d\mu \leq (S)\int_B f d\mu$;
(5) $\mu(A \cap L_\alpha f) \leq \alpha \Rightarrow (S)\int_A f d\mu \leq \alpha$;
(6) if $\mu(A) < \infty$*, then* $\mu(A \cap L_\alpha f) \geq \alpha \Leftrightarrow (S)\int_A f d\mu \geq \alpha$;
(7) when $A = X$*, then* $(S)\int_A f d\mu = \vee_{\alpha \geq 0}(\alpha \wedge \mu(L_\alpha f))$;
(8) if $\alpha \leq \beta$*, then* $L_\beta f \subseteq L_\alpha f$;
(9) $(S)\int_A f d\mu \geq 0$.

Theorem 17.4 ([9], p. 135) *Here* $f \in \mathcal{F}_+$*, the class of all finite nonnegative measurable functions on* (X, Σ, μ)*. Then*
(1) if $\mu(A) = 0$*, then* $(S)\int_A f d\mu = 0$*, for any* $f \in \mathcal{F}_+$;
(2) if $(S)\int_A f d\mu = 0$*, then* $\mu(A \cap \{x | f(x) > 0\}) = 0$;
(3) $(S)\int_A f d\mu = (S)\int_X f \cdot \chi_A d\mu$*, where* χ_A *is the characteristic function of* A;
(4) $(S)\int_A (f + a) d\mu \leq (S)\int_A f d\mu + (S)\int_A a d\mu$*, for any constant* $a \in [0, \infty)$.

Corollary 17.5 ([9], p. 136) *Here* $f, f_1, f_2 \in \mathcal{F}_+$*. Then*
(1) $(S)\int_A (f_1 \vee f_2) d\mu \geq (S)\int_A f_1 d\mu \vee (S)\int_A f_2 d\mu$;
(2) $(S)\int_A (f_1 \wedge f_2) d\mu \leq (S)\int_A f_1 d\mu \wedge (S)\int_A f_2 d\mu$;
(3) $(S)\int_{A \cup B} f d\mu \geq (S)\int_A f d\mu \vee (S)\int_B f d\mu$;
(4) $(S)\int_{A \cap B} f d\mu \leq (S)\int_A f d\mu \wedge (S)\int_B f d\mu$.

In general we have

$$(S)\int_A (f_1 + f_2) d\mu \neq (S)\int_A f_1 d\mu + (S)\int_A f_2 d\mu,$$

and

$$(S)\int_A a f d\mu \neq a (S)\int_A f d\mu, \text{ where } a \in \mathbb{R},$$

see [9], p. 137.

Lemma 17.6 ([9], p. 138) $(S)\int_A f d\mu = \infty$ *iff* $\mu(A \cap L_\alpha f) = \infty$ *for any* $\alpha \in [0, \infty)$.

We need

Definition 17.7 ([4]) A fuzzy measure μ is subadditive iff $\mu(A \cup B) \leq \mu(A) + \mu(B)$, for all $A, B \in \Sigma$.

We mention

Theorem 17.8 ([4]) *If* μ *is subadditive, then*

$$(S)\int_X (f + g) d\mu \leq (S)\int_X f d\mu + (S)\int_X g d\mu, \tag{17.2}$$

for all measurable functions $f, g : X \to [0, \infty)$.

Moreover, if (17.2) holds for all measurable functions $f, g : X \to [0, \infty)$ *and* $\mu(X) < \infty$, *then* μ *is subadditive.*

Notice here in (17.1) we have that $\alpha \in [0, \infty)$.

We have

Corollary 17.9 *If* μ *is subadditive,* $n \in \mathbb{N}$, *and* $f : X \to [0, \infty)$ *is a measurable function, then*

$$(S)\int_X nfd\mu \leq n\,(S)\int_X fd\mu, \tag{17.3}$$

in particular it holds

$$(S)\int_A nfd\mu \leq n\,(S)\int_A fd\mu, \tag{17.4}$$

for any $A \in \Sigma$.

Proof By (17.2). ■

A very important property of Sugeno integral follows.

Theorem 17.10 *If* μ *is subadditive measure, and* $f : X \to [0, \infty)$ *is a measurable function, and* $c > 0$, *then*

$$(S)\int_A cfd\mu \leq (c+1)\,(S)\int_A fd\mu, \tag{17.5}$$

for any $A \in \Sigma$.

Proof Let the ceiling $\lceil c \rceil = m \in \mathbb{N}$, then by Theorem 17.3 (3) and (4) we get

$$(S)\int_A cfd\mu \leq (S)\int_A mfd\mu \leq m\,(S)\int_A fd\mu \leq (c+1)\,(S)\int_A fd\mu,$$

proving (17.5). ■

From now on in this chapter we work on the fuzzy measure space $(Q, \mathcal{B}, \mu)$, where $Q \subset \mathbb{R}^k$, $k \geq 1$ is a convex compact subset, $\mathcal{B}$ is the Borel σ-algebra on Q, and μ is a finite fuzzy measure on $\mathcal{B}$. Typically we take it to be subadditive.

The functions f we deal with here are continuous from Q into $\mathbb{R}_+$.

We make

Remark 17.11 Let $f \in C(Q, \mathbb{R}_+)$, and μ is a subadditive fuzzy measure such that $\mu(Q) > 0$, $x \in Q$. We will estimate

$$E(x) := \left|(S)\int_Q f(t)\,d\mu(t) - \mu(Q) \wedge f(x)\right| \tag{17.6}$$

(by Theorem 17.3 (2))

$$= \left| (S) \int_Q f(t) \, d\mu(t) - (S) \int_Q f(x) \, d\mu(t) \right| .$$

We notice that

$$f(t) = f(t) - f(x) + f(x) \leq |f(t) - f(x)| + f(x),$$

then (by Theorem 17.3 (3) and Theorem 17.4 (4))

$$(S) \int_Q f(t) \, d\mu(t) \leq (S) \int_Q |f(t) - f(x)| \, d\mu(t) + (S) \int_Q f(x) \, d\mu(t), \quad (17.7)$$

that is

$$(S) \int_Q f(t) \, d\mu(t) - (S) \int_Q f(x) \, d\mu(t) \leq (S) \int_Q |f(t) - f(x)| \, d\mu(t). \quad (17.8)$$

Similarly, we have

$$f(x) = f(x) - f(t) + f(t) \leq |f(t) - f(x)| + f(t),$$

then (by Theorem 17.3 (3) and Theorem 17.8)

$$(S) \int_Q f(x) \, d\mu(t) \leq (S) \int_Q |f(t) - f(x)| \, d\mu(t) + (S) \int_Q f(t) \, d\mu(t),$$

that is

$$(S) \int_Q f(x) \, d\mu(t) - (S) \int_Q f(t) \, d\mu(t) \leq (S) \int_Q |f(t) - f(x)| \, d\mu(t). \quad (17.9)$$

By (17.8) and (17.9) we derive that

$$\left| (S) \int_Q f(t) \, d\mu(t) - (S) \int_Q f(x) \, d\mu(t) \right| \leq (S) \int_Q |f(t) - f(x)| \, d\mu(t). \quad (17.10)$$

Consequently it holds

$$E(x) \overset{\text{(by (17.6), (17.10))}}{\leq} (S) \int_Q |f(t) - f(x)| \, d\mu(t), \quad (17.11)$$

where $t = (t_1, ..., t_k)$, $x = (x_1, ..., x_k)$.

We will use (17.11).

17.3 Main Results

We make

Remark 17.12 Here $Q := \prod_{i=1}^{k} [a_i, b_i]$, where $a_i < b_i$; $a_i, b_i \in \mathbb{R}$, $i = 1, ..., k$; $x = (x_1, ..., x_k) \in \prod_{i=1}^{k} [a_i, b_i]$ is fixed, and $f \in C^1\left(\prod_{i=1}^{k} [a_i, b_i], \mathbb{R}_+\right)$. Consider $g_t(r) := f(x + r(t - x))$, $r \geq 0$. Note that $g_t(0) = f(x)$, $g_t(1) = f(t)$. Thus

$$f(t) - f(x) = g_t(1) - g_t(0) = g_t'(\xi)(1 - 0) = g_t'(\xi), \tag{17.12}$$

where $\xi \in (0, 1)$.

I.e.

$$f(t) - f(x) = \sum_{i=1}^{k} (t_i - x_i) \frac{\partial f}{\partial t_i}(x + \xi(t - x)). \tag{17.13}$$

Hence

$$|f(t) - f(x)| \leq \sum_{i=1}^{k} |t_i - x_i| \left| \frac{\partial f}{\partial t_i}(x + \xi(t - x)) \right|$$

$$\leq \sum_{i=1}^{k} |t_i - x_i| \left\| \frac{\partial f}{\partial t_i} \right\|_\infty. \tag{17.14}$$

By (17.11) we get

$$\left| (S) \int_{\prod_{i=1}^{k} [a_i, b_i]} f(t) \, d\mu(t) - \mu\left(\prod_{i=1}^{k} [a_i, b_i]\right) \wedge f(x) \right| \leq$$

$$(S) \int_{\prod_{i=1}^{k} [a_i, b_i]} |f(t) - f(x)| \, d\mu(t) \overset{(17.14)}{\leq}$$

$$(S) \int_{\prod_{i=1}^{k} [a_i, b_i]} \left(\sum_{i=1}^{k} |t_i - x_i| \left\| \frac{\partial f}{\partial t_i} \right\|_\infty \right) d\mu(t) \overset{(17.2)}{\leq}$$

$$\sum_{i=1}^{k} (S) \int_{\prod_{i=1}^{k} [a_i, b_i]} |t_i - x_i| \left\| \frac{\partial f}{\partial t_i} \right\|_\infty d\mu(t) \overset{(17.5)}{\leq}$$

$$\sum_{i=1}^{k}\left(\left\|\frac{\partial f}{\partial t_i}\right\|_{\infty}+1\right)\left((S)\int_{\prod_{i=1}^{k}[a_i,b_i]}|t_i-x_i|\,d\mu(t)\right). \tag{17.15}$$

Here μ is a fuzzy subadditive measure with $\mu\left(\prod_{i=1}^{k}[a_i,b_i]\right)>0$.

Therefore we get

$$\left|\frac{1}{\mu\left(\prod_{i=1}^{k}[a_i,b_i]\right)}(S)\int_{\prod_{i=1}^{k}[a_i,b_i]}f(t)\,d\mu(t)-\left(1\wedge\frac{f(x)}{\mu\left(\prod_{i=1}^{k}[a_i,b_i]\right)}\right)\right|\overset{(17.15)}{\leq} \tag{17.16}$$

$$\sum_{i=1}^{k}\left(\frac{\left\|\frac{\partial f}{\partial t_i}\right\|_{\infty}+1}{\mu\left(\prod_{i=1}^{k}[a_i,b_i]\right)}\right)\left((S)\int_{\prod_{i=1}^{k}[a_i,b_i]}|t_i-x_i|\,d\mu(t)\right).$$

Notice here $\left(1\wedge\frac{f(x)}{\mu\left(\prod_{i=1}^{k}[a_i,b_i]\right)}\right)\leq 1$, and

$\frac{1}{\mu\left(\prod_{i=1}^{k}[a_i,b_i]\right)}(S)\int_{\prod_{i=1}^{k}[a_i,b_i]}f(t)\,d\mu(t)\overset{\text{(by Thm. 17.3 (1))}}{\leq}1,$

where $(S)\int_{\prod_{i=1}^{k}[a_i,b_i]}f(t)\,d\mu(t)\geq 0$.

If $f:\prod_{i=1}^{k}[a_i,b_i]\to\mathbb{R}_+$ is a Lipschitz function of order $0<\alpha\leq 1$, i.e. $|f(x)-f(y)|\leq K\|x-y\|_{l_1}^{\alpha}$, $\forall\, x,y\in\prod_{i=1}^{k}[a_i,b_i]$, $K>0$, where $\|x-y\|_{l_1}:=\sum_{i=1}^{k}|x_i-y_i|$, denoted by $f\in Lip_{\alpha,K}\left(\prod_{i=1}^{k}[a_i,b_i],\mathbb{R}_+\right)$, then by (17.11) we get

$$\left|(S)\int_{\prod_{i=1}^{k}[a_i,b_i]}f(t)\,d\mu(t)-\mu\left(\prod_{i=1}^{k}[a_i,b_i]\right)\wedge f(x)\right|\leq \tag{17.17}$$

$$(S)\int_{\prod_{i=1}^{k}[a_i,b_i]}|f(t)-f(x)|\,d\mu(t)\leq$$

$$(S)\int_{\prod_{i=1}^{k}[a_i,b_i]}K\|t-x\|_{l_1}^{\alpha}\,d\mu(t)\overset{(17.5)}{\leq}$$

$$(K+1)\,(S)\int_{\prod_{i=1}^{k}[a_i,b_i]} \|t-x\|_{l_1}^{\alpha}\, d\mu(t).$$

We have proved

$$\left| \frac{1}{\mu\left(\prod_{i=1}^{k}[a_i,b_i]\right)} (S)\int_{\prod_{i=1}^{k}[a_i,b_i]} f(t)\, d\mu(t) - \left(1 \wedge \frac{f(x)}{\mu\left(\prod_{i=1}^{k}[a_i,b_i]\right)}\right) \right| \leq \tag{17.18}$$

$$\frac{(K+1)}{\mu\left(\prod_{i=1}^{k}[a_i,b_i]\right)} (S)\int_{\prod_{i=1}^{k}[a_i,b_i]} \|t-x\|_{l_1}^{\alpha}\, d\mu(t).$$

We have established the following multivariate Ostrowski–Sugeno inequalities.

Theorem 17.13 *Here μ is a fuzzy subadditive measure with $\mu\left(\prod_{i=1}^{k}[a_i,b_i]\right) > 0$, $x \in \prod_{i=1}^{k}[a_i,b_i]$.*

(1) Let $f \in C^1\left(\prod_{i=1}^{k}[a_i,b_i], \mathbb{R}_+\right)$, then

$$\left| \frac{1}{\mu\left(\prod_{i=1}^{k}[a_i,b_i]\right)} (S)\int_{\prod_{i=1}^{k}[a_i,b_i]} f(t)\, d\mu(t) - \left(1 \wedge \frac{f(x)}{\mu\left(\prod_{i=1}^{k}[a_i,b_i]\right)}\right) \right| \leq \tag{17.19}$$

$$\sum_{i=1}^{k} \left(\frac{\left\| \frac{\partial f}{\partial t_i} \right\|_{\infty} + 1}{\mu\left(\prod_{i=1}^{k}[a_i,b_i]\right)} \right) \left((S)\int_{\prod_{i=1}^{k}[a_i,b_i]} |t_i - x_i|\, d\mu(t) \right).$$

(2) Let $f \in Lip_{\alpha,K}\left(\prod_{i=1}^{k}[a_i,b_i], \mathbb{R}_+\right)$, $0 < \alpha \leq 1$, then

$$\left| \frac{1}{\mu\left(\prod_{i=1}^{k}[a_i,b_i]\right)} (S)\int_{\prod_{i=1}^{k}[a_i,b_i]} f(t)\, d\mu(t) - \left(1 \wedge \frac{f(x)}{\mu\left(\prod_{i=1}^{k}[a_i,b_i]\right)}\right) \right| \leq \tag{17.20}$$

$$\frac{(K+1)}{\mu\left(\prod_{i=1}^{k}[a_i,b_i]\right)}(S)\int_{\prod_{i=1}^{k}[a_i,b_i]}\|t-x\|_{l_1}^{\alpha}\,d\mu(t).$$

We make

Remark 17.14 Let Q be a compact and convex subset of $\mathbb{R}^k$, $k \geq 1$. Let $f \in \left(C(Q,\mathbb{R}_+) \cap C^{n+1}(Q)\right)$, $n \in \mathbb{N}$ and $x \in Q$ is fixed such that all partial derivatives $f_\alpha := \frac{\partial^\alpha f}{\partial t^\alpha}$, where $\alpha = (\alpha_1, \ldots, \alpha_k)$, $\alpha_i \in \mathbb{Z}^+$, $i = 1, \ldots, k$, $|\alpha| = \sum_{i=1}^{\kappa} \alpha_i = j$, $j = 1, \ldots, n$ fulfill $f_\alpha(x) = 0$.

By [2], p. 513, we get that

$$|f(t)-f(x)| \leq \frac{\left[\left(\sum_{i=1}^{k}|t_i-x_i|\left\|\frac{\partial}{\partial t_i}\right\|_\infty\right)^{n+1}f\right]}{(n+1)!}, \forall t \in Q. \tag{17.21}$$

Call

$$D_{n+1}(f) := \max_{\alpha:|\alpha|=n+1}\|f_\alpha\|_\infty. \tag{17.22}$$

For example, when $k = 2$ and $n = 1$, we get that

$$\left[\left(\sum_{i=1}^{2}|t_i-x_i|\left\|\frac{\partial}{\partial t_i}\right\|_\infty\right)^2 f\right] =$$

$$(t_1-x_1)^2\left\|\frac{\partial^2 f}{\partial t_1^2}\right\|_\infty + 2|t_1-x_1||t_2-x_2|\left\|\frac{\partial^2 f}{\partial t_1 \partial t_2}\right\|_\infty + (t_2-x_2)^2\left\|\frac{\partial^2 f}{\partial t_2^2}\right\|_\infty, \tag{17.23}$$

and

$$D_2(f) = \max_{\alpha:|\alpha|=2}\|f_\alpha\|_\infty. \tag{17.24}$$

Clearly, it holds

$$\left[\left(\sum_{i=1}^{2}|t_i-x_i|\left\|\frac{\partial}{\partial t_i}\right\|_\infty\right)^2 f\right] \leq D_2(f)\left(|t_1-x_1|+|t_2-x_2|\right)^2. \tag{17.25}$$

Consequently, we derive that

$$\left[\left(\sum_{i=1}^{k}|t_i-x_i|\left\|\frac{\partial}{\partial t_i}\right\|_\infty\right)^{n+1} f\right] \leq D_{n+1}(f)\|t \quad x\|_{l_1}^{n+1}, \forall t \subset Q. \tag{17.26}$$

By (17.11) we get

$$\left|(S)\int_Q f(t)\,d\mu(t) - \mu(Q)\wedge f(x)\right| \leq$$

$$(S)\int_Q |f(t) - f(x)|\,d\mu(t) \overset{(17.21)}{\leq} \tag{17.27}$$

$$(S)\int_Q \frac{\left[\left(\sum_{i=1}^{k}|t_i - x_i|\left\|\frac{\partial}{\partial t_i}\right\|_\infty\right)^{n+1} f\right]}{(n+1)!}d\mu(t) \overset{(17.26)}{\leq}$$

$$(S)\int_Q \frac{D_{n+1}(f)\,\|t-x\|_{l_1}^{n+1}}{(n+1)!}d\mu(t) \overset{(17.5)}{\leq} \tag{17.28}$$

$$\left(\frac{D_{n+1}(f)}{(n+1)!}+1\right)(S)\int_Q \|t-x\|_{l_1}^{n+1}\,d\mu(t).$$

Here μ is a fuzzy subadditive measure with $\mu(Q) > 0$.

By (17.27) and (17.28) we obtain

$$\left|\frac{1}{\mu(Q)}(S)\int_Q f(t)\,d\mu(t) - \left(1\wedge\frac{f(x)}{\mu(Q)}\right)\right| \leq$$

$$\frac{\left(\frac{D_{n+1}(f)}{(n+1)!}+1\right)}{\mu(Q)}(S)\int_Q \|t-x\|_{l_1}^{n+1}\,d\mu(t). \tag{17.29}$$

We have established the following multivariate Ostrowski–Sugeno general inequality:

Theorem 17.15 *Let Q be a compact and convex subset of $\mathbb{R}^k$, $k \geq 1$. Let $f \in \left(C(Q,\mathbb{R}_+)\cap C^{n+1}(Q)\right)$, $n \in \mathbb{N}$, $x \in Q$ be fixed: $f_\alpha(x) = 0$, all $\alpha : |\alpha| = j$, $j = 1, ..., n$. Here μ is a fuzzy subadditive measure with $\mu(Q) > 0$. Then*

$$\left|\frac{1}{\mu(Q)}(S)\int_Q f(t)\,d\mu(t) - \left(1\wedge\frac{f(x)}{\mu(Q)}\right)\right| \leq$$

$$\frac{\left(\frac{D_{n+1}(f)}{(n+1)!}+1\right)}{\mu(Q)}(S)\int_Q \|t-x\|_{l_1}^{n+1}\,d\mu(t). \tag{17.30}$$

Corollary 17.16 *All as in Theorem 17.15. Then*

$$\left| \frac{1}{\mu(Q)} (S) \int_Q f(t)\, d\mu(t) - \left(1 \wedge \frac{f(x)}{\mu(Q)}\right) \right| \leq$$

$$\frac{\left(1 + \frac{1}{(n+1)!}\right)}{\mu(Q)} (S) \int_Q \left[\left(\sum_{i=1}^{k} |t_i - x_i| \left\| \frac{\partial}{\partial x_i} \right\|_\infty \right)^{n+1} f \right] d\mu(t) . \quad (17.31)$$

Next we take again $Q := \prod_{i=1}^{k} [a_i, b_i]$, we set $a := (a_1, \ldots, a_k)$, $b := (b_1, \ldots, b_k)$, and $\frac{a+b}{2} = \left(\frac{a_1+b_1}{2}, \ldots, \frac{a_k+b_k}{2}\right) \in \prod_{i=1}^{k} [a_i, b_i]$.

Corollary 17.17 *Let* $f \in \left(C\left(\prod_{i=1}^{k} [a_i, b_i], \mathbb{R}_+ \right) \cap C^{n+1}\left(\prod_{i=1}^{k} [a_i, b_i] \right) \right)$, $n \in \mathbb{N}$, *such that* $f_\alpha\left(\frac{a+b}{2}\right) = 0$, *all* $\alpha : |\alpha| = j$, $j = 1, \ldots, n$. *Here* μ *is a fuzzy subadditive measure with* $\mu\left(\prod_{i=1}^{k} [a_i, b_i] \right) > 0$. *Then*

$$\left| \frac{1}{\mu\left(\prod_{i=1}^{k} [a_i, b_i] \right)} (S) \int_{\prod_{i=1}^{k} [a_i, b_i]} f(t)\, d\mu(t) - \left(1 \wedge \frac{f\left(\frac{a+b}{2}\right)}{\mu\left(\prod_{i=1}^{k} [a_i, b_i] \right)} \right) \right| \leq$$

$$\frac{\left(\frac{D_{n+1}(f)}{(n+1)!} + 1 \right)}{\mu\left(\prod_{i=1}^{k} [a_i, b_i] \right)} (S) \int_{\prod_{i=1}^{k} [a_i, b_i]} \left\| t - \frac{a+b}{2} \right\|_{l_1}^{n+1} d\mu(t) . \quad (17.32)$$

Proof By Theorem 17.15. ■

We make

Remark 17.18 By multinomial theorem we have that

$$\|t - x\|_{l_1}^{n+1} = \left(\sum_{i=1}^{k} |t_i - x_i| \right)^{n+1} =$$

$$\sum_{r_1 + r_2 + \ldots + r_k = n+1} \binom{n+1}{r_1, r_2, \ldots, r_k} |t_1 - x_1|^{r_1} |t_2 - x_2|^{r_2} \ldots |t_k - x_k|^{r_k} , \quad (17.33)$$

where

$$\binom{n+1}{r_1, r_2, ..., r_k} = \frac{(n+1)!}{r_1!r_2!...r_k!}. \tag{17.34}$$

By (17.27), (17.28) we get

$$\left|(S)\int_Q f(t)\,d\mu(t) - \mu(Q) \wedge f(x)\right| \leq$$

$$(S)\int_Q \frac{D_{n+1}(f)}{(n+1)!}\|t-x\|_{l_1}^{n+1}\,d\mu(t) \overset{\text{(by (17.33), (17.34))}}{=}$$

$$(S)\int_Q \left[\sum_{r_1+r_2+...+r_k=n+1}\left(\frac{D_{n+1}(f)}{r_1!r_2!...r_k!}\right)\left(\prod_{i=1}^k |t_i - x_i|^{r_i}\right)\right] d\mu(t) \overset{(17.2)}{\leq}$$

$$\sum_{r_1+r_2+...+r_k=n+1}(S)\int_Q \left(\frac{D_{n+1}(f)}{r_1!r_2!...r_k!}\right)\left(\prod_{i=1}^k |t_i - x_i|^{r_i}\right) d\mu(t) \overset{(17.5)}{\leq}$$

$$\sum_{r_1+r_2+...+r_k=n+1}\left(\frac{D_{n+1}(f)}{r_1!r_2!...r_k!}+1\right)(S)\int_Q \left(\prod_{i=1}^k |t_i - x_i|^{r_i}\right) d\mu(t). \tag{17.35}$$

We have proved the following multivariate Ostrowski–Sugeno general inequality:

Theorem 17.19 *Here all as in Theorem 17.15. Then*

$$\left|\frac{1}{\mu(Q)}(S)\int_Q f(t)\,d\mu(t) - \left(1 \wedge \frac{f(x)}{\mu(Q)}\right)\right| \leq$$

$$\sum_{r_1+r_2+...+r_k=n+1}\left(\frac{\left(\frac{D_{n+1}(f)}{r_1!r_2!...r_k!}+1\right)}{\mu(Q)}\right)(S)\int_Q \left(\prod_{i=1}^k |t_i - x_i|^{r_i}\right) d\mu(t). \tag{17.36}$$

We make

Remark 17.20 In case $k = 2$, $n = 1$, by (17.27), (17.28) we get

$$\left|(S)\int_Q f(t)\,d\mu(t) - \mu(Q) \wedge f(x)\right| \leq$$

$$(S)\int_Q \frac{D_2(f)}{2}\|t-x\|_{l_1}^2\,d\mu(t) =$$

$$(S)\int_Q \frac{D_2(f)}{2}\left[(t_1-x_1)^2 + 2|t_1-x_1||t_2-x_2| + (t_2-x_2)^2\right] d\mu(t) \leq \tag{17.37}$$

$$(S)\int_Q \frac{D_2(f)}{2}(t_1-x_1)^2\,d\mu(t)+(S)\int_Q D_2(f)\,|t_1-x_1|\,|t_2-x_2|\,d\mu(t)$$

$$+(S)\int_Q \frac{D_2(f)}{2}(t_2-x_2)^2\,d\mu(t)\leq$$

$$\left(1+\frac{D_2(f)}{2}\right)(S)\int_Q (t_1-x_1)^2\,d\mu(t)+(1+D_2(f))\,(S)\int_Q |t_1-x_1|\,|t_2-x_2|\,d\mu(t)$$

$$+\left(1+\frac{D_2(f)}{2}\right)(S)\int_Q (t_2-x_2)^2\,d\mu(t).$$

We have proved

Corollary 17.21 *Let Q be a compact and convex subset of $\mathbb{R}^2$. Let $f\in(C(Q,\mathbb{R}_+)\cap C^2(Q))$, $x=(x_1,x_2)\in Q$ be fixed: $\frac{\partial f}{\partial t_1}(x_1,x_2)=\frac{\partial f}{\partial t_2}(x_1,x_2)=0$. Here μ is a fuzzy subadditive measure with $\mu(Q)>0$. Then*

$$\left|\frac{1}{\mu(Q)}(S)\int_Q f(t)\,d\mu(t)-\left(1\wedge\frac{f(x)}{\mu(Q)}\right)\right|\leq$$

$$\frac{\left(1+\frac{D_2(f)}{2}\right)}{\mu(Q)}(S)\int_Q (t_1-x_1)^2\,d\mu(t)+\frac{(1+D_2(f))}{\mu(Q)}(S)\int_Q |t_1-x_1|\,|t_2-x_2|\,d\mu(t) \tag{17.38}$$

$$+\frac{\left(1+\frac{D_2(f)}{2}\right)}{\mu(Q)}(S)\int_Q (t_2-x_2)^2\,d\mu(t).$$

References

1. G.A. Anastassiou, Multivariate Ostrowski type inequalities. Acta Math. Hung. **76**(4), 267–278 (1997)
2. G.A. Anastassiou, *Quantitative Approximations* (Chapman & Hall/CRC, Boca Raton, 2001)
3. G.A. Anastassiou, Multivariate Ostrowski-Sugeno fuzzy inequalities (2018). Submitted
4. M. Boczek, M. Kaluszka, On the Minkowaki-Hölder type inequalities for generalized Sugeno integrals with an application. Kybernetica **52**(3), 329–347 (2016)
5. A. Ostrowski, Über die Absolutabweichung einer differentiebaren Funktion von ihrem Integralmittelwert. Comment. Math. Helv. **10**, 226–227 (1938)
6. E. Pap, *Null-Additive Set Functions* (Kluwer Academic, Dordrecht, 1995)
7. D. Ralescu, G. Adams, The fuzzy integral. J. Math. Anal. Appl. **75**, 562–570 (1980)
8. M. Sugeno, Theory of fuzzy integrals and its applications. Ph.D. thesis, Tokyo Institute of Technology (1974)
9. Z. Wang, G.J. Klir, *Fuzzy Measure Theory* (Plenum, New York, 1992)

Printed by Printforce, the Netherlands